Cell death in biology and pathology

Cell death in biology and pathology

edited by
I. D. BOWEN
Department of Zoology, University College, Cardiff

and
R. A. LOCKSHIN
Department of Biological Sciences, St John's University, New York

CHAPMAN AND HALL
LONDON   NEW YORK

First published 1981
by Chapman and Hall Ltd
11 New Fetter Lane, London EC4P 4EE
Published in the USA by
Chapman and Hall
in association with Methuen, Inc.
733 Third Avenue, New York, NY 10017

Printed in the United States of
America

ISBN 0 412 16010 2

British Library Cataloguing in Publication Data

Cell death in biology and pathology.

   1. Pathology, Cellular   2. Cytology
   I. Bowen, I D   II. Lockshin, R A
   574.876'5   QH671   80-41769

   ISBN 0-412-16010-2

mns 2-18-83

Contents

List of contributors

Luigi Aloe — Laboratorio di Biologia Cellulare, CNR, Roma, Italy

D. Bellamy — Department of Zoology, University College, Cardiff CF1 1XL, UK

Irene K. Berezesky — University of Maryland School of Medicine, Department of Pathology, 10 South Pine Street, Baltimore, Maryland 21201, USA

Ivor D. Bowen — Department of Zoology, University College, Cardiff CF1 1XL, UK

Timothy Carter — Department of Biological Sciences, St John's University, Jamaica, NY 11439, USA

Gerald R. Crabtree — Department of Pathology, Dartmouth Medical School, Hanover, New Hampshire 03755, USA

P. B. Gahan — Biology Department, Queen Elizabeth College, Campden Hill Road, London W8 7AH, UK

Gale A. Granger — Department of Molecular Biology and Biochemistry, University of California, Irvine, California 92717, USA

Leonard Hayflick — Children's Hospital Medical Center, Bruce Lyon Memorial Research Laboratory, 51st and Grove Streets, Oakland, California 94609, USA

J. R. Hinchliffe — Zoology Department, University College of Wales, Aberystwyth SY23 3DA, Dyfed, UK

S. M. Hinsull — Department of Zoology, University College, Cardiff CF1 1XL, UK

Michael Joesten — Department of Biological Sciences, St John's University, Jamaica, NY 11439, USA

Rita Levi-Montalcini — Laboratorio di Biologia Cellulare, CNR, Roma, Italy

Richard A. Lockshin — Department of Biological Sciences, St John's University, Jamaica NY 11439, USA

Allan Munck — Department of Physiology, Dartmouth Medical School, Hanover, New Hampshire 03755, USA

Monica W. Ross — Department of Molecular Biology and Biochemistry, University of California, Irvine, California 92717, USA

Moira Royston — Department of Biological Sciences, St John's University, Jamaica, NY 11439, USA

Alvaro R. Osornio-Vargas — Instituto Nacional de Cardiologia, Departamento de Patologia, Mexico 22, DF, Mexico

Benjamin F. Trump — University of Maryland School of Medicine, Department of Pathology, 10 South Pine Street, Baltimore, Maryland 21201, USA

N. A. Wright — Department of Histopathology, Royal Postgraduate Medical School, Hammersmith Hospital, Ducane Road, London W12 0HS, UK

A. H. Wyllie — Department of Pathology, University of Edinburgh Medical School, Teviot Place, Edinburgh, UK

Robert S. Yamamoto — Department of Molecular Biology and Biochemistry, University of California, Irvine, California 92717, USA

Preface

The subject of cell death includes several seemingly diverse fields of inquiry, and there is to date no common source of information for the researcher or advanced student. Dispersion of the information on the subject also hinders the development of a theoretical or historical sense of the significance of cell death as a biological phenomenon. This book is an attempt to collate the thrust of current research on cell death and also to point out the direction of future lines of research.

Cell death may be accidental or induced or may be genetically programmed (Chapter 1). As a matter of principle, programmed cell death should be regarded as a mechanism for survival rather than destruction. It is a paradox that cell death is intimately involved in the birth and development of life as we know it, occurring during the formation of eggs and sperm and at many vital stages during embryological development (Chapter 2), and forming the basic mechanism underlying morphogenesis and metamorphosis (Chapter 3). It exerts a balancing or homeostatic function in relation to tissue kinetics and may be an important factor in the control of size and shape in normal tissues, organs, appendages and whole organisms (Chapter 4). Cell death is also an important element in the control of abnormal growth such as tumours (Chapter 6) and many therapeutic agents exploit this factor to induce regression.

An understanding of the mechanisms of cell death is fundamental to an understanding of cell biology. Genetically programmed cell death may involve differential gene activity, differential synthetic activity in terms of enzymes and proteins and finally, differential cell lysis or destruction (Chapters 5, 9 and 13). All these steps are controlled in ways which we are now only beginning to understand.

Cell death is of considerable medical significance (Chapters 6, 8, 10, 11 and 12) and is of obvious import to pathologists who have to interpret histological patterns of cell death. This book deals with the causes of induced cell death (Chapter 7) including disease and also outlines recent techniques for demonstrating cell death (Chapter 13).

It is evident that cell death can be broadly classified into two categories: lysis or coagulative necrosis, in which cells osmotically swell and rupture: and apoptosis or related forms of physiological cell death, in which the cell undergoes more elaborate conversions before succumbing (Chapters 1, 2, 3, 10, 11, 13). The mechanisms of the latter are neither as obvious nor as well understood as are those of the former, but substantial progress has been made in the last few years, and the field is burgeoning.

It is clear that lysosomal enzymes often play a role in the destruction of the cytoplasm, but very few authorities feel that they initiate the process (Chapters 1, 2, 3, 5-8, 12, 13). The cells show many forms of damage, and sometimes even complete destruction, before lysosomes become a dominant part of the environment. What initiates the process is still unclear, although in several instances it appears that the death of a cell may arise from any one of several pathways (Chapters, 10, 11). It is rather interesting that evolution has chosen to achieve the same goal by different means. Apparently no one point is exceptionally or preferentially vulnerable, though a common pathway, such as permeability of the plasma membrane to calcium (Chapter 7), might currently be too subtle for routine identification. Factors which affect membrane stability and which induce membrane bending can lead to blebing, cell fragmentation and death. Thus, more work on the changing chemistry of the plasma membrane in relation to environmental fluctuations would be welcomed.

Space requirements and the major orientation of the book forced the exclusion of several very interesting topics: an evolutionary treatment of the advantages of cell death as a means of eliminating vestigial organs or embryonic scaffolding; or consideration of the merits of body sculpting by cell death rather than cell growth. These ideas have been thoughtfully examined by the founders of the field, to whom reference is made in many chapters. It appears that if cell death can be evoked as a physiological command, then it is far less expensive to change shape by cutting away than by adding. It is presumably easier to call forth existing genes, such as those regulating catabolism, than to eliminate genes for particular structures — especially if the tissues destined for evolutionary loss serve as scaffolding for other tissues.

Finally, there are vast areas that deserve to be considered but which are excluded from this work, either because of the immenseness of the field or lack of information. Destruction of cells by chemotherapeutic agents or by viruses are areas of great interest, but most of the literature focusses on the behaviour of the virus rather than the host cell, and the literature relating to the invocation of self-destruction pathways by the host cell remains diffuse. Also, in spite of several excellent studies in the field (see especially Chapters 8, 9, and 11) the relation of nuclear control to cytoplasmic destruction is not well understood. The discussions emphasize our need to know to what extent cytoplasmic destruction is a spontaneous, perhaps physico-chemical, response to specific conditions, to what extent it is brought about by prior restrictions on nuclear activity, and to what extent it is a direct result of specific gene function.

These and several other questions need to be resolved, but it is gratifying to see how similar are the thoughts from the different fields. The latent unanimity is itself justification for presenting this work at this time, and we hope that our effort will serve to stimulate, provoke, and crystallize the thoughts on the subject.

R. A. L., New York, March, 1981

I. D. B., Cardiff, March, 1981

Acknowledgements

My wife Janine has been of great personal help as well as being extremely tolerant and understanding, and my parents, parents-in-law, and children have also been of great help at various times during the preparation of this book. Professionally, Jacques Beaulaton has made many of the contributions with which I have dealt, and my recent crop of graduate students, especially (in alphabetical order) Arlene Colon, Agnes Dorsey, Michael Joesten, and Moira Royston, have kept the laboratory together (and moving!) while I disappeared under piles of paper. I remain deeply indebted to my original mentor, Carroll M. Williams. Finally, I wish also to thank the members, mostly anonymous, of the various committees and panels of the National Science Foundation (USA) who have not only supported my work but through criticism and discussion have helped me grow.

R.A.L.

I should like to thank my wife, Dr. S. M. Bowen, for her help with the reference work and in indexing the book. Acknowledgement is due to Dr. N. R. Smith, Library Information Officer, University College, Cardiff, for his extensive and thorough literature searches. My thanks also go to Mr. G. W. Jones, Mrs. J. den Hollander and Mr. G. H. J. Lewis for their assistance and for providing useful research data. I am grateful to Dr. W. T. Coakley and Dr. B. W. Staddon for the many stimulating discussions we had on the subject of cell death, and to Professor D. Bellamy who provided me with the impetus to help edit the book.

I.D.B.

Introduction

The death of cells is a phenomenon often mentioned, the importance of which is recognized, and which attracts the attention of researchers in numerous fields of the biological sciences. In searching through the literature, one finds the term referenced as a keyword in papers deriving from the separate fields of pathology, neurology, development, genetics, radiation, teratology, immunology and oncology, bacteriology and aging. In spite of the interest, however, several problems become apparent to anyone who takes the time to try to understand the process and mechanism of cell death: the term may refer to several different phenomena; the vast majority of papers in the area note the existence of dead cells without further analysis of mechanism or causality; and there exists no compendium attempting to provide guidelines or a source of comparison or unification of ideas. Thus the time is ripe for such a treatise.

The first and most obvious difficulty is that there is, to our knowledge, no adequate definition as to what constitutes a dead cell. For those seeking to recognize, in histological sections, dark condensed cells or fragments of cells, with dense compacted chromatin, lysed completely extracted cells or dark condensed cells residing within vacuoles of phagocytes, there is no particular problem: surely these cells are dead — although the questions of why they condense, lyse, or are attacked by phagocytes do not seem to have been examined closely. It is more difficult to cope with the question if one is seeking to understand how cell death occurs: is a cell dead because it has ceased to synthesize DNA? Neurons may live more than a century in a post-mitotic state. Is a cell dead because it has ceased to synthesize RNA? In many systems the death of a cell may be prevented by inhibition of RNA synthesis. In some experimental systems, such as developing embryos, metamorphosing insects and involuting thymocytes, cells can be shown to follow a programme toward their own death, the programme being reversible by specific experimental manoeuvres up to a specific time before the cell can be seen to deteriorate rapidly; should an operational definition be employed to define cell death as a point of experimental irreversibility?

Although several authors suggest different definitions, our initial assumption will be that the most accurate judgement is based on the cytology of the cell. Accordingly, it will be most useful to imagine that a cell is dead when the nucleus condenses or lyses. Presumably this event correlates with the final cessation of

DNA synthesis, if the cell was previously capable of synthesizing DNA, or with the cessation of synthesis of RNA. Condensation of the nucleus occurs simultaneously or nearly simultaneously with the collapse of membrane permeability barriers in at least one insect system (Lockshin and Beaulaton, 1979); otherwise, the question seems not to have been greatly explored, and gross biochemical analyses, such as synthesis or lack of synthesis of total RNA, cannot at present be considered to be sufficiently precise to permit their use as definitions.

Such biochemical analyses may ultimately prove to be of value in studies involving metamorphosing animals, although, even in these models, tissue heterogeneity and lack of synchrony may well limit the usefulness of definitions based on such phenomena. For those studying random isolated instances of cell death, as occurs in cell turnover or at higher frequency in many pathological conditions, or the gradual dropout of cells from cultures of fibroblasts, at present only the histological definitions seem sufficiently precise, although radioautographic analysis of certain synthetic events may prove of value. Nevertheless, we may almost certainly assume that unequivocal cytological alteration is secondary to the event we should properly term as cell death and which we would like to understand. Our excuse for accepting cytological alteration operationally is that, as our expertise in the subcellular location and characterization of the alteration increases, the alteration will help us to define the border.

That our definitions remain in this unsatisfactory state derives in part from the lack of focus in the field. In spite of the outstanding efforts of Glücksmann (1951) to classify and organize the field, the excellent review by Saunders (1966) and a further effort to develop a theme by Lockshin and Beaulaton (1974) most authors view the death of cells as a landmark or end point for experiments directed toward other goals, and many are unaware of studies or interpretations deriving from distantly related fields. We hope that this book will bring together some of these ideas so that, even if ultimately our guesses, predictions and themes prove to be wrong, we at least can provide a starting point.

The second reason for the unsatisfactory state of our definitions can best be described in tabular form. Although many authors mention cell death in passing or by using other terms, one can get a general picture of the field by compiling a list of those who have considered the phenomenon of sufficient importance to include it in a title or keyword. A rough and personal classification of such articles, based on a computer search which turned up 668 items from 1871 to 1978, is shown in Table 0.1; some references are included under more than one category. Of the types of tissue examined, and especially recently, those dealing with mammalian neural and circulatory tissue lead the list by far; most represent observations of cell death in embryonic tissues such as sympathetic ganglia, either under normal conditions or when deprived of a source of nerve growth factor; or they are observations of loss of cells in aging animals. The majority of papers dealing with cell death in the circulatory system referred to acute anoxic death of cardiac muscle in simulations of infarcts; and, under the rubric 'connective tissue', most of the studies referred to the death of fibroblasts in long-term culture, or immunological killing. Many of the latter studies refer simply to the lysis of a target cell, without reference to the mode of death. The mechanisms involved in all of the subjects are dealt with in the essays that follow.

Table 0.1 Range of studies described as 'cell death, necrosis, deletion, autolysis or apoptosis', categorized by phylogenetic classification, type of tissue and research interest (British Library Automated Information Service search)

Tissue or type of analysis	Phylogenetic affinity					Total
	Bacteria	Plants	Invertebrates	Vertebrates		
				Lower	Birds and Mammals	
Tissues						
Neural	0	0	2	3	29	34
Circulatory	0	0	0	0	30	30
Epithelial	0	0	1	1	14	16
Mesodermal (connective)	0	0	2	3	11	16
Muscle	0	0	2	1	10	13
Other (cellular)	3	8	5	4	23	43
Research interests						
Pathology	5	22	3	0	50	80
Aging	0	0	1	0	6	7
Pharmacology	5	7	3	1	19	35
Radiation	2	7	3	0	25	37
Genetics	5	6	9	0	14	34
Development	4	3	12	10	30	59

The research interests in which the term attracts the most attention are, first of all, pharmacology and pathology, in which a chemical or other vector leads to the death of the cell. In most of these studies, the mode of death was observed, generally by cytological means, as a means of analysing the vector. The radiation experiments refer most often to cells isolated in culture or to bacteria. Development, another area which includes a large number of papers, again refers to primarily observational phenomena, such as the observation of cells being phagocytosed in various organs at different stages of development. Most of the studies involved embryological situations; metamorphosis did not figure heavily. Very few of the studies could be described as evincing an interest in how the cells died.

The lack of interest in mechanism is more strikingly illustrated in Table 0.2, in which the same articles are classified according to the major research technique used in the study. The bulk of the papers are treatises of anatomy or histology — that is to say, an observation that dead cells are present in a given location at a given time or circumstance. A lesser number of articles employ histochemistry as a major technique, the demonstration, usually, of acid phosphatase in a lysosomal derivative, thus demonstrating the destruction of one or more elements of the cytoplasm. Many researchers direct their attention to the origin of the lysosome and its enzymes; except for the papers considered to be pharmacological in nature, none reflect on the alteration of the organelle attacked, the reason for the attack, the selectivity of the attack, or the intracellular signalling that led to that specific outcome. The biochemical studies tend to be a bit more diversified, although the studies on cellular respiration and synthetic processes include, predominantly, studies of bacteria or cell cultures exposed to radiation or specific pharmacological agents. By far the greatest interest in the mechanisms controlling cell death is in the extracellular signal rather than in the intracellular response to that signal; and most of these papers in this survey referred to the response of nerve cells to the loss of nerve growth factor or peripheral target stimuli. The basic experimental technique was the observation of the size of a ganglion *in situ* or *in vitro*, or the observation of dead or phagocytosed cells in similar situations.

Table 0.2 Range of studies described as 'cell death, necrosis, deletion, autolysis or apoptosis', categorized by tissue and research technique or discipline

Type of study	Tissue							
	Neural	Liver	Epithel-ial	Digestive	Circulat-ory	Connec-tive	Other	Total
Anatomical	7	6	0	3	12	3	6	37
Histological	21	7	14	3	13	9	22	89
Histochemical	1	2	0	1	4	1	5	14
Biochemical								
Lytic enzymes	1	2	0	0	1	0	12	16
Metabolism	3	18	5	4	10	7	28	75
Nucleic acids (syntheses)	2	4	1	4	2	4	12	29
Proteins (syntheses)	0	1	1	3	0	1	5	11
Membranes	0	2	0	1	3	1	2	9
Control mechanisms								
Extracellular	14	2	4	1	4	1	6	32
Intracellular	3	1	0	1	0	1	2	8
Genetic	3	0	0	0	1	1	5	10
	55	45	25	21	50	29	115	321

Many other studies which treat these questions, do of course exist, but they do not include cell death in their titles or keywords. In bringing these subjects together then under the title 'Cell Death', we hope to be of service to the scientific community. In doing so, we would like to draw attention to the following questions, which this book represents a first step in trying to answer:

1. *What is the proper definition of cell death?* We do not know at what point in the physiological evolution of a cell we can consider it to be truly and irreversibly dead. Does the passage of this point commonly or frequently refer to the failure of a specific organelle, such as the cell membrane, or to activation or inactivation of a specific class of genes; or is the failure more diversified?

2. *To what extent is atrophy related to cell death?* Surprisingly, most researchers

focus on the formation of autophagic vacuoles or the collapse of cytoplasmic organelles, assuming that the loss of cytoplasmic mass is equivalent to, or synonymous with, cell death. Such a conclusion is not really warranted; many types of tissue including unused muscle and sexual tissue in seasonal reproducers, may regress considerably without loss of cell number, as may most glands considered to be targets of pituitary, secondary, or other hormones. What distinguishes cell atrophy, in which the cell retires to a hypoplastic quiescent state but nevertheless remains capable of reawakening, from cell death, in which a cell permanently loses its capacity for resuscitation, with or without preliminary loss of cytoplasm? In what manner is the destruction of the cytoplasm linked or related to the collapse of nuclear potential? The latter might be achieved through an active blockage of regenerative mechanisms, or the nucleus might fail, in a sense accidentally, because the cytoplasm is no longer capable of maintaining an ionic or organic milieu in which the nucleus can function. Also, does the death of a traumatized cell ensue from the same causes of physiological events as does the death of a cell responding to non-pathological, physiological, commands? Physiological control mechanisms may exploit identifiable vulnerabilities of the cell, and our knowledge of these vulnerabilities will aid us in controlling cell death, for experimental or medical purposes.

3. *To what extent are cell and organelle membranes maintained in dying cells?* The presumed failure of synthetic processes may derive from an alteration of ionic concentrations, which itself would reflect failure of one or more membranes. Some dying cells lose water and condense, while others swell and lyse. Our understanding of the osmotic forces involved in the movement of water does not provide simultaneously satisfactory explanations for the two opposing developments. Furthermore, we have no satisfactory understanding of the selectivity involved in the formation and choice of target of autophagic vacuoles, or even, in specific instances of the extremely high specificity of the autophagic vacuoles.

4. *What role do nucleic acids play in cell death?* Many types of cells, in pathological and non-pathological states, undergo an elaborate programme leading to their own destruction. This programme can be demonstrated to involve the synthesis of nucleic acid and protein. We have no meaningful knowledge of the nature and significance of the materials being synthesized — whether they are enzymes, genotropic substances, inhibitors of key metabolic pathways, or messengers of an unknown nature. The existence of such entities suggests but does not prove that all cells have the capability of destroying themselves; we have no idea how often such a mechanism is evoked, how it is evoked, or what kinds of metabolic signals trigger it.

5. *How is the extracellular signal translated?* It is self-evident that throughout the biological world many kinds of endocrine, neural and embryonic inducer signals can command the survival or death of cells. Do these disparate signals nevertheless operate through one or a few common pathways? How are these extracellular stimuli transduced within the cell? Specific pores of the cell membrane might be blocked, or, alternatively, permeability barriers might be damaged. Genes

might be activated or inactivated; metabolic pathways might be inhibited; or intracellular signalling molecules might be activated.

6. *What is the function of cell death?* The seemingly obvious functions of cell death have been hypothesized but not experimentally demonstrated. For instance, metamorphosing tadpoles can survive amputation of the tail. The recycling of cell materials has been presumed, but the size and state of the materials is a matter of conjecture. An undefined amount of substances escapes from the cell and presumably attracts phagocytes. The turnover of organelles does not appear to relate intimately to the turnover of cells, but documentation is incomplete.

The loss of cells in cell turnover appears to be a random event, but it may occur in response to random injury to cells; or the cell may embark on a programme of self-destruction in response to a specific internal or external signal. Loss of cells during aging is considered to contribute to the gradual deterioration observed in aging animals: can one manipulate the process? Is the death of fibroblasts in long-term culture part of the same phenomenon as physiological cell death?

If cells contain an automatic self-destruct mechanism, then selection for such a mechanism implies a hazard as a consequence of uncontrolled survival. The hazard might consist of the release of toxins or vectors in an uncontrolled manner, teratomatous growths or malformations, malignancies or related complications. Malignancy might arise in any of three manners as a result of the failure of a system for self-destruction: the tumorous cell might lack the internal mechanism for programmed cell death and thus, by the force of its own reproduction, expand beyond its borders; it might fail to recognize, or directly inactivate, a call by neighbouring cells or the immune surveillance system for its suicide; or it might advance its invasive character by evoking in surrounding cells calls for their own self-destruction, in a predatory exploitation of a common code of behaviour.

7. Finally, *can we identify and exploit mutants for cell death?* We already know various types of congenital malformations, such as brachyury in the mouse or cleft palate, which involve abnormal regulation of a normal process of localized cell death; and of course many genetic defects such as retinal degeneration result from the death (undoubtedly for many biochemical reasons) of the affected cells, but can we identify any mutations in which the mechanism of cell death itself would be altered? Undoubtedly such a mutation would be catastrophic, but in what manner? How could we identify it, search for it, or select for it?

The answers to some of these questions are suggested in the chapters which follow, and directions will be suggested in the final chapter. But perhaps the greatest contribution we can hope that this book will make would be to suggest that the answers can be obtained, and to encourage the search for those answers.

References

Glücksman, A. (1951), Cell deaths in normal vertebrate ontogeny. *Biol. Rev. Cambridge. Philos. Soc.*, **26**, 59–86.

Lockshin, R. A. and Beaulaton, J. (1974), Programmed cell death. *Life Sci.*, **15**, 1549–1565.

Lockshin, R. A. and Beaulaton, J. (1979), Programmed cell death. Cytological studies of dying muscle fibers of known physiological parameters. *Tissue Cell*, **11**, 803–809.

Saunders, J. W., Jr. (1966), Death in embryonic systems. *Science*, **154**, 604–612.

1 Cell death: a new classification separating apoptosis from necrosis

A. H. WYLLIE

1.1 Introduction

Studies of cell death in the past 20 years have reflected differing philosophies. Some workers saw it as a strictly pathological process, arising only in abnormal circumstances and producing manifestations of disease (see reviews by Trump and Mergner, 1974; Jennings, Ganote and Reimer, 1975). Their experimental models included disruption of metabolic regulation by various injurious agents, and their conclusions were orientated towards expanding our understanding of cellular events in certain diseases. Others, however, regarded cell death as a physiological phenomenon, necessary for the normal development and maintenance of tissue shape (see reviews by Glücksmann, 1951; Saunders, 1966). These workers drew their experimental material largely from embryonic tissues, and their conclusions sought to align cell death with other phenomena of differentiation. Still others were impressed by the frequency of cell death within tumours, as judged by indirect kinetic measurements (see Steel, 1977). This spontaneous cell death in tumours engendered particular interest amongst workers seeking novel approaches to the control of malignancy, as it gave some support to the view that tumour tissues are subject to residual homoeostatic mechanisms whose function in normal tissues would be to regulate cell numbers by balancing cell gain and loss (Laird, 1969).

The question arises whether conclusions from these different experimental systems and philosophies can be synthesized and superimposed, or whether cells die in several fundamentally different ways. And if the latter is true, how can the different types of death be recognized and classified?

Bessis (1964), a pioneer in the study of cell death, commented that what he termed 'accidental' death (i.e. death due to environmental perturbation) might be very different in mechanism from the physiological 'natural' cell death of normal cell turnover. The distinction is important but at that time could not form the basis of a practicable classification, as it depended upon interpretation of the circumstances of death rather than events within the dying cell itself. It is therefore not surprising that investigators turned to ultrastructural study in the hope that it might segregate major types of death and provide some clue to their mechanisms. Unfortunately death seemed to involve such different structural

9

changes that some writers concluded that even in cells of the same tissue and animal, dying as a result of similar initiating circumstances, many different forms of death might exist (Beckingham-Smith and Tata, 1976).

There are reasons to believe, however, that cell death is not so diverse. First, it is probable that many of the observed changes, although associated with injury, may be reversible and hence are not sufficient to define death at all (see e.g. McLean, McLean and Judah, 1965). Secondly, appearances which are distinguishable from each other in static electron micrographs may merely represent different phases in a single process. Awareness of these possibilities permits some simplification and suggests that two major types of death are recognizable on morphological grounds.

The first type, called *necrosis* in this chapter, is characterized by cellular oedema and culminates in rupture of plasma and internal membranes, and leakage of cellular contents into the extracellular space. The second, called *apoptosis*, involves progressive contraction of cellular volume, widespread chromatin condensation, and preservation of the integrity of cytoplasmic organelles. The affected cells separate into membrane-bounded fragments which are rapidly phagocytosed by adjacent cells. This classification has been presented elsewhere (Kerr, Wyllie and Currie, 1972; Wyllie, Kerr and Currie, 1980) but is still relatively new; this chapter seeks to outline its potential value. Like all new classifications in biology, this one carries immediate semantic problems. It should be pointed out that *necrosis* has sometimes been used, particularly in toxological and pathological literature, to embrace all cell death, without regard to type, whilst the constituent phases of *apoptosis* attracted many names when observed in a variety of different circumstances, prior to recognition of its widespread significance in biology. The ensuing sections deal in turn with the circumstances of incidence, morphology and mechanisms of necrosis and apoptosis. Finally the validity of the classification is judged in terms of its scope, practicability and relevance to biological realities beyond the structural changes which form its basis.

1.2 Necrosis

1.2.1 *Incidence*

With very few exceptions necrosis occurs exclusively in circumstances of wide departure from physiological conditions. It has been observed in cells injured by hypoxia (Saladino and Trump, 1968; Kloner, Ganote, Whalen and Jennings, 1974; Ganote, Seabra-Gomes, Nayler and Jennings, 1975; Jennings *et al.*, 1975), complement (Prieto, Kornblith and Pollen, 1967; Hawkins, Ericsson, Biberfeld and Trump, 1972), inhibition of oxidative phosphorylation, glycolysis or Krebs cycle enzymes (McDowell, 1972a, 1972b; Laiho and Trump, 1975), hyperthermia (Buckley, 1972) or exposure to a variety of toxins whose intracellular point of action is not precisely known (McLean *et al.*, 1965; Evan and Dail, 1974). It is also the mode of cell death in autolysis *in vitro* (Trump, Goldblatt and Stowell, 1965; Finlay-Jones and Papadimitriou, 1973).

Growing tumours frequently include both necrosis and apoptosis. The zones of necrosis almost certainly result from hypòxia and deprivation of other substrates. Thus in one carefully studied mouse mammary carcinoma, cords of viable tumour cells surrounding blood vessels lie interspersed with zones of necrosis; the thickness of the viable cell cord from blood vessel wall to the interface with necrotic tissue corresponds closely to the calculated diffusion distance of oxygen carried by the blood (Thomlinson and Gray, 1955). Similar conclusions on the importance of oxygen diffusion are reached from cell culture systems (Franko and Sutherland, 1979). *In vivo* also, the proportion of necrotic tissue in the rat Walker sarcoma is increased by lowering the oxygenation of the blood (Schatten, 1962) or by decreasing the perfusing blood pressure (Schatten and Burson, 1965).

There is no evidence that necrosis ever mediates homoeostatically regulated cell death, in normal or neoplastic tissues, with possible exceptions only amongst the invertebrates, discussed later (see Section 1.4.2).

1.2.2 *Morphology*

There are several good descriptions of the morphology of necrosis (McLean *et al.*, 1965; Trump and Ginn, 1969; Buckley, 1972; McDowell, 1972a, 1972b; Trump, Valigorsky, Dees, Mergner, Kim, Jones, Pendergrass, Garbus and Cowley, 1973; Finlay-Jones and Papadimitriou, 1973; Evan and Dail, 1974; Kloner *et al.*, 1974; Jennings *et al.*, 1975; Ganote *et al.*, 1975) and the major findings are summarized in Fig. 1.1 and exemplified in Figs. 1.2 and 1.3. The earliest changes include a mild degree of cytoplasmic oedema, dilation of endoplasmic reticulum, slight mitochondrial swelling, disaggregation of polysomes, and the appearance of multiple small aggregates of condensed chromatin around the nuclear periphery. These changes are probably all reversible, but in dying cells are followed by a further set of changes that are irreversible; it is the initiation of these which defines the 'point of no return' in the route to death (Trump and Ginn, 1969; Laiho and Trump, 1975). This second set includes 'high amplitude' swelling of mitochondria — a florid dilation with rupture of internal cristae and, usually, development of matrix densities of flocculent or granular types. Later lethal changes include extensive cytoplasmic swelling, dissolution of cytoplasmic organelles and rupture of plasma membranes. Occasionally bizarre local convolutions of plasma membranes are also observed. Lysosomes appear intact until cytoplasmic degradation is advanced (Ginn, Shelburne and Trump, 1968; Buckley, 1972; Hawkins *et al.*, 1972). Intercellular junctions do eventually shear, but cells undergoing necrosis frequently remain adherent to each other or to adjacent basal laminae until apparently irreversible mitochondrial changes are evident. Moreover, remnants of junctional complexes may still be discerned in greatly degraded cells (Saladino and Trump, 1968), a finding consistent with the known chemical stability of these structures (Campbell and Campbell, 1971). In contrast to the cytoplasmic changes, nuclear changes are relatively unremarkable. There is persistence of the increased chromatin margination already noted as a 'sublethal' change. The nucleus swells and the nucleoplasm

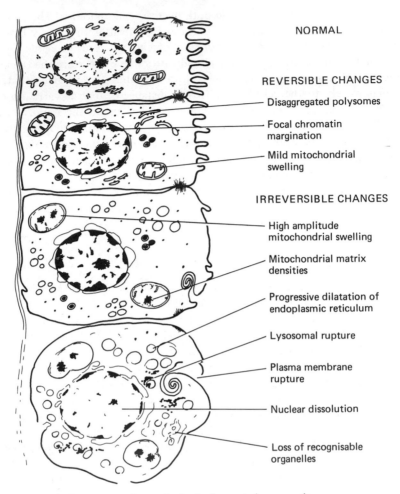

NORMAL

REVERSIBLE CHANGES
— Disaggregated polysomes
— Focal chromatin margination
— Mild mitochondrial swelling

IRREVERSIBLE CHANGES
— High amplitude mitochondrial swelling
— Mitochondrial matrix densities
— Progressive dilatation of endoplasmic reticulum
— Lysosomal rupture
— Plasma membrane rupture
— Nuclear dissolution
— Loss of recognisable organelles

Fig. 1.1 Summary of morphological events in necrosis.

becomes increasingly electron lucent, containing scattered small coarse chromatin masses. Nuclear pores in normal disposition can be observed in cells with the characteristic mitochondrial lesions (Fig. 1.2). It appears that the nucleus may rupture around the same time as the plasma membrane, or it may escape, with other organelles, when the plasma membrane bursts.

Corresponding to these ultrastructural changes, cells undergoing necrosis, when observed in the light microscope (Fig. 1.3), show a uniform, eosinophilic cytoplasm and a nucleus which may show substantially normal staining, some hyperchromatism (pyknosis), although with nucleoli still evident, or complete dissolution (karyolysis). Necrosis usually affects sheets of cells; their ghostly outlines are still evident, even late after the event of death, but at the edges of such sheets neutrophil polymorphonuclear leucocytes and other inflammatory cells are usually observed.

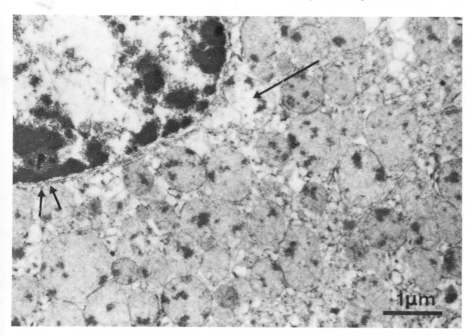

Fig. 1.2 Ultrastructural features of necrosis. This hepatocyte lay in the partially necrotic wall of a fluke track within rat liver. The mitochondria are distended, some showing generally lucent profiles with ruptured cristae (long arrow), and all bearing focal flocculent matrix densities. The normally plentiful rough endoplasmic reticulum is replaced by dilated membranous profiles without attached ribosomes. Nuclear chromatin is aggregated focally, but pores are intact (short arrows). (By courtesy of Dr. John R. Foster, University College, Cardiff.)

1.2.3 Mechanism

The dominant morphological event in necrosis is cellular swelling and there is now much evidence for the theory that the essential defect is loss of control of cell volume (see Chapter 7 and Trump and Mergner, 1974, for reviews, and Fig. 1.4). Loss of selective permeability of the membrane of dying cells is the basis of many tests currently used to assess 'viability' of cells in suspension; such tests include vital dye exclusion and the ability to retain previously internalized radioactive chromium. Freeze-fractured preparations of necrotic cell membranes have demonstrated trans-membrane defects (Ashraf and Halverson, 1977) and studies of cell permeability *in vivo* to colloidal electron-dense material have shown that cells with the morphology of necrosis are abnormally permeable to the marker molecules (Hoffstein, Gennaro, Fox, Hirsch, Streuli and Weissmann, 1975). Certain cells showing only the earlier morphological changes, preceding high amplitude mitochondrial swelling, also take up the marker. This is consistent with the view that

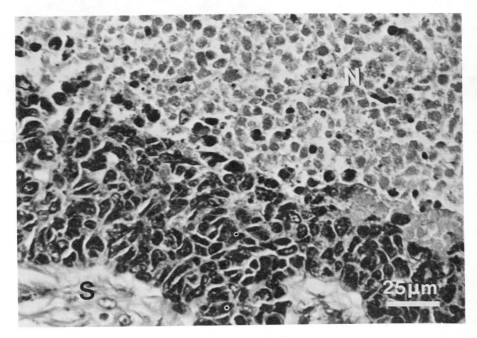

Fig. 1.3 Light microscopic features of necrosis. This poorly differentiated human ovarian carcinoma shows an extensive sheet of necrosis (N) encircled by bands of apparently viable cells, close to the vascular stroma (S). Note the ghostly outlines of the necrotic cells, most of which are completely devoid of chromatin staining.

the abnormal permeability precedes and probably causes the morphological changes, and this is confirmed by the early disappearance of membrane ion-pumping activity in cells of several different types prior to the development of morphologically recognizable necrosis. Membrane function may be rendered defective in this way as a result of the action of diverse injurious agents; injury to the membrane may be direct (for example by complement) or secondary to cellular energy depletion (for example in hypoxia or in the presence of respiratory poisons). Once membrane damage exists, it appears that secondary injury is inflicted on the mitochondria, thus removing the source of oxidative metabolism and the possibility of recovery of membrane ion-pumping activity (Farber and El-Mofty, 1975). There is an obvious parallel between this secondary mitochondrial damage, which heralds irreversibility, and the 'point of no return', alluded to above and associated with morphological change in the mitochondria. At present we do not know exactly how irreversible mitochondrial damage is effected — it may be the result of abnormally high local concentrations of simple material like calcium ions (El-Mofty, Scrutton, Serroni, Nicolini and Farber, 1975; Chien, Abrams, Pfau and Farber, 1977). The granular mitochondrial matrix densities observed in necrosis are in fact rich in calcium, but the flocculent densities, which are more consistently observed,

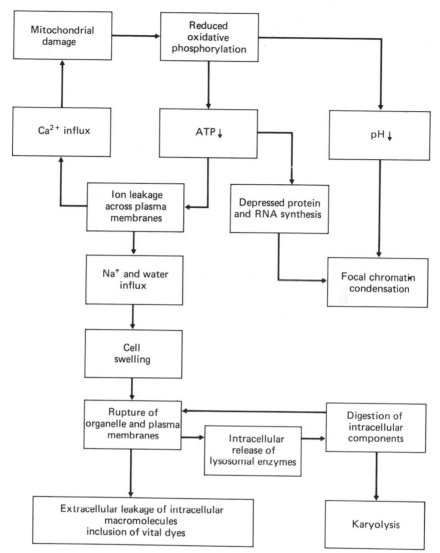

Fig. 1.4 Summary diagram of mechanisms in necrosis (after Trump and Mergner, 1974).

probably represent denatured matrix protein (Buja, Dees, Harling and Willerson, 1976).

Perhaps surprisingly, the role of the lysosome is poorly defined in necrosis. It is certain that lysosomal hydrolases are released in the last stages of cellular dissolution, but irreversible structural changes characteristic of necrosis can occur prior to lysosomal rupture. This has been demonstrated by Hawkins *et al.* (1972) using both iodoacetate (an inhibitor of Krebs cycle) and complement to kill cells in tissue culture, and establishing the integrity of lysosomes by their retention of previously supplied ferritin particles or Acridine Orange.

1.3 Apoptosis

1.3.1 *Incidence*

Apoptosis occurs in normal tissue turnover (Kerr and Searle, 1973; Wyllie, Kerr, Macaskill and Currie, 1973a; Wyllie, Kerr and Currie, 1973b), embryogenesis (Bellairs, 1961; Saunders and Fallon, 1966; Farbman, 1968; Manasek, 1969; Hammar and Mottet, 1971; Schweichel, 1972; O'Connor and Wyttenbach, 1974), metamorphosis (Goldsmith, 1966; Kerr, Harmon and Searle, 1974; Decker, 1976) and endocrine-dependent tissue atrophy (Kerr and Searle, 1973; Wyllie *et al.*, 1973b; Hopwood and Levison, 1976; O'Shea, Hay and Cran, 1978; Sandow, West, Norman and Brenner, 1979), but is apparently reduced in incidence in endocrine-stimulated hyperplasia and hypertrophy (Wyllie *et al.*, 1980). Endothelial cells of blood vessels in endocrine tissues during atrophy also undergo apoptosis (O'Shea, Nightingale and Chamley, 1977). Most growing tumours include apoptotic cells, and their number is increased in tumour regression (Kerr *et al.*, 1972; Searle, Lawson, Abbott, Harmon and Kerr, 1975) and in models of such regression *in vitro* which preclude hypoxia as a causative agent (Robertson, Bird, Waddell and Currie, 1978). Nevertheless, degrees of hypoxia inadequate to produce necrosis do enhance apoptosis in some tissues (Kerr, 1971). Cell killing mediated by T cells has the morphology of apoptosis (Russell, Rosenau and Lee, 1972; Battersby, Egerton, Balderson, Kerr and Burnett, 1974; Slavin and Woodruff, 1974; Sanderson, 1976; Matter, 1979). Low doses of X-irradiation and radiomimetic cytotoxic agents can cause apoptosis (Searle *et al.*, 1975; Potten, 1977; Potten, Al-Barwari and Searle, 1978). In all the circumstances in which apoptosis occurs *in vivo* large numbers of cells may be deleted in a short time, yet leaving intact the overall stromal–parenchymal organization of the tissue.

In concluding this section it should be pointed out that not all the authors whose work is referred to used the term apoptosis, but their descriptions leave no doubt as to the identity of the morphological changes they observed.

1.3.2 *Morphology*

The major morphological changes in apoptosis are summarized in Fig. 1.5. In the light microscope (Fig. 1.6) apoptotic cells in tissues are inconspicuous, appearing singly or in small groups and consisting of portions of intensely staining cytoplasm, usually with a smooth contour and sometimes including in the plane of section small densely basophilic nuclear fragments. *Pyknosis* is thus one of the features of apoptosis in the light microscope, but as we have seen it may also be used to describe hyperchromatism in the nuclei of necrotic cells. Although apoptotic cells are frequently digested by macrophages (*vide infra*), neutrophil polymorphs do not appear to ingest or be attracted by them.

FORMATION OF APOPTOTIC CELLS

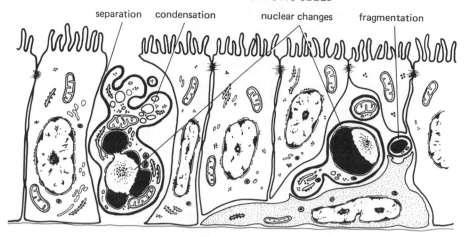

FATE OF APOPTOTIC CELLS

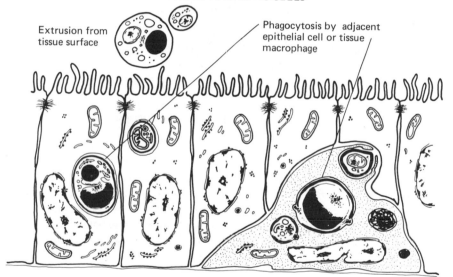

Fig. 1.5 Summary diagram of morphological events in apoptosis.

Ultrastructurally the earliest changes of apoptosis include the loss of cell junctions and other specialized plasma membrane structures such as microvilli. At the same time the cytoplasm becomes condensed and nuclear chromatin marginates into one or several large masses, which initially may 'blister' the nuclear membrane outwards and then coalesce to form crescentic caps around half or more of the nucleus (Fig. 1.7 and 1.8). Nuclear pores are seldom observed adjacent to the masses of condensed chromatin, although they remain clearly visible in parts of the membrane adjacent to uncondensed chromatin. The nucleolus disaggregates

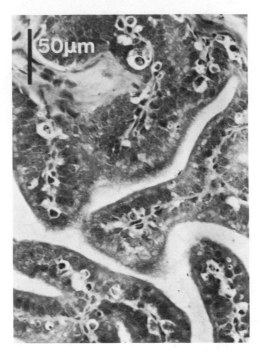

Fig. 1.6 Light microscopic features of apoptosis in rat prostate 4 days after orchidectomy. Numerous apoptotic bodies are present in the epithelium, several including nuclear fragments.

to form a cloud of coarse osmiophilic granules. As the process continues the nuclear membrane adopts complex internal folds and the nucleus may split into several fragments (Fig. 1.9). Eventually the nuclear membrane develops discontinuities, and coarse granular chromatin masses and short fragments of pore-bearing nuclear membrane lie free in the cytoplasm. Meanwhile progressive changes have been occurring in the cytoplasm. There is contraction of cytoplasmic volume apparently associated with loss of intracellular fluid. The cell transiently adopts a deeply convoluted outline (Fig. 1.10). Cytoplasmic organelles – notably mitochrondria and ribosomes – become arranged in densely compacted formations, with little intervening cytosol (Fig. 1.11). Endoplasmic reticulum sometimes dilates, forming prominent vesicles, some of which appear to fuse with the cell surface, and this may be one mechanism whereby intracellular fluid is lost. Subsequently the convoluted cell breaks up into several membrane-bounded smooth-surfaced 'apoptotic bodies' containing a variety of cytoplasmic organelles, and some including nuclear fragments. Even at this stage mitochondrial structure is frequently intact, without any evidence of high amplitude swelling or granular or flocculent matrix densities.

Apoptotic bodies vary greatly in size, the smallest consisting merely of plasma membrane surrounding electron-dense cytoplasm perhaps containing a few ribosomes.

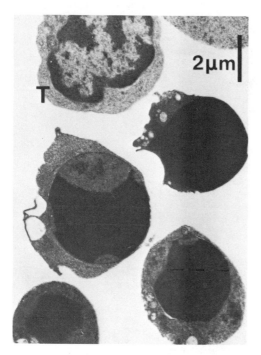

Fig. 1.7 Apoptosis in rat thymocytes *in vitro*, 5h after exposure to gluco-corticoid (10^{-5} mol/l methylprednisolone). Note the sharp distinction between the chromatin of the normal thymocyte (T) and its apoptotic neighbours.

The tendency of the dying cell to produce multiple apoptotic bodies appears to be a function of its size: small lymphocytes of the thymus cortex undergo typical apoptosis but in our experience mostly remain as single bodies, whereas multiple fragments are commonly observed when larger cells such as hepatocytes, tumour cells and adrenal and prostatic epithelial cells undergo apoptosis.

In vivo, apoptotic bodies either are extruded into an adjacent lumen or – more commonly – undergo phagocytosis by nearby cells (Fig. 1.12). The engulfing cells are frequently members of the mononuclear-phagocyte system, but apoptotic bodies may also be ingested by vascular endothelium, by adjacent normal epithelial cells or in the case of carcinomas by nearby tumour cells. Apoptotic bodies apparently provide a potent stimulus for phagocytosis and are seldom observed in the extracellular space without some approaching or adherent phagocyte processes.

Once ingested, the apoptotic body undergoes autolytic destruction within the phagosome; initially mitochondria retain their compacted profiles but flocculent densities appear in their matrices. Eventually through further degradation and condensation the appearance of the engulfed apoptotic body becomes that of any advanced secondary lysosome. Those apoptotic bodies which do not undergo phagocytosis (as may occur *in vivo* after extrusion into a gland lumen, or regularly

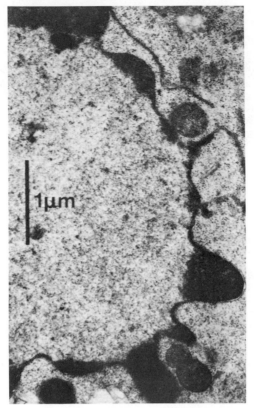

Fig. 1.8 Early nuclear changes of apoptosis in murine lymphoma cell (L5178YC3) following treatment *in vitro* with methylprednisolone. Dense chromatin aggregates form blisters under the nuclear membrane which is still intact. (By courtesy of Alison L. Smith.)

in vitro) show progressive dilation and degradation of cytoplasmic organelles, a process that has been called 'secondary necrosis'.

Observation of apoptosis in cultured cells by time-lapse cinematography has indicated that the process of cellular fragmentation occurs extremely rapidly (Russell *et al.*, 1972; Mullinger and Johnson, 1976), and this concords with the relative rarity with which the convoluted stage is observed *in vivo*, even in tissues in which many cells are known to be undergoing apoptosis. The most commonly observed appearances are various phases of intraphagosomal degradation. After their formation apoptotic cells may remain detectable in tissue sections for over 12h (Wyllie *et al.*, 1980), a time close to that required by phagocytes to destroy other large inclusions (Perkins and Makinodan, 1965; Gordon and Cohn, 1973).

1.3.3 *Mechanism*

Cells undergoing apoptosis within tissues are usually surrounded and outnumbered

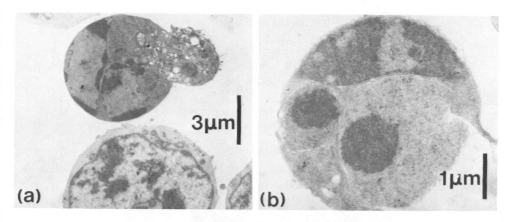

Fig. 1.9 Early apoptosis in a murine lymphoma cell (S49) following treatment with methylprednisolone. In (a) the nucleus has fragmented, and shows caps of dense chromatin. The cell has an unusually lobular outline, with cytoplasmic organelles packed towards one pole. Compare the features of the adjacent normal cell. (b) shows an apoptotic body, at a later stage, consisting largely of nuclear fragments.

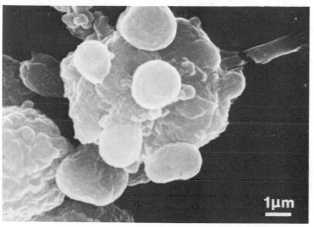

Fig. 1.10 Apoptosis in a human lymphoid cell (from the cell line BLA$_1$) *in vitro* following treatment with methylprednisolone, showing numerous smooth-surfaced cytoplasmic protrusions. (From Robertson *et al.*, 1978, by courtesy of the editor of the Journal of Pathology.)

by viable cells. For this reason few biochemical studies on regressing tissues have told us much about apoptosis itself, although they do provide a quantitative measurement of related tissue change (such as the increase in lysosomal enzyme activity attributable to macrophages engaged in phagocytosis of apoptotic bodies). In only a few cell systems can apoptosis be initiated sufficiently synchronously and in adequate numbers of cells to allow direct study of biochemical processes within

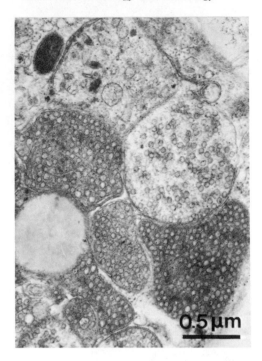

Fig. 1.11 Cytoplasmic compaction within a rat adrenocortical epithelial cell during apoptosis induced *in vivo* by ACTH withdrawal. Although densely stacked together, the mitochondria are structurally intact; in particular, flocculent and granular matrix densities are absent.

them. It is from this limited material that the generalizations below are drawn, and it is of course an assumption that mechanisms pertaining to apoptosis in one cell system apply in another.

Apoptosis requires energy. At least some apoptotic cells retain normal ATP levels, despite major morphological changes in chromatin (Waddell, Nicholson, Durie, Robertson and Wyllie, unpublished work). This contrasts with necrosis, where failure of ATP supply to membrane ionic pumps is regarded as a central feature, but it is entirely consistent with data showing that apoptotic cells are metabolically active, exclude sodium and retain potassium normally (Gullino and Lanzerotti, 1972; Walker and Lucas, 1972; Rosenau, Goldberg and Burke, 1973), and exclude vital dyes (Waddell, Wyllie, Robertson, Mayne, Au and Currie, 1979).

Possibly relevant to this argument is the role of adenosine $3':5'-$cyclic monophosphate (cyclic AMP) in cell death. Cyclic AMP or its dibutyryl derivative causes regression of some primary carcinogen-induced animal tumours, several transplantable tumours and many cell lines *in vitro* (see review by Cho-Chung, 1979). In some

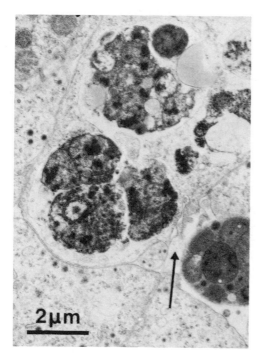

Fig. 1.12 Phagocytosis of apoptotic bodies deriving from rat adrenocortical epithelial cells. The tissue, taken from the cortico-medullary boundary, shows a macrophage process extending towards a condensed apoptotic body lying between two medullary cells (arrow). Within the macrophage cytoplasm are several phagosomes, including cortical epithelial apoptotic bodies in various stages of degradation, but recognizable by their characteristic mitochondria. Note the presence at this stage of prominent flocculent matrix densities in the mitochondria of the apoptotic bodies.

instances the regression is known to be mediated by cell death with the morphology of apoptosis. Dibutyryl cyclic AMP also causes premature deletion of embryonic palatal shelf tissue (Pratt and Martin, 1975), where again the morphology is that of apoptosis, and Epidermal Growth Factor, which blocks this deletion, is thought to exert its effect through depletion of endogenous cyclic AMP (Hassell and Pratt, 1977). The mode of action of cyclic AMP in cell death is not fully understood, but in regressing mammary tumour models, where it has been extensively studied, a necessary step is binding to a cytoplasmic protein which then translocates to the nucleus. This protein is very closely associated with, and probably identical to, cyclic AMP-dependent protein kinase, and the result of its arrival in the nucleus, together with cyclic AMP, is the appearance of new species of phosphorylated nuclear proteins (Cho-Chung, 1979). As cyclic AMP generation is dependent upon ATP, it seems unlikely that this mechanism could operate in cells sufficiently depleted of ATP to produce failure of membrane pumping mechanisms.

Apoptosis requires macromolecular synthesis. Inhibitors of translation suppress apoptosis in several situations (Weber, 1965; Tata, 1966; Lieberman, Verbin, Landay, Liang, Farber, Lee and Starr, 1970; Ben-Ishay and Farber, 1975; Pratt and Greene, 1976; Wyllie, unpublished work). In some of these – but not all (Pratt and Greene, 1976) – inhibitors of transcription exert a similar effect, but it is uncertain whether this is due to blockade of RNA synthesis or to the concurrent depression of protein synthesis.

Apoptosis may result from new gene expression. Glucocorticoids, thyroxine and ecdysone induce new gene expression in many cell types on which they exert important biological effects. All three hormones can also cause apoptosis in the appropriate target cells (respectively, lymphoid cells, anuran tail tissues and insect imaginal disc cells). It is very tempting to conclude that apoptosis – in at least these cell systems – requires new gene expression also. Indirect support for this conclusion derives from the effectiveness of cortexolone in inhibiting glucocorticoid-induced apoptosis in thymus cells (Wyllie, unpublished work); this drug competes with active glucocorticoid for the cytoplasmic receptors involved in the early steps of hormone translocation to the nucleus. Further, RNA polymerase B activity increases in two very different cell types under conditions known to lead to death (Kunitomi, Rosenau, Burke and Goldberg, 1975; Borthwick and Bell, 1978) by a process morphologically identifiable as apoptosis. It is not known whether this activity directs the synthesis of functional message, but sensitive methods such as two-dimensional electrophoresis should eventually provide a clear answer as to whether newly synthesized protein species are consistently associated with apoptosis.

In contrast, there is already strong evidence that apoptosis is under *remote* genetic control. Both avian and insect mutants exist (Fristrom, 1969; Hinchliffe and Ede, 1973; Hinchliffe and Thorogood, 1974; Murphy, 1974) in which apoptosis is either excessive or deficient in certain sites.

Apoptosis is associated with endogenous endonuclease activation. Recent evidence from glucocorticoid-treated thymocytes and lymphoid cell lines indicates that the chromatin condensation of apoptosis is associated with endogenous endonuclease activity which excises nucleosome chains from nuclear chromatin (Wyllie, 1980). The enzyme responsible has not yet been isolated and we know little about its properties. It is possible that it may be a cation-sensitive enzyme similar to that described by Hewish and Burgoyne (1973) and now known to be a component of chromatin (Ishida, Akiyoshi and Takahashi, 1974). At any rate it is unlikely to be a lysosomal enzyme (Wyllie, 1980). The appearance of this new activity in apoptotic cells is consistent with the proposition that the nuclease itself – or perhaps its activators – may be amongst the macromolecules on whose synthesis apoptosis depends in certain cell systems. Alternatively, activation of *pre-existing* nuclease might result from arrival in the nucleus of a cyclic AMP-dependent protein kinase of the type already discussed. Apoptosis involves several changes in the cell other than those in chromatin (for example contraction, blebbing and surface alterations permitting recognition by phagocytes), so it is likely that these propositions at best represent only part of the story. One of their attractions is that they can be tested by direct experiments.

1.4 Validity of the classification

The aim of this section is to examine the validity of the morphological classification of cell death presented above, in terms of its relevance, scope and practicability.

1.4.1 *Relevance*

A strong case can be made that the proposed classification, although based on structure, delineates processes which differ from each other in their fundamental biology. We have seen that whilst necrosis occurs in many non-physiological circumstances and is hardly ever involved in homoeostatic regulation of cell numbers, apoptosis is frequently physiological, and regulated by changes in the levels of recognized trophic hormones or by lesser known but undoubtedly physiological factors in embryonic development. Further, necrosis apparently represents loss of plasma membrane volume control, initiated, or at least perpetuated by collapse of cellular energy supply, but the initiation of apoptosis in several different cell types has been shown to require continuing macromolecular synthesis and perhaps new gene activation and involves the early appearance of non-lysosomal nuclease activity in the nucleus.

A stringent test of the relevance of this classification to events undetectable by morphological study alone would be demonstration of its predictive value. Does the finding of apoptosis imply that intracellular initiating processes involving synthetic activity should always be discovered, however improbable the circumstances? One challenge to the classification – at present not fully answered – is the frequent finding of apoptosis after X-irradiation and treatment with radiomimetic anti-cancer drugs (Trowell, 1966; Searle *et al.*, 1975; Potten, 1977). These agents represent 'environmental perturbation' and have long been known to damage important intracellular macromolecules such as DNA; it would be reasonable to expect the mode of death to be necrosis. However, recent evidence shows that the lethal effects of X-irradiation and some of the radiomimetic drugs are substantially influenced by the nature of the target cell: cell populations (such as gut crypt epithelium) which normally proliferate rapidly are sensitive to unexpectedly low doses (Potten, 1977). Further, at a given treatment dose, cell populations of normally high proliferation rate remain more sensitive than normally slow-turnover populations which have been stimulated to divide at an equivalent rate (Farber and Baserga, 1969; Alison and Wright, 1979). It is possible that DNA damage, which in slow-turnover populations is sublethal (and presumably repaired), triggers apoptosis in cells of high-turnover populations. At a teleological level one may postulate that in these rapidly proliferating populations, deletion of damaged cells by apoptosis affords a long-term advantage over the consequences of their survival after error-prone repair. Hopefully, continuing study of the intracellular effects of X-irradiation will answer the question whether this cell death does indeed depend – as predicted from its morphology – on the activation of intrinsic processes.

1.4.2 Scope

A different type of test of the validity of the classification is posed by challenging its ability to encompass all or even most of the cell death which occurs. There are obvious constraints in attempting to meet this challenge. Apoptosis itself, which is now documented in a variety of tissues and circumstances (Wyllie *et al.*, 1980), was largely unobserved until recently and it is theoretically possible that other morphological forms of death at present suffer similar disregard. Our own experience, however, and review of the descriptions of others, suggests that apoptosis and necrosis together account for most cell death, at least in the phylogenetically higher animals. In the paragraphs below, three of the more obvious challenges to the comprehensiveness of the classification are discussed.

Anuran metamorphosis is one of the circumstances in which claims that cells may die in many different ways have received some documentation (see Beckingham Smith and Tata, 1976, for bibliography). As indicated earlier, however, many of the appearances can be interpreted as stages in the phagocytosis and degradation of apoptotic cells, which have also been clearly identified (Kerr *et al.*, 1974; Decker, 1976). Some of the difficulty in interpreting changes in the regressing anuran tail arises from the fact that apoptosis in the striped muscle cells is modified by the cells' own structural peculiarities; the cardinal features of apoptosis are all present, however: nuclear condensation, cytoplasmic fragmentation to membrane-bounded bodies, and swift phagocytosis.

A type of reaction which poses more difficulty in classification is the appearance of '*Type B*' or '*very dark*' cells, as described in both physiological and pathological circumstances (Johannisson, 1968; Nussdorfer, 1970; Kosek, Mazze and Cousins, 1972). These cells show electron-dense cytoplasm and nucleus, dilated endoplasmic reticulum, conserved intercellular junctions and nuclear pores and, within the dense nucleus, retention of nucleolar structure. Electron radioautography suggests that they are unable to undertake synthetic functions seen in adjacent cells with normal morphology (Nussdorfer, 1970). Such 'very dark' cells appear to share some of the features described in injured cells prior to the development of necrosis (Reimer, Ganote and Jennings, 1972; Brown, 1977). It is not certain, however, that the changes in 'very dark' cells are irreversible. Further, some writers doubt whether all such dark cells are a true representation of a biological process at all; some may be artefacts of fixation (Ganote and Moses, 1968; Rhodin, 1971; Brown, 1977). Under these circumstances it seems reasonable not to include 'very dark' cells as a separate mode of death in this classification.

In *lower animals*, the situation appears more complicated. 'Regulated' or 'programmed' cell death occurs, in so far as this event can ever be recognized by circumstances alone, but the morphology is sometimes difficult to classify, or possesses prominent features which differ from those of vertebrate apoptosis. Saturniid moth intersegmental muscles during metamorphosis, for example, show chromatin condensation and subdivision of the sarcoplasm by expansion of the transverse tubular system, to produce multiple membrane-bounded cytoplasmic bodies. Degeneration of contractile elements appears to be initiated without lysosomal activation (Beaulaton and Lockshin, 1977). These features closely resemble those of apoptosis in vertebrate muscle, but the insect fibres also show very widespread autophagic activity which gives them an unusual appearance. In the developing nematode larva, the highly predictable deletion of cells in the ventral nerve cord shares some but not all of the features of vertebrate apoptosis (A.M.G. Robertson, personal communication) In particular, the chromatin condensation is less dramatic, although rapid phagocytosis and cellular fragmentation do occur. Finally, the dying cells in starved and aging planaria adopt a morphology differing widely from apoptosis, and more akin to necrosis (Bowen and Ryder, 1974; Bowen, Ryder and Dark, 1976).

In summary, classification of death into apoptosis and necrosis appears to provide a fairly comprehensive coverage of the types of death currently recognized in the phylogenetically higher animals, and in these animals 'programmed' death is associated with apoptosis. In lower animals, however, 'programmed' death shows morphological features which differ somewhat from those of mammalian apoptosis, or may be more similar to necrosis.

1.4.3 *Practicability*

In general, necrosis and apoptosis can be distinguished readily, although recourse to ultrastructure may be necessary. This section deals with two circumstances in which the distinction between the two modes of death may be blurred. First, the late stages of apoptosis, whether or not intraphagosomal, involve cytoplasmic degradation and can be confused with necrosis. Cells showing such late degradative changes are sometimes unclassifiable. Since the characteristic early changes in apoptosis are of short duration, the investigator may be confronted with material in which all, or at any rate the great majority of, dying cells show these unclassifiable changes. In such circumstances there is little that can be done other than redesigning the experiment to focus on the earlier and more definitive changes. Secondly, ingested apoptotic bodies are sometimes mistaken for autophagic vacuoles. Here, the presence of organelles atypical of the ingesting cell, or of nuclear material within its phagosomes, clearly differentiates between autophagic activity and apoptosis, but the size of the ingested body may itself provide some guide. It is probable that most 'autophagic vacuoles' larger than a cell nucleus in size are phagocytosed apoptotic bodies.

1.5 Conclusions and summary

Cell death is not a single entity but is heterogeneous in structure, mechanisms, circumstances of initiation and biological function. A classification is offered, based on clearly recognizable morphological features, which appears to distinguish death resulting from pathological deviation in the environment (necrosis) from that occurring in more physiological circumstances (apoptosis). The morphology of each concords with what is known of its mechanisms. Thus necrosis shows cytoplasmic swelling and mitochondrial damage consistent with the hypothesis that it results from failure in osmotic regulation, caused or perpetuated by loss of cellular energy supplies. Apoptosis shows cytoplasmic condensation and prominent nuclear changes, the latter closely associated with endonuclease activation. In many cell types apoptosis fails to proceed in the absence of intact macromolecular synthesis. These observations are compatible with the hypothesis that apoptosis represents a cellular function activated from within.

These two modes of death appear to account for most cell death in higher organisms, but in phylogenetically lower animals the connotations of necrosis may be different, including homoeostatic regulation, and morphological forms deviating somewhat from apoptosis may also exist.

References

Alison, M. R. and Wright, N. A. (1979), Differential lethal effects of both cytosine arabinoside and hydroxyurea on jejunal crypt cells and testosterone-stimulated accessory sex glands. *Cell Tissue Kinet.*, **12**, 477–491.

Ashraf, M. and Halverson, C. A. (1977), Structural changes in the freeze-fractured sarcolemma of ischaemic myocardium. *Am. J. Pathol.*, **88**, 583–594.

Battersby, C., Egerton, W. S., Balderson, G., Kerr, J. F. and Burnett, W. (1974), Another look at rejection in pig liver homografts. *Surgery (St. Louis)*, **76**, 617–623.

Beaulaton, J. and Lockshin, R. A. (1977), Ultrastructural study of the normal degeneration of the intersegmental muscles of *Antheraea polyphemus* and *Manduca sexta* (Insecta, Lepidoptera) with particular reference to cellular autophagy. *J. Morphol.*, **154**, 39–58.

Beckingham-Smith, K. and Tata, J. R. (1976), Cell death. Are new proteins synthesised during hormone-induced tadpole tail regression? *Exp. Cell Res.*, **100**, 129–146.

Bellairs, R. (1961), Cell death in chick embryos as studied by electron microscopy. *J. Anat.*, **95**, 54–60.

Ben-Ishay, Z. and Farber, E. (1975), Protective effects of an inhibitor of protein synthesis, cycloheximide, on bone marrow damage induced by cytosine arabinoside or nitrogen mustard. *Lab. Invest.*, **33**, 478–490.

Bessis, M. (1964), Studies on cell agony and death: an attempt at classification. In *Ciba Foundation Symposium on Cellular Injury* (eds. A. V. S. de Reuck and J. Knight), Churchill, London, pp. 287–316.

Borthwick, N. M. and Bell, P. A. (1978), Glucocorticoid regulation of rat thymus RNA polymerase activity: the role of RNA and protein synthesis. *Mol. Cell. Endocrinol.*, **9**, 269–278.

Bowen, I. D. and Ryder, T. A. (1974), Cell autolysis and deletion in the planarian *Polycelis tenuis* Iijima. *Cell Tissue Res.*, **154**, 265–274.

Bowen, I. D., Ryder, T. A. and Dark, C. (1976), The effects of starvation on the planarian worm *Polycelis tenuis* Iijima. *Cell Tissue Res.*, **169**, 193–209.

Brown, A. W. (1977), Structural abnormalities in neurones. *J. Clin. Pathol.*, **30**, Suppl. 11 (Royal College of Pathologists Symposium on Hypoxia and Ischaemia), 155–169.

Buckley, I. K. (1972), A light and electron microscopic study of thermally injured cultured cells. *Lab. Invest.*, **26**, 201–209.

Buja, L. M., Dees, J. H., Harling, D. F. and Willerson, J. T. (1976), Analytical electron microscopic study of mitochondrial inclusions in canine myocardial infarcts. *J. Histochem. Cytochem.*, **24**, 508–516.

Campbell, R. D. and Campbell, J. H. (1971), Origin and continuity of desmosomes.

In *Origin and Continuity of Cell Organelles* (*Results and Problems in Cell Differentiation*, vol. 2) (eds. J. Reinert and H. Ursprung), Springer-Verlag, Berlin, Heidelberg, New York, pp. 261–298.

Chien, K. R., Abrams, J., Pfau, R. G. and Farber, J. L. (1977), Prevention by chlorpromazine of ischaemic liver cell death. *Am. J. Pathol.*, **88**, 539–558.

Cho-Chung, Y. S. (1979), Cyclic AMP and tumour growth *in vivo*. In *Influences of Hormones in Tumour Development*, vol. 1 (eds. J. A. Kellen and R. Hilf), CRC Press, Florida, pp. 55–93.

Decker, R. S. (1976), Influence of thyroid hormones on neuronal death and differentiation in larval *Rana pipiens. Dev. Biol.*, **49**, 101–118.

El-Mofty, S. K., Scrutton, M. C., Serroni, A., Nicolini, C. and Farber, J. L. (1975), Early reversible plasma membrane injury in galactosamine-induced liver cell death. *Am. J. Pathol.*, **79**, 579–596.

Evan, A. P. and Dail, W. G. (1974), The effects of sodium chromate on the proximal tubules of the rat kidney. Fine structural damage and lysozymuria. *Lab. Invest.*, **30**, 704–715.

Farber, E. and Baserga, R. (1969), Differential effects of hydroxyurea on survival of proliferating cells *in vivo. Cancer Res.*, **29**, 136–139.

Farber, J. L. and El-Mofty, S. K. (1975), The biochemical pathology of liver cell necrosis. *Am. J. Pathol.*, **81**, 237–250.

Farbman, A. I. (1968), Electron microscope study of palate fusion in mouse embryos. *Dev. Biol.*, **18**, 93–116.

Finlay-Jones, J.-M. and Papadimitriou, J. M. (1973), The autolysis of neonatal pulmonary cells *in vitro*: an ultrastructural study. *J. Pathol.*, **111**, 125–134.

Franko, A. J. and Sutherland, R. M. (1979), Oxygen diffusion distance and development of necrosis in multicell spheroids. *Radiat. Res.*, **79**, 439–453.

Fristrom, D. (1969), Cellular degeneration in the production of some mutant phenotypes in *Drosophila melanogaster. Mol. Gen. Genet.*, **103**, 363–379.

Ganote, C. E. and Moses. H. L. (1968), Light and dark cells as artifacts of liver fixation. *Lab. Invest.*, **18**, 740–745.

Ganote, C. E., Seabra-Gomes, R., Nayler, W. G. and Jennings, R. B. (1975), Irreversible myocardial injury in anoxic perfused rat hearts. *Am. J. Pathol.*, **80**, 419–450.

Ginn, F. L., Shelburne, J. D. and Trump, B. F. (1968), Disorders of cell volume regulation. I. Effects of inhibition of plasma membrane adenosine triphosphatase with ouabain. *Am. J. Pathol.*, **53**, 1041–1071.

Glücksmann, A. (1951), Cell deaths in normal vertebrate ontogeny. *Biol. Rev. Cambridge Philos. Soc.*, **26**, 59–86.

Goldsmith, M. (1966), The anatomy of cell death. *J. Cell Biol.*, **31**, 41 abstr.

Gordon, S. and Cohn, Z. A. (1973), The macrophage. *Int. Rev. Cytol.*, **36**, 171–214.

Gullino, P. M. and Lanzerotti, R. H. (1972), Mammary tumour regression. II. Autophagy of neoplastic cells. *J. Nat. Cancer Inst.*, **49**, 1349–1356.

Hammar, S. P. and Mottet, N. K. (1971), Tetrazolium salt and electron microscopic studies of cellular degeneration and necrosis in the interdigital areas of the developing chick limb. *J. Cell Sci.*, **8**, 229–251.

Hassell, J. R. and Pratt, R. M. (1977), Elevated levels of cAMP alters the effect of epidermal growth factor *in vitro* on programmed cell death in the secondary palate epithelium. *Exp. Cell. Res.*, **106**, 55–62.

Hawkins, H. K., Ericsson, J. L. E., Biberfeld, P. and Trump, B. F. (1972), Lysosome and phagosome stability in lethal cell injury. *Am. J. Pathol.*, **68**, 255–288.

Hewish, D. R. and Burgoyne, L. A. (1973), Chromatin sub-structure. The digestion of chromatin DNA at regularly spaced sites by a nuclear deoxyribonuclease. *Biochem, Biophys. Res. Commun.*, **52**, 504–510.

Hinchliffe, J. R. and Ede, D. A. (1973), Cell death in the development of limb form and skeletal pattern in normal and *wingless* (*ws*) chick embryos. *J. Embryol. Exp. Morphol.*, **30**, 753–772.

Hinchliffe, J. R. and Thorogood, P. V. (1974), Genetic inhibition of mesenchymal cell death and the development of form and skeletal pattern in the limbs of *talpid*[3] (*ta*[3]) mutant chick embryos. *J. Embryol. Exp. Morphol.*, **31**, 747–760.

Hoffstein, S., Gennaro, D. E., Fox, A. C., Hirsch, J., Streuli, F. and Weissmann, G. (1975), Colloidal lanthanum as a marker for impaired plasma membrane permeability in ischaemic dog myocardium. *Am. J. Pathol.*, **79**, 207–218.

Hopwood, D. and Levison, D. A. (1976), Atrophy and apoptosis in the cyclical human endometrium. *J. Pathol.*, **119**, 159–166.

Ishida, R., Akiyoshi, H. and Takahashi, T. (1974), Isolation and purification of calcium and magnesium dependent endonuclease from rat liver nuclei. *Biochem. Biophys. Res. Commun.*, **56**, 703–710.

Jennings, R. B., Ganote, C. E. and Reimer, K. A. (1975), Ischaemic tissue injury. *Am. J. Pathol.*, **81**, 179–198.

Johannisson, E. (1968), The foetal adrenal cortex in the human. *Acta Endocrinol. (Copenhagen)*, **58**, Suppl. **130**, 7–107.

Kerr, J. F. R. (1971), Shrinkage necrosis: a distinct mode of cellular death. *J. Pathol.*, **105**, 13–20.

Kerr, J. F. R., Harmon, B. and Searle, J. (1974), An electron-microscope study of cell deletion in the anuran tadpole tail during spontaneous metamorphosis with special reference to apoptosis of striated muscle fibres. *J. Cell Sci.*, **14**, 571–585.

Kerr, J. F. R. and Searle, J. (1973), Deletion of cells by apoptosis during castration-induced involution of rat prostate. *Virchows Arch. Abt. B.*, **13**, 87–102.

Kerr, J. F. R., Wyllie, A. H. and Currie, A. R. (1972), Apoptosis: a basic biological phenomenon with wide-ranging implications in tissue kinetics. *Br. J. Cancer*, **26**, 239–257.

Kloner, R. A., Ganote, C. E., Whalen, D. A. and Jennings, R. B. (1974), Effect of a transient period of ischaemia on myocardial cells II. Fine structure during the first few minutes of reflow. *Am. J. Pathol.*, **74**, 399–422.

Kosek, J. C., Mazze, R. I. and Cousins, M. J. (1972), The morphology and pathogenesis of nephrotoxicity following methoxyflurane (penthrane) anaesthesia. An experimental model in rats. *Lab. Invest.*, **27**, 575–580.

Kunitomi, G., Rosenau, W., Burke, G. C. and Goldberg, M. L. (1975), Alterations in RNA synthesis in lymphotoxin-treated target cells. *Am. J. Pathol.*, **80**, 249–260.

Laiho, K. U. and Trump, B. F. (1975), Studies on the pathogenesis of cell injury. Effect of inhibitors of metabolism and membrane function on the mitochondria of Ehrlich ascites tumor cells. *Lab. Invest.*, **32**, 163–182.

Laird, A. K. (1969), Dynamics of growth in tumors and in normal organisms. In *Human Tumor Cell Kinetics (National Cancer Institute Monograph 30)* (ed. S. Perry), National Cancer Institute, Bethesda, pp. 15–28.

Lieberman, M. W., Verbin, R. S., Landay, M., Liang, H., Farber, E., Lee, T.-N. and Starr, R. (1970), A probable role for protein synthesis in intestinal epithelial cell damage induced *in vivo* by cytosine arabinoside, nitrogen mustard or X-irradiation. *Cancer Res.*, **30**, 942–951.

Manasek, F. J. (1969), Myocardial cell death in the embryonic chick ventricle. *J. Embryol. Exp. Morphol.*, **21**, 271–284.

Matter, A. (1979), Microcinematographic and electron microscopic analysis of target cell lysis induced by cytotoxic T lymphocytes. *Immunology*, **36**, 179–190.

McDowell, E. M. (1972a), Light and electron microscopic studies of the rat kidney after administration of inhibitors of the citric acid cycle *in vivo* I. Effects of sodium fluoroacetate on the proximal convoluted tubule. *Am. J. Pathol.*, **66**, 513–542.

McDowell, E. M. (1972b), Light- and electron-microscope studies of the rat kidney after administration of inhibitors of the citric acid cycle *in vivo*: changes in the proximal convoluted tubule during fluorocitrate poisoning. *J. Pathol.*, **108**, 303–318.

McLean, A. E. M., McLean, E. and Judah, J. D. (1965), Cellular necrosis in the liver induced and modified by drugs. *Int. Rev. Exp. Pathol.*, **4**, 127–157.

Mullinger, A. M. and Johnson, R. T. (1976), Perturbation of mammalian cell division. III. The topography and kinetics of extrusion subdivision. *J. Cell Sci.*, **22**, 243–285.

Murphy, C. (1974), Cell death and autonomous gene action in lethals affecting imaginal discs in *Drosophila melanogaster. Dev. Biol.*, **39**, 23–36.

Nussdorfer, G. G. (1970), The fine structure of the newborn rat adrenal cortex II. Zona juxtamedullaris. *Z. Zellforsch. Mikrosk. Anat.*, **103**, 398–409.

O'Connor, T. M. and Wyttenbach, C. R. (1974), Cell death in the embryonic chick spinal cord. *J. Cell Biol.*, **60**, 448–459.

O'Shea, J. D., Hay, M. F. and Cran, D. G. (1978), Ultrastructural changes in the theca interna during follicular atresia in sheep. *J. Reprod. Fertil.*, **54**, 183–187.

O'Shea, J. D., Nightingale, M. G. and Chamley, W. A. (1977), Changes in the small blood vessels during cyclical luteal regression in sheep. *Biol. Reprod.*, **17**, 162–177.

Perkins, E. H. and Makinodan, T. (1965), The suppressive role of mouse peritoneal phagocytes in agglutinin response. *J. Immunol.*, **94**, 765–777.

Potten, C. S. (1977), Extreme sensitivity of some intestinal crypt cells to X- and γ-irradiation. *Nature (London)*, **269**, 518–521.

Potten, C. S., Al-Barwari, S. E. and Searle, J. (1978), Differential radiation response amongst proliferating epithelial cells. *Cell Tissue Kinet.*, **11**, 149–160.

Pratt, R. M. and Greene, R. M. (1976), Inhibition of palatal epithelial cell death by altered protein synthesis. *Dev. Biol.*, **54**, 135–145.

Pratt, R. M. and Martin, G. R. (1975), Epithelial cell death and cyclic AMP increase during palatal development. *Proc. Nat. Acad. Sci. U.S.A.*, **72**, 874–877.

Prieto, A., Kornblith, P. L. and Pollen, D. A. (1967), Electrical recordings from meningioma cells during cytolytic action of antibody and complement. *Science*, **157**, 1185–1187.

Reimer, K. A., Ganote, C. E. and Jennings, R. B. (1972), Alterations in renal cortex following ischaemic injury III. Ultrastructure of proximal tubules after ischaemia or autolysis. *Lab. Invest.*, **26**, 347–363.

Rhodin, J. A. G. (1971), The ultrastructure of the adrenal cortex of the rat under normal and experimental conditions. *J. Ultrastruct. Res.*, **34**, 23–71.

Robertson, A. M. G., Bird, C. C., Waddell, A. W. and Currie, A. R. (1978), Morphological aspects of glucocorticoid-induced cell death in human lymphoblastoid cells. *J. Pathol.*, **126**, 181–187.

Rosenau, W., Goldberg, M. L. and Burke, G. C. (1973), Early biochemical alterations induced by lymphotoxin in target cells. *J. Immunol.*, **111**, 1128–1135.

Russell, S. W., Rosenau, W. and Lee, J. C. (1972), Cytolysis induced by human lymphotoxin. Cinematographic and electron microscopic observations. *Am. J. Pathol.*, **69**, 103–118.

Saladino, A. J. and Trump, B. F. (1968), Ion movements in cell injury; effects of inhibition of respiration and glycolysis on the ultrastructure and function of the epithelial cells of the toad bladder. *Am. J. Pathol.*, **52**, 737–776.

Sanderson, C. J. (1976), The mechanism of T cell mediated cytotoxicity II. Morphological studies of cell death by time-lapse microcinematography. *Proc. R. Soc. London Ser. B*, **192**, 241–255.

Sandow, B. A., West, N. B., Norman, R. L. and Brenner, R. M. (1979), Hormonal control of apoptosis in hamster uterine luminal epithelium. *Am. J. Anat.*, **156**, 15–36.

Saunders, J. W. (1966), Death in embryonic systems. *Science*, **154**, 604–612.

Saunders, J. W. and Fallon, J. F. (1966), Cell death in morphogenesis. In *Major Problems in Developmental Biology* (25th Symposium of the Society for Developmental Biology) (ed. M. Locke), Academic Press, New York, pp. 289–314.

Schatten, W. E. (1962), An experimental study of necrosis in tumors. *Cancer Res.*, **22**, 286–290.

Schatten, W. E. and Burson, J. L. (1965), Effect of altering blood flow on necrosis in tumors. *Neoplasma*, **12**, 435–440.

Schweichel, J. U. (1972), Electron microscopical studies on the degradation of the apical ridge during the development of limbs in rat embryo. *Z. Anat. Entwicklungsgcs.*, **136**, 192–203.

Searle, J., Lawson, T. A., Abbott, P. J., Harmon, B. and Kerr, J. F. R. (1975), An electron-microscope study of the mode of cell death induced by cancer-chemotherapeutic agents in populations of proliferating normal and neo-plastic cells. *J. Pathol.*, **116**, 129–138.

Slavin, R. E. and Woodruff, J. M. (1974), The pathology of bone marrow transplantation. *Pathol. Annu.*, **9**, 291–344.

Steel, G. G. (1977), *Growth Kinetics of Tumours*. Clarendon Press, Oxford.

Tata, J. R. (1966), Requirement for RNA and protein synthesis for induced regression of tadpole tail in organ culture. *Dev. Biol.*, **13**, 77–94.

Thomlinson, R. H. and Gray, L. H. (1955), The histological structure of some human lung cancers and the possible implications for radiotherapy. *Br. J. Cancer*, **9**, 539–549.

Trowell, O. A. (1966), Ultrastructural changes in lymphocytes exposed to noxious agents *in vitro*. *Q. J. Exp. Physiol.*, **51**, 207–220.

Trump, B. F. and Ginn, F. L. (1969), The pathogenesis of subcellular reaction to lethal injury. In *Methods and Achievements in Experimental Pathology*, vol. 4 (eds. E. Bajusz and G. Jasmin), Karger, Basle, pp. 1–29.

Trump, B. F., Goldblatt, P. J. and Stowell, R. E. (1965), Studies on necrosis of mouse liver *in vitro*. Ultrastructural alterations in the mitochondria of hepatic parenchymal cells. *Lab. Invest.*, **14**, 343–371.

Trump, B. F. and Mergner, W. J. (1974), Cell Injury. In *The Inflammatory Process*, vol. 1, 2nd ed. (eds. B. W. Zweifach, L. Grant and R. T. McCluskey), Academic Press, New York, San Francisco, London, pp. 115–257.

Trump, B. F., Valigorsky, J. M., Dees, J. H., Mergner, W. J., Kim, K. M., Jones, R. T., Pendergrass, R. E., Garbus, J. and Cowley, R. A. (1973), Cellular change in human disease. A new method of pathological analysis. *Hum. Pathol.*, **4**, 89–109.

Waddell, A. W., Wyllie, A. H., Robertson, A. M. G., Mayne, K., Au, J. and Currie, A. R. (1979), Lethal and growth-inhibitory actions of glucocorticoids. In *Glucocorticoid Action and Leukaemia* (7th Tenovus Workshop) (eds. P. A. Bell and N. M. Borthwick), Alpha Omega, Cardiff, pp. 75–83.

Walker, S. M. and Lucas, Z. J. (1972), Cytotoxic activity of lymphocytes II. Studies on mechanism of lymphotoxin-mediated cytotoxicity. *J. Immunol.*, **109**, 1233–1244.

Weber, R. (1965), Inhibitory effect of actinomycin D on tail atrophy in *Xenopus* larvae at metamorphosis. *Experientia*, **21**, 665–666.

Wyllie, A.H. (1980), Glucocorticoid induces in thymocytes a nuclease-like activity associated with the chromatin condensation of apoptosis. *Nature (London)*, **284**, 555–556.

Wyllie, A. H., Kerr, J. F. R. and Currie, A. R. (1973b), Cell death in the normal neonatal rat adrenal cortex. *J. Pathol.*, **111**, 255–261.

Wyllie, A. H., Kerr, J. F. R. and Currie, A. R. (1980), Cell death: the significance of apoptosis. *Int. Rev. Cytol.*, **68**, 251–306.

Wyllie, A. H., Kerr, J. F. R., Macaskill, I. A. M. and Currie, A. R. (1973a), Adreno-cortical cell deletion: the role of ACTH. *J. Pathol.*, **111**, 85–94.

2 Cell death in embryogenesis

J. R. HINCHLIFFE

2.1 Introduction

Cell death is a paradox of development which is usually considered to be a process of growth and diversification. Why should an embryo destroy its resources of energy, nutrition and information? Nevertheless, it is now clear that substantial cell death is a normal part of embryonic development (reviewed in Saunders, 1966; Saunders and Fallon, 1967), and its frequent incidence is charted by Glücksmann (1951). Glücksmann distinguishes three types of cell death in development: (i) morphogenetic, involved in alterations in form, (ii) histogenetic, involved in tissue or organ differentiation and (iii) phylogenetic, removing larval and vestigial structures. In fact, these categories overlap, and the mechanisms of cell death do not appear to differ according to category. The process of rudimentation – organ or tissue regression (whether phylogenetic or morphogenetic) and its control – has recently been extensively reviewed in a congress report (Raynaud, 1977a).

The examples reviewed here are chosen to illustrate the positive contribution of cell death to the emergence of embryonic form and also to illustrate the variety of control mechanisms now revealed by experimental approaches. The developing amniote limb is considered in detail, since it demonstrates dramatically the contribution of interdigital cell death to limb shaping, as well as positional and genetic control mechanisms and phylogenetic differences in pattern (Saunders and Fallon, 1967; Hinchliffe, 1974b; Hinchliffe and Johnson, 1980). The importance of cell death in developmental abnormality (reviewed by Gruenwald, 1975) is also briefly illustrated by studies on teratogenic or genetic effects on limb morphogenesis. The important concept of peripheral loading as a control of neuronal cell number (Prestige, 1970) is introduced, as well as the hormonally controlled cell death of the oviduct in the male embryo (Scheib, 1977). Many examples have to be omitted, such as the intervention of cell death in eye development (Silver and Hughes, 1973) and heart development (Pexieder, 1975; Ojeda and Hurle, 1975; Hurle, Lafarga and Ojeda, 1977). Cell death should be borne in mind in quantitative attempts to assess embryonic proliferation rates, since, even in the early mammal embryo with its small cell number at the time of inner cell mass differentiation, cell death is already taking a significant toll (El-Shershaby and Hinchliffe, 1974; Copp, 1978). Clearly, embryonic cell death shades off into larval cell death at metamorphosis, and this is considered (especially in insects and amphibians) in Chapter 3 by Lockshin.

Finally, the question of lysosomal and acid hydrolase participation — whether primary or secondary — in the control of cell death is discussed.

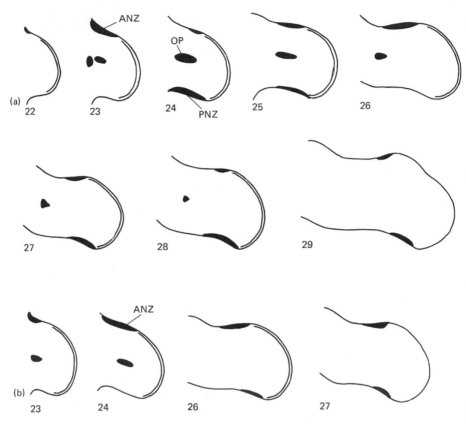

Fig. 2.1 Areas of mesenchymal cell death in the development of (a) the chick wing bud (Hamburger–Hamilton staging) and (b) the Herring gull. Stage 26 is at 5 days of incubation in the chick and approximately 7 days in the Herring gull. ANZ and PNZ, anterior and posterior necrotic zones; OP, opaque patch. (a, After Saunders *et al.*, 1962, and Hinchliffe and Johnson, 1980; b, original, J.R.H.).

2.2 Limb development and cell death

2.2.1 *The areas of cell death*

The occurrence of cell death is a prominent feature of limb development in the amniotes (reptiles, birds and mammals). We shall refer principally to Hamburger and Hamilton stages of the chick limb, on which most experimental analysis has been focused, but in addition give briefer consideration to other amniote species

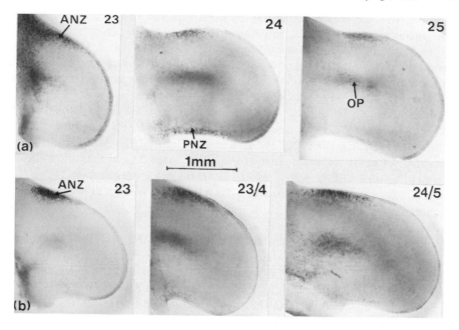

Fig. 2.2 Vitally stained chick limb buds (a, wing bud; b, leg bud — Hamburger–Hamilton staging) showing the pattern of mesenchymal cell death in anterior and posterior necrotic zones (ANZ, PNZ) and in the opaque patch (OP).

(reviewed in Saunders and Fallon, 1967; Hinchliffe, 1974b; Hinchliffe and Johnson, 1980). The areas of cell death are conveniently mapped out by vital staining *in ovo* of the embryo using Neutral Red or Nile Blue sulphate which is selectively concentrated in moribund cells and in the macrophages which are engaged in mopping up the dead and dying cells (Saunders, Gasseling and Saunders, 1962). A simple formol-calcium fixation method is available for preserving the vitally stained limb bud as a permanent preparation (Hinchliffe and Ede, 1973).

While most attention has been given to mesenchymal cell death, the cell death in the apical ectodermal ridge (AER) should also be briefly considered. The AER is a thickening of the ectoderm at the distal end of the limb bud, and is considered to induce the outgrowth of the underlying mesoderm (Saunders, 1977). Paradoxically, there is considerable cell death along the distal edge of the AER throughout its period of activity in the chick (Jurand, 1965; Hinchliffe and Ede, 1967) and in other amniotes (Figs. 2.2 and 2.5) (Milaire 1967b, 1970; Kelly and Fallon, 1976). Some dead cells are taken up by their neighbours and digested, while others are probably lost into the amniotic fluid. This loss of cells does not flatten the AER, and curiously enough much later when the ridge finally flattens out at the end of its inductive life, it does so without the intervention of cell death. The existence of cell death in the AER is probably best explained by reference to the work of Amprino and Ambrosi (1973), who showed that the ectoderm 'slides' in a distal direction over the mesoderm during limb growth. On this view the dead cells represent a migrant population of ectodermal cells in the process of being sloughed off the limb bud. Recent work by Saunders (Errick and Saunders, 1976; Saunders, Gasseling and Errick, 1976) suggests, however, that only a proportion of AER cells can be lost in this way, since there is cell continuity over a considerable time in grafts of quail AER to the apex of a chick limb bud.

In the mesenchyme of the early chick limb bud, there are initially three areas of cell death: the anterior and posterior necrotic zones (ANZ, PNZ) and the centrally located opaque patch (Figs. 2.1 and 2.2). In the wing bud, the ANZ appears as two successive waves of cell death, first proximally at stage 23 and then more distally at stages 25–26. The PNZ which appears later, reaches a peak at stage 24 and then disappears for a time. In later stages (27–31) ANZ and PNZ are still present distally, at either end of the AER. The chick leg bud has a very prominent ANZ at stages 23 and 24.

In other avian species, the pattern of cell death in ANZ and PNZ differs from that in the chick, the amount of cell death being somewhat reduced. In the duck wing bud, the ANZ is prominent from stages 23 to 29, and in the leg bud from stages 25 to 28. In the Herring gull wing bud, there is an ANZ from stages 23 to 31 (Fig. 2.1), but little cell death in the leg bud until stage 26 (Hinchliffe, 1981a). In the mouse and rat, however, both fore and hind limb buds lack persistent and substantial ANZs and PNZs, although in both there is a small anterior area of cell death (present at 11 days in the mouse – comparatively later than stage 23) adjacent to the prospective first digit and at the anterior end of the AER (Fig. 2.5) (Milaire, 1971, 1977a, 1977b; Rooze, 1977). At least one reptile, the lizard *Calotes*, has an ANZ and small PNZ also at a relatively late stage (stage 32; Goel and Mathur, 1977).

A further area of cell death is the 'opaque patch', so-called because it is opaque to transmitted light (Fell and Canti, 1935). This occurs in wing and leg buds in a number of avian species, including the chick (Dawd and Hinchliffe, 1971). It appears at the distal end of the femur or humerus and is most prominent at the time when radius and ulna (or tibia and fibula) blastemata first differentiate.

The most dramatic intervention of cell death occurs later, in the interdigital zones. These appear at 8 days in the chick where they form in the mesenchymal areas separating the chondrifying digits of the developing footplate (Figs. 2.3 and 2.4) (Menkes, Deleanu and Ilies, 1965; Saunders and Fallon, 1967). The necrosis follows a precise developmental programme, different areas appearing at specific stages. Thus, between chick toes 2 and 3, necrosis begins proximally, then a second area appears distally under the AER and finally the two areas coalesce (Pautou, 1975). Such interdigital zones (INZ) have been found in all amniote species investigated, or at least in those with non-webbed digits. INZs are found in human (Kelley, 1970), mouse and rat (Ballard and Holt, 1968) embryos, in several avian species and in reptiles (lizards and turtles: Goel and Mathur, 1977; Fallon and Cameron, 1977). Once cell death in the interdigital tissue has ceased, it continues along the anterior and posterior margins of the digits, where it removes further undifferentiated mesenchyme.

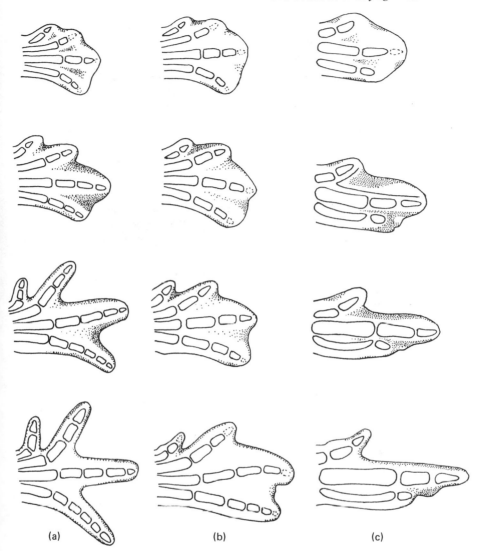

Fig. 2.3 Interdigital cell death in development of (a) chick leg (7½, 8¼, 9¼ and 10 days; partly from Pautou, 1974), (b) duck leg (8½, 9½, 10½ and 11½ days; partly from Pautou, 1974) and (c) chick wing (7¼, 7¾, 8 and 8¾ days, original J.R.H.). Note the shaping by interdigital cell death (dots) of the digital contours of the chick foot, the relative inhibition of interdigital cell death during webbed-foot development, and the intervention of cell death to limit the growth of digit 4 in the wing.

Perhaps because the INZs of the leg are dramatic and their morphogenetic role is clear, the INZs of the wing have received less attention. While cell death intervenes between digits 2 and 3 in a way similar to that of the leg INZs, the cell death

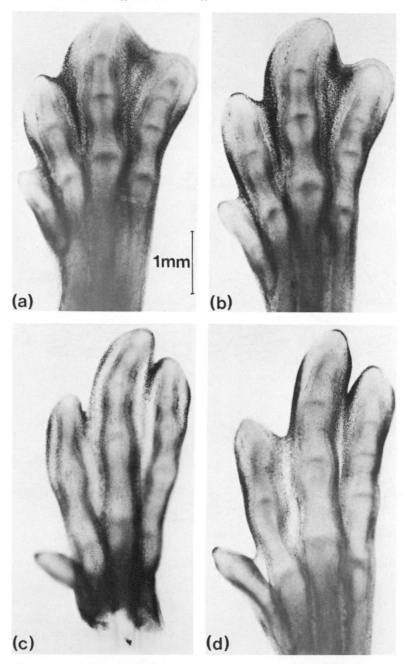

Fig. 2.4 Interdigital cell death in the development of the chick leg, between 8 and 9 days, shown by vital staining [permanent preparations of stages 33 (a), 34 (b), 34–5 (c) and 35 (d) obtained by the method of Hinchliffe and Ede, 1973].

between digits 3 and 4 follows a different pattern (Fig. 2.3). Here cell death begins proximally and spreads distally, but then the INZ thus formed and the PNZ located posteriorly to digit 4 spread out and overlap, thus removing undifferentiated mesenchyme distal to the last formed phalangeal element of digit 4 (Hinchliffe, unpublished observation). As the wing continues to develop, digit 3 differentiates a further phalangeal element, while digit 4, initially of similar size to digit 3, becomes relatively shorter and fails to add a further phalangeal element. These observations are consistent with the idea that cell death, by limiting the quantity of mesenchyme available, is exercising a controlling influence on the differentiation of skeletal pattern. We shall now consider further the morphogenetic role of cell death.

2.2.2 *The role of cell death*

There is compelling evidence that the role of the INZ in the amniotes is to separate and shape the digits (Menkes and Deleanu, 1964; Saunders and Fallon, 1967). This view is confirmed by studies on web footed birds which provide a natural experiment. In the duck (Fig. 2.3) (Pautou, 1974), Herring gull (Hinchliffe, 1981a) and turtle (Fallon and Cameron, 1977) considerable interdigital cell death is still present, but it is of shorter duration and less extensive in area than in the chick. This relative inhibition of cell death accounts for the survival of the interdigital webs, the remaining cell death presumably accounting for the thinning out of the mesenchyme of the interdigital webs. In both the web-footed avian species there is no web between digits 1 and 2 where cell death is as intensive as in this position in the chick. A number of mutant embryos also support the INZ-shaping theory. In the *talpid* embryos in the chick (Hinchliffe and Thorogood, 1974), and in the polysyndactylous mutant of the mouse (Johnson, 1969), interdigital cell death is completely or relatively inhibited, with the consequence that the mutant limbs show soft-tissue syndactyly, their digits being joined by webs of mesenchymal tissue.

Further confirmation comes from studies in which Janus Green, injected into the 6½-day chick amniotic sac, resulted in soft-tissue syndactyly of the foot. Janus Green, which is taken up in the mesenchyme by the mitochondria which are thereby damaged and whose population is temporarily diminished (Fallon, 1972), has the effect of greatly reducing interdigital cell death (Menkes and Deleanu, 1964; Saunders and Fallon, 1967; Kieny, Pautou and Sengel, 1976).

Little attention has been paid to the later stages of the INZ which persists for a considerable time along the anterior and posterior margins of the now separated digits, whose form is presumably sculptured by the process. The cell death eventually spreads to the distal tips of the digits where its appearance coincides with the final disappearance of the AER. It is tempting to suggest that the AER regression is a consequence of cell death in the subjacent mesenchyme (Kieny *et al.*, 1976).

The role of ANZ and PNZ is more puzzling. At first it seemed likely that these zones too played a role in shaping the emerging limb, e.g. by removing elbow angle material. But this interpretation – for the PNZ at least – is not supported by experiments of Saunders (1966) in which he suppressed the chick wing bud PNZ by grafting central wing mesoderm over the prospective PNZ. A normal wing formed, and Saunders was forced to conclude that the PNZ was dispensible for normal limb development. But if we adopt a comparative approach, there is other evidence that ANZ and PNZ may play a role in controlling skeletal pattern. As we have already seen, mouse and rat which have broad limb buds lack ANZ and PNZ at early stages, and these animals have the full number of digits — five — of the penta-dactyl limb. In birds, however, the digit number is reduced, the wing having essentially only three, and the leg four main digits (Hinchliffe, 1977c). This suggests that in the narrow limb buds of birds, ANZ and PNZ may limit the quantity of distal mesenchyme available for digit formation (Hinchliffe, 1977a). Other obser-vations support this interpretation.

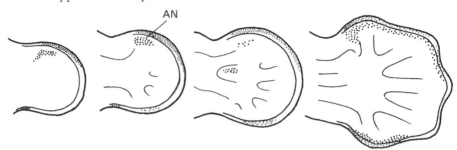

Fig. 2.5 Cell death in the development of the mouse forelimb between 10½ and 12 days of development (after Milaire, 1977a). Note the small anterior necrotic area (AN) at early stages, and the cell death in the preaxial AER.

In the mole (*Talpa*) the minor anterior area of necrosis found (Fig. 2.5) in the rat and mouse is absent (Milaire, 1977b). This absence is correlated with the forma-tion in the mole of a sixth (falciform) digit, while the first true digit is longer than the short first digit of the rat and mouse. In various mutants also, increases or decreases in areas of cell death are correlated with increases or decreases in digital number (Hinchliffe, 1974b, Figs. 10 and 11). Thus in the *talpid* mutant, ANZ and PNZ are absent, and the resulting polydactylous limbs have up to eight digits (Hinchliffe and Ede, 1967; Cairns, 1977). The increased size of the ANZ in the *wingless* wing bud appears to be responsible for the *wingless* phenotype which ranges from total winglessness to preaxial digital deficiencies.

One possible role of the ANZ and PNZ which is suggested by these observations is that areas of cell death are a device by which a correctly sized limb bud is formed by regulating mesodermal cell number. An analogy is provided by the spinal ganglia which regulate their cell number in relation to 'peripheral loading' as discussed in Section 2.3.2. It is known that chick limb buds up to stage 22–23 have the power of regulating their size, making good mesenchymal cell deficiency or adjusting for cellular excess (Wolff and Kahn, 1947; Kieny, 1964; Kieny and Pautou, 1976; though for a contrary view see Wolpert, Lewis and Summerbell, 1975, and Wolpert, 1978).

The cell-death-regulation hypothesis has been tested by removing wedges or circles of tissue from the proximal part of stage 18–21 chick wing buds. Such limbs had regulated for the deficiency judged by macroscopic observation of size and skeleton 3 days after operation, and the pattern of cell death in such regulating limbs was examined by vital staining 24h and 48h after the operation (Hinchliffe and Gumpel-Pinot, unpublished observations). In no case was there any reduction in the ANZ and PNZ in the part of the limb not affected by the operation. For example, when the anterior wedge was removed, the PNZ which formed later was normal (Fig. 2.6).

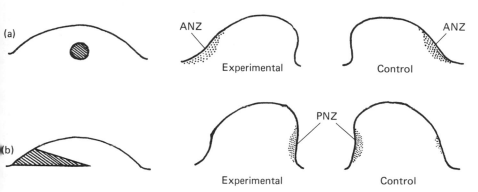

Fig. 2.6 When limb buds regulate for deficiency, they do not do so at the expense of the prospective PNZ or ANZ. In (a), a central 0.4 or 0.5 mm diameter disc is removed. Subsequent limb development is normal, and there is no decrease in ANZ 24h after operation. In (b), an anterior wedge (shaded) is removed and the flap pinned back. Limb development is again normal and there is no decrease in the PNZ at more than 24h after operation.

These experiments show clearly that when regulation of deficiency takes place it does not do so at the expense of the prospective PNZ (pPNZ) and ANZ, which are in theory a potential source of additional cells. More probably it requires adjustment in rates of mesenchymal cell division, though this has not yet been examined. One possibility is that the pANZ and pPNZ cells are already determined at the time of the experiment. The evidence here is equivocal, since Saunders *et al.* (1962) found that pPNZ was determined for cell death in flank grafts at stage 17, but that this was not irreversible until stage 22 when the determination is no longer reversible in associations of pPNZ with central wing mesoderm (see Section 2.2.4).

The role of the opaque patch also remains obscure, but one model of skeletal development proposes it as a factor involved in initial separation of radius and ulna (Waddington, 1962; Dawd and Hinchliffe, 1971). The opaque patch is largest at the stage when the condensations of radius and ulna become defined. Condensation formation requires increased cell adhesion, while in dying cells this adhesion is lost and the cells round up. Thus, given interaction between a process of condensation along the central axis of the limb and the intervention of cell death distally, the axial condensation will become bifurcated or Y-shaped, as are the blastemata representing the limb long bones.

2.2.3 *Phylogenetic aspects*

There are major differences in the incidence of cell death in limb development between the major tetrapod groups. As already pointed out, separation of the digits by the INZ is the rule throughout the amniotes. In the amphibia, however, there is no INZ. In urodeles (e.g. the axolotl) digits are formed one by one (Hinchliffe and Johnson, 1980), while in anurans (e.g. *Xenopus*) as in amniotes, a footplate is formed from which digital rays grow out. In neither amphibian group is there interdigital cell death (Hinchliffe, 1972; Cameron and Fallon, 1977), and the sculpturing of the limb must emerge from differential growth patterns.

The early intervention and persistence of ANZ and PNZ seem to be restricted to birds, and, even there, to be less marked in species other than the chick. Although a general survey is lacking, a minor anterior zone appears briefly in some mammals and a reptile, but such zones are absent from amphibians. In Section 2.2.2 it is argued that these zones may be associated with the digital reduction characteristic of birds, with their highly specialized forms of the pentadactyl limb.

Several groups of tetrapods are limbless, although the only systematic study is that by Raynaud (1972, 1977b, 1977c) of the reptiles. The expectation that increased cell death might be implicated has been borne out. The evidence may be briefly considered in a legless lizard (e.g. the slow worm, *Anguis fragilis*), in which limb buds are formed which then regress, and a snake (e.g. the python), which has a small hindlimb bud which forms a vestigial hindlimb, but has no forelimb bud at all.

In the slow worm forelimb bud, there is a deficiency in the somitic processes which contribute to the somatopleure (which forms the mesodermal component of the limb bud). The cells of the distal part of the somitic processes degenerate (Raynaud and Adrian, 1975). Normally in the chick the somitic processes contribute the musculature (Christ, Jacob and Jacob, 1977), and here and in the lizard *Lacerta viridis* they may in addition have a stimulatory morphogenetic effect on the somatopleure (Pinot, 1970; Raynaud, 1977c). In Raynaud's view, in the slowworm, the defective somitic processes are the initial cause of regression. Nonetheless a small limb bud, capped with an AER, forms and for a brief period grows normally. The ridge soon degenerates, but for a further period the mesoderm continues to proliferate. Then its growth too ceases and cell death begins in the proximal parts and spreads distally (Raynaud and Vasse, 1972), so that the limb bud disappears. This cell death is not, however, due to an extension of existing areas.

In the python, a similar story is found. The AER of the small hindlimb bud soon degenerates (Raynaud, 1972), and further growth ceases. Even those chondrogenic elements, such as the femur and pelvic elements which have formed fail to grow and many of their cells die (Renous, Raynaud, Gasc and Pieau, 1976). Raynaud emphasizes that in apodous reptiles there is a pattern of first somitic deficiency and then AER regression, and he argues that these deficiencies are genetically controlled.

2.2.4 *The programming of cell death*

We have seen how cell death in some instances helps to shape the developing limb, and we can now consider how these areas of cell death become determined and whether cell death is a form of morphogenesis lying at the end of a pathway of differentiation.

In a classic study, Saunders *et al.* (1962) analysed the process of determination of the PNZ in the chick wing bud. The PNZ is first recognizable at stage 24 as a scattering of individual dead cells and macrophages, but an apparently healthy pPNZ indistinguishable from the neighbouring viable mesenchyme can be located at the posterior margin of the wing bud when this appears at stage 17. Already at this stage, the pPNZ is determined for cell death at stage 24. If pPNZ cells of stage 17 are excised and flank-grafted, the cells die on schedule after the appropriate interval. Control cells similarly grafted survive. But the determination is not irreversible at stage 17 and can be modified by the local cellular environment. If pPNZ of stages 17–22 is grafted to the dorsal surface of a host wing where it is in contact with the central wing mesoderm, the 'death clock' is stopped and the cells survive. But after stage 22 their determination is irreversible and they die even if grafted as just described.

This inhibition of cell death by central wing mesoderm was further analysed by Saunders and Fallon (1967) using organ culture. Isolated in culture, pPNZ dies at the same time as the control pPNZ of the controlateral wing bud of the host. But, as with grafting, pPNZ grown with central wing mesoderm (though not with somitic mesoderm) survives. If a Millipore filter is inserted, the inhibition is mediated by pore size: $0.45 \mu m$ pores permit the inhibition, but a reduction to $0.05 \mu m$ prevents it, and the pPNZ cells die. Saunders concluded that a diffusible factor exerted control, but following recent work (Saxen and Wartiovaara, 1976) on filter-pore penetration by cell processes, direct contact cannot be ruled out. Whatever the precise mechanism of control, this evidence should be considered as further experimental evidence showing that this position in the limb bud is a factor controlling differentiation (Wolpert *et al.*, 1975; reviewed by Hinchliffe and Johnson, 1980).

Although cell death in the pPNZ is clearly programmed by stage 22, the cells do not differ visibly from their viable neighbours until the processes of autophagy and autolysis begin the cell-fragmentation process at stage 24 (Hurle and Hinchliffe, 1978 – see Section 2.6). In this gap between determination and expression, the cell division rate as indicated by radioautography with tritiated thymidine begins to drop at stage 22 in the pPNZ compared with neighbouring areas (Saunders and Fallon, 1967; Pollak and Fallon, 1974). RNA synthesis declines in the pPNZ by stage 22 (Pollak and Fallon, 1974) and protein synthesis by stage 23 (Pollak and

Fallon, 1976). The pPNZ cells appear to shut down their genome before cytological changes become detectable.

A position effect is also demonstrated in recent experiments (Hinchliffe, Garcia-Porrero and Gumpel-Pinot, 1981) which show that the anterior part of the limb bud can neither survive nor differentiate without the influence of the posterior half. In chick wing buds at stages 19 and 20, when the anterior half of the bud was removed, the posterior half survived without any change in the normal pattern of PNZ, and after 3 days formed its normal complement of skeletal elements. Following removal of the posterior half, anterior halves did not survive well and showed massive cell death beginning anteriorly at 18h, and moving distally where it was still prominent after 24h and 48h (Fig. 2.7). This pattern of necrosis may well be due to extension of the existing ANZ. Leaving a part of the ZPA attached to the anterior half prevented such anterior necrosis (Hinchliffe, Garcia-Porrero, and Gumpel-Pinot, 1981).

To determine precisely what skeletal elements were lost during this anterior regression it was necessary to map out the prospective areas of the limb bud. Using the chick–quail cell marker system we replaced stage 18–21 chick wing bud parts (e.g. the anterior half) with corresponding parts of the quail wing bud. The chimeric bud was allowed to develop for 3 days until the digits were clearly defined and histological analysis could be carried out. The chimeric wing buds frequently developed normally, and it was possible to determine that the anterior half of the wing bud contributed part of the humerus, the radius, radiale and digit 2 to the skeleton (Fig. 2.8). These were also the parts of the skeleton often missing when the anterior half of the wing bud was excised, and which the isolated anterior half wing bud was unable to form.

Recent work on the zone of polarizing activity (ZPA) goes some way to explaining this result. The ZPA is an area of mesenchyme which has the property, when grafted from its normal posterior position to an anterior site in the limb bud apex, to initiate a second limb outgrowth, and to determine in this second limb the antero-posterior axis (A-P axis) (Saunders, 1972, 1977; Summerbell and Tickle, 1977; Wolpert, 1978). Opinions differ as to whether the ZPA removal interferes with antero-posterior axis of a normal wing (Fallon and Crosby, 1975; Saunders, 1977; Summerbell, 1979). Our own experiments may be interpreted as supporting the view of Summerbell and Wolpert that the ZPA determines the A-P axis in normal development, and they suggest in addition that its presence, even at a distance, is necessary to maintain the anterior part of the limb bud.

A formal model (modified from Ede, 1977) which supports this idea may be proposed (Fig. 2.9), based on the view of Summerbell (1979) that the ZPA acts as a source of diffusible morphogen which specifies differentiation along the antero-posterior axis. Thus a high level of morphogen specifies digit 4, a much lower level digit 2. In our model, limb mesenchyme dies below a certain threshold of morphogen. Normally only the extreme anterior mesenchyme (the ANZ) falls below this threshold but when the posterior half of the limb and thus the morphogen source is removed, the level anteriorly falls so that the remaining anterior

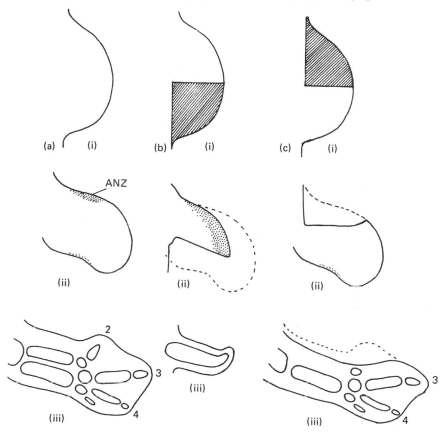

Fig. 2.7 After excision of the posterior half of the wing bud (b), the remaining anterior half following massive cell death (perhaps due to extension of ANZ), which begins 18h after operation, develops only a short cartilaginous spike. When the anterior half is excised (c), the remaining posterior half survives and differentiates normally. (a) Control wing bud (i) 24h later with normal ANZ (ii), and 72h later with normal cartilaginous skeleton (iii). (b) Posterior half deletion (i), followed after 24h by massive cell death (ii), the remaining limb bud forming little of the skeleton (iii). (c) Anterior half deletion (i), 24h later (ii), the remaining limb bud making its normal skeletal contribution (iii) (See Fig. 2.8). (Hinchliffe, Garcia-Porero and Gumpel-Pinot, 1981.)

mesenchyme is below the critical level and dies. This model has the advantage that it accommodates other observations (Hinchliffe, 1981a). The pattern of cell death following removal of the posterior half resembles a posteriorwards extension of the ANZ in the wing buds of the wingless (*ws*) mutant (Hinchliffe and Ede, 1973; Hinchliffe, 1977b). While the pattern of cell death in the Zwilling wingless mutant (*wg*) has not been fully described (Saunders' unpublished observations, quoted in Hinchliffe and Ede, 1973, p.772), it is significant that this mutant lacks a ZPA

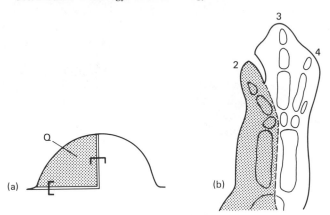

Fig. 2.8 Development of chimeric wing buds. In this experiment (a) a quail anterior half (Q) has replaced the corresponding chick anterior half: over two-thirds of these chimeras gave perfect or well-formed limbs (as in b). The quail cells are histologically distinct. The boundary line indicated represents the mean. Note that the posterior half contributes digits 3 and 4, the anterior half only 2. (Hinchliffe and Gumpel-Pinot, unpublished experiments.)

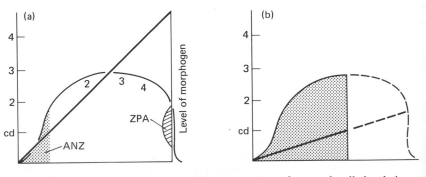

Fig. 2.9 A possible explanation of the extension of normal cell death in an isolated anterior half limb bud. In (a), the ZPA is a source of hypothetical morphogen whose level (thick line) as it declines anteriorly specifies the digits (4, high; 3, mid; 2, low) (based on the results of Summerbell and Tickle, 1977; Wolpert, 1978; Summerbell, 1979). Anterior cells die (ANZ — stippled) below a critical threshold (cd) of morphogen. If the posterior half with the ZPA is excised (b), the morphogen in the anterior half rapidly drops below this critical level, so that the ANZ enlarges and extends posteriorly.

(Saunders, 1972). Clearly this model should be tested by attempting to 'save' wingless (*ws*) wing buds by grafting a foreign ZPA posteriorly.

2.2.5 *Genetic control of cell death in the limb bud*

There is convincing evidence, mainly from chick embryos carrying mutant genes, that the areas of cell death are under direct or indirect genetic control. Such areas may be increased or inhibited and changes in their extent produce changes in limb shape (Hinchliffe, 1974b). In the *talpid*$^{2+3}$ mutants ANZ and PNZ are eliminated (Figs. 2.10 and 2.11), there is excessive elongation of the AER and the resulting polydactylous limbs have up to eight digits (Ede and Kelly, 1964; Hinchliffe and Ede, 1967; Cairns, 1977). Survival of the ANZ and PNZ cells is clearly related to the distal excess of mesenchyme of 5-6 days which gives the limb bud its characteristic mushroom-like profile and which permits the formation of supernumerary digits at 8-9 days. The inhibition of ANZ and PNZ is probably not the only factor contributing to the distal excess in the talpid limb bud, since *talpid*^{3} mesenchyme cells have been shown to be more adhesive and less motile than normal mesenchyme (Ede and Flint, 1975a, 1975b), and in Ede's view this cell behaviour is responsible for the short spade-like talpid limb profile. The link between the talpid mesenchyme movement anomaly, which presumably involves cell surface changes, and the inhibition of cell death is not at present understood. Talpid cells are resistant to treatments, such as removal of the AER (Cairns, 1977) or dissociation (Ede and Flint, 1972), which will kill a proportion of normal mesenchymal cells. The inhibition of cell death in talpid extends to other zones: the opaque patch and the interdigital zones. The opaque patch has been suggested as a factor associated with the initial separation of radius and ulna (Dawd and Hinchliffe, 1971). Its absence in talpid is correlated with the fusion of radius and ulna, while the later inhibition of interdigital cell death produces soft-tissue syndactyly (Hinchliffe and Thorogood, 1974).

In other mutants, the areas of cell death are enlarged. In the sex-linked wingless (*ws*) mutant, the ANZ makes a precocious appearance and is larger than normal. The wide range of phenotypic expression from complete winglessness to relatively normal wings is probably due to the degree to which the ANZ is enlarged. The ANZ in wingless wing buds tends to extend in a posterior direction at 3 and 4 days of development. There is a similar though less pronounced effect in the leg bud ANZ which is often enlarged, so that in severe cases preaxial skeletal elements are reduced or missing (Hinchliffe and Ede, 1973; Hinchliffe, 1977b). In the ametapodia mutant of the chick, the ANZ is suppressed, but the PNZ is enlarged between stages 23 and 25. Later, at stage 27 there is cell death — though unrelated to normal interdigital areas of cell death — throughout the metapodial region which is thought to be responsible for the principal effect of the gene: severely reduced metacarpals in the wing and metatarsals in the leg (Ede, 1968, 1977).

The same point can equally be made with reference to mouse mutants (reviewed Milaire, 1970). Dominant hemimelia (*Dh*) belongs to the luxate group of mutants which have essentially the same syndrome of tibial (or radial) reduction sometimes compensated for by polydactyly in which prospective tibial material appears to be redistributed to the preaxial digits (Fig. 2.12) (reviews Grüneberg, 1963; Hinchliffe and Johnson, 1980). Rooze (1977) showed that in *Dh* embryos at 11 and 12 days the preaxial necrotic area is suppressed in the forelimb bud. In oligosyndactylism, this area is more extensive than normal at 11 days, thus reducing the width of the

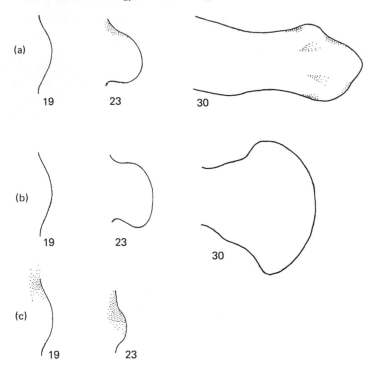

Fig. 2.10 Areas of cell death are under genetic control, and changes in the pattern of cell death are associated with changes in limb stage (Hinchliffe, 1974b). (a) Normal wing bud. (b) *talpid*[3] wing bud: note the excess of distal mesenchyme associated with inhibition of cell death (Hinchliffe and Ede, 1967; Hinchliffe and Thorogood, 1975). (c) *wingless* wingbud; note the precocious and enlarged ANZ (Hinchliffe and Ede, 1973). By stage 30, there is no wing (Hamburger-Hamilton stages: areas of cell death are stippled).

footplate within which digital blastemata may be squashed close together or even be completely missing (Fig. 2.12; Milaire, 1967a).

The relation of such variation in mesenchymal cell death to the AER has thus far been ignored. However, a general pattern can be discerned. Where ANZ (or the corresponding mouse zone) and PNZ are inhibited, the limb bud which is wider distally has an elongated AER (talpid, *Dh*). Where the zones are enlarged, the narrower limb bud has a shortened AER which regresses anteriorly (wingless, oligosyndactylism). Milaire (1967b) has shown that the mouse AER undergoes considerable normal regression anteriorly, but that this is halted in *Dh* embryos, and also in the teratogenic effects of cytosine arabinoside which induces poly-dactylism which may be considered as a phenocopy of *Dh* (Scott, Ritter and Wilson, 1977). But it is difficult with such descriptive studies to decide whether the origin of the anomaly is in mesoderm or AER. The interpretations of the talpid and wingless mutants have been in terms of ridge regression (wingless) or enlargement (talpid) reflecting variation in the supply of maintenance factor (Zwilling, 1974) through changes in the necrotic areas (Hinchliffe, 1977b). This approach has the merit that it accords with the view that ectoderm generally plays a non-specific role in the limb, and indeed, in feather development (Sengel, 1976; Saunders, 1977). Nonetheless, it is still possible that the primary lesion is in the AER.

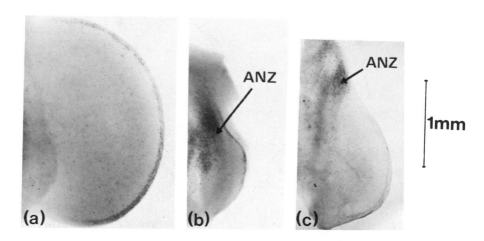

Fig. 2.11 Changes in the pattern of mesenchymal cell death in wingless (*ws*) and *talpid*[3] mutant limb buds. (a) *talpid*[3] forelimb bud at stage 24; (b) wingless wing bud and (c) leg bud both at stage 21. Both show a precocious and enlarged ANZ, but the effect is more severe in the wing bud. (Hinchliffe and Ede, 1967, 1973.) Originally published in *J. Embryol. Exp. Morphol.*

One interpretation of this group of mutants is that the variation in cell death areas is secondary to changes in morphogenetic factors located posteriorly in the limb bud. Much recent work emphasizes the morphogenetic importance of the posterior part, which may be due to the postaxial concentration of the maintenance factor (Zwilling and Hansborough, 1956) or to a morphogen from the ZPA diffusing anteriorly (Summerbell, 1979). Diffusion from cell to cell and from cell to environment may well be affected by gene-controlled properties of the cell surface, which are currently a focus of developmental biological research (Ede, 1976). If cell death is triggered whenever the concentration of the morphogen along the antero-posterior axis drops below a critical threshold (see Section 2.2.4), and the normal concentration gradient is altered by changes in the properties of the mutant cells, we may expect variation also in the extent of mesenchymal cell death. Consistent with this hypothesis are two observations: (i) that the ANZ is frequently affected, and (ii) that in mutants the skeletal anomalies of the limb are more common preaxially than postaxially (reviewed by Hinchliffe and Ede, 1967).

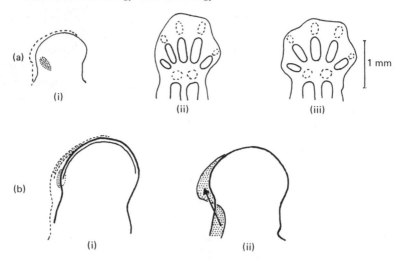

Fig. 2.12 Limb development in mouse mutants shows variation in the pattern of cell death. (a) Oligosyndactyly. The forelimb bud (solid line) is already reduced preaxially (i) compared with the normal at 11 days, and there is an increased area of preaxial mesenchymal cell death (dots). In the narrowed footplate at 13 days (ii), digit 2 is either absent or unusually close to the adjacent digits, as compared with the control (iii) (Grüneberg, 1962; Milaire, 1967a). (b) Dominant hemimelia (a luxate mutant). At 11 days the hindlimb bud (solid line) is narrower than the normal (i). The mutant bud lacks the normal anterior area of mesenchymal cell death (dots), but has an AER free from the preaxial cell death found in controls (dots) (Milaire, 1971; Rooze, 1977). Luxate hindlimb development may be interpreted as involving a shift of mesenchyme from making tibia to making preaxial digits (ii) (Hinchliffe and Johnson, 1980).

2.2.6 Teratogenic modification of cell death patterns in limb morphogenesis

Limb abnormalities (reviews by Gruenwald, 1975; Kochhar, 1977) are also produced by teratogenic agents, the majority of which are not specific to the limbs alone. Because of this, these malformations are not considered in detail here, though it is worth emphasizing the importance of the issue in screening for potentially teratogenic drugs such as thalidomide.

The effect of Janus Green in inhibiting interdigital cell death in the chick hindlimb has already been mentioned (Section 2.2.2). The effect of Janus Green is not restricted to interdigital mesenchyme, however, since it also causes hypophalangy (loss of the most distal elements) in digits 3 and 4. Pautou (1978) believes both these effects are due to Janus Green acting directly on the AER and flattening it during periods of active ectoderm–mesenchyme morphogenetic interaction. S. A. Marsh (unpublished observations) in this laboratory was able to confirm by using the scanning electron microscope the finding of Fallon (1972) that Janus Green induced flattening of the AER (Fig. 2.13). Marsh also found – as did Pautou

(1978) – that the distal mesenchyme of the digits under the Janus Green-flattened ridge becomes necrotic, and both these workers attribute the hypophalangy of digits 3 and 4 to the loss of AER over these digits where normally it persists the longest.

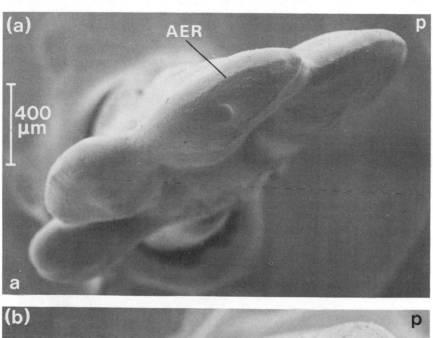

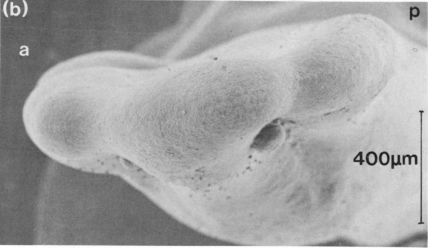

Fig. 2.13 The effect on the chick embryo of Janus Green is to flatten the AER. (a) Control stage 32–33 and (b) experimental stage 31–32 hindlimbs viewed distally by the scanning electron microscope (SEM; 10 μg. Janus Green is injected at 6½ days (micrograph courtesy of S. A. Marsh). a, Anterior; AER, apical ectodermal ridge; p, posterior.

Another example frequently classified as being due to an extension of naturally occurring cell death is that of insulin-induced micromelia (shortening of the limbs). Zwilling (1959) studied this phenomenon in the development of the chick leg and attributed degeneration in the joint region to the extension in area of the 'opaque patch' and to its persistence for a longer period of time than normal. This interpretation is accepted in a number of authoritative reviews of cell death (Zwilling 1964, 1968; Saunders, 1966; Saxen and Rapola, 1969) as a classic example of interference with a control mechanism restricting the spread of a necrotic focus. Later evidence, however, tends to contradict this interpretation. Dawd (1969) found insulin had no effect in increasing the area of the opaque patch at its period of maximum extent (stages 24–26). Hinchliffe (1974a) in a radioautographic analysis of chondrogenesis using $^{35}SO_4$ uptake found that the effect of insulin was to suppress chondrogenesis and create new and unrelated areas of cell death at the centres of the head of 7-day tibia and fibula, and in the distal end of the femur (Fig. 2.14). He attributed his findings to the interference by insulin with the nutrition of the cartilage and joint regions, possibly through hypoglycaemia.

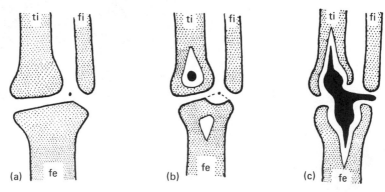

Fig. 2.14 The effect of 3 international units of insulin on chick embryos at 3½ days on leg development at 7 days is to create new areas of cell death in the heads of femur (fe), tibia (ti) and fibula (fi). (a) Normal control; (b) moderate and (c) strong insulin effect. The stippling represents regions of chondroitin sulphate synthesis. Black represents cell death (Hinchliffe, 1974a,)

Other teratogens have effects clearly unrelated to existing patterns of cell death. Some affect specific tissues in the developing limb, while others are cytotoxic agents causing widespread cell death throughout the embryo including the limb.

Amongst teratogens affecting specific tissues should be mentioned retinoic acid which attacks only the chondrogenic cells at the time of their differentiation (Kochhar and Agnish, 1977), and 6-aminonicotinamide which attacks the cartilage of developing limbs (Seegmiller, Overman and Runner, 1972). Jurand (1966, 1977) found that thalidomide selectively damaged the vascular system (specifically the endothelial lining of the axial artery) of the chick limb bud. Subsequently much of the surrounding proximal mesoderm became necrotic, so that proximal skeletal elements were lacking, while more distal mesoderm survived to contribute the digital skeleton.

Other teratogens which have general cytotoxic effects have maximum effect on rapidly proliferating tissue, e.g. undifferentiated limb bud mesenchyme, but less effect once the limb skeleton is present as cartilage which has a lower rate of cell division (Janners and Searls, 1970). Thus a teratogen applied early to limb buds affects proximal elements (which develop first), while later application produces more distal defects. This is illustrated by analysis (Summerbell, 1973; Salzgeber, 1976a, 1976b; Salzgeber and Gumpel-Pinot, 1977) of the effects of nitrogen mustard or X-irradiation on limb development. These agents produce scattered cell death throughout the limb bud mesenchyme, and, presumably as a consequence, the AER frequently regresses. As in the wingless mutant, AER regression brings limb development to a halt. The nitrogen mustard affects particularly the mesodermal component, since if treated (T) limb buds are separated into their two ectodermal (Ec) and mesodermal (Me) components, which are then recombined with the corresponding normal (N) component, the chimeric grafted limb bud frequently produces normal limbs when the ectoderm is the treated components (TEc, NMe), but never does so when the mesoderm has been treated (NEc, TMe) (Salzgeber 1968, 1976a). A further teratogen which has a non-specific toxic effect on proliferating limb bud mesenchyme is cytosine arabinoside (AraC) an inhibitor of DNA synthesis (Ritter, Scott and Wilson, 1973; Kochhar and Agnish, 1977).

Caution should be exercised in interpreting the effects of teratogens on limb development, since the mode of action at molecular and cellular level is in almost all cases not understood, and it is difficult to determine a 'pedigree of causes'. While cell death in these malformed limbs is common, it is clearly in many cases secondary and bears little relationship to the programmed cell death of normal limb morphogenesis.

2.3 Development of the nervous system

2.3.1 *Neural tube closure*

There is cell death during the morphogenesis of the neural plate into neural tube and during its later detachment from the ectoderm (Glücksmann, 1930, 1951; Schlüter, 1973). Such cell death is particularly marked in the dorsal part of the neural tube during closure, in the ectodermal stalk during its separation from the tube, and also in the anterior neuropore area. This cell death seems to be required to achieve continuity of epithelium and of the two sides of the neural tube, and also of the associated mesoderm (McKenzie, 1977).

McKenzie (1977) and Watt (1977) have studied the ontogeny of rumplessness induced by insulin in 24 h chick embryos. In these experiments, they consider that the failure of the neural tube to close dorsally is due to the action of insulin in increasing the normally occurring areas of cell death in the dorsal part of the neural ridges, and also the adjacent mesoderm. The normal cell death is, in their view, caused by conditions of hypoxia in the dorsal part of the closing neural tube, and the effect of insulin is to exaggerate this condition by interfering with oxidative metabolism throughout the embryo. While there is the usual problem in teratogenic experiments of knowing whether increased cell death is due to primary or secondary actions of the teratogen, such normal or experimentally increased patterns of necrosis in the neural tube may well be critically important in the ontogeny of such neural malformations as exencephaly, spina bifida and rumplessness. There is a parallel with the dominant rumplessness mutant investigated by Zwilling (1942), in which there is exaggeration in the tail bud of the chick embryo of an area of normal cell death, which in the mutant enlarges its normal area in the caudal mesoderm and spreads into the neural tube itself, thus producing the rumpless condition.

2.3.2 *The role of the periphery in limb innervation*

A clear demonstration that the periphery plays an important role in controlling the number of cells in neural centres in the chick embryo has been made by Hamburger and Levi-Montalcini (1949). As the neural tube develops, motor cells originating in the ventral part of the tube spin out their axons which migrate outwards and are gathered together in compact motor columns on a segmental basis [Fig. 2.15(a)]. The cell bodies of these motor neurons remain in the spinal cord where they form the ventral horn. In the dorsal root, the cell bodies are contained in ganglia arranged segmentally. There is cell death in all the spinal ganglia, but it takes place differentially so as to control the final and characteristic size of the ganglion. Thus there is substantial cell death in the dorsal root ganglia at cervical and thoracic levels, and much of the size difference between these ganglia and the larger ganglia at the brachial and lumbar–sacral level is directly due to this cell loss.

Experimental work on the chick embryo confirms the role of the periphery in controlling the extent of cell death. If the developing leg bud is extirpated at 2½ days of incubation, the pioneering fibres once they have reached the scar have their further growth blocked. Within 3 days of limb bud removal many of the motor neurons degenerate. If, on the other hand, an additional wing bud is transplanted, this becomes innervated, and the ganglia which supply this additional wing become larger, since more of their neurons survive (Fig. 2.15b, Hamburger, 1939; Hamburger and Levi-Montalcini, 1949).

Prestige (1965, 1967a, 1967b, 1970) has clearly demonstrated a corresponding process in the tadpoles of an anuran amphibian, *Xenopus*. In an experiment on tadpoles closely paralleling that just described, *Xenopus* limb amputations were followed by degeneration affecting about half the neurons in the dorsal root ganglion (Prestige, 1967a). In the ventral horn also, there proved to be sufficiently few cells to count. During development, the number of cells in each ventral horn decreases sharply (from 5000–6000 initially to 1200 at 60 days later), though the surviving cells increase in size. The dying cells are clearly identifiable within the developing horn. The rate of cell death at the leg level is highest during the time when movement is being developed (i.e. larvae aged 26–45 days — stages 54–59 in *Xenopus*). The two processes of increase in cell size and decrease in cell number are separable experimentally. Cell loss continues and differentiation ceases in larvae starved of thyroxine. While the anterior part of the horn differentiates earlier, contributing the earliest neurons to the developing limb, the growing horn is recruiting further cells at its posterior end.

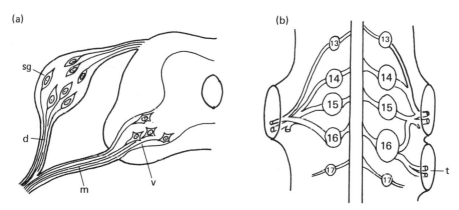

Fig. 2.15 Influence of changes in peripheral load on neuron survival. (a) Embryonic spinal cord and ganglion of 8-day chick embryo. d, dorsal sensory root; m, ventral motor column; sg, spinal ganglion; v, ventral horn. (b) Increase in peripheral load due to transplantation of chick wing bud (t) increases the size of the spinal ganglia (nos. 15 and 16) which innervate it (after Hamburger and Levi-Montalcini, 1949).

Once again the periphery controls the number of neurons in the ventral horn. Amputation of the hindlimb at stages 55–56 causes immediate degeneration of some of the motor cells. Prestige (1967b) has shown that the response of the ventral horn cells depends on their degree of differentiation. In phase I amputation has no effect, but at phase II it causes the horn to degenerate rapidly, while at phase III degeneration is postponed for several weeks.

Such a phase specific response to amputation provides insight into the nature of the 'feedback' between the periphery and neurons. Prestige (1967b) suggests that the cells in phases II and III die after amputation because they are no longer getting from the leg an essential 'maintenance factor' normally carried in the motor axons. Phase III cells have been able to store this substance, and are thus able to postpone their death. The more mature cells which have accumulated larger stores will thus survive when their fellows die. In such a model two steps may be involved: (i) a nerve/peripheral tissue recognition factor and (ii) an axonal information-transfer system. The flow of material from cell body along the axon has previously been clearly demonstrated, and Prestige (1970) argues that there is convincing evidence for a retrograde transfer at both the chemical and informational level.

Such a model might also provide an explanation of the normal cell death in the differentiation of the nervous system (see Pittman and Oppenheim, 1979). Cowan (1973) suggests that a proportion of neurons die because their axons fail to find appropriate peripheral 'targets', implying competition between outgrowing axons for contact sites. The best evidence for this is provided by Prestige and Wilson (1972), who found an approximately 1:1 correspondence between numbers of axons in the ventral roots of *Xenopus* and numbers of motor neurons in the corresponding spinal segment. The number of fibres in the ventral root decreases in development (e.g. in the ninth root from 2000 at stage 54 to 500 at stage 55), paralleling the decrease in number of motor neuron cell bodies.

The significance of such overproduction followed by degeneration and the nature of the mechanisms involved is discussed in great detail in the reviews by Cowan (1973) and Prestige (1970).

Although the role of the periphery has been discussed only with reference to limb innervation, there is a substantially similar body of evidence concerning eye development. Cowan (1973) reviews the evidence that optic vesicle removal is followed by cell death in the eye motor nuclei of the brainstem.

2.4 Differentiation of the reproductive system

The sexual differentiation in vertebrates of the accessory reproductive structures (such as oviduct and Wolffian duct) provides an interesting example of hormonal control of cell death. Initially before the gonads begin their secretion the reproductive system passes through an indifferent stage: both male and female embryos contain the same rudiments. In the male, the Wolffian duct differentiates forming epididymus and vas deferens, while the Müllerian duct (the oviduct) regresses. In the female, the Müllerian duct differentiates in birds as an oviduct with shell gland (Fig. 2.16 – only the left survives) or as a uterus in mammals. In the male chick embryo, gonad differentiation at 7–8 days precedes the regression of the Müllerian duct between 9 and 12 days.

It has long been known that the stabilization of one or other of the systems is controlled by the sex hormones (review, Jost, 1971; Finn and Porter, 1975). Injection of oestrogens (female sex hormones) at the indifferent stage feminizes a male chick embryo, which retains its Müllerian duct. Conversely, androgens (male sex hormones) injected into female embryos result in the breakdown of the Müllerian duct. Pioneering work by Jost (1947) on rabbit embryos showed that there was a critical period for the rudiment during which its fate could be influenced by gonad removal. Removal of the embryonic testis during this period resulted in the retention of the Müllerian duct by the male embryo, while with the removal of the female gonads the rabbit embryo continued to develop as a female. (In birds, however, the differentiation of the oviduct requires positive intervention of oestrogen.) Grafting a male gonad into a female rabbit embryo close to a Müllerian duct results in its regression, but (since the other Müllerian duct shows

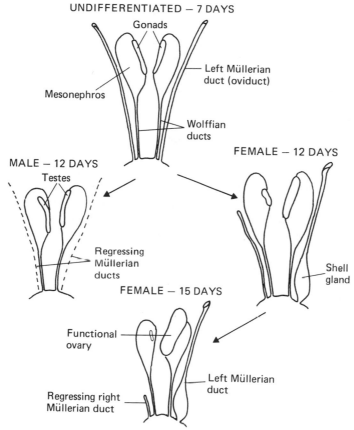

UNDIFFERENTIATED – 7 DAYS

Gonads

Left Müllerian
duct (oviduct)

Mesonephros

Wolffian
ducts

MALE – 12 DAYS

Testes

FEMALE – 12 DAYS

Regressing
Müllerian
ducts

Shell
gland

FEMALE – 15 DAYS

Functional
ovary

Left Müllerian
duct

Regressing right
Müllerian duct

Fig. 2.16 Sexual differentiation in the chick embryo. The Müllerian duct (oviduct) regresses in male embryos, but survives and differentiates on the left in female embryos.

no response) the response is localized. From these results it is clear that the testis is producing a substance (or substances) causing Müllerian duct regression. The obvious candidate is clearly testosterone. However, castration plus implantation of crystalline testosterone, while stimulating the Wolffian duct, does not affect the development of the Müllerian duct. In Jost's view, the foetal testis is secreting at least two hormones: testosterone and a Müllerian duct inhibitor.

The regression of the Müllerian duct in male embryos has been studied in the mouse (Price, Donahoe, Ito and Hendren, 1977) and in particular detail in the chick (Scheib, 1977). In the epithelium of the Müllerian duct of the male chick at 9 days, ultrastructural studies reveal that acid phosphatase-positive autophagic bodies containing deteriorating mitochondria appear in the apex of the cells. Later, autolysis begins, the chromatin of the nuclei becomes condensed and the cells break up. The view that the regression is the consequence of activation of the lysosomal system is supported by biochemical studies. As a proportion of the total protein of the duct, the specific activity of acid phosphatase and other lysosomal enzymes increases in male Müllerian ducts (as compared with the female duct) during their regression between 8 and 10 days (Scheib and Wattiaux, 1962). Moreover, the proportion which is 'unsedimentable' increases, suggesting that the lysosomal enzymes may be 'leaking' into the remainder of the cell.

Such a clear target–trigger system has raised hopes of improving our understanding of the mechanism of programmed cell death, by analogy with the discovery of oestrogen-receptor translocation during Müllerian duct differentiation in female chick embryos (Teng and Teng, 1976). However, the experimental culture of Müllerian ducts with various hormones *in vitro* and the treatment of lysosomal preparations with hormones have produced conflicting results. In some experiments, Müllerian ducts cultured *in vitro* regressed when androgens were added (Lutz-Ostertag, 1974), but in others, differentiation or regression depended not on the addition of oestrogen or androgen respectively, but rather on the concentration of hormone used (Hamilton and Teng, 1965). Another possibility explored is that androgens act directly on the lysosomes of the Müllerian duct, releasing acid hydrolases into the cell cytoplasm. Homogenates of undifferentiated Müllerian ducts, treated with androgens, showed increased free activity of acid phosphatase while the total activity of the enzyme remained unchanged (Scheib, 1963). But since oestrogens equally increase free activity, this evidence is not compelling. The precise role of lysosomes in Müllerian duct regression remains obscure, and may involve either release of pre-existing acid hydrolases from lysosomes by hormone action or neosynthesis of lysosomal enzymes, or a combination of the two. Equally, it is not clear whether lysosomes play a primary role – directly activated by hormones – or make secondary responses to deterioration in the cytoplasm caused by hormone action on other organelles.

2.5 Epithelial cell death during fusion of the secondary palate in mammalian development

The secondary palate, a structure which separates oral and nasal cavities, develops by the growth, rotation and fusion of the left and right palatal shelves (reviewed by Greene and Pratt, 1976). The fusion process has attracted attention because of the frequency with which cleft palate arises spontaneously or through teratogenic action (Saxen and Rapola, 1969). Palate fusion involves initial adhesion between the apposed epithelial surfaces involving extracellular glycoproteins. The epithelial cells in the midline then undergo autolysis enabling the mesenchyme of the two shelves to fuse and complete the secondary palate (Fig. 2.17). The epithelial cells destined to die first cease DNA synthesis in the rat 36h before contact, although synthesis of extracellular material continues. Then lysosomes appear in the form of acid phosphatase-positive autophagic vacuoles (Angelici and Pourtois, 1968); the epithelial cells and their basal lamina become disorganized and macrophages clear up the degenerating cells.

Culture *in vitro* of palatal material (Konegni, Chan, Moriarty, Weinstein and

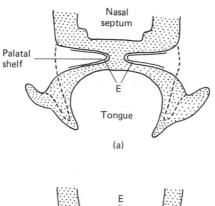

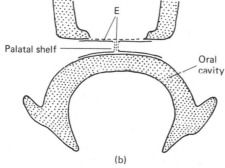

Fig. 2.17 Diagram illustrating palatal-shelf rotation and fusion, during which the edge epithelial cells die. The diagrams represent frontal sections of the anterior hard palate area of the rat embryo at (a) 16 days 8 h and (b) 17 days 8 h. In (a) the prerotation position of the shelves is indicated by dashes. In (b) the palatal shelf edge and nasal epithelial surfaces have adhered, and the midline epithelium (dashes) is autolysing. E, edge epithelium of the palatal shelves (after Greene and Pratt, 1976).

Gibson, 1965) has permitted detailed analysis of the fusion process. Pourtois (1969), on the basis of recombinations of palatal epithelium and mesenchyme, has suggested that the death of the palatal edge epithelium is dependent on a mesenchymal factor. The death of the edge epithelium has also been shown to be programmed and not due to the physical pressure of fusion, since it occurs on schedule when single palatal processes are cultured in isolation (Smiley and Koch, 1971).

Levels of cyclic AMP increase in the period just before epithelial contact, though it is not clear whether this change is restricted to the epithelial cells (Pratt and Martin, 1975). Pratt argues that the developmental programme of the epithelial cells (decreased proliferation, increased adhesiveness and death) may be mediated by cyclic nucleotides (see Chapter 3). Treatment *in vitro* of palatal processes with 6-diazo-5-oxo-L-norleucine (DON), which is a glutamine analogue and an inhibitor of glycoprotein synthesis, prevents both cell death and adhesion (Greene and Pratt, 1977; Pratt and Greene, 1976). The effect on cell death is explained through inhibition by DON of those lysosomal enzymes which are also glycoproteins. Since lysosomes do not appear directly to cause cell death (see Chapter 3), the treated epithelial cells should be examined for 'blocked' initial stages of deterioration.

2.6 Lysosomes and the control of embryonic cell death at the cellular level

The above account emphasizes that cell death in development occurs either autonomously, or as a response to extracellular mechanisms, but the underlying mechanisms at the cellular level remain largely unknown. The emergence of the lysosome concept provided an attractive model for the control of cell death, in which the integrity of the lysosome and its contained acid hydrolases could be related to the survival of cells. But it is still not clear whether such a lysosomal mechanism initiates cell death.

Instances were soon found in which cell death involved early lysosomal participation, which took the form of first autophagy and then autolysis as acid hydrolase leaked out of the autophagic vacuoles and damaged the cytoplasm. Epithelial cell death provided clear examples of this type of mechanism (Ericsson, 1969), as in mesonephros regression in the chick (Salzgeber and Weber, 1966; Haffen and Salzgeber, 1977), insect salivary gland regression (Jurand and Pavan, 1975) and epithelial deterioration in the uterus during implantation of the mouse egg (El-Shershaby and Hinchliffe, 1975), or in Müllerian duct regression (Scheib, 1977). Other examples are known which fail to show evidence of early lysosomal intervention, though it is clear that disintegrating cells are frequently digested in the heterophagosomes of macrophages which are often differentiated locally. An example of this is muscle regression in metamorphic tadpole tails (Weber, 1977), where, though acid hydrolases rise sharply, their source seems to be mainly the connective tissue macrophages rather than the muscle itself. This illustrates the problem that the many biochemical studies showing correlation between increases in lysosomal enzyme activity and regression do not indicate the cellular site of the increase. A further example of non-lysosomal cell deterioration appeared, until recently, to be the mesenchyme in interdigital, PNZ or 'opaque patch' cell death in rat and chick embryos (Saunders and Fallon, 1967; Ballard and Holt, 1968; Pautou and Kieny, 1971; Dawd and Hinchliffe, 1971; Litvac and Litvac, 1973). But a closer look at the processes of cell death in the PNZ illustrates the point that lysosomally controlled autophagy–autolysis or macrophage digestion of dead cells are not necessarily alternatives.

From the studies just quoted on mesenchymal cell death in the limb bud emerged a picture of firstly, cytoplasmic deterioration, followed by chromatopyknosis, cell fragmentation and a 'mopping up' process by local viable mesenchyme cells which became transformed into macrophages (Figs. 2.18 and 2.21). The increase in acid phosphatase (AP) activity was considered to be due to synthesis by the macrophages, which digested the dead cells in AP-rich heterophagic vacuoles. However, a recent reinvestigation (Hurle and Hinchliffe, 1977, 1978) suggested the following sequence: (1) appearance in the cytoplasm of small AP-positive autophagic vacuoles

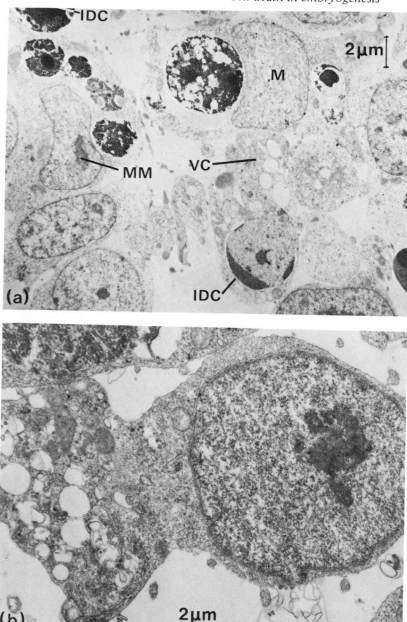

Fig. 2.18 (a) Cell death in the enlarged ANZ of the *wingless* wing bud at stage 21, show-ing various stages of cell death and macrophage digestion. Note isolated dead cells (IDC) with nuclei showing condensation of the chromatin and with vacuolated cytoplasm (VC). Phases of phagocytosis of the dead cells are illustrated by a mesenchyme cell (M), con-taining a single dead cell, and by a mature macrophage (MM). (b) Dying PNZ cell with vacuolated cytoplasm probably detaching from the nucleus and its adjacent cytoplasm. Originally published in *J. Embryol. Exp. Morphol.*

[Fig. 2.19(b)] enclosing organelles including mitochondria; (2) an increase in size of these vacuoles [Fig. 2.20(a)], which then leak AP activity so that the cell becomes autolytic [Fig. 2.20(b)]; (3) cell fragmentation, clearly seen in SEM preparations [Fig. 2.22(c) — Hurle and Hinchliffe, 1978] beginning with separation of a large nucleus-containing fragment (frequently little deteriorated) from the cytoplasmic portion; (4) chromatic condensation becomes more marked while the deteriorating and frequently AP-rich cytoplasmic portion [Fig. 2.20(c)] itself fragments; (5) the fragments [Figs. 2.21(b) and 2.22(b)] are phagocytosed by mesenchyme cells or macrophages which come to contain large numbers of dead cells [Fig. 2.18(a)] or their fragments in AP-rich vacuoles. Earlier analysis had perhaps focused on the more obvious but mainly AP-negative nuclear fragment [Fig. 2.18(b)] than on the small AP-positive cytoplasmic fragments. The source of the AP in the dead mesenchyme cell thus remains unresolved: it may be released from pre-existing lysosomes during cell death, or be synthesized either in the dying cell, or later via the macrophage cytoplasm. Neosynthesis is indicated by evidence from other examples that synthesis of RNA and protein is required for cell deterioration to take place (Tata, 1966; Weber, 1977).

The picture just presented has much in common with the 'apoptosis' scheme of cell fragmentation followed by uptake of the fragments by other cells, as proposed by Kerr, Wyllie and Currie (1972). Apoptosis is considered as an active and inherently programmed cell activity: a general self-destruct mechanism common to both embryonic cell death, and cell turnover in healthy tissues. It should be noted, however, that the fragmentation in apoptosis is defined as being non-lysosomal, while in limb buds it appears to be correlated with autophagy and autolysis.

Early lysosomal activation thus seems a common feature of developmental cell death, but there is little evidence that it is the *primary* cause. It is likely that the process represents an exaggeration or a loss of control of the normal intracellular digestion system responsible for destruction of unwanted organelles in discrete autophagic vacuoles. If the lysosomal role is not primary, the problem of identification of the initiating mechanism remains, but few of the developmental studies have considered other than lysosomal mechanisms, though Webster and Gross (1970) consider that cross-linking of DNA blocking the normal progress of the cell through its cycle is a possibility. Possible models which should be considered have advanced further in analysis of insect muscle regression in metamorphosis (see Chapter 3).

We understand more of the extrinsic controls of developmental cell death than of the self-destruct mechanism itself. Some of the controls can be clearly defined in terms of a trigger–target–response complex. As Saunders and Fallon (1967) stress, programmed cell death may be regarded, using the classical concepts of embryology, as the prospective fate of a group of competent cells progressively determined by a hierarchy of genetic, spatial and temporal factors.

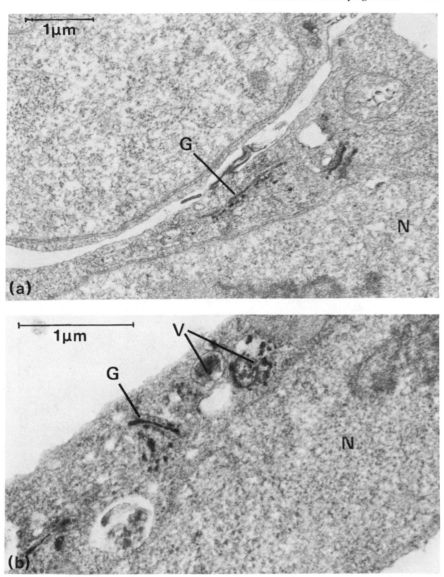

Fig. 2.19 Cell death in the mesenchyme of the PNZ of the chick limb bud. Acid phosphatase (AP) has been localized by a modified Gomori method (Hurle and Hinchliffe, 1978). (a) Healthy non-PNZ cell with AP activity localized in the Golgi (G) complex. (b) Early cytoplasmic lesions take the form of small autophagic vacuoles (V). The nucleus (N) and cell profile remain normal. Originally published in *J. Embryol. Exp. Morphol.*

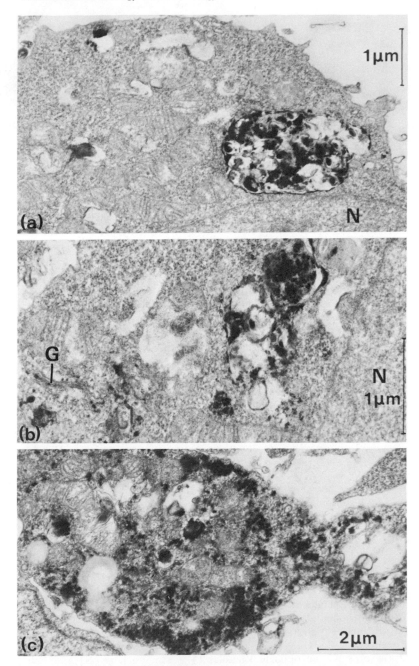

Fig. 2.20 Cell death in the PNZ of the chick limb bud. AP activity localised by the modified Gomori method. (a) Cell with large AP-positive autophagic vacuole. (b) The same cell in a different section: AP activity has spread from the vacuole in (a) into the surrounding damaged cytoplasm. The nucleus (N) remains normal. G, Golgi complex. (c) AP-positive cytoplasmic fragment. Originally published in *J. Embryol. Exp. Morphol.*

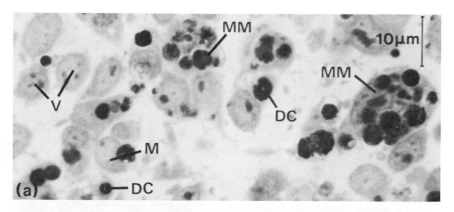

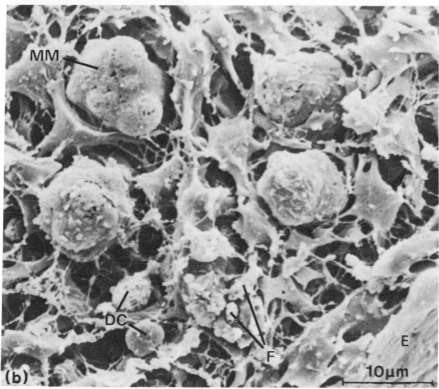

Fig. 2.21 Cell death in the PNZ of the chick limb bud. (a) Light micrograph of plastic section of cell death as in Fig. 2.18(a), showing dead cells (DC) and viable (V) cells, early phases of ingestion by mesenchyme cells (M) and mature macrophages (MM). (b) SEM micrograph giving panoramic view of PNZ. Stellate mesenchyme cells, dead cells and fragments (F) and mature macrophages are visible (E, ectoderm). Originally published in *J. Embryol. Exp. Morphol.*

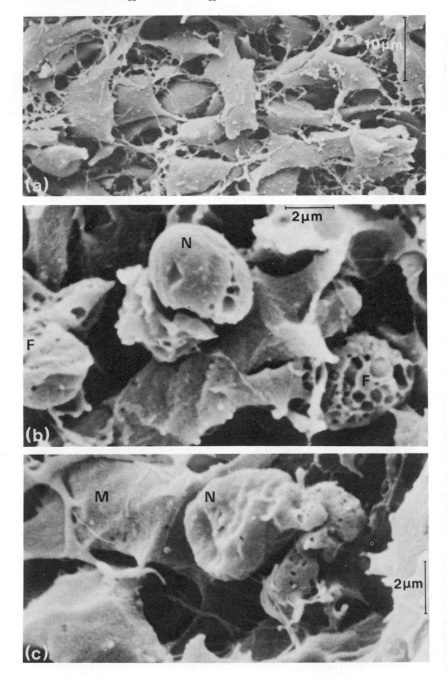

Acknowledgements

The author gratefully acknowledges Wellcome Foundation support during the joint work (described in Section 2.4) with Dr M. Gumpel-Pinot at the Institut d'Embryologie, Nogent-sur-Marne, France. The majority of the electron micrographs are published by courtesy of Dr J. Hurle.

Fig. 2.22 Cell death in the PNZ of the chick limb bud. (a) SEM micrograph showing viable mesenchyme cells with stellate profile and prominent filopodia. (b) SEM micrograph of isolated dead cell fragments (F) lying between normal mesenchymal cells. Note the pitted surfaces, and a probable nuclear fragment (N). (c) SEM micrograph of fragmenting cell. Note probable nuclear fragment (N) and small cytoplasmic pieces with pitted surfaces (M, viable mesenchymal cell). Orginally published in *J. Embryol. Exp. Morphol.*

References

Amprino, R. and Ambrosi, G. (1973), Experimental analysis of the chick embryo limb bud growth. *Arch. Biol. (Bruxelles)*, **84**, 35–86.
Angelici. D. and Pourtois, M. (1968), The role of acid phosphatase in the fusion of the secondary palate. *J. Embryol. Exp. Morphol.*, **20**, 15–23.
Ballard, K. J. and Holt, S. J. (1968), Cytological and cytochemical studies on cell death and digestion in foetal rat foot: the role of macrophages and hydrolytic enzymes. *J. Cell Sci.*, **3**, 245–262.
Cairns, J. M. (1977), Growth of normal and talpid[2] chick wing buds: an experimental analysis. In *Vertebrate Limb and Somite Morphogenesis* (eds. D. A. Ede, J. R. Hinchliffe and M. Balls), Cambridge University Press, Cambridge, pp. 123–137.
Cameron, J. A. and Fallon, J. F. (1977), The absence of cell death during development of free digits in amphibians. *Dev. Biol.*, **55**, 331–338.
Christ, B., Jacob, H. J. and Jacob, M. (1977), Experimental analysis of the origin of the wing musculature in avian embryos. *Anat. Embryol.*, **150**, 171–186.
Copp, A. J. (1978), Interaction between inner cell mass and trophectoderm of the mouse blastocyst. I. A study of cellular proliferation. *J. Embryol. Exp. Morphol.*, **48**, 109–125.
Cowan, W. M. (1973), Neuronal death as a regulative mechanism in the control of cell number in the nervous system. In *Development and Aging in the Nervous System* (ed. M. Rockstein), Academic Press, New York, pp. 19–41.
Dawd, D. S. (1969), Studies on the 'opaque patch' of cell death in the mesenchyme of the developing chick limb. Ph. D. thesis, University College of Wales, Aberystwyth.
Dawd, D. S. and Hinchliffe, J. R. (1971), Cell death in the 'opaque patch' in the central mesenchyme of the developing chick limb: a cytological, cytochemical and electron microscope analysis. *J. Embryol. Exp. Morphol.*, **26**, 401–424.
Ede, D. A. (1968), Abnormal development at the cellular level in *talpid* and other mutants. In *The Fertility and Hatchability of the Hen's Egg* (eds. T. C. Carter and B. M. Freeman), Oliver & Boyd, Edinburgh, pp. 71–83.
Ede, D. A. (1976), Cell interactions in vertebrate limb development. In *Cell Surface in Animal Embryogenesis and Development* (eds. G. Poste and G. L. Nicholson), Elsevier, Amsterdam, pp. 495–543.
Ede, D. A. (1977), Limb skeleton deficiencies in some mutants of the chick embryo. In *Méchanismes de la Rudimentation des Organes chez les Embryons de Vertébrés*, Colloques CNRS 266, Paris, pp. 187–191.
Ede D. A. and Flint, O. (1972), Patterns of cell division, cell death and chondrogenesis in cultured aggregates of normal and *talpid*[3] mutant chick limb mesenchyme cells. *J. Embryol. Exp. Morphol.*, **27**, 245–260.

Ede, D. A. and Flint, O. P. (1975a), Intercellular adhesion and formation of aggregates in normal and *talpid³* mutant chick limb mesenchyme. *J. Cell Sci.,* **18**, 97-111.

Ede, D. A. and Flint, O. P. (1975b), Cell movement and adhesion in the developing chick wing bud: studies on cultured mesenchyme cells from normal and *talpid³* mutant embryos. *J. Cell Sci.,* **18**, 301-313.

Ede, D. A. and Kelly, W. A. (1964), Developmental abnormalities in the trunk and limbs of the *talpid³* mutant of the fowl. *J. Embryol. Exp. Morphol.,* **12**, 339-356.

El-Shershaby, A. M. and Hinchliffe, J. R. (1974), Cell redundancy in the zona-intact preimplantation mouse blastocyst: a light and electron microscopic study of dead cells and their fate. *J. Embryol. Exp. Morphol.,* **31**, 643-654.

El-Shershaby, A. M. and Hinchliffe, J. R. (1975), Epithelial autolysis during implantation of the mouse blastocyst: an ultrastructural study. *J. Embryol. Exp. Morphol.,* **33**, 1067-1080.

Ericsson, J. L. E. (1969), Mechanism of cellular autophagy. In *Lysosomes in Biology and Pathology,* vol. 2, (eds. J. T. Dingle and H. B. Fell), North-Holland, Amsterdam, pp. 354-394.

Errick, J. and Saunders, J. W., Jr. (1976), Limb outgrowth in the chick embryo induced by dissociated and reaggregated cells of the apical ectodermal ridge. *Dev. Biol.,* **50**, 26-34.

Fallon, J. F. (1972), The morphology and fate of the apical ectodermal ridge in the normal and Janus Green B-treated chick foot. *Am. Zool.,* **12**, 701-702.

Fallon, J. F. and Cameron, J. A. (1977), Interdigital cell death during limb development of the turtle and lizard with an interpretation of evolutionary significance. *J. Embryol. Exp. Morphol.,* **40**, 285-289.

Fallon, J. F. and Crosby, G. M. (1975), Normal development of the chick wing following removal of the polarizing zone. *J. Exp. Zool.,* **193**, 449-455.

Fell, H. B. and Canti, R. G. (1935), Experiments on the development *in vitro* of the avian knee-joint. *Proc. R. Soc. London Ser. B,* **116**, 316-351.

Finn, C. A. and Porter, D. G. (1975), *The Uterus.* Elek Science, London.

Goel S. C. and Mathur, J. K. (1977), Morphogenesis in reptilian limbs. In *Vertebrate Limb and Somite Morphogenesis* (eds. D. A. Ede, J. R. Hinchliffe and M. Balls), Cambridge University Press, Cambridge, pp. 387-404.

Glücksmann, A. (1930), Über die Bedeutung von Zellvorgängen für die Formbildung epitheliales Organe. *Z. Anat. Entwicklungsges.,* **93**, 35-92.

Glücksmann, A. (1951), Cell deaths in normal vertebrate ontogeny. *Biol. Rev. Cambridge Philos. Soc.,* **26**, 59-86.

Greene, R. M. and Pratt, R. M. (1976), Developmental aspects of secondary palate formation. *J. Embryol. Exp. Morphol.,* **36**, 225-245.

Greene, R. M. and Pratt, R. M. (1977), Inhibition by diazo-oxonorleucine (DON) of rat palatal glycoprotein synthesis and epithelial cell adhesion *in vitro. Exp. Cell Res.,* **105**, 27-37.

Gruenwald, P. (1975), Embryological basis of abnormal development with special reference to cell death. In *The Mammalian Foetus* (ed. E. S. Hafez), Chas. C. Thomas, Springfield, pp. 251-281.

Grüneberg, H. (1963), *The Pathology of Development,* Blackwell, Oxford.

Haffen, K. and Salzgeber, B. (1977), Les modalités de la régression du mésonephros chez les amniotes. In *Mécanismes de la Rudimentation des organes chez les Embryons de Vertébrés,* Colloques CNRS 266, Paris, pp. 251-262.

Hamburger, V. (1939), The development and innervation of transplanted limb primordia of chick embryos. *J. Exp. Zool.,* **80**, 347-385.

Hamburger, V. and Levi-Montalcini, R. (1949), Proliferation, differentiation and degeneration in the spinal ganglia of the chick embryo under normal and experimental conditions. *J. Exp. Zool.*, **111**, 457–502.

Hamilton, T. H. and Teng, C. S. (1965), Sexual stabilisation of Müllerian ducts in the chick embryo. In *Organogenesis* (eds. R. L. de Haan and H. Ursprung), Holt, Rinehart and Winston, New York, pp. 681–700.

Hinchliffe, J. R. (1972), Cell death in relation to the genesis of form and pattern in the developing chick and amphibian limb. In *Colloque International sur le Développement du Membre* (ed. P. Sengel), Grenoble, p. 26.

Hinchliffe, J. R. (1974a), Experimental modification of patterns of cell death and chondrogenesis in insulin-induced micromelia of the developing chick limb: an autoradiographic analysis of $^{35}SO_4$ uptake into chondroitin sulphate. *Teratology*, **9**, 263–274.

Hinchliffe, J. R. (1974b), The patterns of cell death in chick limb morphogenesis. *Lib. J. Sci.*, **4A**, 23–32.

Hinchliffe, J. R. (1977a), 'Rudimentation', reduction and specialisation in the development and evolution of the bird wing. In *Mécanismes de la Rudimentation des Organes chez les Embryons de Vertébrés*, Colloques CNRS 266, Paris, pp. 411–414.

Hinchcliffe, J. R. (1977b), The development of winglessness (*ws*) in the chick. In *Mécanismes de la Rudimentation des Organes chez les Embryons de Vertébrés*, Colloques CNRS 266, Paris, pp. 173–185.

Hinchliffe, J. R. (1977c), The chondrogenic pattern in chick limb morphogenesis. In *Vertebrate Limb and Somite Morphogenesis* (D. A. Ede, J. R. Hinchliffe and M. Balls), Cambridge University Press, Cambridge, pp. 293–309.

Hinchliffe, J. R. (1981a). Control by the posterior border of cell death patterns in limb bud development of amniotes: evidence from experimental amputations and from mutants. 4th Symposium on Pre-natal Development, Berlin (in press).

Hinchliffe, J. R. (1981b), Cell death and the development of limb form in the developing wing and webbed foot of the Herring Gull, *Larus argentatus*. *J. Embryol. Exp. Morphol.*, (in press).

Hinchliffe, J. R. and Ede, D. A. (1967), Limb development in the polydactylous *talpid*[3] mutant of the fowl. *J. Embryol. Exp. Morphol.*, **17**, 385–404.

Hinchliffe, J. R. and Ede, D. A. (1973), Cell death and the development of limb form and skeletal pattern in normal and wingless (*ws*) chick embryos. *J. Embryol. Exp. Morphol.*, **30**, 753–772.

Hinchliffe, J. R., Garcia-Porrero, J. A. and Gumpel-Pinot, M. (1981), The role of the zone of polarising activity (ZPA) in controlling the maintenance and antero-posterior differentiation of the apical mesenchyme of the chick wing bud: Histochemical techniques in the analysis of a developmental problem. *Histochem. J.* (in press).

Hinchliffe, J. R. and Johnson, D. R. (1980), *The Development of the Vertebrate Limb*, Oxford University Press, Oxford.

Hinchliffe, J. R. and Thorogood, P. V. (1974), Genetic inhibition of mesenchymal cell death and the development of form and skeletal pattern in the limbs of *talpid*[3] mutant chick embryos. *J. Embryol. Exp. Morphol.*, **31**, 747–760.

Hurle, J. and Hinchliffe, J. R. (1977), Cytological and stereoscan studies of cell death in the posterior necrotic zone of the embryonic chick wing-bud. *Proc. R. Microsc. Soc.*, **12**, 214.

Hurle, J. and Hinchliffe, J. R. (1978), Cell death in the posterior necrotic zone (PNZ) of the chick wingbud: a stereoscan and ultrastructural survey of autolysis and cell fragmentation. *J. Embryol. Exp. Morphol.*, **43**, 123–136.

Hurle, J. M., Lafarga, M. and Ojeda, J. L. (1977), Cytological and cytochemical studies of the necrotic area of the bulbus of the chick embryo heart. *J. Embryol. Exp. Morphol.*, **41**, 161–173.

Janners, M. Y. and Searls, R. L. (1970), Changes in the rate of cellular proliferation during the differentiation of cartilage and muscle in the mesenchyme of the embryonic chick wing. *Dev. Biol.*, **23**, 136–165.

Johnson, D. R. (1969), Polysyndactyly, a new mutant gene in the mouse. *J. Embryol. Exp. Morphol.*, **21**, 285–294.

Jost, A. (1947), Recherches sur la différentiation sexuelle de l'embryon de lapin. *Arch. Anat. Microsc. Morphol. Exp.*, **36**, 271–315.

Jost, A. (1971), Hormonal factors in the sex differentiation of the mammalian foetus. *Philos. Trans. R. Soc. London Ser. B*, **259**, 119–130.

Jurand, A. (1965), Ultrastructural aspects of early development of the forelimb buds in the chick and the mouse. *Proc. R. Soc. London Ser. B*, **162**, 387–405.

Jurand, A. (1966), Early changes in limb buds of chick embryos after thalidomide treatment. *J. Embryol. Exp. Morphol.*, **16**, 289–300.

Jurand, A. (1977), Ultrastructural study of the effects of thalidomide on limb buds of chick embryos. In *Mécanismes de la Rudimentation des Organes chez les Embryons de Vertébrés*, Colloques CNRS 266, Paris, pp. 281–290.

Jurand, A. and Pavan, C. (1975), Ultrastructural aspects of histolytic processes in the salivary gland cells during metamorphic stages in *Rhynchosciara hollaenderi* (Diptera, Sciaridae). *Cell differ.*, **4**, 219–236.

Kelley, R. O. (1970), An electron microscopic study of mesenchyme during development of interdigital spaces in man. *Anat. Res.*, **168**, 43–54.

Kelley, R. O. and Fallon, J. R. (1976), Ultrastructural analysis of the apical ectodermal ridge during vertebrate limb morphogenesis. I. The human forelimb with special reference to gap junctions. *Dev. Biol.*, **51**, 241–256.

Kerr, J. F. R., Wyllie, A. H. and Currie, A. R. (1972), Apoptosis: a basic biological phenomenon with wide ranging implications in tissue kinetics. *Br. J. Cancer*, **26**, 239–257.

Kieny, M. (1964), Étude du mécanisme de la régulation dans le développement du bourgeon de membre de l'embryon de poulet. I. Régulation des excédents. *Dev. Biol.*, **9**, 197–229.

Kieny, M. and Pautou, M. P. (1976), Régulation des excédents dans le développement du bourgeon de membre de l'embryon d'oiseau. Analyse expérimentale de combinaisons xenoplastiques caille/poulet. *Wilhelm Roux Arch. Entw-Mech Org.*, **179**, 327–338.

Kieny, M., Pautou, M. P. and Sengel, P. (1976), Limb morphogenesis as studied by Janus Green B and vinblastine-induced malformations. In *Tests of Teratogenicity in vitro*, North-Holland, Amsterdam, pp. 389–415.

Kochhar, D. M. (1977), Abnormal organogenesis in limbs. In *Handbook of Teratology* (eds. J. G. Wilson and F. C. Fraser), Plenum Press, New York, pp. 453–479.

Kochhar, D. M. and Agnish, N. D. (1977), 'Chemical surgery' as an approach to study morphogenetic events in embryonic mouse limb. *Dev. Biol.*, **61**, 388–394.

Konegni, J. S., Chan, B. C., Moriarty, T. M., Weinstein, S. and Gibson, R. D. (1965), A comparison of standard organ culture and standard transplant techniques in the fusion of palatal processes of rat embryos. *Cleft Palate J.*, **2**, 219–228.

Litvac, B. and Litvac, E. (1973), Early structural changes during the morphogeneti-cal necrosis of the mesenchyme in the interdigital membrane of the chick embryo. *Rev. Roum. d'Embryol.*, **10**, 127–133.

Lutz-Ostertag, Y. (1974), Nouvelles preuves de l'action de la testosterone sur la développement des canaux de Müller de l'embryon d'oiseau en culture in vitro. *C.R. Hebd. Séances Acad. Sci. Ser. D*, **278**, 2351–2353.

McKenzie, J. (1977), Ontogeny of rumplessness and its genetic determination. In *Méchanismes de la Rudimentation des Organes chez les Embryons de Ver-tébrés*, Colloques CNRS 266, Paris, pp. 37–45.

Menkes, B. and Deleanu, M. (1964), Leg differentiation and experimental syn-dactyly in chick embryo. *Rev. Romaine Embryol. Cytol.*, **1**, 69–77.

Menkes, B., Deleanu, M. and Ilies, A. (1965), Comparative study of some areas of physiological necrosis of the embryo of man, laboratory mammalians and fowl. *Rev. Romaine Embryol. Cytol.*, **2**, 161–172.

Milaire, J. (1967a), Histochemical observations on the developing foot of normal, oligosyndactylous (*Os*/+) and syndactylous (*sm/sm*) mouse embryos. *Arch. Biol. (Liège)*, **78**, 223–288.

Milaire, J. (1967b), Evolution des processus dégeneratifs dans le cape apicale au cours du développement des membres chez le rat et la souris. *C. R. Hebd. Séances Acad. Sci., Ser. D*, **265**, 137–140.

Milaire, J. (1970), Evolution et déterminisme des dégenerescences cellulaires au cours de la morphogenèse des membres et leurs modifications teratologiques. In *Malformations Congènitales des Mammifères* (ed. H. Tuchmann-Duplessis), Colloques Pfizer, Amboise, pp. 131–149.

Milaire, J. (1971), Étude morphogénétique de la syndactylie postaxiale provoquée chez le rat par l'hadacine. II. Les bourgeons de membres chez les embryons de 12 a 14 jours. *Arch. Biol. (Liège)*, **82**, 253–322.

Milaire, J. (1977a), Histochemical expression of morphogenetic gradients during limb morphogenesis (with particular reference to mammalian embryos). *Birth Defects*, **13**, 37–67.

Milaire, J. (1977b), Rudimentation digitale au cours du développement normale de l'autopod chez les mammifères. In *Mécanismes de la Rudimentation des Organes chez les Embryons de Vertébrés*, Colloques CNRS 266, Paris, pp. 221–233.

Ojeda, J. L. and Hurle, J. M. (1975), Cell death during the formation of tubular heart of the chick embryo. *J. Embryol. Exp. Morphol.*, **33**, 1–12.

Pautou, M. M. (1974), Evolution comparé de la nécrose morphogène interdigitale dans le pied de l'embryon de poulet et de canard. *C. R. Hebd. Séances Acad. Sci., Ser. D*, **278**, 2209–2212.

Pautou, M. P. (1975), Morphogenèse de l'autopode chez l'embryon de poulet. *J. Embryol. Exp. Morphol.*, **34**, 511–529.

Pautou, M. P. (1978), Cessation de l'activité de la crête apicale ectodermique au course de la morphogenèse de l'autopode chez l'embryon de poulet: analyse histologique. *Arch. Biol. (Bruxelles)*, **89**, 27–66.

Pautou, M. P. and Kieny, M. (1971), Sur les mécanismes histologiques et cytologiques de la nécrose morphogène interdigitale, chez l'embryon de poulet. *C. R. Hebd. Séances Acad. Sci., Ser. D*, **272**, 2025–2028.

Pexieder, T. (1975), Cell death in the morphogenesis and teratogenesis of the heart. *Adv. Anat. Embryol. Cell Biol.*, **51**, 1–100.

Pinot, M. (1970), Relations entre le mésenchyme somitique et la plaque latérale dans l'organogenèse précoce des membres, chez le poulet. *Ann. Biol.*, **9**, 277–284.

Pittman, R. and Oppenheim, R. W. (1979), Cell death of motoneurons in the chick embryo spinal cord. IV. Evidence that a functional neuromuscular interaction is involved in the regulation of naturally occurring cell death and the stabilization of synapses. *J. Comp. Neurol.*, **187**, 425–446.

Pollak, R. D. and Fallon, J. F. (1974), Autoradiographic analysis of macromolecular synthesis in prospectively necrotic cells of the chick limb bud. I. Protein synthesis. *Exp. Cell Res.*, **86** 9–14.

Pollak, R. D. and Fallon, J. F. (1976), Autoradiographic analysis of macromolecular synthesis in prospectively necrotic cells of the chick limb bud. II. Nucleic acids. *Exp. Cell Res.*, **100**, 15–22.

Pourtois, M. (1969), Study on the fusion of palatal processes *in vitro* and *in vivo*. *J. Dent. Res.*, **48**, 1135.

Pratt, R. M. and Greene, R. M. (1976), Inhibition of palatal epithelial cell death by altered protein synthesis. *Dev. Biol.*, **54**, 135–145.

Pratt, R. M. and Martin, G. R. (1975), Epithelial cell death and cyclic AMP increase during palatal development. *Proc. Natl. Acad. Sci., U.S.A.*, **72**, 874–877.

Prestige, M. C. (1965), Cell turnover in the spinal ganglia of *Xenopus laevis* tadpoles. *J. Embryol. Exp. Morphol.*, **13**, 63–72.

Prestige, M. C. (1967a), The control of cell number in lumbar spinal ganglia during the development of *Xenopus laevis* tadpoles. *J. Embryol. Exp. Morphol.*, **17**, 453–471.

Prestige, M. C. (1967b), The control of cell number in the lumbar ventral horns during the development of *Xenopus laevis* tadpoles. *J. Embryol. Exp. Morphol.*, **18**, 359–387.

Prestige, M. C. (1970), Differentiation, degeneration and the role of the periphery: quantitative considerations. In *The Neurosciences* (ed. F. O. Schmitt), Rockefeller University Press, New York, pp. 73–82.

Prestige, M. C. and Wilson, M. A. (1972), Loss of axons from ventral roots during development. *Brain Res.*, **41**, 467–470.

Price, J. M., Donahoe, P. K., Ito, Y. and Hendren, W. H. (1977), Programmed cell death in the Müllerian duct induced by Müllerian inhibiting substance. *Am. J. Anat.*, **149**, 353–376.

Raynaud, A. (1972), Morphogenèse des membres rudimentaires chez les reptiles: un problème d'embryologie et d'évolution. *Bull. Soc. Zool. Fr.*, **97**, 469–485.

Raynaud, A. (1977a) (ed.), *Mécanismes de la Rudimentation des Organes chez les Embryons de Vertébrés*, Colloques CNRS 266, Paris.

Raynaud, A. (1977b), Les modalités de la rudimentation des membres chez les embryons de reptiles serpentiformes. In *Mécanismes de la Rudimentation des Organes chez les Embryons de Vertébrés*, Colloques CNRS 266, Paris, pp. 201–219.

Raynaud, A. (1977c), Somites and early morphogenesis of reptile limbs. In *Vertebrate Limb and Somite Morphogenesis* (eds. D. A. Ede, J. R. Hinchliffe and M. Balls), Cambridge University Press, Cambridge, pp. 373–385.

Raynaud, A. and Adrian, M. (1975), Caractéristiques ultrastructurales des divers constituants des ébauches des membres, chez les embryons d'Orvet (*Anguis fragilis* L.) et de Lézard vert (*Lacerta viridis* Laur). *C. R. Hebd. Séances Acad. Sci., Ser. D*, **280**, 2591–2594.

Raynaud, A. and Vasse, J. (1972), Les principales étapes du développement de l'ébauche du membre anterieur de l'Orvet (*Anguis fragilis* L.) etudiées au moyen de l'autoradiographie. *C. R. Hebd. Séances Acad. Sci., Ser. D*, **274**, 1938–1941.

Renous, S., Raynaud, A., Gasc, J.-P. and Pieau, C. (1976), Caractères rudimentaires, anatomiques et embryologiques, de la ceinture pelvienne et des appendices postérieurs du Python réticulé (*Python reticulatus* Schneider, 1801). *Bull. Mus. Natln. Hist. Nat., Paris* (3e sér), **379**, 547–584.

Ritter, E. J., Scott, W. J. and Wilson, J. G. (1973), Relationship of temporal patterns of cell death and development to malformations in the rat limb. Possible mechanisms of teratogenesis with inhibitors of DNA synthesis. *Teratology,* **7**, 219–226.

Rooze, M. A. (1977), The effects of the *Dh.* gene on limb morphogenesis in the mouse. *Birth Defects,* **13**, 69–95.

Salzgeber, B. (1968), Étude sur la genèse de malformations expérimentales des membres chez l'embryon de poulet. *Ann. Embryol. Morphol.,* **1**, 313–331.

Salzgeber, B. (1976a) Effects of nitrogen mustard on the development of limb buds. Studies on the genesis of limb malformations. In *Tests of Teratogenicity in vitro*, North-Holland, Amsterdam, pp. 417–434.

Salzgeber, B. (1976b), Limb malformation production in chicken embryos treated with nitrogen mustard at 48 hours of incubation. In *Drugs and Foetal Development* (eds. M. A. Klingberg, A. Abramovici and J. Chemke), Plenum, New York.

Salzgeber, B. and Gumpel-Pinot, M. (1977), Étude des effets de l'ypérite azotée et des rayons X dans la production expérimentale de la micromélie, de la phocomélie et de l'ectromélie chez l'embryon de poulet. In *Mécanismes de la Rudimentation des Organes chez les Embryons de Vertébrés*, Colloques CNRS 266, Paris.

Salzgeber, B. and Weber, R. (1966), Le régression du mésonéphros chez l'embryon de poulet. Étude des activités de la phosphatase acide et des cathepsines. Analyse biochimique, histochimique et observations au microscope électronique. *J. Embryol. Exp. Morphol.,* **15**, 397–419.

Saunders, J. W., Jr. (1966), Death in embryonic systems. *Science,* **154**, 604–612.

Saunders, J. W., Jr. (1972), Developmental control of three-dimensional polarity in the avian limb. *Ann. N.Y. Acad. Sci.,* **193**, 29–42.

Saunders, J. W., Jr. (1977) The experimental analysis of chick limb bud development. In *Vertebrate Limb and Somite Morphogenesis* (eds. D. A. Ede, J. R. Hinchliffe and M. Balls), Cambridge University Press, Cambridge, pp. 1–24.

Saunders, J. W., Jr. and Fallon, J. F. (1967), Cell death in morphogenesis. In *Major Problems in Developmental Biology* (ed. M. Locke), Academic Press, New York, pp. 289–314.

Saunders, J. W., Jr., Gasseling, M. T. and Errick, J. (1976), Inductive activity and enduring cellular constitution of a supernumerary apical ectodermal ridge grafted to the limb bud of the chick embryo. *Dev. Biol.,* **50**, 16–25.

Saunders, J. W., Jr., Gasseling, M. T. and Saunders, L. C. (1962), Cellular death in morphogenesis of the avian wing. *Dev. Biol.,* **5**, 147–178.

Saxen, L. and Rapola, J. (1969), *Congenital defects*, Holt, Rinehart and Winston, New York.

Saxen, L. and Wartiovaara, J. (1976), Embryonic induction. In *The Developmental Biology of Plants and Animals* (eds. C. F. Graham and P. F. Wareing), Blackwell, Oxford, pp. 127–140.

Scheib, D. (1963), Properties and role of acid hydrolases of the Müllerian ducts during sexual differentiation in the male chick embryo. In *Lysosomes* (eds. A.V.S. de Reuck and M.P. Cameron), Ciba Foundation Symposium, Churchill, London.

Scheib, D. (1977), Mécanismes cellulaires et déterminisme hormonal de la régression des canaux de Müller chez l'embryon de poulet: étude cytologique et bio-chemique. In *Mécanismes de la Rudimentation des Organes chez les Embryons de Vertébrés*, Colloques CNRS 266, Paris, pp. 59–70.

Scheib, D. and Wattiaux, R. (1962), Étude des hydrolases acides des canaux de Müller de l'embryon de poulet. Activités totales et solubles des canaux d'embryons de 8 à 10 jours d'incubation. *Dev. Biol.*, **5**, 205–217.

Schlüter, G. (1973), Ultrastructural observations on cell necrosis during formation of the neural tube in mouse embryos. *Z. Anat. Entwicklungs ges.*, **141**, 251–264.

Scott, W. J., Ritter, E. J. and Wilson, J. G. (1977), Delayed ectodermal cell death as a mechanism of polydactyly induction. *J. Embryol. Exp. Morphol.*, **42**, 93–104.

Seegmiller, R. E., Overman, D. O. and Runner, M. N. (1972), Histological and fine structural changes during chondrogenesis in micromelia induced by 6-amino-nicotinamide. *Dev. Biol.*, **28**, 555–574.

Sengel, P. (1976), *Morphogenesis of skin*, Cambridge University Press, Cambridge.

Silver, J. and Hughes, A. F. W. (1973), The role of cell death during morphogenesis of the mammalian eye. *J. Morphol.*, **140**, 159–170.

Smiley, G. R. and Koch, W. E. (1971), Fine structure of mouse secondary palate development *in vitro*. *J. Dent. Res.*, **59**, 1671–1677.

Summerbell, D. (1973), Growth and regulation in the development of the chick limb. Ph. D. thesis, London University.

Summerbell, D. (1979), The zone of polarising activity: evidence for a role in normal chick morphogenesis. *J. Embryol. Exp. Morphol.*, **50**, 217–233.

Summerbell, D. and Tickle, C. (1977), Pattern formation along the anterio-posterior axis of the chick limb bud. In *Vertebrate Limb and Somite Morphogenesis* (eds. D. A. Ede, J. R. Hinchliffe and M. Balls), Cambridge University Press, Cambridge, pp. 41–57.

Tata, J. R. (1966), Requirement for RNA and protein synthesis for induced re-gression of the tadpole tail in organ culture. *Dev. Biol.*, **13**, 77–94.

Teng, C. S. and Teng, C. T. (1976), Studies on sex-organ development. Oestrogen-receptor translocation in the developing chick Müllerian duct. *Biochem. J.*, **154**, 1–9.

Waddington, C. H. (1962), *New Patterns in Genetics and Development*, Columbia University Press.

Watt, D. J. (1977), The mechanism of spina bifida in the chick. Ph. D. thesis, Aberdeen University.

Weber, R. (1977), Biochemical aspects of tail atrophy during anuran metamorphosis. In *Mécanismes de la Rudimentation des Organes chez les Embryons de Vertébrés*. Colloques CNRS 266, Paris, pp. 137–146.

Webster, D. A. and Gross, J. (1970), Studies on possible mechanisms of programmed cell death in the chick embryo. *Dev. Biol.*, **22**, 123–136.

Wolff, E. and Kahn, J. (1947), Le régulation de l'ébauche des membres chez les oiseaux. *C. R. Séances Soc. Biol.*, **141**, 915.

Wolpert, L. (1978), Pattern formation in biological development. *Sci. Am.*, **239**, 124–137.

Wolpert, L., Lewis, J. and Summerbell, D. (1975), Morphogenesis of the vertebrate limb. In *Cell patterning*, Ciba Foundation Symposium 29, new series, Associ-ated Scientific Publishers, Amsterdam, pp. 95–119.

Zwilling, E. (1942), The development of dominant rumplessness in chick embryo. *Genetics*, **27**, 641–656.

Zwilling, E. (1959), Micromelia as a direct effect of insulin. Evidence from *in vitro* and *in vivo* experiments. *J. Morphol.*, **104**, 159–179.

Zwilling, E. (1964), Controlled degeneration during development. In *Cellular injury* (eds. A. V. S. de Reuck and J. Knight), Ciba Foundation Symposium, Churchill, London, pp. 352–362.

Zwilling, E. (1968), Abnormal morphogenesis in limb development. In *Limb development and deformity: problems of evaluation and rehabilitation* (ed. C. A. Swinyard), Charles C. Thomas, Springfield, pp. 100–113.

Zwilling, E. (1974), Effects of contact between mutant (wingless) limb buds and those of genetically normal chick embryos: confirmation of a hypothesis. *Dev. Biol.*, **39**, 37–48.

Zwilling, E. and Hansborough, L. A. (1956), Interaction between limb bud ectoderm and mesoderm in the chick embryo. III. Experiments with polydactylous limbs. *J. Exp. Zool.*, **132**, 219–39.

3 Cell death in metamorphosis

RICHARD A. LOCKSHIN

3.1 Introduction

Metamorphosing animals provide the most obvious models for the study of cell death. In insects and amphibia, large masses of relatively homogeneous tissue degenerate rapidly and predictably. In many instances, the extracellular controls are well understood and the control manipulatable at the experimenter's convenience; the response may be evocable *in vitro*. In other instances of rapid development – the metamorphosis of embryonic invertebrates (molluscs, echinoderms or ascidians) or tissue metamorphosis (Müllerian and Wolffian ducts, ovulation, involution of the corpus luteum) – there are equal opportunities for experimentation, but they have not been as thoroughly studied (Glücksmann, 1951).

Published information on the death of cells in metamorphosing animals encompasses whole-animal physiology, enzymology of macromolecular syntheses of lytic enzymes, cytochemistry and cytology, and subcellular physiology. Themes and mechanisms for the control of cell death remain to some extent inchoate, partly because of the diversity of the disciplines and techniques involved. To some extent also, our knowledge suffers from the fact that different laboratories use different materials and tissues. A biochemist's report on metamorphosis of the tail of a bullfrog might be compared with a cytologist's observations concerning the involution of the tail of the African clawed frog, in which metamorphosis occurs at a markedly different pace; or the collapse of muscles in a Saturniid moth is assumed to be essentially similar to that in a Sphingiid, and may be compared to that of a horsefly. One may nevertheless identify, without stretching the point, a surprising unity in the overall conclusions. The mechanisms of most of the rapid involutionary processes fall into a limited number of categories, and the mechanisms and sites of failure, where astutely guessed at, seem to be similar. These similarities and themes may be most clearly understood if we first review the observations and then attempt to draw functional conclusions. Thus the first part of the sections on amphibians and insects consists of a rapid description of involution in specific organs and tissues of favourite research organisms. The second part, in a more theoretical vein, is devoted to common elements in the various studies, with some extrapolation of the generalities, mechanisms and meanings of studies using different techniques. The section on the maintenance of nerves, being more unified, treats the two subjects together. Finally, a model of the mechanisms of cell death is sketched, with indications of directions for future lines of research.

3.2 Amphibian metamorphosis

3.2.1 *Anuran tail fins*

By far the great bulk of studies of anuran metamorphosis have focused on tail tissues, primarily muscle. Although the reasons for this preoccupation are self-evident, interpretation of the sequence of events remains somewhat less certain than might be hoped. A major complication is an apparent variation among species in the cytological sequence of events, which occur over time scales varying 20- to 50-fold. [The length of the tail of *Rana catesbeiana* regresses at approximately 12% per day; that of *Bufo americana*, 10% per hour (Lockshin, 1980; Lockshin, Colon and Dorsey, 1980).] Nevertheless the corpus of information bears an identifiable structure, so that an outline of the involution of the tail may be sketched.

(a) *Role of thyroxine* That rising levels of thyroxine initiate resorption of the tail is unequivocal. Not only has tail regression been a landmark assay for the hormone, the development of the isolated tail-tip preparation (see Weber, 1969a, for discussion) has demonstrated that thyroxine alone can provoke the involution of the cultured remnant. The response of the tail to thyroxine or tri-iodothyronine has been observed in several studies, focusing especially on the appearance of lysosomal enzymes and on various synthetic activities.

The first prominent sign of collapse of the tail is the resorption of the ventral tail fin, followed shortly thereafter by the dehydration and beginning collapse of the dorsal tail fin (Etkin, 1964; Dodd and Dodd, 1976). The fins are relatively simple structures, consisting primarily of epidermis (including melanocytes), fibroblasts, a few macrophages and a highly hydrated collagen matrix. The release of proline from collagen is markedly increased when thyroxine is injected into the tadpoles *(Rana catesbeiana)*, and injection of prolactin reduces this response (Yamaguchi and Yasumasu, 1978). Thus attention directed to this tissue resulted in the identification of collagenase which resides within the tissue as an inactive procollagenase and, when activated, initiates a proteolytic cascade, leading to the destruction of the tail fin collagen (Gross and LaPiere, 1962; Gross and Bruschi, 1971; Harper, Bloch and Gross, 1971; Harper and Gross, 1972; Davis, Jeffrey, Eisen and Derby, 1975). The collagen is degraded within macrophages (Fox, 1972b, 1973a), but there is no systematic cytological description of the fate of the resident fibroblasts. Several of the researchers working in the field feel that the destruction of collagen is a necessary prelude to the resorption of the rest of the tail, resting their argument on both the primacy of the collagenolytic reaction and the temperature sensitivity. [Tadpole collagen denatures at 37°C and becomes susceptible to attack by various proteases; tadpole tail will not degenerate if the animal is maintained below 15°C (Gross and LaPiere, 1972; Davis *et al.*, 1975), but the point is somewhat disputed by Dodd and Dodd in a recent review (1976).] The temperature sensitivity of other enzymes and substrates has not been documented and the argument, though provocative, remains unproved. Destruction of the ground substance could of course allow muscles to relax, a condition known to enhance muscle breakdown in other systems (Li and Goldberg, 1976; Li and Jefferson, 1978; Li, Higgins and Jefferson, 1979; Etlinger, Kameyama, Toner, van der Westhuyzen and Matsumoto, 1980).

Several studies describe other biochemical responses of tail fin to thyroxine. Lysosomal hydrolases increase in quantity and concentration as the tail fin involutes; the K_m values and pH optima remain the same. Simultaneously, other enzymes such as lactate dehydrogenase maintain constant specific activity, decreasing in total activity as the tail tissue is lost (Greenfield and Derby, 1972).

Insofar as synthetic events are concerned, actinomycin D, an inhibitor of DNA-dependent RNA synthesis, can block the breakdown of the isolated tail tip, appearing to interfere with the degradation of the connective tissue (Weber, 1977). The synthesis at metamorphosis of a specific or particular lytic protein has not been detected (Beckingham-Smith and Tata, 1976). Thyroxine rapidly inhibits uptake of [^{14}C] proline into the fin; when the tail is preloaded with proline, prolactin stimulates its incorporation into acid-insoluble material; when the two hormones are administered together, the retention of proline is reduced from that of the prolactin-alone preparation, but it is not eliminated. In the bullfrog, tri-iodothyronine rapidly inhibits the transport and incorporation of L-site amino acids, apparently by direct competition at the transport site (Frieden and Campbell, 1978). The authors note, however, that the doses used were in excess of physiologically effective doses, and that thyromimetic compounds could be found which did not inhibit transport. These latter compounds nevertheless depressed incorporation of L-site amino acids, an effect which may be physiologically important (Frieden and Campbell, 1978). Thus, in complex systems, thyroxine appears ultimately to stimulate protein synthesis and degradation, at least partially by effects on the transport of amino acids. Nevertheless, the mechanism still appears to be uncertain (Perriard and Weber, 1971; Little, Atkinson and Frieden, 1973; Little, Garner and Pelley, 1979; Wanatabe, Khan, Sasaki and Iseki, 1978a; Wanatabe, Sasaki and Khan, 1978b). Since it has recently become possible to establish primary cultures of liver cells of *Xenopus* (Wahli and Weber, 1977) and since cultures of fibroblasts of homoeotherms have long been favoured experimental material for studies of aging, cell growth and, in general, cell metabolic function, it would appear to be an obvious experimental direction to collect fibroblasts from dissociated tail fins and to subject these homogeneous lines to thyroxine *in vitro*. For that matter, and in an entirely different vein, it would be of interest to compare the longevity in culture of lines of fibroblasts derived from the tail (destined to die) and of lines derived from hindlimbs or other adult tissue (potential lifespan of the frog 10–20 years) in order to establish 'Hayflick limits' for these two types of cells (see Chapter 8). Culture of amphibian cells, however, still involves several problems not completely resolved (Rafferty, 1976).

The studies on isolated tail tips have also highlighted two anomalies which may provide keys to the causal sequence of events: if the isolated tail is not allowed to heal at relatively cool temperatures, lytic activities at the wound site progress and ultimately involve and destroy the entire tail tip, in the absence of added hormones (Davis *et al.*, 1975); and newly regenerated tail is insensitive to the action of thyroxine (Weber, 1969b). These rather subtle curiosities suggest that the effect of thyroxine is mediated through intracellular communication mechanisms, which themselves are subject to alternative metabolic control, and further investigation will undoubtedly prove rewarding.

3.2.2 *Anuran tail muscle*

Within the tail, undifferentiated biochemical analyses refer in essence to changes within the muscle mass, and most cytological assays are directed to this tissue. The death of the muscle cell is a unique and spectacular event which has been recognized, though perhaps not fully understood, since the end of the 19th century (Eeckhout, 1969; Dodd and Dodd, 1976). The massive muscle fragments into small pieces, approximately 5–20 sarcomeres in length and 5–20 myofibrils in width, termed 'sarcolytes', which are then ingested by macrophages and ultimately degraded within the heterophagic vacuoles of these cells (Gona, 1969; Weber, 1969b; Fox, 1973b; Kerr, Harmon and Searle, 1974). The sarcolytes are formed by the development of longitudinal clefts within the fibres, presumably deriving from the sarcoplasmic reticulum. The fibre is promptly invaded by macrophages, which then segregate off rather impressive bites of sarcolytes. Although the sarcoplasmic reticulum dilates, the other organelles retain a relatively normal appearance, apart from an overall condensation of the cell and increase in electron density (Kerr, 1973; Kerr *et al.*, 1974). These latter authors analogize the process to a complex ballet of dying cells which they term apoptosis — essentially, a fragmentation of a mortally wounded but not lysed cell — drawing particular attention to the preservation of the organelles, the condensation of the cell, and the belated but definite nuclear pyknosis (see Chapter 1). They also emphasize that autophagic lysosomes do not play a significant role in the destruction of the muscle, a fact with which the other authors agree. Other authors, however, claim that some erosion of myofilaments occurs prior to the engulfment of sarcolytes by the macrophages (Fox, 1973b; Weber, 1969b). We will return to the question of the degradation of the myofilaments in a later section, and further note here that the condensation of the cell must reflect either the extrusion of water from the cell or in precipitation of the proteins *in situ*, by mechanisms which have not been suggested for this context (but see Bank and Mazur, 1972). This point also is discussed later.

3.2.3 *Other tail tissues*

Fox has made a systematic study of several non-muscle tissues in the involuting tail of *Rana temporaria*: tail skin (Fox, 1972a), notochord (Fox, 1973a) and notochordal and subnotochordal collagen (Fox, 1972b). In these studies, he has come to several general conclusions. First, in this species at least, he notes that degeneration begins at the tail tip and progresses anteriorly along a rather sharply defined frontier, with more anteriorward tissues presenting an apparently normal undamaged appearance. Inattention to such a detail, if true in other species, could of course be a source of much confusion for the unwary investigator. Second, acid phosphatase-containing lysosomes (autophagic vacuoles) play a prominent part in the destruction of mitochondria and other intracellular organelles. Epidermal cells furthermore flatten and keratinize before being sloughed off. Third, macrophages play a prominent role, ingesting the partially degraded collagen and perhaps initiating the fragmentation of the notochord. The macrophages contain large heterophagic vacuoles, in which collagen and cell debris are found, and may also possess lysosomal enzymes on the outer surfaces of their cell membranes. The latter hypothesis, however, is based on depositions of lead phosphate which Fox concedes potentially to be artifacts. There is some evidence that liver lysosomes may contain enzymes on both sides of the membrane (Schneider, Burnside, Gorga and Nettleton, 1977).

The origin of the macrophages remains somewhat controversial. They appear in degenerating tail, even when degeneration is provoked *in vitro*, and hence are presumed to differentiate from previously existing resident precursors (Weber, 1969a, 1969b). Their differentiation is presumed to account for the rise in lysosomal enzymes, since lysosomes are prominent features of their cytoplasm (Hickey, 1971). The ability of inhibitors of RNA and protein synthesis (actinomycin D and cycloheximide respectively) to block tail resorption has been attributed to the action by these drugs of preventing the presumed differentiation of the macrophages (Tata, 1966; Weber, 1969b). Similar conclusions have been reached for insect systems (Lockshin, 1969); and the initiation of proteolysis may require the synthesis of proteins (Gunn, 1978). A latent period during which explanted tail fin discs will not respond to thyroxine may be reduced if the discs are nestled against discs which have been previously treated with thyroxine, and it is presumed that some induced material is transmitted from the exposed tissue to the previously unexposed tissue (Kim, Hamburgh, Frankfort and Etkin, 1977). It is possible that the material transmitted is in fact activated macrophages, or collagenase, as described below.

The action of the antibiotics has itself received some attention, since it appears that metamorphosis of isolated tail tips is completely blocked by the drugs. In a perhaps related manner, the inhibitor of cathepsin D, pepstatin, completely inhibits resorption (Seshimo, Ryuzaki and Yoshizato, 1977). Thus several authors have recently begun to speculate on a possible primary role for the collagenase (Weber, 1977), or another proteinase, the lysosomal cathepsin D (Seshimo *et al.*, 1977). Both groups of authors think along similar lines that the synthesis of a proteinase plays a primary role in initiating the collapse of the tail. The Japanese group speculates that, since cathepsin D appears to block the entire metamorphosis of the tail and not simply proteolysis, cathepsin D might be a genetic regulator. The question is not yet resolved, but it would appear that sentiment is gaining in favour of a cascade of events triggered by an initial and relatively subtle shift. Insofar as the activation of the macrophages is concerned, this event might well be triggered by the release of some product of degradation, perhaps a fragment of proteolysis, and hence be itself secondary. A similar phenomenon is seen in insects (see below). Whether the concept of the cascade is biologically accurate or is merely a reflection of the preferences and psychology of human thinking remains to be seen.

3.2.4 *Anuran tissues other than tail*

The metamorphosis of anurans is of course profound and, although naturally the destruction of the tadpole tail has attracted the greatest attention of biologists, the involution of other tissues has not escaped notice. Following several involutional changes during embryonic development (van Evercooren and Picard, 1978), virtually every tissue in the amphibian body undergoes extensive remodelling in preparation for the new life of the organism: intestines, lungs and gills, liver, pancreas, intestinal and epidermal epithelia, haematopoietic system, nervous system including Mauthner's cells, and many others for which no recent reports are readily available. One relatively simple preparation which appears not to have been studied is the epidermis overlying the forearms; this presumably easily cultured region regresses rapidly during metamorphic climax. Comparable epidermal regions, located elsewhere, respond to thyroxine but do not resorb, and it would seem a matter of some interest to know more about it. It first deviates in appearance from other epidermis at stage XV (prometamorphosis), and the response of the tissue appears to depend on the underlying gill and operculum, and to a limited extent on pressure from the limb (Reichel, 1976). Its sensitivity to thyroxine differs from that of other epidermis (Dodd and Dodd, 1976).

Table 3.1 Comparison of stages of development of *Xenopus laevis* [Nieuwkoop–Faber (1967) stages] and of *Rana pipiens* [Taylor-Kollros (1946) stages]

Stages	*Xenopus*	*Rana*	Distinguishing features
Premetamorphosis	47–51	II–VI	Young feeding tadpole
	52	VI	Foot paddle
Prometamorphosis	55–57	XI–XVII	Growth of hindlimbs
Climax*	58	XX	Forelimbs erupt
	62	XX–XXI	Rapid tail resorption
	66	XXV	Miniature adult

*Climax is defined as beginning at stage 58 in *Xenopus*; in *Rana*, it is defined as beginning at forelimb eruption (Etkin, 1968), or, earlier, at stage XVII, resorption of the cloacal tailpiece (Taylor and Kollros, 1946). Eruption of the forelimbs occurs relatively earlier in *Xenopus* than in *Rana* (Dodd and Dodd, 1976).

Several tissues evolve in function with neither massive cell death nor cell replacement. In the bullfrog liver, for instance, total DNA does not change, although there is a rise in total RNA and protein (Eaton and Frieden, 1968; Kistler, Yoshizato and Frieden, 1975). Measurements of rates of synthesis and degradation are complicated by fluctuations in the sizes of intracellular pools of amino acids, presumably deriving both from changes in permeability barriers and from changing rates of protein degradation (Kistler *et al.*, 1975). No changes in composition or relative rates of synthesis were noted in the metamorphosing liver (Morris and Cole, 1978).

The haematopoietic system may also be a site of cell death. The difference between adult and larval haemoglobins has long been known, and derives from the appearance of a new class of cells. The transition is relatively slow: in *Xenopus*, conversion from larval to adult haemoglobin requires at least 4 weeks, deriving apparently from a preferential synthesis of adult haemoglobin (Just, Schwager and Weber, 1977). In bullfrogs, however, there is a decided increase in the rate of destruction of larval red blood cells, beginning at stage XVIII and reaching a maximum at stage XXI–XXII (Osaki, James and Frieden, 1974). Whether the difference reflects species differences or simply different orientations on the part of the experimenters is not clear. *Xenopus* is an aquatic frog which is capable of absorbing a large percentage of its oxygen through its skin, but the rate of metamorphosis of the haemoglobin does not appear to depend on the habitat assumed by the frog or toad (Dodd and Dodd, 1976). The metamorphosis of the erythropoietic system is controlled by thyroxine, though at different concentrations from those necessary for morphological metamorphosis. When *Xenopus* tadpoles are treated with thiourea, morphological transformation is blocked, but the haemoglobin transition still occurs (Just *et al.*, 1977), and thyroxine can effect a conversion of axolotl haemoglobin from larval type to adult type, in the absence of anatomical metamorphosis (Dulcibella, 1974a). In the latter instance, the conversion starts with the inhibition of the formation of larval red blood cells (Dulcibella, 1974b). A word of caution should, however, be expressed: the effect of thyroxine and thiourea on amphibian metabolism is equivocal, and the chemicals are at least potentially capable of altering metabolic rate and therefore oxygen extraction by the tissues. Since the affinity for oxygen of larval and adult haemoglobins differs, the switch from one to the other may well derive from a feedback relating to the oxygen demand, rather than directly as a result of endocrine stimulation. If studies exist of the stem cells themselves, either in isolation or identified by immunological markers, I am unaware of them.

One potentially very interesting tissue, study of which has just begun, is the pancreas. As the organ metamorphoses, there is first a period of cell death, followed by the regeneration of the adult organ. During the period of regression, incorporation of nucleotides into RNA shows two peaks. The first peak derives from an apparent increase in the specific activity of the precursor pool, without an overall increase in synthesis. The mechanism for the increase in pool specific activity is not known; it may relate to tissue heterogeneity or changes in subcellular uptake or compartmentation. Just before the period of maximum regression, there is a disproportionate increase in low- molecular-weight (4S) RNA, perhaps reflecting degradation of the ribosomes (Bollin, Carlson and Kim, 1973). The pancreatic system should eventually prove to be extremely interesting, not only from the standpoint of analysis of the nucleic acids and endocrine control of the pancreas, but also because of its high potential for analysis. As studies of the embryonic pancreas have so elegantly demonstrated (Rutter, Pictet and Morris, 1972). the presence of several enzymes, for which antibodies exist or can be made, permits very fine resolution of the biological activities of several specific types of cells. The biological activities may then be correlated with cytological appearance, histochemical activities and rates of synthesis and degradation of nucleic acids. From these, a highly elaborate composite of cell death may be constructed. The evolution of some of the enzymes has been documented (Cohen, 1966; Mori and Cohen, 1978), and the evolution of the others may be presumed from the drastic change of diet which some anurans undergo. In this system, however, *Xenopus* is likely to be less interesting than Ranids such as *R. catesbeiana*, the bullfrog, since the diet of *Xenopus* changes less markedly.

Another tissue which has been recently subjected to closer analysis is the gill system, which atrophies rapidly as the lungs develop. Involution of the gills may be followed either *in vivo* (Atkinson and Just, 1975) or *in vitro* (Derby, Jeffrey and Eisen, 1979). Biochemically, natural metamorphosis and thyroxine-forced metamorphosis are similar, but the morphology is different. The major difference involves the development of the lungs, which probably depends on a finely honed regulation of thyroxine levels. The involuting gills manifest a sharp increase in incorporation of amino acids and thymidine, coupled with a decrease in gill vascularity and blood flow. The gills appear to darken because the melanocytes are lost less rapidly than the epidermal cells (Atkinson and Just, 1975). *In vitro*, involuting bullfrog gills release collagenase into the medium, which increases roughly in proportion to the amount of gill resorbed and the amount of collagen lost from the explant. Collagenase is present in freshly explanted gills and declines gradually in the absence of hormone (Derby *et al.*, 1979). It appears that the collagenase may be synthesized or activated in response to the act of wounding; the relation of induction by thryoxine to induction of tail atrophy by wounding or maintenance at too high a temperature (Davis *et al.*, 1975) is not known. Resorption of the gill in these experiments was accompanied by an increase in specific activity of acid phosphatase, but the entire increase could be attributed to the differential loss of protein; there was no net increase in the activity of this presumably lysosomal enzyme (Derby *et al.*, 1979).

The epidermis also undergoes metamorphosis, which may be induced *in vivo* and *in vitro* by thyroxine and identified by the appearance of adult keratin (Reeves, 1977). The larval epidermis is lost by molting; the cells of the stratum corneum flatten before they are shed. Prior to this point, when *Rana ridibunda* tadpoles have four legs (equivalent to *R. pipiens* stage XX – Table 3.1), the stratum corneum contains both 'light' and 'dark' cells; the latter eventually develop vacuoles. Whether they are already moribund is not clear, since younger larvae were not studied by electron microscopy (Rosenberg and Warburg, 1978). None of the studies attempt to correlate the fate of the epidermis from the several body regions with that overlying the gills.

The intestinal epithelium undergoes a rapid involution resulting in the replacement of the primary (larval) epithelium by a secondary (adult) epithelium. The death of the primary epithelium is characterized by the condensation of cells, pyknosis of the nuclei, and formation of vacuoles in the cytoplasm of the cells. Lysosomal enzymes are purportedly released directly into the cytoplasm (Hourdry, 1971). The physical release of lysosomal enzymes, as opposed to their release during preparation of tissue, is an idea which attracts less enthusiasm today than previously, but, since the fate of these cells differs from that of many other cells (they are discarded into the intestinal lumen) the concept should not be automatically discarded. Free acid phosphatase is still considered to be a sign of cell death in several other systems (Lewis, Bowen and Bellamy, 1979; Bowen, 1980). An early sign of the involution of the primary intestinal epithelium is a sudden decrease in incorporation of tritiated thymidine, which occurs by stage XVII (mid-prometamorphosis) in *Alytes obstetricans*, whereas the epithelium is not actually discarded until after climax begins (stage XX – Dauça and Hourdry, 1978b). These workers have recently established a technique for the isolation of viable intestinal epithelium, an achievement which promises to render the tissue much more accessible to experimental manipulation (Dauça and Hourdry, 1978a).

The pronephros also degenerates at metamorphosis, in a process which involves the formation of large autophagic vacuoles in the tissue. The focus of the most recent study was the controversy over the direct or indirect effects of the thyroid hormones. By means of a series of alterations of the concentrations of thyroid hormones and the aquarium temperature, Fox (1971) opted for the argument of a direct effect of the hormone.

Metamorphosis of course includes the extensive reorganization of the amphibian nervous system, involving both cell proliferation and cell death. Thyroxine, long known to influence the development and function of the nervous system in mammals, evokes both types of changes in brain cells of *Rana pipiens*; some cells grow rapidly, whereas others degenerate and are scavenged by phagocytic microglia (Decker, 1976). In the dying cells, the Golgi dilates and then fragments into vesicles; mitochondria also show early degenerative changes such as an increase in matrix density followed by distention of the cristae; fragmentation of the nucleus is a rather later event. Approximately two-thirds of the neurons die. Decker provokes metamorphosis by adding thyroxine to the aquarium, and makes a very interesting observation: at low doses of the hormone, dying neurons are filled with lysosomes, whereas, if the tadpoles are subjected to higher doses of hormone ($50 \mu g/l$ as opposed to $10 \mu g/l$) no lysosomes are formed, but the cells nevertheless die (Decker, 1976). He relates this observation to his concept that the predominance of lysosomes is proportional to the degree of differentiation of the cell (Decker, 1974a, 1974b), but there is another aspect which likewise should be emphasized: in the high-dose situation, either the lysosomal enzymes exist in the absence of morphologically identifiable lysosomes – an idea which is open to examination and deserves attention – or lysosomes are not causal of, or synonymous with, the death of the cell. We will return to this point in the final section of the chapter. When thyroxine is administered to *Xenopus* tadpoles, there is a 3- to 6-fold increase in incorporation of [^{3}H] uridine into brain cells (Stadler and Weber, 1970), but these authors make no comments on the numbers of cells involved or the frequency with which involuting cells or non-incorporating cells are seen. It is possible to survey simultaneously for dividing and dying cells (Lewis *et al.*, 1979).

Of particular interest among neurons is the fate of the Mauthner's cell, for the Mauthner's cell is a giant (100 μm in diameter) easily accessible neuron involved in the lateral-line sensory mechanism of aquatic vertebrates. Located in the medulla of fish and larval amphibians, it acts as an integrative centre for several sensory inputs, and initiates a startle response (Zottoli, 1978). It atrophies at metamorphic climax, but is still present, though small, in *Xenopus* frogs 2 months after metamorphosis. It has not been found in considerably older adults (Zottoli, 1978). Since the lateral line persists in the aquatic adult *Xenopus*, it is possible that the Mauthner cell does not disappear, but merely regresses to a size at which it is not easily identifiable, (Kimmel and Model, 1978). In the arboreal *Hyla*, Mauthner cells disappear rapidly after climax. A similar situation presumably obtains for the less aquatic frogs such as *Rana pipiens* and *Rana temporaria*, and for the Bufonid toads.

In spite of the extensive possibilities for research on the Mauthner's cell, relatively little attention has been paid to its involution. Attempts have been made to determine the role of thyroxine in its disappearance, but the effect is as yet unclear. Earlier studies purported to demonstrate that thyroxine provokes involution, but more recent experiments argue that the cell hypertrophies in the presence of thyroxine and finally involutes when thyroxine levels decrease following metamorphosis (Pesetsky, 1966). By this interpretation, the cell acquires dependence on thyroxine when it is exposed to the hormone, much in the manner that embryonic urodele limbs become dependent on innervation for regeneration only after the growing nerves invade the limb bud (Yntema, 1959). Since the effects of implanted hormones are relatively small, however, it is also possible that the changes seen result from other factors (Kimmel and Model, 1978). The question of survival of the nerve to synaptic contacts made upon it is a fascinating subject worthy of treatment in its own right (Landmesser and Pilar, 1978).

The cytological appearance of the degenerating Mauthner's cell has been observed (Moulton, Jurand and Fox, 1968). A preliminary slight hypertrophy of the cell is suggested by an increase in rough endoplasmic reticulum; later, there is dispersion of the reticulum, coupled with the appearance in the cytoplasm of lysosomes and other, clear, vesicles. Although both *Xenopus* and *Rana temporaria* tadpoles were used in these studies, the cells were not followed into the later stages of involution. Clearly, the subject is ripe for further work.

3.2.5 *Cell death in amphibian metamorphosis: summary*

Thus, if one were to characterize the state of the literature on cell death in amphibian metamorphosis, one could say that the microscopy is generally good, though with some inconsistencies. The story of the collagenase cascade has been most productive, but descriptions of other proteolytic systems is less satisfactory. Our knowledge of the state and dynamics of nucleic acids in the tail could be vastly improved, and much tighter attention to the cellular site of action of thyroxine would be welcomed. Questions such as the similarity of response to wounding and to hormone treatment deserve more serious attention, as does the lack of responsiveness of regenerate tissue. Many of the complications of interpretation derive from heterogeneity of tissue, even in *in vitro* preparations. Combined analytical and observational studies on single homogeneous tissues would be valuable, and several candidates present themselves: intestinal and epidermal epithelia, Mauthner's cells and, potentially, fibroblasts. When these tissues are under study, meaningful interpretations of nucleic acid metabolism and enzymology will be possible. From a physiological standpoint, it would be useful also to obtain information such as to what extent tail circulation changes as a primary event of metamorphosis. Warren Chin (1980) has noted, for instance that amputation of the tail of *Xenopus* causes much less haemorrhage shortly after the beginning of metamorphic climax.

3.3 Metamorphosis in invertebrates

3.3.1 *General comments*

By comparison with the vertebrates, the invertebrates can offer at least one major advantage: their tissues are frequently far simpler in construction and are more homogeneous. Insect ectodermal tissues, for instance, are typically one cell layer deep and, since blood does not course through the tissue, contaminant blood cells are much less of a problem. Of greater concern in such tissues are cells such as insect tracheoblasts, which support the tracheal respiratory system and are to be found deeply embedded among the other cells of the tissue. Although the concentrated and massive effort often directed toward questions in mammalian physiology is not characteristic of studies on invertebrates, several tissues have been examined with a thoroughness capable of providing an overall picture, and to a large extent the concept described at the end of this chapter derives from the wealth of detail available from these sources. Many tissues have been examined at least occasionally (Lockshin and Beaulaton, 1974c), and at least three have been studied by several techniques: muscle, fat body (a massive insect tissue functioning as an equivalent to both liver and adipose) and salivary or silk glands. In the latter instance especially, large extrapolations are necessary among species, genera and orders, but since there appears to be an overall similarity, the approach seems tolerable for at least theoretical purposes.

3.3.2 *Insect muscle*

Muscles are probably the most thoroughly studied of insect metamorphosing tissue. The subject has been reviewed several times in recent years (Lockshin and Beaulaton,

1974c; Lockshin, 1980; Lockshin *et al.*, 1980) and will be only briefly summarized below, primarily in reference to those questions which have not been adequately treated or analysed.

An interesting and illustrative study, which well illuminates the distinctions involved, is a series of papers by Wissocq, concerning the evolution of muscles not in insects but in Nereid worms. These animals, it will be recalled, undergo stolonization, or epitoky, in which several segments differentiate as a motile reproductive unit. The muscles in this unit differ in several respects from the muscles in the asexual form. The old myofilaments are resorbed and new ones take their place. Wissocq carefully differentiates between this dedifferentiation, in which the fibres evolve in a new direction (Wissocq, 1970a, 1970b; 1977), and degeneration, in which the fibres die (Wissocq, 1970c). Dedifferentiation, which occurs in both asexual and sexual muscles, occurs as a rapid loss of thick filaments, which are replaced by a homogeneous granular matrix. The membranes remain intact, and the fibres appear to lose contact with one another when approximately two-thirds of the myofilaments are lost. Some vacuoles form within the fibres. The myofilaments erode rapidly in a centripetal direction from the coelom (Wissocq, 1970c, 1977, 1978). In images of an often spectacular clarity, Wissocq demonstrates that lysosomes as morphologically identifiable bodies are not involved in the loss of the myofilaments, nor do phagocytes play a major role.

In degeneration, as opposed to dedifferentiation, the myofilaments persist but become disorganized. Once this disorganization is seen, the fibre fragments into sarcolytes (Wissocq, 1970c). The sarcolytes are ultimately phagocytosed. At that time, they are dense condensed cells. This form of degeneration occurs only in asexual fibres and is attributed by Wissocq to loss through aging. Thus muscle, with its massive cytoplasm, becomes a tissue *par excellence* for which the question may be asked, 'To what extent is cell death related to destruction of the cytoplasm?'

In our own laboratory we have attempted, if not to deal directly with this question, at least to analyse some of its components. For technical reasons, our tissue of choice has been the intersegmental muscles of large Saturniid and Sphingiid moths, which degenerate during the 48h immediately following the emergence of the moth. In *Manduca sexta*, the tobacco hornworm, the intersegmental muscles extend anteriorly one or two segments immediately prior to the approaching adult ecdysis; these new fibres as well degenerate (Colon, 1980; Fig. 3.1).

The involution of these muscles begins with an endocrine step, the beginning of adult development. In the Saturniids and Sphingiids, development of the pupa to the eclosion (ecdysis) of the adult takes approximately 3 weeks. The actual collapse of the muscle, as measured by physiological parameters, is triggered by a neural event, itself triggered by the release of eclosion hormone and perhaps bursicon (Taylor and Truman, 1974; Reynolds, Taghert and Truman, 1979). The involvement of the neural system is rather easily demonstrated in some of the Saturniids, in which the nerves cease spontaneous activity but at least temporarily survive (Lockshin and Williams, 1965a, 1965b). In Sphingiids, the motor neurons die within 2 days (Truman and Reiss, 1976); presumably for this reason, efforts to preserve the muscles by pharmacologically maintaining the motor activity have proved unsuccessful (Lockshin, unpublished work).

The role of the nervous system is more equivocal in other instances of metamorphic cell death, including the degeneration of the proleg retractor in metamorphosing *Galleria* larvae (Runion and Pipa, 1970) and the reproduction-controlled degeneration of flight muscles of *Ips* and other bark beetles (Sahota, 1975) and crickets (Srihari, Gutmann and Novak, 1975; Finlayson, 1975; Chudakova and Gutmann, 1978) as well as, more expectedly, the involution of silk and salivary glands in various insects. These differences are perhaps disconcerting but not totally surprising, since, it should be remembered, the intersegmental muscles are, in origin, larval muscles which are retained throughout the pupal phase to serve in the ecdysis of the adult, whereas in the other situations the larval muscles disappear at metamorphosis or are related to the endocrinology of reproduction. It is presumably the interpolation of the delay mechanism that, in Saturniids at least, involves the nervous system. It is also relevant to note that the morphology of denervation atrophy of the proleg retractor is similar to, if less co-ordinated than, that of programmed cell death (Randall and Pipa, 1969; Randall, 1970).

The muscle may be in trouble prior to the triggering event. In doing a survey of easily assayable enzymes, we recently found, to our surprise, a marked increase in an M-type lactate dehydrogenase, starting approximately 2 days prior to the onset of rapid degradation at the ecdysis of the moth (Bidlack and Lockshin, 1976). We also noted a slow but definitely increased net loss of muscle protein at this time (Lockshin, Schlichtig and Beaulaton, 1977). Subsequently, we have demonstrated that very-short-term exposure of the insect to anoxia results in the rapid activation of this enzyme (Lockshin and Finn, 1978) and even that forced continual activity of the larva can provoke its appearance (Finn and Lockshin, 1980). We do not have direct proof, but we tentatively interpret these results to mean that, unlike the flight muscles, the intersegmental muscles are not endowed with an abundantly expansive oxygen-delivery system, and that they can become hypoxic during chronic exercise and that, when they do, they rapidly activate a glycolytic system; furthermore, shortly prior to adult ecdysis, the muscles become hypoxic and the lactate dehydrogenase system is activated. This latter event appears to be a physiological response, well within the capacity of the insect at any time, but which may relate to the impending destruction of the muscle. There is also, during this period, a minor but definite increase in lysosomes, as detected cytologically (dense bodies: Lockshin and Beaulaton, 1974a) and biochemically (Lockshin and Williams, 1965c; Dorsey, 1980). The lysosomes contain acid phosphatase, cathepsin D and cathepsin B. The appearance of these enzymes is correlated with periods of breakdown of the muscle but occurs at slightly different times for the various enzymes (Lockshin *et al.*, 1976; Dorsey, 1980; Fig. 3.2). The appearance of cathepsin B slightly precedes the initiation of rapid degradation, while that of cathepsin D is nearly simultaneous with that event. A premonitory rise in lysosomes in other muscles has not been noticed or sought; indeed, in *Calliphora* (Crossley, 1968, 1972a, 1972b) and *Rhodnius* (Auber-Thomay, 1967) easily identifiable lysosomes appear to be absent from the degenerating muscle.

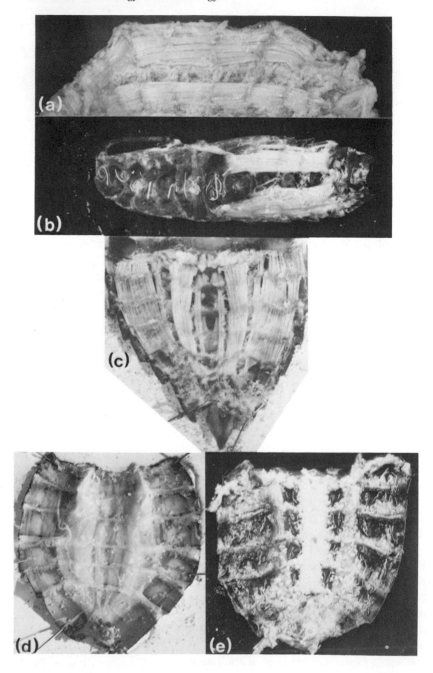

The muscle at emergence is fully functional and in fact it works vigorously during this period. As examined by transmission electron microscopy, it appears to be essentially normal in appearance. However, an ambiguous alteration in the structure of the Z-line led us to wonder whether we were truly fixing the muscles at appropriate rest lengths, and we have recently been rapidly freezing the insects, dissecting out the muscles in alcohol, and measuring sarcomere lengths by phase-contrast microscopy. Our early results suggest that the sarcomere length of the emerged moth is approximately 25 per cent shorter than that from the unemerged insect (Lockshin and Chin, unpublished work), an observation which would be consistent with the shorter set that the adult abdomen ultimately assumes. Since the abdomen is ultimately held by recently formed cutaneous muscles (when the intersegmental muscles have degenerated), we cannot at present say to what extent the muscles are under load, but it is evident that, if the physiological demands on the muscle influence the status of its myofilaments, as is seen in vertebrate muscles (Li and Goldberg, 1976; Vandenburgh and Kaufman, 1979; Etlinger *et al.*, 1980), this physiological change may also be considered to be a relevant alteration potentially leading to cell death. In all instances of muscle involution in relation to moulting, the insect changes size and hence the stretch on the muscle alters just before the breakdown of the muscle. A larva just prior to pupation may be half the size it was, and after pupation it contracts further. The stretch factor, however, may not apply to the resorption of flight muscles of beetles in response to reproductive events.

At the emergence of the insect, the muscles, though reduced somewhat in total protein content (Lockshin *et al.*, 1977; Dorsey, 1980; Fig. 3.2) are fully functional; they have resting potentials as negative as −70 mV, and they contract vigorously.

Fig. 3.1 Fate of the intersegmental muscles in *Manduca sexta*. The ventral and posterior ends of the insects are indicated. The preparations have been fixed in alcohol. The photographs are not to scale. (a) Fifth instar larva. The intersegmental muscles extend the entire length of the body. The bulk of the tissue is destroyed at pupation (Fig. 3.2) but for technical reasons this period of destruction is not readily accessible for study. (b) Left half of young pupa. The intersegmental musculature was reduced, during the first 3 days of pupal life, to three pairs of bands extending from the fourth or fifth to the seventh abdominal segments. (c) Abdomen of pharate (near-emergence) adult. The musculature has extended anteriorally to the second segment, and several new bands of muscles underlie the intersegmental muscles. They are easily distinguished from these latter and persist after the intersegmental muscles have degenerated. Here they are visible near the mid-ventral line. (d) Similar preparation, 15h after emergence. The tracheae have filled with fluid and no longer refract light. This event, coupled with the considerable loss of myofilament protein, makes even fixed muscles somewhat translucent. It is approximately at this time that the muscle depolarizes. (e) Similar preparation, 30h after emergence. All intersegmental muscles, including the new anteriorward extensions, have degenerated. Remnants can still be found; these consist essentially of pyknotic nuclei (Beaulaton and Lockshin, 1977).

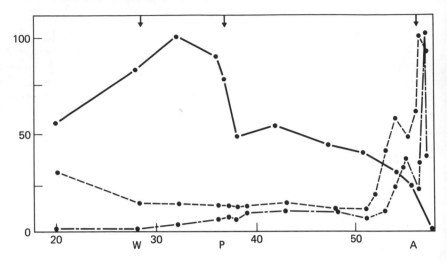

Fig. 3.2 Proteolytic enzymes in the intersegmental muscles of *Manduca sexta*. Abscissa: days from hatching of egg. Since there is some variability in the time at which the larvae pass through the various instars, the beginning of the fifth (final larval) instar is normalized to day 20. The larva leaves its food and begins wandering (*W*) on day 28, pupates (*P*) on day 32, and emerges as an adult (*A*) on day 56. Ordinate: percentage of maximal level of the indicated parameter. Solid line: total protein in homogenate of muscle. Dashed line: cathepsin B, measured as liberation of naphthylamine per unit of protein from benzyloxycarbonyl-Ala-Arg-Arg-β-naphthylamine. Dot-dash: cathepsin D, measured as production of trichloroacetic acid-soluble ^{3}H per unit of protein from ^{3}H-labelled haemoglobin. The relative activities of both enzymes rise sharply for the breakdown of the muscle after the emergence of the adult, the rise in cathepsin B slightly preceding the rise in cathepsin D. Although the total activities of both enzymes are high when the anteriormost and posteriormost extensions of the muscles are lost at pupation (the maximum for cathepsin B again preceding the maximum for cathepsin D), the relative activities are much lower. Note that there is some loss of protein throughout the last days of pupal life. In general, the activity of cathepsin B appears to precede the period of maximal lysis, while the activity of cathepsin D correlates closely with that period. The data are from Dorsey (1980).

Together they can generate a hydraulic pressure in the abdomen of over 1 atm, presumably by this means forcing haemolymph into the thorax and helping the wings to expand (Lockshin and Williams, 1964). At this time also several changes of potential significance to the fate of the muscle occur: spontaneous motor activity to the muscles appears to decrease markedly (Lockshin and Williams, 1965a). Owing to loss of fluid to the wings, loss of meconium (excretory material from the pupal stage) and loss of body fluid through evaporation, the rest length of the abdomen decreases and an intracellular commitment to degeneration is enacted in a process that can be blocked by cycloheximide and actinomycin D (Lockshin, 1969; Lockshin and Beaulaton, 1974b). The shift in motor activity occurs within the central nervous system of the moth, and is provoked by the hormones which stimulate the actual eclosion of the insect (Truman and Riddiford, 1970). In Saturniids, one may block the installation of the change by treating the insect, within the first hours after it emerges, with cholinergic drugs, these acting on the central nervous system (Lockshin and Williams, 1965b).

During the next few hours the muscle retains an appearance similar to that seen just prior to emergence, with two exceptions: the slightly altered appearance of the Z-line, the significance of which is not understood at present; and the gradual increase in dense-body lysosomes (Lockshin and Beaulaton, 1974a). In some of our preparations, we see elongated trails of lead phosphate precipitates (acid phosphatase) which other authors have interpreted as evidence for the origin of the lysosomes (Trout, Stauber and Schottelius, 1979). Between 7 and 10 h after emergence, the frequency of autophagic vacuoles increases sharply, and these organelles tend to contain, rather specifically, mitochondria. Later they will contain glycogen or ribosomes; mixed contents are not common (Beaulaton and Lockshin, 1977). Although there are signs of muscle damage – arginine kinase, the insect equivalent of creatine kinase, appears in the haemolymph (Lockshin, 1975a) – and a steady increase in the effective input resistance reflects a continual loss of muscle mass (Lockshin, 1973), the fibres remain contractile, and resting potentials remain within normal limits (Lockshin, 1973; Lockshin and Beaulaton, 1979a). In electron micrographs, the thickness of the peripheral sarcoplasm appears to have diminished (Beaulaton and Lockshin, 1977, 1978) and myofilaments are lost somewhat randomly throughout the fibre (Beaulaton and Lockshin, 1977). The degenerating proleg retractor of *Galleria* larvae differs in that miniature endplate potentials can be shown to continue, and there appears to be a gradual depolarization of the fibre (Runion and Pipa, 1970). Also, erosion appears to take place from the end inward (Randall, 1970).

Approximately 10–15 h after ecdysis, the pace of erosion increases rapidly. The apparent order of events is as follows. First, the tracheae fill with fluid, thereby eliminating the major source of oxygen to the muscle; whether the collapse of the tracheae is a spontaneous process or results secondarily from the cessation of respiration is not known. Resting potential persists briefly, but as the pace of erosion of the myofilaments increases, the T-system swells and dissects the fibre, the number of autophagic vacuoles in the cytoplasm skyrockets (glycogen, mitochondria and, slightly later, ribosomes are seen in them); the nucleus shows signs of pyknosis, resting potential collapses, and contractility is lost. At this time, approximately half the cytoplasm contains myofilaments, and these remaining myofilaments are lost in the ensuing 2–5 h. Membranous materials are cast off into the haemolymph, and phagocytes consume the remnant basement membrane (Beaulaton and Lockshin, 1977; Lockshin and Beaulaton, 1979a). The cell is dark and condensed. Vesicles apparently deriving from the destruction of the post-synaptic membrane are common, but presynaptic nerves are relatively unperturbed (Beaulaton and Lockshin, 1978).

The rapidity of loss of the myofilaments raises questions as to the enzymes involved, since myofilaments are not seen within autophagic vacuoles. Primary choices would be the lysosomal proteases cathepsins B and D, but proof of their involvement would have to meet the following experimental qualifications: in general, thick filaments appear to be eroded more rapidly than thin, with the last identifiable remnant of the contractile apparatus being small I-Z-I brushes. When degeneration is blocked by the rather toxic inhibitor of protein synthesis, puromycin, thick filaments remain (Lockshin and Beaulaton, 1974b). Also, the lysosomotropic drug chloroquine inhibits the resorption of the muscle, as might be expected, but not the dissolution of the myofilaments. The dead cell remains filled, not with myofilaments, but with protein-containing and other vacuoles (Lockshin and Beaulaton, 1979b). Finally, the stoichiometric inhibitor of the lysosomal enzyme cathepsin D, pepstatin, fails to prevent loss of birefringence even when ionophoresed into the fibre (Lockshin, 1975b). These observations indicate that enzymes other than lysosomal proteases play some role in the dissolution of the myofilaments.

Two lysosomal proteases, cathepsins B and D, are found in the muscle, and both rise toward the time when the muscle will involute (Fig. 3.2). Cathepsin B rises slightly prior to the beginning of involution; it also rises in total amount, but not in concentration, shortly prior to the loss of muscle during the first 3 days of pupal life. At the adult moult, the rise in cathepsin B is more spectacular than that of cathepsin D. The rise of cathepsin D is concurrent with the beginning of rapid involution. We do not have these enzymes sufficiently pure or concentrated to evaluate their effect on the muscle proteins, and can only speculate as to their role. As the muscle breaks down rapidly, a protein weighing approximately 60–70 000 daltons accumulates in the muscle. Although its origin is not yet certain, the quantity of protein suggests that it is a fragment of myosin, and indeed, we have occasionally seen it liberated from iodinated myosin (Dorsey, 1980; Colon, 1980). It is soluble in low-salt buffers which are used to precipitate myofilaments. The fragment is present in chloroquine-treated muscles (Lockshin and Beaulaton, 1979b; Fig. 3.3). Our hypothesis, therefore, is that the myosin is initially cleaved to this fragment, the myofilament dissociates, and the soluble proteins are digested within autophagic vacuoles. The soluble proteins would of course be unrecognizable in electron micrographs without the use of labelling techniques or immunological labelling, neither of which are within our competence as yet.

Rat liver cathepsins B and D cleave insect myosin into quite distinctive fragments which are not, however, the same as the 70 000-dalton fragment we observe (Dorsey, 1980). However, when radioiodinated myosin is stored for several weeks, it spontaneously degrades to a slightly lower-molecular-weight compound, which is not readily degraded by the mammalian enzymes. Our attempts to isolate the enzyme responsible for cleavage to the 70 000-dalton fragment have so far proved unsuccessful, as we have only sporadically and non-reproducibly seen the cleavage when we mix substrate myosin with extracts.

Our working hypothesis is that there exists among the myofibrils an enzyme or group or enzymes capable of cleaving native myosin to a 70 000-dalton fragment, and that the enzyme is either highly labile or subject to quite effective inhibition, or that the tertiary and quaternary structure of the substrate, as affected by ions, ATP and regulatory molecules, exerts a marked effect on its susceptibility to attack by the enzyme. Such an hypothesis is of course fully consistent with modern concepts of enzyme action, but may be doubly valid for large insoluble structural proteins such as the myofilament proteins. The relation of the cleavage enzyme to the cathepsins awaits concentration and purification of one or more of the enzymes — a project for the immediate future.

In general, the breakdown of muscles in other insects is similar, in that lysosomes are seen but are not prominent in the destruction of the myofilaments (Jones, Davis, Hung and Vinson, 1978); phagocytes appear very late in the process or not at all [with the exception of the higher Diptera, and then only in a muscle destined for reconstruction (Crossley, 1968, 1972a)]. Erosion appears not to be localized, the cells condense, and nuclear pyknosis is a late phenomenon.

In other systems, a few studies other than those reviewed relatively recently (Lockshin and Beaulaton, 1974c; Lockshin, 1980; Lockshin *et al.*, 1980) should be mentioned: hormones have been shown to determine the fate of other flight muscles, ecdysterone retarding the degeneration of the flight muscles of *Acheta domestica*, the house cricket (Srihari *et al.*, 1975), and juvenile hormone increasing lysosomal enzymes in the flight muscles of the Douglas fir beetle, *Dendroctonus pseudotsugae* (Sahota, 1975). Juvenile hormone was also reported to have the same effect *in vitro*, an idea which seems unlikely. Eserine, an inhibitor of acetylcholinesterase, had no effect on the flight muscles of *Dendroctonus*, indicating a lack of neural involvement (Sahota, 1975). There is a suggestion that, at least in *Carausius*, the walking stick, and *Rhodnius*, an assassin bug, neurosecretion may be controlled by cholinergic nerves (Finlayson, Osborne and Anwyl, 1976), although cholinergic motor activity could be differentiated from neurosecretion in the moth *Antheraea* (Williams, 1969; Lockshin, 1971). The growth of flight muscles in *Calliphora* does not require the presence of air-filled tracheae (Houlihan and Newton, 1979); nor, from studies on mutants, does there appear to be an equivalent of disuse atrophy in insects (Deak, 1976). These latter experiments illustrate the utility of using mutants as analytical tools, as is considered in the discussion of the salivary glands, below.

Preoccupation with the cytoplasm, although interesting in its own right, does not answer the question of the death of the cell. The cytoplasm appears to be destroyed as a physiological response to immediate physiological problems which, in a somewhat risky extrapolation, we would describe as hypoxia and release of tension on the muscle. (It is fair to note that maintaining the muscle under stretch has not been shown directly to protect it.) What, however, causes the muscle to develop these problems, or to fail to respond in a corrective manner? We have only two observations available at this point: that protein synthesis is required during the first few hours after ecdysis, and that the nucleus shows signs of pyknosis at approximately 15 h. For several technical reasons the muscle is not very suited to the study of synthesis of nucleic acids and proteins at the time of ecdysis, and so this aspect of cell death — the nuclear events — is best studied in other accessible

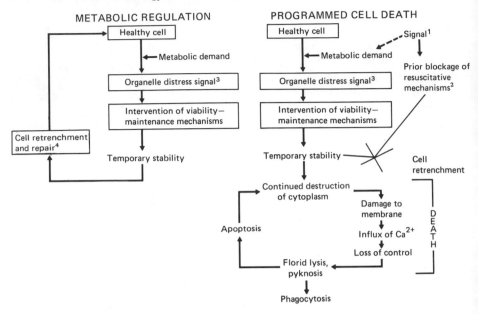

Fig. 3.3 Hypothetical relation between cell death and metabolic demand. The model is presented for heuristic purposes and does not assume that all steps have been documented in any one system. The distinction between reversible and irreversible regression of a cell is presumed to be the absence of regenerative mechanisms. The ideas to which the notes refer are elaborated in the text. (1) The signal is most often endocrine, but may also be neural, local chemical, or other. It is non-toxic in that its presence does not discomfort non-responding cells. Evidence of metabolic distress, prior to the irreversible stage, has been documented for the intersegmental muscles. (2) Blockage of resuscitative mechanisms is suggested by the similarity of the early stages of cell death to non-lethal situations of metabolic demand. It is probable that the documented involvement in cell death of nucleic acid synthesis operates either at this level or at the level of synthesis of lytic enzymes. (3) The concept of organelle distress signal refers to evidence of demand on specific organelles, such as specific clearance of organelles, or a rise in cellular lactate dehydrogenase. Presumably, accumulated metabolites evoke a feedback response, leading to gene-controlled efforts at restoration. A role for nucleotide cyclases has been adumbrated but not well analysed (Greene and Pratt, 1979). The cellular response to the organelle in distress depends on the origin of the cell: ectodermal cells isolate the organelle in isolation membranes, whereas mesodermal cells frequently do not (Ericsson, 1969), and the formation of the autophagic vacuole may depend on the speed at which regression is forced. (4) In metabolic regulation, before the cell reaches a point of agony or morbidity, repair mechanisms have begun, as is documented by the synthesis of nucleic acids in muscles undergoing denervation atrophy. This sequence apparently does not intervene in cell death.

The similarity of the two schemes may be justified, in spite of differing morphological appearances (Locke and Sykes, 1975; Wissocq, 1977, 1978) by differing rates and magnitudes of the response (Decker, 1976; Morimoto systems. It is of note that a rise in lactate dehydrogenase activity is seen in the salivary gland as well as in the muscle (Bidlack and Lockshin, 1976).

3.3.3 *Other invertebrate tissues*

Of the numerous other sources of experimental material, several have been looked at briefly, and only a few have been looked at in any significant detail. Information ranges from the bizarre and somewhat defeatist self-destruction of roughly handled holothurians (Hill, 1966) through occasional studies of various tissues such as the epidermis (Bautz, 1975) and nervous system (Pipa, 1978) of metamorphosing insects. A wider range of effort has been directed toward the evolution of the fat body, the prothoracic glands, and the silk or salivary glands of various insects.

Epidermal tissues, whether being replaced in the course of metamorphosis (Bautz, 1975) or as a developmental phenomenon deriving from a micropterous state (Lavenseau and Surleve-Bazeille, 1974; Thangavelu, 1978), involve the massive participation of autophagic lysosomes. During the degeneration of larval epidermis of the fleshfly *Calliphora*, apoptosis is a prominent phenomenon. Acid phosphatase is seen along cell boundaries and in Golgi-derived membranes; and segregated parts of the cell, cast off into the haemolymph, are ultimately phagocytosed (Bautz, 1975). Similarly, during the destruction of the wings of the micropterous female moth *Orgya antiqua*, large autolytic vacuoles are formed, apparently from a degranulated rough endoplasmic reticulum (Lavenseau and Surleve-Bazeille, 1974). The origin of the isolation membranes, which are quite prominent in insects, is a subject of considerable dispute in insects as in other organisms (Locke and Collins, 1965; Scharrer, 1965, 1966; Beaulaton, 1964, 1967a, 1967b; Matsuura, Morimoto, Nagata and Tashiro, 1968; Morimoto, Matsuura, Nagata and Tashiro, 1968; Tashiro, Morimoto, Matsuura and Nagata, 1968). It is not appropriate to enter into the argument here, except to note that two major alternatives — origin from the Golgi-endoplasmic reticulum–lysosome system (GERL) and origin by degranulation of the rough endoplasmic reticulum, with somewhat mysterious origin of the enzymes — still find favour in different circles. The major focus of these arguments is research

---

et al., 1976). Some workers feel that, during the cell's agony, a small but irreparable lesion has damaged the plasmalemma or other membrane, thus creating a chronic drain on resources or influx of calcium (Bank and Mazur, 1972; Humbert, 1978; Schanne *et al.*, 1979). The point which should be defined as death remains ambiguous, as is indicated by the brackets enclosing a positive feedback cycle. The concept of apoptosis, as described by Kerr (1973), indicates a controlled shrinkage and segregation of the cytoplasm; but data of Bank and Mazur (1972) and Humbert (1978) indicate that the shrunken 'dark' cell is permeable to lanthanum and to ethidium bromide. Thus the point at which the cell actually loses control remains to be defined. Several authors consider that the collapse of barriers to calcium represents that moment (see, for instance, Schanne *et al.*, 1979).

on the exocrine glands and fat body of metamorphosing insects, which provide other viewpoints of interest, as is discussed in a later section.

In the nervous system of metamorphosing and embryonic insects, cell death occurs as it does in the nervous system of vertebrates, but unfortunately little is known about the cells themselves (Heywood, 1965; Tung and Pipa, 1972; Robertson and Pipa, 1973; Hohnson, Saum, McDaniel and Berry, 1976; Truman and Reiss, 1976; Pipa, 1978). The nervous system responds even *in vitro* to hormones (Robertson, 1974); and acid phosphatase (Buser, 1972) and autophagic vacuoles (Tung and Pipa, 1972) are present. There is a considerable increase in high-molecular-weight RNA in the metamorphosing nervous system of *Galleria* (Ishikawa and Newburgh, 1971). Unlike vertebrate sensory neurons, insect sensory neurons can differentiate in the absence of their normal synaptic targets (Sanes, Hildebrand and Prescott, 1976), a fact which renders doubtful the existence of a factor analogous to nerve growth factor.

Other tissues of potential interest include oocytes which, in insects like roaches, may be resorbed (Bell, 1971; Kai and Hesegawa, 1973; Barton Browne, van Gerwen and Williams, 1979). In the silkworm, an esterase may digest the yolk cell membrane at the termination of diapause (Kai and Hesegawa, 1973). Haemocytes also involute as a developmental event, responding to hormones (Seligman and Doy, 1973; Beel and Feir, 1977), perhaps being reprogrammed by juvenile hormone (Hwang-Hsu, Reddy, Kumaran, Bollenbacher and Gilbert, 1979). The appearance of chromatin droplets or the uptake of Trypan Blue appear to be better means of estimating cell death than does assay of phosphatase by p-nitrophenol (Feir and Pantle, 1971; Beel and Feir, 1977). This substrate has been used effectively at an ultrastructural level to evaluate cell death (Bowen and Lewis, 1980; Bowen, 1980). Fluorescent compounds such as ethidium bromide or fluorescein diacetate have also been used to measure cell death in other systems (Bank, Buchner and Hunt, 1979a; Bank, Emerson, Buchner and Hunt, 1979b; Miller and Selverston, 1979). For biochemical assay, p-nitrophenol provides a sensitive but not specific assay (Aidells, Lockshin and Cullin, 1971).

The midgut also degenerates, in a collembolan at every moulting cycle, by the formation of autophagic vacuoles and the expulsion of the old cell into the lumen of the gut (Humbert, 1978, 1979).

Among the most interesting tissues examined at metamorphosis is the fat body, less on the assumption that the cells actually die than for the reason that microscopy has been quite careful and revealing. Briefly, three important conclusions come from the studies of several insects: first, large masses of proteins and other resource material may be sequestered in the fat body. This material is generally isolated by the formation of isolation membranes, deriving (according to most authors) from degranulated endoplasmic reticulum (Locke and Collins, 1965, 1968; Locke and Sykes, 1975; Dean, 1978; de Priester and van der Molen, 1979; Collins, 1979). Material thus isolated may be fully degraded or stored; acid phosphatase is present but is difficult to isolate, depending on the pH of the homogenization medium (Van Pelt-Verkuil, 1979) and because of the known contamination and membranolytic properties of fat in homogenization media. Second, organelles are cleared in a highly sequential and specific manner; autophagic vacuoles are seen to contain only one type of organelle at a time and, shortly thereafter, these organelles are rare in the cytoplasm (Locke and Sykes, 1975; Dean, 1978; de Priester and van der Molen, 1979). The specificity and precision of this phenomenon bespeaks an alteration of the target organelle rather than an unprovoked attack by the isolation membranes, but much work remains to be done in this fascinating field. Third, there are involutionary changes at each moult as well as at metamorphosis; but the clearance of organelles differs according to the ultimate fate of the cell. Thus microbodies are cleared by unknown means during a larval moult in the skipper moth *Calpodes ethlius*, but autophagic vacuoles isolate them before the fat body breaks down at metamorphosis (Locke and Sykes, 1975). Whether the enzymic mechanisms are nevertheless similar is as yet untested, but the observations highlight the subtleties less likely to be taken into consideration in less favourable tissues. There is, of course, in addition to the changes in synthetic and lytic activities coincident with the moulting cycles, a metabolic regulation of protein synthesis in the fat body, perhaps deriving from a signal related proteolytic release of endogenous leucine (Bosquet, 1979).

Finally, the most promising of the tissues for the study of cell death are the insect exocrine glands, the prothoracic glands and the silk or salivary glands. In these tissues, including especially glands from *Drosophila, Chironomus, Bombyx* and *Manduca*, a unique combination of knowledge of nucleic acid activity, good genetic information, availability of genes, awareness of physiological activity and good electron microscopy and cytochemistry exists. There is as yet, however, no unifying concept. Briefly, all or most of the salivary or silk glands involute at pupation, an anterior portion sometimes being retained for adult function (Hakim and Kafatos, 1976). The prothoracic glands, source of the moulting hormone, degenerate during the final moult (Beaulaton, 1967a, 1967b).

Activity of nucleic acids in the silk glands of the silkworm, *Bombyx mori*, has been well-studied since attempts were first made to isolate fibroin messenger RNA (Suzuki and Brown, 1972; Suzuki, 1977), and current techniques are quite sophisticated (Chavancy and Fournier, 1979; Chevallier and Garel, 1979; Cornet, Chavancy and Daillie, 1978; Coulon, 1979; Daillie, 1979; Fournier, 1979; Pavé, 1979; Prudhomme and Couble, 1979). Synthesis of nucleic acids ceases rather early, endomitoses terminating after one or two cycles in the fifth instar (Perdrix-Gillot, 1978, 1979). The protein synthetic machinery is accumulated during the first half of the larval stage, and the silk is synthesized during the last half (Prudhomme and Couble, 1979). Similar observations, based on the weight of nucleic acids in the glands, were made by Matsuura and Tashiro (1976). The rate of synthesis declines markedly during the last 2 days before the animal begins to spin its cocoon (Matsuura *et al.*, 1968). The RNA may then be used during the period of spinning, but, as the gland begins to degenerate, some is released in autophagic vacuoles to the haemolymph (Matsuura *et al.*, 1968; Sridhara and Levenbook, 1973). Other workers, however, note that ribosomal RNA is rapidly degraded with a gradual appearance of soluble-like RNA, perhaps a breakdown product of the ribosomes (Okabe, Koyanagi and Koga, 1975).

DNA is lost from the cells much less rapidly than is RNA (Matsuura et al., 1968). Chinzei (1975a, 1975b) feels that intact DNA, smaller than normal DNA, is released and stored in the fat body; perhaps this DNA represents the chromatin blocks released by apoptosis. It is in any case evident that the synthesis of nucleic acids ceases rather early in the process. The subject is discussed in further detail in Chapter 9.

The synthesis of proteins continues apace and indeed is most active once stimulated by ecdysone. However, the same hormone which initiates the phase of rapid synthesis also probably terminates it (Chinzei, 1975a, 1975b; Kawai, 1978): such is the case in Drosophila salivary glands (Poels, de Loof and Berendes, 1971; Eeken, 1977) and in the salivary glands of the tobacco hornworm, Manduca sexta (Epstein and Lockshin, 1977; Epstein, 1979). Equivalent sentiments have been expressed for the salivary glands of Chironomus, which like Drosophila has giant chromosomes (Agapova and Kiknadze, 1979). As for the fat body, there is a similar but less profound degenerative cycle at the fourth larval instar (Morimoto et al., 1968) and earlier instars (Matsuura and Tashiro, 1976). Again, the existence of regressive, as opposed to lethal, states raises the question of what distinguishes the two. Are the differences quantitative or qualitative? In this instance, the cytological appearance of the earlier phases in every way resembles that of the later phases.

Involution, whether at a larval moult or metamorphosis, is characterized by the formation of large autophagic vacuoles and autolysosomes. The autophagic vacuoles encompass mostly rough endoplasmic reticulum, rarely mitochondria (Tashiro et al., 1968; Morimoto et al., 1968). The isolation membranes appear to form from the rough endoplasmic reticulum, which loses its ribosomes. The forming vacuoles fuse, forming larger and larger vacuoles which ultimately sequester the cytoplasm; meanwhile, free ribosomes remain in the cytoplasm. The nucleus is partitioned in a similar way, releasing chromatin blocks to the haemolymph, as Chinzei (1975a, 1975b) reported (Matsuura et al., 1968). Related to the cytological evidence for the appearance of autophagic vacuoles, less thoroughly documented evidence for a dramatic increase in lysosomal enzymes has been reported for homologous glands in Chironomus (Diptera; cathepsin D — Henrikson and Clever, 1972), Galleria (Lepidoptera; acid phosphatase — Aidells et al., 1971) and Pericallia (Lepidoptera; acid phosphatase — Vishnoi and Narain, 1979).

The autophagic vacuoles appear while the gland is still secreting, even during its phase of highest activity, and it is to be emphasized that destruction begins while the gland is highly functional. A similar situation obtains in the intersegmental muscles (see Section 3.3.2), but in the instance of the silk glands evidence suggests that the synthesis of nucleic acids may have been shut down prior to the beginning of the formation of autophagic vacuoles. Analysis of nucleic acids is, however, too crude at present to allow more than speculation, and it is to be hoped that, in the two situations for which we have good knowledge of nuclear activity — the silk gland of the silkworm, Bombyx mori and the salivary gland of Drosophila or

Chironomus – information will soon be forthcoming. An earlier report by Ashburner (1970) of a puffing pattern related to involution does not seem to have been pursued.

Studies of the involution of the prothoracic glands give essentially similar results, to wit, that epidermal tissues form impressive autophagic vacuoles, whereas mesodermal tissues such as muscle usually do not. Of course, evidence for the activity of nucleic acids is missing; but, again, degeneration begins while the gland is still functioning (Glitho, Delbecque and Delachambre, 1979). In the cockroach *Nauphoeta* at least, the death of the gland is programmed in a several-step process (Lanzrein, 1975); probably it is also in other prothoracic glands, as is the death of *Bombyx* silk glands (Chinzei, 1975a, 1975b; Kawai, 1978) and *Antheraea* intersegmental muscles (Lockshin and Beaulaton, 1974c). Carefully prepared electron micrographs have revealed dramatic autophagic vacuoles in the prothoracic glands of *Leucophaea*, another roach (Scharrer, 1965, 1966) and *Antheraea* (Beaulaton, 1967a, 1967b). Beaulaton, noting again that the vacuoles form in secreting cells, points out that the isolation membranes form by degranulation of the rough endoplasmic reticulum, creating multiple membrane layers around the vacuoles, in which the organelles are broken down. These membranes contain several enzymes, which he suggests are activated *in situ*.

3.4 A model of cell death in metamorphosis

These several stories suggest a model of cell death which is of course grossly oversimplified and generalized, but which has the virtue of attempting to face several questions, and which to a large extent can be tested. As is illustrated in Fig. 3.3, the basic tenet is as follows: that the mechanism leading to programmed or physiological cell death is the same as that involved in day-to-day metabolic regulation. The difference is that at some point the negative feedback loop is blocked, so that restorative mechanisms cannot come into play; thus the cell continues to disassemble itself until it loses control of its function. The point of death is somewhat equivocally defined as occurring during this loss of control, but the actual order of events at that moment is not yet known. It may include morphologically unidentified shutdown of the nucleus; other workers feel that subtle damage to the membrane, creating a slow leak formally akin to a slow haemorrhage and allowing calcium to enter, is the defining event (Bank and Mazur, 1972; Bank *et al.*, 1979a, 1979b; Bank, 1979; Schanne, Kane, Young and Farber, 1979; Katz and Reuter, 1979; Kameyama and Etlinger, 1979). As is described in the following section, there is some evidence that the cyclic nucleotides control dramatic shifts in metabolic direction, such as one might expect to encounter in this situation. The morphology of ectodermal and mesodermal involution differs, primarily in the clarity of definition of autophagic vacuoles (Ericsson, 1969) and isolation membranes.

The function of the model, which is drawn from the examples cited above, is to call attention to several questions, in particular the difference between cell retrenchment and cell death. The suggestion that blockage is at the nuclear level is merely a guess; also, the deliberate ambiguity of the point of death illustrates the fact that we know little of the physiology of pyknosis.

Metamorphosis provides a unique opportunity to study the process of cell death. Through appropriate experimentation, many more specific questions may be answered, and the directions are in view. A suitable format for summarizing current thinking is to work within the context of the questions themselves.

To what extent is the mechanism of death innate? The process is innate insofar as involution and death may be spontaneously initiated in isolated tissues ('programmed cell death': Saunders, 1966; Lockshin and Beaulaton, 1974c) and insofar as in all reported cases the initial alterations are seen within the involuting cells. The initial alterations range from relatively minor cytologically visible shifts, such as swelling in the endoplasmic reticulum, to the entire destruction of the cell, and in no case are phagocytes or other aggressive cells seen to unleash an unprovoked attack against previously unchanged cells. The dying cell frequently undergoes an extraordinarily elaborate dismantling process, all of which is assumed to be under its own physiological control (Kerr, 1973; Wyllie in Chapter 1 of this book) but which from other data (Bank and Mazur, 1972; Humbert, 1978) may be questioned (see below).

The timing of cell death is nevertheless controlled externally in the sense that, with the possible exception of metamorphoses in early embryology, such as the tail of ascidians, the event is never a function of time or biological time alone, but always relates to a specific developmental event: a moult, metamorphic climax, a sexual reproductive cycle. The death of the cells concerned, like the event itself, is commonly timed by the endocrine system; less commonly, it is timed by the nervous system or locally diffusing chemicals. Thus cell death in metamorphosis is a developmental process similar to growth processes, with the tissue-specific responses of the cells having a rather different outcome. The chemical nature of the signal is therefore relatively unimportant, except insofar as we would like to know how it achieves the status of a signal. As for any developmental event, control of the timing of the signal, its binding to, or other interaction with, the target tissue, and subsequent communication, are of interest. These questions are in fact presumably the same, since the same concentrations of thyroxine that provoke growth responses in leg muscles provoke the involution of the tail.

How is the signal transmitted within the cell? Surprisingly little effort has been expended to answer this question. In general, the assumption has been that endocrine-transmitting mechanisms are identical with those which provoke the growth of tissues. The most thorough studies have been on the involution of thymocytes exposed to glucocorticoids, a process related to metamorphosis in that the thymus involutes at puberty. Glucocorticoids bind to thymocyte nuclei and stimulate the synthesis of as-yet-to-be-identified materials, which in some manner interfere with glucose uptake by the cell (see Chapter 11). Synthesis of unknown species of RNA and protein have likewise been suggested in the involution of tadpole tail (Weber, 1969b; Tata, 1966) and moth intersegmental muscles (Lockshin, 1969; Lockshin and Beaulaton, 1974b). A nuclear site of action remains one of the popular ideas for the mode of action of thyroxine (Hulbert, 1978); in the insect muscle, degeneration is brought about by a neural event following potentiation by the water-soluble steroid ecdysone. In this latter instance, the communication between the post-synaptic membrane and the nucleus is unknown. There have been very few serious efforts to measure cyclic AMP or cyclic GMP or their respective synthetic or hydrolytic enzymes in involuting tissues, although the assays are now readily available (Taylor and Newburgh, 1978). Such studies are clearly warranted.

The phosophorylation of proteins, involving the mediation of cyclic AMP and cyclic GMP and fluctuations in intracellular calcium, is considered to play an important role in an ever-widening array of physiological regulations (Greengard, 1978; Standaert and Dretchen, 1979); and modifications of levels of the cyclic nucleotides are considered relevant in at least two instances of cell death. In hereditary retinal degeneration in setter dogs, the cyclic GMP phosphodiesterase fails to mature from an activator-dependent to an activator-independent form (Liu, Krishna, Aguirre and Chader, 1979), but whether this failure indicates a causal relationship or merely the involvement of the cyclase mechanism, like other metabolic pathways, in the disease process is not known. More pertinent is the observation by Greene and Pratt (1979) of a sharp rise in adenylate cyclase activity on the plasma membrane of cells of the secondary palate of rats, which are undergoing or about to undergo degeneration. Viable cells did not display this activity, and the authors considered the cyclase activity to be correlated with programmed cell death. Surely this is a fruitful field, and one deserving of considerably more attention than it has received. Adenylate cyclase can be measured in relatively small bits of insect tissue (Taylor and Newburgh, 1978).

To what extent does cell death differ from atrophy? The question of the definition of cell death is not a trivial one, for many cells and tissues can atrophy to a completely quiescent state, in some instances consisting of barely more than a cytoplasmic membrane, a few mitochondria and a nucleus, and promptly regenerate under appropriate conditions (Wigglesworth, 1956; Auber-Thomay, 1967). Similarly, muscle tissue in vertebrates and invertebrates serves as a protein reservoir, a large percentage of which is labile and readily subjected to degradation under conditions of starvation, disuse or loss of innervation (Goldberg and Dice, 1974; Goldberg and St. John, 1976). The cytological and biochemical appearance of dying muscle cells does not differ strikingly from these instances of retrenchment. The two physiological situations are often interchanged, compared and confused in the literature. In fact, many of the early events of degeneration appear to be the same as those obtaining in atrophy, to the extent that cell death often appears to be a form of atrophy or retrenchment carried to the point of no return (see below). Is then cell death an exaggerated form of atrophy, a form of atrophy in which the expected regenerative mechanisms do not appear, or a fundamentally different process? Attention has been directed to the cytoplasm, which admittedly is easier to visualize and which is better understood than the nucleus, but the identification of distinctions between atrophy and death remains poorly understood but critical.

The most helpful example, the atrophy, regrowth and subsequent death of the intersegmental muscles of the Hemipteran *Rhodnius prolixus* is intriguing (Wigglesworth, 1956), but tissue supply is too limited for easy investigation except by electron microscopy. A study of the lytic and synthetic events in tadpole tail muscle regressing as a result of denervation, starvation or wounding (see above) might prove very interesting; likewise, denervation or starvation atrophy of the intersegmental muscles might be compared with the status of the muscles at metamorphosis (Randall, 1970).

Precisely what role does the nucleus play in the death of the cell? It is self-evident that programmed cell death is encoded in the genes, but a role for the nucleus has been proposed in diametrically opposed mechanisms: that a particular product of the nucleus, blockable by antibiotics, initiates cell death (Tata, 1966; Weber, 1969b; Lockshin, 1969; Leung and Munck, 1975) or that failure of the nucleus to incorporate precursors of RNA (more rarely, DNA) is one of the earliest signs of impending cell death (Perdrix-Gillot, 1978, 1979; and Chapter 9 of this book). The former hypothesis presumes that a nuclear product interferes with cytoplasmic function, but the two hypotheses may ultimately be related. For instance, it would be possible that a nuclear product blocks other genes, these latter being responsible for metabolic maintenance or restorative functions. Thus the challenged cell would fail because it could not adapt to the new physiological demand. Although some progress is being made in the field (see Chapter 11), it is evident that much more can be done. Among the well-known metamorphosing tissues, the nucleic acids are best known for the silk gland of *Bombyx* and the salivary gland of *Drosophila*. Thus we can expect to learn, as analyses move beyond the crude stage, which types of nucleic acid are synthesized in the functioning and early-involuting silk or salivary gland, which types of syntheses are blocked and what effect this type of blockage has on the cell. As we refine further our understanding of the timing of these events, we will be able to ask how the blockage comes about: perhaps a nuclear product or a membrane lesion engenders the blockage. At a later time, we may be able to link the metamorphosis of the nucleic acids with a specific functional alteration of the cell — the lack of a critical enzyme of intermediary metabolism, or an alteration of membrane function. At present, knowledge of general metabolism of nucleic acids is helpful but not really satisfying.

To what extent can the transmission of the signal be compared with physiological communications which lead to atrophy? As we have attempted to unravel the sequence of events in cell death in muscle, we have been struck by intimations of changes of general physiological consequence across which we have stumbled. These include, first, an apparent hypoxia in the tissue, as judged by the appearance of an activatable form of lactate dehydrogenase (Bidlack and Lockshin, 1976; Lockshin and Finn, 1978; Finn and Lockshin, 1980; but see Marty and Weber, 1968) and, second, suggestion that muscle shortening and perhaps unloading occurs just prior to the involution of the muscle (Lockshin and Beaulaton, 1979a). Muscle work is known to affect myofibril turnover in mammals (Li and Goldberg, 1976; Etlinger *et al.*, 1980). Whether or not the observations in insects prove to have merit, they highlight the argument that the signalling may be indirect, setting up those conditions in which a cell is programmed to enter into a catabolic mode. By this interpretation, the death of the cell would result from the failure of restorative mechanisms or from an unremitting stranglehold, blocking the escape of the agonizing cell from these commands. Since cells may possess a programme to jettison damaged material (Chapter 1), multi-function use of an extant capacity would be evolutionarily conservative.

What metabolic events initiate the destruction of the cytoplasm? The foregoing discussion suggests that initial metabolic changes may evoke the destruction of the cytoplasm, whether the cell is wounded or hypoxic; the organism requires substrates; or the tissue is regressing in response to developmental stimuli. Unfortunately we have very little information available on this subject. A vast literature exists for metabolic control in mammals, especially focusing on the turnover of muscle protein in various physiological states (Goldberg and Dice, 1974; Goldberg and St. John, 1976; Bird, 1975), the suppression of lymphoid and connective tissues by glucocorticoids (Chapter 11), the catabolic state created by excess thyroid hormone (Hulbert, 1978) and the privation of insulin or amino acids from perfused rat liver (Schworer and Mortimer, 1979). The reader is referred to these several studies.

Information equivalent to the observations described in the reviews cited above should be obtained for dying cells. The preparations should be in culture if possible, in order to maintain better control. Then basic observations on metabolic parameters should be taken: oxygen consumption and release of carbon dioxide; uptake of glucose and other substrates; uptake of amino acids and rate of protein synthesis; uptake of phosphate and uridine, and rate of synthesis of nucleic acids; content of carbohydrate and fat, and turnover of these resources; rates of breakdown of labile and non-labile protein reserves; intracellular content of cyclic AMP and GMP; intracellular activity of the respective cyclases and diesterases; prostaglandins; movement of calcium and fluorescent markers into and out of cells and organelles. When this information is available, we will be able to evaluate the similarities and differences between cells which are dying and cells which are merely retreating.

What is the significance of the 'dark cell'? In general, dying cells assume one of two forms. When traumatized beyond all possibility of metabolic control, they lyse (coagulative necrosis, Chapter 7). When injured, or in several forms of physiological cell death, they become condensed and dark, frequently breaking off pieces of cytoplasm in a process termed apoptosis (Bank and Mazur, 1972; Kerr, 1973; Unnithan, Nair and Bowers, 1977; Bradley and Edwards, 1979). It is surprising that the condensation of the cell has not attracted more attention, since the event presumably tells us something about the process of death. Kerr (1973) assumes that apoptosis is under the control of the cell (Wyllie in Chapter 1), in which case the removal of water from the cell implies an unaccounted-for osmotic shift or contraction of the cytoskeleton. Alternatively, Humbert (1978) reports that dark cells can be penetrated by lanthanum, and Bank and co-workers (Bank *et al.*,

1979a, 1979b) emphasize that cells injured by freezing manifest permeability to ethidium bromide and fluorescein at times when they still are capable of maintaining some semblance of function. In the latter instance, according to Bank (1979) the cells are doomed and agonizing, the membrane permeability being irreversible, but the various parameters of function are lost over periods ranging from hours to a few days. Our observations with insects would not contradict this hypothesis, since the beginning of obvious condensation is roughly synchronous with the collapse of resting potential (Lockshin and Beaulaton, 1979a), but the sequence has to be worked out much more carefully. Since the first deduction from the hypothesis that the condensed cell is osmotically permeable is that apoptosis is a complex organic reaction which does not require the organized participation of the cell, it is obvious that the mechanism of cell condensation and the status of the condensed cell are worthy of further study. An ideal preparation would be silk glands or salivary glands of insects, the cells of which are large enough to be penetrated by microelectrodes, and which to some extent can be manipulated in culture.

Do lysosomes cause cell death? It may seem pugnacious to insist upon the point, but most emphatically they do not. The correlation has often been made in rather too facile a manner. There are several reasons for believing that lysosomes (generally autophagic vacuoles) appear secondarily to primary lytic mechanisms, that there are several lysosomal and non-lysosomal mechanisms for the destruction of cell organelles or soluble substances and that in any case the elimination of specific or non-specific segments of the cytoplasm may be symptomatic of a metabolic problem but does not 'explain' the death of the cell. (It need not be emphasized that the formation of an autophagic vacuole requires energy and at least a minimal sense of order; it thus is a process under the control of the yet-living cell.) First, of course, many types of cells, muscles in particular, can involute without the formation of classical forms of autophagic vacuoles (Auber-Thomay and Srihari, 1973; Crossley, 1968, 1972a; see also Borgers and Thoné, 1976; Bowen and Ryder, 1976; Bowen and Bryant, 1978) or lysosomal enzymes in other intracellular organelles. Even in muscles in which autophagic vacuoles form, these organelles are occupied with the destruction of other formed organelles such as mitochondria, while the myofilaments erode without being necessarily confined within limiting membranes (Lockshin and Beaulaton, 1974a; Beaulaton and Lockshin, 1977). Although some researchers argue for the existence of enzymes on the outer — cytoplasmic — face of lysosomal membranes (Schneider *et al.*, 1977) or the presence of lysosomes in unconventional and hence cytologically unrecognized forms (Bird, 1975; Bird, Schwartz and Spanier, 1977; see also Chapter 13), at the very least the common conception of the marauding autophagic vacuole must be revised, for there are certainly occult proteolytic and potentially other hydrolytic processes in the cell. Second, even where primary alteration of the cytoplasm is not cytologically identifiable, the demonstrated selectivity of lysosomal attack argues for alteration of the target organelle rather than the lysosome. In the fat body of the metamorphosing Lepidopteran *Calpodes*, individual classes of organelles are synchronously swept from the cytoplasm in a highly ordered and reproducible rhythm (Locke and Collins, 1965; Locke and Sykes, 1976). In other systems, the same phenomenon is observed but in a less spectacular manner. The membrane (Bank and Mazur, 1972; Humbert, 1978) and other organelles may alter without the change being detectable by transmission electron microscopy. Furthermore, similar and parallel events may occur with the facultative rather than obligatory participation of lysosomes (Locke and Sykes, 1976; Matsuura and Tashiro, 1976); the appearance of cytologically identifiable autophagic vacuoles may depend on the state of differentiation of the cell or the speed at which it is regressing (Decker, 1976). Whether Decker's observations mean that packaging, but not the synthesis of lysosomal enzymes, has been bypassed, or that the 'death' of the cell is not directly related to lysosomes, is not clear, but the observation remains intriguing. Finally, although a few reports have indicated that, when degeneration is blocked by antimetabolites such as actinomycin D or cycloheximide, lysosomes are absent from the treated tissue (Weber, 1969b; Lockshin and Beaulaton, 1974b), the observation does not prove any causal relationship; the lysosomes may form in response to an earlier metabolic change within the cell, which itself was blocked by the drug.

3.5 Cell death in metamorphosis: the future

These several comments and questions indicate that much remains to be learned on the subject of cell death. Judging from the status of our knowledge, however, in 1951 (Glücksmann, 1951), 1966 (Saunders, 1966) and 1974 (Lockshin and Beaulaton, 1974c), we have come a fair distance. At least by this point we are asking rather fundamental questions in a manner that indicates that we can obtain the answers. As the questions are broached, metamorphosing animals promise to play a larger and larger role in the elucidation of the answers. Perhaps at some point the disparate, sometimes confusing and contradictory, information will hold the key to a unifying hypothesis.

Acknowledgements

This research has been supported by grants from the National Science Foundation, USA. Much of the information derives from the research of several graduate students, most conspicuously Arlene Colon, Agnes Dorsey, David Epstein, Michael Joesten and Moira Royston.

References

Agapova, O. A. and Kiknadze, I. I. (1979), Puffing and tissue-specific function of *Chironomus thummi* salivary gland cells. III. The cell ultrastructure during larval molt. *Tsitologiya,* **21**, 508–513.

Aidells, B., Lockshin, R. A. and Cullin, A-M. (1971), Breakdown of the silk glands of *Galleria mellonella*. Acid phosphatase in involuting glands. *J. Insect Physiol.,* **17**, 857–870.

Ashburner, M. (1970), Function and structure of polytene chromosomes during insect development. *Adv. Insect Physiol.,* **7**, 1–95.

Atkinson, B. G. and Just, J. J. (1975), Biochemical and histological changes in the respiratory system of *Rana catesbeiana* larvae during normal and induced metamorphosis. *Dev. Biol.,* **45**, 151–165.

Auber-Thomay, M. (1967), Modifications ultrastructurales au cours de la dégénérescence et de la croissance de fibres musculaires chez un insecte. *J. Microsc.,* **6**, 627–638.

Auber-Thomay, M. and Srihari, T. (1973), Evolution ultrastructurale de fibres musculaires intersegmentaires chez *Pieris brassicae* (L.) pendant le dernier stade larvaire et la nymphose. *J. Microsc.,* **17**, 27–36.

Bank, H. L. (1979), Private communication.

Bank, H. L., Buchner, L. and Hunt, H. (1979a), A statistical design for estimating functional survival. *Cryobiology,* **16**, 481–491.

Bank, H. L., Emerson, D., Buchner, L. and Hunt, H. (1979b), Cryogenic preservation of rat polymorphonuclear leukocytes. *Blood Cells*, in the press.

Bank, H. and Mazur, P. (1972), Relation between ultrastructure and viability of frozen–thawed chinese hamster tissue-culture cells. *Exp. Cell Res.,* **72**, 441–454.

Barton Browne, L., van Gerwen, A. C. M. and Williams, K. L. (1979), Oocyte resorption during ovarian development in the blowfly *Lucilia cuprina*. *J. Insect Physiol.,* **25**, 147–154.

Bautz, A-M. (1975), Growth and degeneration of the larval abdominal epidermis in *Calliphora erythrocephala* (Meig.) (Diptera: Calliphoridae) during larval life and metamorphosis. *Int. J. Insect Morphol. Embryol.,* **4**, 495–515.

Beaulaton, J. (1964), Evolution du chondriome dans la glande prothoracique du ver à soie tussor (*Antheraea pernyi* Guér.) au cours du cycle secretoire pendant les 4me et 5me stades larvaires. *J. Microsc.,* **3**, 167–186.

Beaulaton, J. (1967a), Localisation d'activités lytiques dans la glande prothoracique du ver à soie du chêne (*Antheraea pernyi* Guér.) au stade prénymphale. I. Structures lysosomiques, appareil de golgi, et ergastoplasm. *J. Microsc.,* **6**, 179–200.

Beaulaton, J. (1967b), Localisation d'activités lytiques dans la glande prothoracique du ver à soie du chêne (*Anthéraea pernyi* Guér.). II. Les vacuoles autolytiques (cytolysomes). *J. Microsc.*, **6**, 349–370.

Beaulaton, J. and Lockshin, R. A. (1977), Ultrastructural study of the normal degeneration of the intersegmental muscles of *Antheraea polyphemus* and *Manduca sexta* (Insecta, Lepidoptera) with particular reference to cellular autophagy. *J. Morphol.*, **154**, 39–58.

Beaulaton, J. and Lockshin, R. A. (1978), Programmed cell death. Ultrastructural study of neuromuscular relations during degeneration of the intersegmental muscles. *Biol. Cell.*, **33**, 169–174.

Beckingham-Smith, K. and Tata, J. R. (1976), Cell death. Are new proteins synthesized during hormone-induced tadpole tail regression? *Exp. Cell Res.*, **100**, 129–146.

Beel, C. and Feir, D. (1977), Effect of juvenile hormone on acid phosphatase activity in six tissues of the milk-weed bug. *J. Insect Physiol.*, **23**, 761–763.

Bell, W. J. (1971), Starvation-induced oocyte resorption and yolk protein salvage in *Periplaneta americana*. *J. Insect Physiol.*, **17**, 1099–1112.

Bidlack, J. M. and Lockshin, R. A. (1976), Evolution of LDH isozymes during programmed cell death. *Comp. Biochem. Physiol.*, **55B**, 161–166.

Bird, J. W. C. (1975), Skeletal muscle lysosomes. In *Lysosomes in Biology and Pathology*, vol. 4, (eds. J. T. Dingle and R. T. Dean), American Elsevier, New York, pp. 75–109.

Bird, J. W. C., Schwartz, W. N. and Spanier, A. M. (1977), Degradation of myofibrillar proteins by cathepsins B and D. *Acta Biol. Med. Germ.*, **36**, 1587–1604.

Bollin, E., Jr., Carlson, C. A. and Kim, K-H. (1973), Anuran pancreas development during thyroxine-induced metamorphosis of *Rana catesbeiana*: RNA metabolism during the regressive phase of pancreas development. *Dev. Biol.*, **32**, 185–191.

Borgers, M. and Thoné, F. (1976), Further characterization of phosphatase activities using non-specific substrates. *Histochem. J.*, **8**, 301–317.

Bosquet, G. (1979), Occurrence of an active regulatory mechanism of protein synthesis during starvation and re-feeding in *Bombyx mori* fat body. *Biochimie*, **61**, 165–170.

Bowen, I. D. (1980), Private communication.

Bowen, I. D. and Bryant, J. A. (1978), The fine structural localization of *p*-nitrophenyl phosphatase activity in the storage cells of pea (*Pisum sativum* L.) cotyledons. *Protoplasma*, **97**, 241–250.

Bowen, I. D. and Lewis, G. H. J. (1980), Acid phosphatase and cell death in mouse thymus. *Histochemistry*, **65**, 173–179.

Bowen, I. D. and Ryder, T. A. (1976), Use of the *p*-nitrophenyl phosphate method for the demonstration of acid phosphatase during starvation and cell autolysis in the planarian *Polycelis tenuis* Iijima. *Histochem. J.*, **8**, 319–329.

Bradley, J. T. and Edwards, J. S. (1979), Ultrastructure of the corpus cardiacum and corpus allatum of the house cricket *Acheta domesticus*. *Cell Tissue Res.*, **198**, 201–208.

Buser, S. M. (1972), Distribution and concentration of acid phosphatases in the nerve cord of the moth, *Galleria mellonella*, during metamorphosis. *J. Insect Physiol.*, **18**, 211–222.

Chavancy, G. and Fournier, A. (1979), Effect of starvation on tRNA synthesis, amino acid pool, tRNA charging levels and aminoacyl-tRNA synthetase activities in the posterior silk gland of *Bombyx mori* L. *Biochimie*, **61**, 229–243.

Chevallier, A. and Garel, J-P. (1979), Studies on tRNA adaptation, tRNA turnover, precursor tRNA and tRNA gene distribution in *Bombyx mori* by using two-dimensional polyacrylamide gel electrophoresis. *Biochimie,* **61**, 245–262.

Chin, W. (1980), A study of the proteolytic enzyme system involved in the tail resorption of *Xenopus* tadpoles. M.S. Thesis, St. John's University, New York.

Chinzei, Y. (1975a), Induction of histolysis by ecdysterone *in vitro*: Breakdown of anterior silk gland in the silkworm *Bombyx mori* (Lepidoptera: Bombycidae). *Appl. Ent. Zool.,* **10**, 136–138.

Chinzei, Y. (1975b), Biochemical evidence of DNA transport from the silk gland to the fat body of the silkworm, *Bombyx mori. J. Insect Physiol.,* **21**, 163–171.

Chudakova, I. and Gutmann, E. (1978), Developmental changes of succinate dehydrogenase, ATPase, and acid phosphatase activity in flight muscles of the normal and allatectomized adult cricket, *Acheta domestica* (Orthoptera). *Zool. J. Physiol.,* **82**, 1–15.

Cohen, P. P. (1966), Biochemical aspects of metamorphosis: Transition from ammonotelism to ureotelism. *Harvey Lect. Ser.,* **60**, 119–154.

Collins, J. V. (1979), Acid hydrolase activity and control of autophagy and heterophagy in larval fat body of *Calpodes ethlius* Stoel (Lepid.). *Comp. Biochem. Physiol.,* **62B**, 317–324.

Colon, A. D. (1980), Proteolysis and degradation of myosin during the involution (programmed cell death) of the intersegmental muscles of *Manduca sexta.* Ph. D. Thesis, St. John's University, New York.

Cornet, P., Chavancy, G. and Daillie, J. (1978), Preparation and properties of nuclei from the posterior silk glands of the silkworm *Bombyx mori. Dev. Growth and Differ.,* **20**, 251–259.

Coulon, M. (1979), *Bombyx mori* development: the context for its endocrine system. *Biochimie,* **61**, 147–152.

Crossley, A. C. (1968), The fine-structure and mechanism of breakdown of larval intersegmental muscles in the blowfly *Calliphora erythrocephala. J. Insect Physiol.,* **14**, 1389–1407.

Crossley, A. C. (1972a), Ultrastructural changes during transition of larval to adult intersegmental muscle at metamorphosis in the blowfly *Calliphora erythrocephala.* I. Dedifferentiation and myoblast fusion. *J. Embryol. Exp. Morphol.,* **27**, 43–74.

Crossley, A. C. (1972b), Ultrastructural changes during transition of larval to adult intersegmental muscle at metamorphosis in the blowfly *Calliphora erythrocephala.* II. The formation of adult muscle. *J. Embryol. Exp. Morphol.,* **27**, 75–101.

Daillie, J. (1979), Juvenile hormone modifies larvae and silk gland development in *Bombyx mori. Biochimie,* **61**, 275–281.

Dauça, M. and Hourdry, J. (1978a), Analyse quantitative des remaniements intestinaux et évolution de la synthèse de l'ADN. Étude chez les larves d'*Alytes obstetricans* et de *Discoglossus pictus* (Amphibiens Anoures) en métamorphose spontaneé. *Biol. Cell.,* **31**, 277–286.

Dauça, M. and Hourdry, J. (1978b), Isolement de l'épithelium intestinal chez le crapaud accoucheur à l'état larvaire ou juvénile. *Biol. Cell.,* **33**, 85–88.

Davis, B. P., Jeffrey, J. J., Eisen, A. Z. and Derby, A. (1975), The induction of collagenase by thyroxine in resorbing tadpole tailfin *in vitro. Dev. Biol.,* **44**, 217–222.

Deak, I. I. (1976), Use of *Drosophila* mutants to investigate the effect of disuse on the maintenance of muscle. *J. Insect Physiol.,* **22**, 1159–1165.

Dean, R. L. (1978), The induction of autophagy in isolated insect fat body by beta-ecdysone. *J. Insect Physiol.,* **24**, 439–447.

Decker, R. S. (1974a), Lysosomal behaviour during lateral motor column neurogenesis. *Dev. Biol.,* **41**, 146–161.

Decker, R. S. (1974b), Lysosomal packaging in differentiating and degenerating anuran lateral motor column neurons. *J. Cell Biol.,* **61**, 599–612.

Decker, R. S. (1976), Influence of thyroid hormones on neuronal death and differentiation in larval *Rana pipiens. Dev. Biol.,* **49**, 101–118.

de Priester, W. and van der Molen, L. C. (1979), Premetamorphic changes in the ultrastructure of *Calliphora* fat cells. *Cell Tissue Res.,* **198**, 79–93.

Derby, A., Jeffrey, J. J. and Eisen, A. Z. (1979), The induction of collagenase and acid phosphatase by thyroxine in resorbing tadpole gills *in vitro. J. Exp. Zool.,* **207**, 391–398.

Dodd, M. H. I. and Dodd, J. M. (1976), The biology of metamorphosis. In *Physiology of the Amphibia,* vol. 3 (ed. B. Lofts), Academic Press, New York, pp. 467–600.

Dorsey, A. M. (1980), The role of lysosomal proteinases in the programmed cell death of intersegmental muscles in *Manduca sexta.* Ph. D. Thesis, St. John's University, New York.

Dulcibella, T. (1974a), The occurrence of biochemical metamorphic events without anatomical metamorphosis in the axolotl. *Dev. Biol.,* **38**, 175–186.

Dulcibella, T. (1974b), The influence of L-thyroxine on the change in red blood cell type in the axolotl. *Dev. Biol.,* **38**, 187–194.

Eaton, J. E., Jr. and Frieden, E. (1968), Molecular changes during anuran metamorphosis; early effects of tri-iodothyronine on nucleotide and RNA metabolism in the bullfrog tadpole liver. *Gunma Symp. Endocrinol.,* **5**, 43–55.

Eeckhout, Y. (1969), Étude biochimique de la métamorphose caudale des amphibiens anoures. *Acad. R. Bel. Cl. Sci. Mem. Collect.,* **38**, 1–113.

Eeken, J. C. J. (1977), Ultrastructure of salivary glands of *Drosophila lebanonensis* during normal development and after *in vivo* ecdysterone administration. *J. Insect Physiol.,* **23**, 1043–1055.

Epstein, D. S. (1979), Ecdysone-controlled secretion in the salivary gland of *Manduca sexta.* Ph. D. thesis, St. John's Unversity, New York.

Epstein, D. S. and Lockshin, R. A. (1977), Ecdysone-controlled secretion from the salivary gland of *Manduca sexta. J. Cell Biol.,* **75**, 179a.

Ericsson, J. L. E. (1969), Mechanism of cellular autophagy. In *Lysosomes in Biology and Pathology,* vol. 2 (eds. J. T. Dingle and H. B. Fell), American Elsevier, New York, pp. 345–394.

Etkin, W. (1964), Metamorphosis. In *Physiology of the Amphibia,* vol. 1 (ed. J. A. Moore), Academic Press, New York, pp. 427–468.

Etkin, W. (1968), Hormonal control of amphibian metamorphosis. In *Metamorphosis, A Problem in Developmental Biology* (eds. W. Etkin and L. I. Gilbert), Appleton, New York, pp. 313–348.

Etlinger, J. D., Kameyama, T., Toner, K., van der Westhuyzen, D. and Matsumoto, K. (1980), Calcium and stretch-dependent regulation of protein turnover and myofibrillar disassembly in muscle. In *Plasticity of Muscle* (ed. D. Pette), Walter de Gruyter, Berlin and New York, in the press, pp. 541–557.

Feir, D. and Pantle, C. R. (1971), *In vitro* studies of insect haemocytes. *J. Insect Physiol.,* **17**, 733–738.

Finlayson, L. H. (1975), Development and differentiation. In *Insect Muscle* (ed. P. N. R. Usherwood), Academic Press, New York, pp. 75–149.

Finlayson, L. H., Osborne, M. P. and Anwyl, R. (1976), Effects of acetylcholine, physostigmine, and hemicholinium-3 on spontaneous electrical activity of neurosecretory nerves in *Carausius* and *Rhodnius*. *J. Insect Physiol.*, **22**, 1321-1326.

Finn, A. F. Jr. and Lockshin, R. A. (1980). Activation by anoxia of latent lactic acid dehydrogenase isozymes in insect intersegmental muscles (*Manduca sexta*, Sphingidae). *Comp. Biochem. Physiol.*, **68c**, 1-7.

Fournier, A. (1979), Quantitative data on the *Bombyx mori* L. silkworm: a review. *Biochimie*, **61**, 283-320.

Fox, H. (1971), Cell death, thyroxine, and the development of *Rana temporaria* larvae with special reference to the pronephron. *Exp. Gerontol.*, **6**, 173-177.

Fox, H. (1972a), Tissue degeneration: An electron microscopic study of the tail skin of *Rana temporaria* during metamorphosis. *Arch. Biol. (Liège)*, **83**, 373-394.

Fox, H. (1972b), Sub-dermal and notochordal collagen degeneration in the tail of *Rana temporaria*: An electron microscopic study. *Arch. Biol. (Liège)*, **83**, 395-405.

Fox, H. (1973a), Degeneration of the tail notochord of *Rana temporaria* at metamorphic climax. Examination by electron microscopy. *Z. Zellforsch. Mikrosk. Anat.*, **138**, 371-386.

Fox, H. (1973b), Ultrastructure of tail degeneration in *Rana temporaria* larvae. *Fol. Morphol.*, **21**, 109-112.

Frieden, E. and Campbell, J. A. (1978), The effect of tri-iodothyronine on the transport and incorporation of amino acids by bullfrog tadpole tail fin cells. *Gen. Comp. Endocrinol.*, **36**, 215-222.

Glitho, I., Delbecque, J. P. and Delachambre, J. (1979), Prothoracic gland involution related to moulting hormone levels during the metamorphosis of *Tenebrio molitor* L. *J. Insect Physiol.*, **25**, 187-192.

Glücksmann, A. (1951), Cell deaths in normal vertebrate ontogeny. *Biol. Rev. Cambridge Philos. Soc.*, **26**, 59-86.

Goldberg, A. L. and Dice, J. F. (1974), Intracellular protein degradation in mammalian and bacterial cells. *Ann. Rev. Biochem.*, **43**, 835-869.

Goldberg, A. L. and St. John, A. C. (1976), Intracellular protein degradation in mammalian and bacterial cells. Part 2. *Ann. Rev. Biochem.*, **45**, 747-803.

Gona, A. G. (1969), Light and electron microscopic study on thyroxine-induced *in vitro* resorption of the tadpole tail fin. *Z. Zellforsch. Mikrosk. Anat.*, **95**, 483-494.

Greene, R. M. and Pratt, R. M. (1979), Correlation between cyclic-AMP levels and cytochemical localization of adenylate cyclase during development of the secondary palate. *J. Histochem. Cytochem.*, **27**, 924-931.

Greenfield, P. and Derby, A. (1972), Activity and localization of acid hydrolases in the dorsal tail fin of *Rana pipiens* during metamorphosis. *J. Exp. Zool.*, **179**, 129-142.

Greengard, P. (1978), Phosphorylated proteins as physiological effectors. *Science*, **199**, 146-152.

Gross, J. and Bruschi, A. B. (1971), The pattern of collagen degradation in cultured tadpole tissues. *Dev. Biol.*, **26**, 36-41.

Gross, J. and LaPiere, C. M. (1962), Collagenolytic activity in amphibian tissues: A tissue culture assay. *Proc. Nat. Acad. Sci. U.S.A.*, **48**, 1014-1022.

Gunn, J. M. (1978), Does the regulation of intracellular protein degradation require protein synthesis? *Exp. Cell Res.*, **117**, 448-451.

Hakim, R. S. and Kafatos, F. C. (1976), Cellular metamorphosis. II. The larval labial gland duct and its prospective adult fates in the tobacco hornworm. *Dev. Biol.*, **49**, 369-380.

Harper, E., Bloch, K. J. and Gross, J. (1971), The zymogen of tadpole collagenase. *Biochemistry*, **10**, 3035–3041.

Harper, E. and Gross, J. R. (1970), Collagenase, procollagenase and activator relationships in tadpole tissue cultures. *Biochem. Biophys. Res. Commun.*, **48**, 1147–1152.

Henrikson, P. A. and Clever, U. (1972), Protease activity and cell death during metamorphosis in the salivary gland of *Chironomus tentans*. *J. Insect Physiol.*, **18**, 1981–2004.

Heywood, R. B. (1965), Changes occurring in the central nervous system of *Pieris brassicae* L. (Lepidoptera) during metamorphosis. *J. Insect Physiol.*, **11**, 413–430.

Hickey, E. D. (1971), Behaviour of DNA, protein and acid hydrolases in response to thyroxine in isolated tail tips of *Xenopus* larvae. *Roux' Arch. Entwicklungsmech. Org.*, **166**, 303–330.

Hill, R. B. (1966), Propylene phenoxytol as a 'preservative' for living holothurians. *Nature (London)*, **211**, 304–305.

Hohnson, E., Saum, T., McDaniel, C. N. and Berry, S. J. (1976), Methyl-xanthine inhibition of neurosecretion in the brain and corpus cardiacum of *Cecropia*. *J. Insect Physiol.*, **22**, 713–723.

Houlihan, D. F. and Newton, J. R. L. (1979), The tracheal supply and muscle metabolism during muscle growth in the puparium of *Calliphora vomitoria*. *J. Insect Physiol.*, **25**, 33–44.

Hourdry, J. (1971), Action de la thyroxine exogène sur l'épithelium intestinal de la larve d'un Amphibien anoure, *Discoglossus pictus Otth. J. Microsc.*, **12**, 205–224.

Hulbert, A. J. (1978), The thyroid hormones: a thesis concerning their action. *J. Theor. Biol.*, **73**, 81–100.

Humbert, W. (1978), Intracellular and intramitochondrial binding of lanthanum in dark degenerating midgut cells of a collembolan (insect). *Histochemistry*, **59**, 117–128.

Humbert, W. (1979), The midgut of *Tomocerus minor* Lubbock (Insecta, Collembola): Ultrastructure, cytochemistry, ageing and renewal during a moulting cycle. *Cell Tissue Res.*, **196**, 39–57.

Hwang-Hsu, K., Reddy, G., Kumaran, A. K., Bollenbacher, W. E. and Gilbert, L. I. (1979), Correlations between juvenile hormone, esterase activity, ecdysone titre and cellular reprogramming in *Galleria mellonella*. *J. Insect Physiol.*, **25**, 105–111.

Ishikawa, H. and Newburgh, R. W. (1971), Changes in RNA during metamorphosis of the central nervous system of *Galleria*. *J. Insect Physiol.*, **17**, 1113–1124.

Jones, R. G., Davis, W. L., Hung, A. C. F. and Vinson, S. B. (1978), Insemination-induced histolysis of the flight musculature in fire ants (*Solenopsis* spp.): An ultrastructural study. *Am. J. Anat.*, **151**, 603–609.

Just, J. J., Schwager, J. and Weber, R. (1977), Haemoglobin transition in relation to metamorphosis in normal and isogenic *Xenopus*. *Roux' Arch. Dev. Biol.*, **183**, 307–323.

Kai, H. and Hesegawa, K. (1973), An esterase in relation to yolk cell lysis at diapause termination in the silkworm, *Bombyx mori. J. Insect Physiol.*, **19**, 799–810.

Kameyama, T. and Etlinger, J. D. (1979), Calcium-dependent regulation of protein synthesis and degradation in muscle. *Nature (London)*, **279**, 344–346.

Katz, A. M. and Reuter, H. (1979), Cellular calcium and cardiac cell death. *Am. J. Cardiol.*, **44**, 188–190.

Kawai, N. (1978), Degenerative changes and carotenoid uptake in the silk gland of the silkworm *Bombyx mori. J. Insect Physiol.,* 24, 17–24.

Kerr, J. F. R. (1973), Some lysosome functions in liver cells reacting to sublethal injury. In *Lysosomes in Biology and Pathology*, vol. 3 (ed. J. T. Dingle), North-Holland, Amsterdam, pp. 365–394.

Kerr, J. F. R., Harmon, B. and Searle, J. (1974), An electron-microscope study of cell deletion in the anuran tadpole tail during spontaneous metamorphosis with special reference to apoptosis of striated muscle fibres. *J. Cell Sci.,* 14, 571–585.

Kim, Y., Hamburgh, M., Frankfort, H. and Etkin, W. (1977), Reduction in the latent period of the response to thyroxin by tadpole tail discs fused to discs pretreated with thyroxin. *Dev. Biol.,* 55, 387–391.

Kimmel, C. B. and Model, P. G. (1978), Developmental studies of the Mauthner cell. In *Neurobiology of the Mauthner Cell* (eds. D. Faber and H. Korn), Raven Press, New York, pp. 183–220.

Kistler, A., Yoshizato, K. and Frieden, E. (1975), Changes in amino acid uptake and protein synthesis of bullfrog tadpole liver and tail tissues during triiodothyronine-induced metamorphosis. *Dev. Biol.,* 46, 151–159.

Landmesser, L. and Pilar, G. (1978), Interactions between neurons and their targets during *in vivo* synaptogenesis. *Fed. Proc. Fed. Am. Soc. Exp. Biol.,* 37, 2016–2022.

Lanzrein, B. (1975), Programming, induction, or prevention of the breakdown of the prothoracic gland in the cockroach, *Nauphoeta cinerea. J. Insect Physiol.,* 21, 367–390.

Lavenseau, L. and Surleve-Bazeille, J-E. (1974), Aspects ultrastructuraux de l'installation du microtérisme chez la femelle du lépidoptère *Orgya antiqua* L. *J. Microsc.,* 21, 189–192.

Leung, K. and Munck, A. (1975), Peripheral actions of glucocorticoids. *Annu. Rev. Physiol.,* 37, 245–272.

Lewis, G. H. J., Bowen, I. D. and Bellamy, D. (1979), Combined autoradiography and histochemistry: The simultaneous detection of [6-^{3}H] thymidine and acid phosphatase activity in cryostat sections. *J. Microsc.,* 117, 255–259.

Li, J. B. and Goldberg, A. L. (1976), Effects of food deprivation on protein synthesis and degradation in rat skeletal muscles. *Am. J. Physiol.,* 231, 441–448.

Li, J. B., Higgins, J. E. and Jefferson, L. S. (1979), Changes in protein turnover in skeletal muscle in response to fasting. *Am. J. Physiol.,* 236, E222–E228.

Li, J. B. and Jefferson, L. S. (1978), Influence of amino acid availability on protein turnover in perfused skeletal muscle. *Biochim. Biophys. Acta,* 544, 351–359.

Little, G. H., Atkinson, B. G. and Frieden, E. (1973), Changes in the rates of protein synthesis and degradation in the tail of *Rana catesbeiana* tadpoles during normal metamorphosis. *Dev. Biol.,* 30, 366–373.

Little, G. H., Garner, C. W. and Pelley, J. W. (1979), Alanine aminopeptidase activity and autolysis in the tails of *Rana catesbeiana* larvae during metamorphosis. *Comp. Biochem. Physiol.,* 62B, 163–165.

Liu, Y. P., Krishna, G., Aguirre, G. and Chader, G. J. (1979), Involvement of cyclic GMP phosphodiesterase activator in an hereditary retinal degeneration. *Nature (London),* 280, 62–64.

Locke, M. and Collins, J. V. (1965), The structure and formation of protein granules in the fat body of an insect. *J. Cell Biol.,* 26, 857–884.

Locke, M. and Collins, J. V. (1968), Protein uptake into multi-vesicular bodies and storage granules in the fat body of an insect. *J. Cell. Biol.,* 36, 453–483.

Locke, M. and Sykes, A. K. (1975), The role of the golgi complex in the isolation and digestion of organelles. *Tissue Cell,* 7, 143–158.

Lockshin, R. A. (1969), Programmed cell death. Activation of lysis by a mechanism involving the synthesis of protein. *J. Insect Physiol.*, **15**, 1505-1516.

Lockshin, R. A. (1971), Programmed cell death: Nature of the nervous signal controlling breakdown of intersegmental muscles. *J. Insect Physiol.*, **17**, 149-158.

Lockshin, R. A. (1973), Degeneration of insect intersegmental muscles: Electrophysiological studies of populations of fibres. *J. Insect Physiol.*, **12**, 2359-2372.

Lockshin, R. A. (1975a), Degeneration of the intersegmental muscles. Alterations in hemolymph during degeneration. *Dev. Biol.*, **42**, 28-39.

Lockshin, R. A. (1975b), Failure to prevent degeneration in insect muscle with pepstatin. *Life Sci.*, **17**, 403-410.

Lockshin, R. A. (1980), Muscle turnover in invertebrates and lower animals. In *Degradative Processes in Heart and Skeletal Muscle* (ed. K. Wildenthal), Elsevier, Amsterdam, 225-254.

Lockshin, R. A. and Beaulaton, J. (1974a), Programmed cell death. Cytochemical evidence for lysosomes during the normal breakdown of the intersegmental muscles. *J. Ultrastruct. Res.*, **46**, 43-62.

Lockshin, R. A. and Beaulaton, J. (1974b), Programmed cell death. Cytochemical appearance of lysosomes when the death of the intersegmental muscles is prevented. *J. Ultrastruct. Res.*, **46**, 63-78.

Lockshin, R. A. and Beaulaton, J. (1974c), Programmed cell death. *Life Sci.*, **15**, 1549-1565.

Lockshin, R. A. and Beaulaton, J. (1979a), Programmed cell death. Cytological studies of dying muscle fibres of known physiological parameters. *Tissue Cell*, **11**, 803-809.

Lockshin, R. A. and Beaulaton, J. (1979b), Programmed cell death. Chloroquine limits resorption but does not delay its onset – an ultrastructural and biochemical study. *Biol. Cell.*, **36**, 37-42.

Lockshin, R. A., Colon, A. D. and Dorsey, A. M. (1980), Control of muscle proteolysis in insects. *Fed. Proc. Fed. Am. Soc. Exp. Biol.*, **39**, 48-52.

Lockshin, R. A. and Finn, A., Jr. (1978), Rapid activation of lactate dehydrogenase by anoxia in insects. *J. Cell Biol.*, **79**, 195a.

Lockshin, R. A., Schlichtig, R. and Beaulaton, J. (1977), Loss of enzymes in dying cells. *J. Insect Physiol.*, **23**, 1117-1120.

Lockshin, R. A. and Williams, C. M. (1964), Programmed cell death. II. Endocrine potentiation of the breakdown of the intersegmental muscles of silkmoths. *J. Insect Physiol.*, **10**, 643-649.

Lockshin, R. A. and Williams, C. M. (1965a), Programmed cell death. III. Neural control of the breakdown of the intersegmental muscles of silkmoths. *J. Insect Physiol.*, **11**, 605-610.

Lockshin, R. A. and Williams, C. M. (1965b), Programmed cell death. IV. The influence of drugs on the breakdown of the intersegmental muscles of silkmoths. *J. Insect Physiol.*, **11**, 803-809.

Lockshin, R. A. and Williams, C. M. (1965c), Programmed cell death. V. Cytolytic enzymes in relation to the breakdown of the intersegmental muscles of silkmoths. *J. Insect Physiol.*, **11**, 831-844.

Marty, A. and Weber, R. (1968), Das Verhalten der Enzyme des Energiestoffwechsels im Schwanz der metamorphosieren den *Xenopus*larve. *Helv. Physiol. Acta*, **26**, 62-78.

Matsuura, S., Morimoto, T., Nagata, S. and Tashiro, Y. (1968), Studies on the posterior silk gland of the silkworm, *Bombyx mori*. II. Cytolytic processes in posterior silk gland cells during metamorphosis from larva to pupa. *J. Cell Biol.*, **38**, 589-603.

Matsuura, S. and Tashiro, Y. (1976), Ultrastructural changes in the posterior silk gland cells in the early larval instars of the silkworm, *Bombyx mori. J. Insect Physiol.*, **22**, 969–979.

Miller, J. P. and Selverston, A. I. (1979), Rapid killing of single neurons by irradiation of intracellularly injected dye. *Science*, **206**, 702–705.

Mori, M. and Cohen, P. P. (1978), Antipain inhibits thyroxin-induced synthesis of carbamyl phosphate synthetase I in tadpole liver. *Proc. Nat. Acad. Sci. U.S.A.*, **75**, 5339–5349.

Morimoto, T., Matsuura, S., Nagata, S. and Tashiro, Y. (1968), Studies on the posterior silk gland of the silkworm, *Bombyx mori*. III. Ultrastructural changes of posterior silk gland cells in the fourth larval instar. *J. Cell. Biol.*, **38**, 604–614.

Morris, S. M., Jr. and Cole, R. C. (1978), Histone metabolism during amphibian metamorphosis. Isolation, characterization, and biosynthesis. *Dev. Biol.*, **62**, 52–64.

Moulton, J. M., Jurand, A. and Fox, H. (1968), A cytological study of Mauthner's cells in *Xenopus laevis* and *Rana temporaria* during metamorphosis. *J. Embryol. Exp. Morphol.*, **19**, 415–431.

Nieuwkoop, P. D. and Faber, J. (1967), Normal table of *Xenopus laevis* (Daudin), 2nd edn., North-Holland, Amsterdam.

Okabe, K., Koyanagi, R. and Koga, K. (1975), RNA in degenerating silk gland of *Bombyx mori. J. Insect Physiol.*, **21**, 1305–1309.

Osaki, S., James, G. T. and Frieden, E. (1974), Iron metabolism of bullfrog tadpoles during metamorphosis. *Dev. Biol.*, **39**, 158–163.

Pavé, A. (1979), Dynamics of macromolecular populations: A mathematical model of the quantitative changes of RNA in the silkgland during the last larval instar. *Biochimie*, **61**, 263–273.

Perdrix-Gillot, S. (1978), Succession des endomitoses dans la glande séricigène au cours du développement larvaire de *Bombyx mori. Biol. Cell.*, **33**, 103–116.

Perdrix-Gillot, S. (1979), DNA synthesis and endomitoses in the giant nuclei of the silkgland of *Bombyx mori. Biochimie*, **61**, 171–204.

Perriard, J-C. and Weber, R. (1971), Das Verhalten der ninhydrin-positiven Stoffe im Schwanzgewebe von *Xenopus*-Larven während Wachstum und Metamorphose. *Wilhelm Roux' Arch. Entwicklungsmech. Org.*, **166**, 365–376.

Pesetsky, I. (1966), The role of the thyroid in the development of the Mauthner's neuron. A karyometric study in thyroidectomized neuron larvae. *Z. Zellforsch. Mikrosk. Anat.*, **75**, 138–145.

Pipa, R. L. (1978), Patterns of neural reorganization during post-embryonic development of insects. *Int. Rev. Cytol. Suppl.*, **7**, 403–437.

Poels, C. L. M., de Loof, A. and Berendes, H. D. (1971), Functional and structural changes in *Drosophila* salivary gland cells triggered by ecdysterone. *J. Insect Physiol.*, **17**, 1717–1729.

Prudhomme, J-C. and Couble, P. (1979), The adaptation of the silkgland cell to the production of fibroin in *Bombyx mori L. Biochimie*, **61**, 215–227.

Rafferty, K. A., Jr. (1976), The physiology of amphibian cells in culture. In *The Physiology of the Amphibia*, vol. 3 (ed. B. Lofts), Academic Press, New York, pp. 112–162.

Randall, W. C. (1970), Ultrastructural changes in the proleg retractor muscles of *Galleria mellonella* after denervation. *J. Insect Physiol.*, **16**, 1927–1943.

Randall, W. C. and Pipa, R. L. (1969), Ultrastructural and functional changes during metamorphosis of a proleg muscle and its innervation in *Galleria mellonella* (L.) (Lepidoptera: Pyralididae). *J. Morphol.*, **128**, 171–194.

Reeves, R. (1977), Hormonal regulation of epidermis-specific protein and messenger RNA synthesis in amphibian metamorphosis. *Dev. Biol.,* **60**, 163–179.

Reichel, P. (1976), Differentiation of the opercular integument in *Rana pipiens. Differentiation,* **5**, 75–83.

Reynolds, S. E., Taghert, P. T. and Truman, J. W. (1979), Eclosion hormone and bursicon titres and the onset of hormonal responsiveness during the last day of adult development in *Manduca sexta* (L). *J. Exp. Biol.,* **78**, 77–86.

Robertson, J. (1974), *Galleria mellonella* nerve cords *in vitro*: Stage-specific survival and differential responsiveness to beta-ecdysone. *J. Insect Physiol.,* **20**, 545–560.

Robertson, J. and Pipa, R. L. (1973), Metamorphic shortening of interganglionic connectives of *Galleria mellonella* (Lepidoptera) *in vitro*; Stimulation by ecdysone analogues. *J. Insect Physiol.,* **19**, 673–680.

Rosenberg, M. and Warburg, M. R. (1978), Changes in structure of ventral epidermis of *Rana ridibunda* during metamorphosis. *Cell Tissue Res.,* **195**, 111–122.

Runion, H. I. and Pipa, R. L. (1970), Electrophysiological and endocrinological correlates during the metamorphic degeneration of a muscle fibre in *Galleria mellonella* (L) (Lepidoptera). *J. Exp. Biol.,* **53**, 9–24.

Rutter, W. J., Pictet, R. L. and Morris, P. W. (1972), Toward molecular mechanisms of developmental processes. *Ann. Rev. Biochem.,* **42**, 601–646.

Sahota, T. S. (1975), Effect of juvenile hormone on acid phosphatases in the degenerating flight muscles of the Douglas-fir beetle, *Dendroctonus pseudotsugae. J. Insect Physiol.,* **21**, 471–478.

Sanes, J. R., Hildebrand, J. G. and Prescott, D. J. (1976), Differentiation of insect sensory neurons in the absence of their normal synaptic targets. *Dev. Biol.,* **52**, 121–127.

Saunders, J. W., Jr. (1966), Death in embryonic systems. *Science,* **154**, 604–612.

Schanne, F. A. X., Kane, A. B., Young, E. E. and Farber, J. L. (1979), Calcium dependence of toxic cell death: A final common pathway. *Science,* **206**, 700–702.

Scharrer, B. (1965), An ultrastructural study of cellular regression as exemplified by the prothoracic gland of *Leucophaea maderae. Anat. Rec.,* **151**, 411.

Scharrer, B. (1966), Ultrastructural study of the regressing prothoracic gland of Blattarian insects. *Z. Zellforsch. Mikrosk. Anat.,* **69**, 1–21.

Schneider, D. L., Burnside, J., Gorga, F. R. and Nettleton, C. J. (1977), Properties of the membrane proteins of rat liver lysosomes. The majority of lysosomal membrane proteins are exposed to the cytoplasm. *Biochem. J.,* **176**, 75–82.

Schworer, C. M. and Mortimer, G. E. (1979), Glucagon-induced autophagy and proteolysis in rat liver: Mediation by selective deprivation of intracellular amino acids. *Proc. Natl. Acad. Sci. U.S.A.,* **76**, 3169–3173.

Seligman, I. M. and Doy, F. A. (1973), Hormonal regulation of the disaggregation of cellular fragments in the haemolymph of *Lucilia cuprina. J. Insect Physiol.,* **19**, 125–136.

Seshimo, H., Ryuzaki, M. and Yoshizato, K. (1977), Specific inhibition of tri-iodothyronine-induced tadpole tail-fin regression by cathepsin D-inhibitor pepstatin. *Dev. Biol.,* **59**, 96–100.

Sridhara, S. and Levenbook, L. (1973), Extracellular ribosomes during metamorphosis in the blowfly, *Calliphora erythrocephala. Biochem. Biophys. Res. Commun.,* **53**, 1253–1270.

Srihari, T., Gutmann, E. and Novak, V. J. A. (1975), Effect of ecdysterone and juvenoid on the developmental involution of flight muscles in *Acheta domestica. J. Insect Physiol.,* **21**, 1–8.

Stadler, H. and Weber, R. (1970), Autoradiographische Untersuchungen über die Früwirkung von Thyroxin auf die RNS-Synthese im Gehirn von Xenopus-larven. Rev. Suisse Zool., 77, 587–596.

Standaert, F. G. and Dretchen, K. L. (1979), Cyclic nucleotides and neuromuscular transmission. Fed. Proc. Fed. Am. Soc. Exp. Biol., 38, 2183–2192.

Suzuki, Y. (1977), Differentiation of the silk gland: A model system for the study of differential gene action. In Biochemical Differentiation in Insect Glands (ed. W. Beerman), Springer-Verlag, New York, pp. 1–44.

Suzuki, Y. and Brown, D. D. (1972), Isolation and identification of the messenger RNA for silk fibroin from Bombyx mori. J. Mol. Biol., 63, 409–429.

Tashiro, Y., Morimoto, T., Matsuura, S. and Nagata, S. (1968), Studies on the posterior silk gland of the silkworm, Bombyx mori. I. Growth of posterior silk gland cells and biosynthesis of fibroin during the fifth larval instar. J. Cell. Biol., 38, 574–588.

Tata, J. R. (1966), Requirement for RNA and protein synthesis for induced regression of the tadpole tail in organ culture. Dev. Biol., 12, 77–94.

Taylor, A. C. and Kollros, J. J. (1946), Stages in the normal development of Rana pipiens larvae. Anat. Rec., 94, 7–23.

Taylor, D. P. and Newburgh, R. W. (1978), Characteristics of adenyl cyclase of the central nervous system of Manduca sexta. Comp. Biochem. Physiol. C, 61, 73–80.

Taylor, H. M. and Truman, J. W. (1974), Metamorphosis of abdominal ganglion of the tobacco hornworm, Manduca sexta. Changes in populations of identified motor neurons. J. Comp. Physiol., 90, 367–375.

Thangavelu, S. (1978), Private communication.

Trout, J. J., Stauber, W. T. and Schottelius, B. A. (1979), A unique acid phosphatase location: the transverse tubule of avian fast muscle. Histochem. J., 11, 417–423.

Truman, J. W. and Reiss, S. E. (1976), Dendritic reorganization of an identified motoneuron during metamorphosis of the tobacco hornworm moth. Science, 192, 477–479.

Truman, J. W. and Riddiford, L. M. (1970), Neuroendocrine control of ecdysis in silkmoths. Science, 167, 1624–1626.

Tung, A. S.-C. and Pipa, R. L. (1972), Insect neurometamorphosis. V. Fine structure of axons and neuroglia in the transforming interganglionic connectives of Galleria mellonella (L.) (Lepidoptera). J. Ultrastruct. Res., 39, 556–567.

Unnithan, G. C., Nair, K. K. and Bowers, W. S. (1977), Precocene-induced degeneration of the corpus allatum of adult females of the bug Oncopeltus fasciatus. J. Insect Physiol., 23, 1081–1094.

Vandenburgh, H. and Kaufman, S. (1979), In vitro model for stretch-induced hypertrophy of skeletal muscle. Science, 23, 265–268.

Van Evercooren, A. and Picard, J. J. (1978), Surface changes during development and involution of the cement gland of Xenopus laevis. Cell. Tissue Res., 194, 303–313.

Van Pelt-Verkuil, E. (1979), Increase in acid phosphatase in the fat body during larval and prepupal development in Calliphora erythrocephala. J. Insect Physiol., 24, 375–382.

Vishnoi, D. N. and Narain, S. (1979), Cytochemical study of acid phosphatase activity in the functional and degenerating spinning gland cells of Pericallia ricinii (Lepidoptera) Fol. Histochem. Cytochem., 17, 175–180.

Wahli, W. and Weber, R. (1977), Factors promoting the establishment of primary cultures of liver cells from Xenopus larvae. Roux'Arch. Dev. Biol., 182, 347–360.

Wanatabe, K., Khan, M. A., Sasaki, F. and Iseki, H. (1978a), Light and electron microscopic investigation of ATPase activity in musculature during anuran tail resorption. *Histochemistry*, **58**, 13–22.

Wanatabe, K., Sasaki, F. and Khan, M. A. (1978b), Light and electron microscopic study of adenosine triphosphatase activity of anuran tadpole musculature. *Histochemistry*, **55**, 293–305.

Weber, R. (1969a), The tadpole tail as a model system for studies on the mechanism of hormone-dependent tissue involution. *Gen. Comp. Endocrinol.*, Suppl. **2**, 408–416.

Weber, R. (1969b), Tissue involution and lysosomal enzymes during anuran metamorphosis. In *Lysosomes in Biology and Pathology*, vol. 2 (eds. J. T. Dingle and H. B. Fell), North-Holland, Amsterdam, pp. 437–461.

Weber, R. (1977), Biochemical characteristics of tail atrophy during anuran metamorphosis. *Colloq. Int. C.N. R. S.*, **266**, 137–146.

Wigglesworth, V. B. (1956), Formation and involution of striated muscle fibres during the growth and moulting cycles of *Rhodnius prolixus* (Hemiptera). *Q. J. Microsc. Sci.*, **97**, 465–480.

Williams, C. M. (1969), Juvenile hormone insecticides. *Quad. Acc. Naz. Dei Lincei*, **128**, 80–87.

Wissocq, J.-C. (1970a), Evolution de la musculature longitudinale dorsale et ventrale au cours de la stolonisation de *Syllis amica* Quatrefages (Annélide polychète). *J. Microsc.*, **9**, 355–388.

Wissocq, J.-C. (1970b), Evolution de la musculature longitudinale dorsale et ventrale au cours de la stolonisation de *Syllis amica* Quatr. (Annélide polychète). II. La dédifférentiation. *J. Microsc.*, **9**, 1049–1074.

Wissocq, J.-C. (1970c), Evolution de la musculature longitudinale dorsale et ventrale au cours de la stolonisation de *Syllis amica* Quatr. (Annélide polychète). III. La dégénérescence. *J. Microsc.*, **9**, 1075–1080.

Wissocq, J.-C. (1977), Evolution de la musculature des *Néréidiens (Annélides polychètes)* au cours de l'épitoque. III. La dédifférenciation des fibres longitudinales. *Roux' Arch. Dev. Biol.*, **182**, 227–253.

Wissocq, J.-C. (1978), Evolution de la musculature des *Néréidiens* (Annélides polychètes) au cours de l'épitoquie. I et II. Fibres atoques et epitoques des faisceaux longitudinaux. *Arch. Anat. Microsc. Morphol. Exp.*, **67**, 37–63.

Yamaguchi, K. and Yasumasu, I. (1978), Effects of thyroxine and prolactin on collagen breakdown in the thigh bone and tail fin of the *Rana catesbeiana* tadpole. *Dev. Growth Differ.*, **20**, 61–69.

Yntema, C. L. (1959), Regeneration in sparsely innervated and aneurogenic forelimbs of *Amblystoma* larvae. *J. Exp. Zool.*, **140**, 101–123.

Zottoli, S. J. (1978), Comparative morphology of the Mauthner cell in fish and amphibians. In *Neurobiology of the Mauthner Cell* (eds. D. Faber and H. Korn), Raven Press, New York, pp. 13–45.

4 Tissue homeostasis and cell death

S. M. HINSULL and D. BELLAMY

4.1 Introduction

'The alimentary canal, the arterial system including the heart, the central nervous system of the vertebrate, including the brain itself, all begin as simple tubular structures. And with them Nature does just what the glass-blower does, and, we might even say, no more than he. For she can expand the tube here and narrow it there; thicken its walls or thin them; blow off a lateral offshoot or caecal diverticulum; bend the tube, or twist and coil it; and infold or crimp its walls as, so to speak, she pleases'.

D'Arcy W. Thompson (1942).

The expansion and modelling of cellular populations which occurs after fertilization of the ovum not only involves the multiplication of cells by binary fission but also the loss of postmitotic cells by selective deletion and general senescence. At maturity, the volume and shape of most tissues containing mitotic cells is mainly a balance between the rate at which cells are produced and the rate at which they die, death occurring either in the tissue of origin or elsewhere in the body. Following Thompson's analogy of 'Nature's Glassblower', the modelling of a primordial ball of cells results from regional differences in the cellular environment which affect the rate of cell division and cell loss in these local subpopulations. Cell death is therefore an integral part of tissue homeostasis from early embryonic development through to the degenerative and proliferative diseases of senescence. It is also evident that evolution proceeds through modifications of the controls governing the division and death of cells. This broad view of the significance of cell death in the total development of an organism describes the limits of this chapter. The obvious regional aspects of division and death of cells within organs also sets the philosophical approach in that they will be discussed largely as expressions of what might be termed 'cellular ecology'. Following the 'modelling' which takes place during embryogenesis, questions related to the control of growth are posed mainly with regard to two phenomena: the general increase in body size; and the obvious precise weight relationships that exist between organs.

Research strategies concerned with the study of growth control fall into three categories, related to the various kinds of control that are thought to be involved.

Central control processes govern the growth of peripheral tissues; endogenous or intrinsic control is an important feature of growth of certain organs; and finally, selective short-distance regulation is involved between adjacent interdependent organs. In terms of their adaptive significance, central control processes ensure that general body size and the growth of specific tissues are synchronized with environmental variations in food, light and temperature. The corresponding research area falls within the province of endocrinology.

4.2 Growth patterns

The characteristic growth patterns of vertebrates are regulated by hormones, and the pituitary gland together with adjacent neural areas is the site of central control of his hormonal system, including somatotropin, the thyroid hormones and glucocorticoids. The latter two hormones are necessary for growth but, when present in excess amounts, inhibit anabolism and stimulate catabolic processes in various tissues. Release of hormones from the pituitary is controlled from the brain through the hypothalamus, and this ensures that environmental information received from the sensory system, such as duration and intensity of light, evokes the correct growth response.

An important aspect of differential growth is connected with environmental needs, such as in reproduction, where environmental resources fluctuate widely from one part of the year to another. Day length governs the release of trophic hormones from the pituitary which in turn alters the secretion of the sex steroids, in turn altering the growth of sexual tissues.

An integral part of the action of the endocrine system on cellular populations is seen in the cellular death which is associated with fluctuating hormone levels. Hormones are thought to act as triggers of cell death in situations such as involution of the Müllerian duct and during the massive reorganization of the body which takes place during insect and amphibian metamorphosis (Lockshin and Beaulaton, 1974). Many animals are seasonal breeders and, in the vertebrates, the size of both primary and secondary reproductive tissues increases under the influence of increasing sex hormone levels but decreases with their decline. This is particularly evident in the organs involved in sexual display, such as head furnishings in birds and the musculature of herbivorous mammals. The size of secondary sexual tissues, such as the reproductive tract, uterus and seminal vesicles, also waxes and wanes with the hormonal rhythm. In the human cyclical endometrium a vast number of gland epithelial cells are lost by apoptosis (Kerr, Wyllie and Currie, 1972) during the late secretory, premenstrual and menstrual phases, whereas apoptotic bodies are sparsely distributed in the epithelium of proliferative phase endometrium (Hopwood and Levison, 1976). The falling levels of oestrogens and progesterone (Speraff and Van de Wiele, 1971) towards the end of the cycle result in sloughing off of the superficial endometrium (Bartelmez, 1957), with death of cells in large secretory glands playing a vital role in the reorganization of the tissue to the early proliferative phase. In addition Hopwood and Levison (1976) have postulated that cell death may be the mechanism of reversion from late secretory to early proliferative phase endometrium in animals that do not menstruate. Numerous comparable cases of the atrophy and remodelling of tissue, associated with declining hormone levels, have been described, including mammary gland (Helminen and Ericsson, 1968) prostate, (Helminen and Ericsson, 1972; Kerr and Searle, 1973) and corpus luteum.

Changes in hormone levels after birth have also been shown to influence the postnatal organization of their target tissues via cell death. Diminution of ACTH in the neonatal rat is associated with adrenal cell atrophy in the zona fasciculata and apoptosis in the zona reticularis, and consequently it has been postulated that apoptosis may play a major role in the homeostasis of the adrenal gland (Wyllie, Kerr, Macaskill and Currie, 1973b; Wyllie, Kerr and Currie, 1973a).

4.3 Organ growth control

The fact that most organs of hypophysectomized animals do not grow in a haphazard way indicates the presence of growth control systems other than the pituitary gland. We know of the existence of many such systems from the reactions of normal animals. Availability of oxygen in inspired air controls the cellular proliferation which eventually gives rise to red blood cells (Tinsley, Moore, Dubach, Minnich and Grinstein, 1949). Bacterial infection stimulates proliferation of cells in the lymphoid systems (Moller, 1969). The compensatory increase in the size of a paired organ, such as kidney (Shirley, 1976), following surgical removal of one of the pair, is also evidence of systems which result in specific adjustments in growth independently of the secretions of the endocrine system. Each organ has a system by which it adjusts its growth largely according to intrinsic information, with the possibility of specific interactions between adjacent cell populations in complex organ systems − for example, limbs, where bone, musculature, blood vessels and nerves must grow as a unit.

Tissues in the adult animal have been classified, at the histological level, on the basis of their ability to proliferate, by Cowdry (1950), who recognized two major categories: tissues consisting of intermitotic cells which retain the ability to divide, and, secondly, postmitotic cells which do not normally divide. Intermitotic cells include populations which continuously divide in the adult organism, e.g. basal epithelial cells, haematoblasts, spermatogonia, as well as cells such as spermatocytes and immature leucocytes which differentiate and divide to form highly differentiated cells. In the cells which are classified as postmitotic, two subgroups are distinguished: reversibly postmitotic, which are capable of division in certain extreme circumstances such as regeneration and hyperplasia, e.g. liver, kidney, blood vessels, and highly differentiated cells such as neurons, heart fibres, erythrocytes, and skeletal muscle which are incapable of division. However, in the latter case of non-dividing cells, and in animals of constant cell number, such as rotifers and nematodes, the cells are still capable of individual growth and there is a continuous turnover of their cytoplasm in the adult organism.

4.3.1 *Cell numbers and turnover*

The adaptive significance of cell turnover has not been evaluated, except in the intestinal epithelium and skin where it appears to be associated with a high chance of cellular damage. Why certain organs such as the liver retain the ability to regenerate after damage, whereas other organs such as the lungs do not has still to be resolved. Regenerative capacity may be simply a reflection of the type of control of organ mass which has evolved to meet day-to-day needs for the organs function.

In the tissues of the adult organism, where cells are capable of division, cell production must be balanced by cell loss. If cell loss were to be prevented in an organ with only 1 per cent of its cells turning over each day, the organ would double in size every 80 days. Bullough (1952) found a 25 per cent reduction in mitosis in the epidermis of starved mice, even though no reduction was observed in the thickness of the epidermis or the size of the sebaceous glands. He therefore concluded that sebum secretion and desquamation of the epidermis were regulated in relation to the rate of cell proliferation.

However it is by no means established that mature organisms maintain a fixed cell number. Technical problems prevent accurate cell counts being made in even the simplest organisms. Indirect measurements of cell mass using ionic markers such as sodium for extracellular fluid and potassium for the intracellular fluid give a measure of overall cell mass. In man, using this type of marker, cell mass gently rises to maturity and falls into old age.

4.3.2 *Control of cell loss and cell death*

Cell loss may take place by such mechanisms as exfoliation, damage, migration or death. Cell death has long been recognized morphologically in extreme situations such as involution and damage and has accordingly been described in terms such as atrophy, autophagy and coagulative necrosis (Brandes and Groth, 1963; Helminen and Ericsson, 1971). However, over the past decade increasing interest has centred around cell death in healthy tissue, which may represent true steady state regulation, and particular attention has been paid to the previously mentioned phenomenon of apoptosis (Kerr, 1971, 1973; Kerr *et al.*, 1972); Kerr and Searle (1973) have described apoptosis as 'nuclear and cytoplasmic condensation and exuberant budding to produce small membrane-bounded compact cell-fragments which are ingested by adjacent cells and degraded by lysosomal enzymes'. Apoptosis has been observed in a variety of tissues including: prostate (Kerr and Searle, 1973), adrenal cortex (Wyllie *et al.*, 1973a, 1973b), endometrium (Hopwood and Levison, 1976), thymus (Abraham, Morris and Hendy, 1969) and embryonic tissue (Farbman, 1968). This ever increasing amount of evidence for the presence of apoptotic bodies in a wide spectrum of tissue suggests that this type of cell death plays a vital role in tissue homeostasis. However, with the exception of the hormone-modulated apoptosis previously mentioned, the factors which are involved in the control of cell death in healthy tissue remain obscure. The modulation of mitosis has, however, been widely investigated and the harmonious relationship between cell proliferation and cell death suggests some kind of interaction between their control mechanisms.

Most of the work relating to this kind of interaction deals with the control of mitosis. Cell death is a more difficult event both to predict and quantify. The relevant research is concerned with experimental alterations of normal organ mass and the study of the subsequent adjustments.

Many organs have the capacity to regenerate to approximately normal size following injury or surgical interference, indicating perception by the body of total and relative organ mass. The capacity for regeneration or compensatory growth under experimental conditions is limited, almost by definition, to organs that fluctuate in size as a normal reaction to changes in general metabolic state. One of the simplest theories of regeneration is that loss of cellular mass results in loss of inhibitors of growth produced by the lost cells.

Although experimental evidence has not been obtained to support this inhibitor mechanism at the level of total body weight, there is evidence that inhibitors determine the constant relative mass of mammalian liver against total body weight (Obearade, Chen, Ove and Lansing, 1974; Simard, Corneille, Deschamps and Verly, 1974; Rabes, 1976; Tsirel'mkov, 1976, Verly, 1976). Each liver cell theoretically releases a liver growth inhibitor into the extracellular fluid, and the liver is consequently maintained at a steady state where cell division balances cellular death at a particular ratio of cells to body volume available. Removal of part of the liver automatically leads to a stimulation of cellular proliferation in the remaining mass until the inhibitor level is restored by the production of cells equivalent to those lost.

Evidence is available that other tissues, such as skin, contain tissue-specific inhibitors of mitosis in epidermal cells (Bullough, 1972). The term 'chalone' has been coined to define this class of growth inhibitors (Houck, 1976). The higher the mitotic rate in a tissue, the shorter the life expectancy of the cells after division. For example, in the epidermis, with a moderate mitotic rate, the postmitotic lifespan is 14-21 days, whilst in the duodenal mucosa, with a higher mitotic rate, lifespan is only about 2 days (Lamerton, 1969). Also the experimental stimulation of cell division by hormones is often associated with a decrease in the postmitotic lifespan. This coupling phenomenon has been taken to indicate that a postmitotic cell dies because the time of death is dictated according to the circumstances of the moment by the same physiological mechanism that controls the mitotic rate. As a corollary to this, any mechanism that slows down cell division would increase potential postmitotic lifespan. In this sense chalones function generally in this way to slow down the rate of aging of all postmitotic cells (Bullough, 1972).

Growth at the organ level may be controlled as a balance between inhibitors and stimulators. Tissue-specific growth stimulators acting over short distances have been observed in a number of experiments. For example, regeneration of amphibian limbs depends on the presence of the regenerating motor nerve (Thornton, 1968; Mescher and Tassava, 1975). New limbs may be organized at another point of the body surface by moving the nerve to that point (Thornton, 1968). Thus the existence of specific nerve factors that have the properties of both a growth hormone and embryonic inducer may be postulated. The growth of nerves themselves may be stimulated by transplantation of certain tumours which results in the accelerated growth of adjacent sympathetic ganglia. Also, in some situations, the nerves are induced to enter the transplanted tumour and ramify within it. The 'nerve growth factor' is a diffusible substance that acts locally by diffusion from the tumour cells (Vernon, Banks, Banthorpe, Berry, Davies-Hiff, Lamont, Pearce and Kenning, 1969).

Other work points to the existence of intrinsic growth control systems that regulate the overall shape, size and development of organs (Hinsull and Bellamy, 1974). This control may be a special aspect of the chalone idea where master cells in the tissue regulate cell division in adjacent cells via a system of short-distance activators and inhibitors.

These experiments and others show that we are in the initial stages of a new phase of research into differential growth and, at the same time, the work also highlights our ignorance of growth regulation at the organ level and, in particular, the role of cell death in this regulation.

4.4 Model systems – the thymus

An important question is to what extent does this control reside within cells or is operating between them. The behaviour of cells growing packed together is fundamental to this problem and therefore it is to the organ, composed of dividing cells, that we must turn for a solution to questions of pattern in division and death. A suitable model system would have tissue consisting of continuously replicating cells which is easily manipulated *in vivo*.

The thymus gland complies with these criteria and, additionally, it is a particularly intriguing organ with regard to its mechanism of homeostasis. The gland in the young rat is bilobed with a well-defined symmetrical shape encapsulated by connective tissue. The blood vessels to each lobe are supported by an irregular network of connective tissue septa, which divide the lobes into lobules. Within each lobule two distinct regions are distinguished, the outer cortex, consisting mainly of densely packed lymphocytes, and the medulla which has a high proportion of epithelial cells to lymphcytes.

The thymus has a unique pattern of growth and involution. The highest thymus to body weight ratio is observed during early postnatal life, but absolute thymus weight continues to increase to attain its maximum immediately prior to sexual maturity, which in rats occurs at 9 weeks of age (Bellamy, Hinsull and Phillips, 1976). After this age a sudden loss in thymus mass, proceeding in two phases, has been observed. The first phase, taking place between 9 and 20 weeks of age, involves a sharp decline in weight which is followed by a second phase of slower loss up to 80 weeks. The fact that thymic involution generally occurs at the time of sexual maturity has given rise to the suggestion that sex hormones are responsible for the initiation and possibly the maintenance of the process of involution (Weaver, 1955). Prepubertal gonadectomy, in rats, results in either increased thymus growth or a continuation of the initial growth rate for a longer period, and the peak thymus weight attained is increased (Bellamy *et al.*, 1976). Postpubertal gonadectomy results in an immediate increase in thymus weight to levels comparable with those observed in prepubertal gonadectomy. However, following these maxima in weight, involution of the gland commences and proceeds at a faster rate than normal for about 4 weeks, at which time no significant difference in thymus mass is observed between intact and gonadectomized animals. Despite the role of these hormones in determining thymic mass, there appears to be an upper limit to the thymic cell population which is independent of gonadal status. Thus, when this limit is reached after postinvolution gonadectomy, the cell population is depleted for a second time, at a faster rate.

The adrenocorticosteroids are also known to influence thymus mass, and it is well established that adrenalectomy results in increased thymus weight (Perris, Weiss and Whitfield, 1970), but does not affect the rate of age-controlled involution. Conversely, a single injection of a corticosteroid such as cortisol causes inhibition of mitosis and destruction of lymphocytes, resulting in acute involution of the thymus to attain a minimum mass 4 days after injection. From this point the remaining viable lymphocytes overcome the suppressive influence of the steroid so that the mitotic rate is tripled compared with control values. Subsequently the thymus weight increases in a linear fashion to stabilize at the control weight by 14 days (Bellamy and Hinsull, 1977), and growth and involution follows the aforementioned pattern. Many factors producing thymus atrophy, such as cold, starvation and disease, operate via stimulation of the adrenal gland.

4.4.1 *Autonomy of the thymus*

This evidence suggests that a considerable degree of autonomy exists within the thymus and that this basic process governing size may be reversibly modulated in relation to hormonal status. Questions arise as to the role of cellular turnover in both maintenance and age involution. In order to ascertain the nature of thymus autonomy it is necessary to turn to transplantation experiments. Transplantation of the thymus results in massive cell death in graft cortical lymphoid tissue, with apparent selective survival of the reticular-epithelial cells (Dukor, Miller, House and Allman, 1965; Hinsull and Bellamy, 1974); it is not known to what extent cell death is due to tissue damage on transplantation and the host's immunological response. Subsequently the graft is progressively cleared of cell debris and the characteristic thymic architecture restored within 14 days of grafting. Evidence obtained from the regeneration of differently sized transplants indicates that the size and shape attained by the regenerated graft is closely related to the size and shape of the donor tissue. In addition, the transplants do not grow in phase with the host thymus but appear to retain an intrinsic developmental programme. After the initial fall in graft weight immediately following transplantation, the transplants increased in weight to attain a maximum at 6 weeks postoperatively, when the host was 11 weeks old. However, the host's own thymus had attained its maximum weight at 9 weeks of age, i.e. 4 weeks after implantation. Therefore the transplants do not grow in phase with the host thymus but appear to retain an intrinsic developmental programme (Hinsull and Bellamy, 1974).

When transplanted thymus tissue is contained within Millipore chambers, the lymphocyte population does not reappear and only the reticular-epithelial cells remain. However, if these remnants are freely transplanted into a second host, the thymus regenerates normally, suggesting that the lymphocytes in the regenerated gland are derived from the host (Bellamy and Hinsull, 1975). This conclusion has been documented by the use of chromosome markers (Dukor *et al.*, 1965). These findings indicate that the specifications for the relative dimensions and mass of the various parts of the gland emanate from the ill-defined reticular-epithelial complex, and that host lymphocytes repopulate the gland.

Certain cells in the reticular-epithelial complex have cytoplasm packed with material which gives a positive periodic acid Schiff (PAS) reaction and it is said that these cells form a stimulatory microenvironment for lymphopoiesis (Ishidate and Metcalf, 1963; Metcalf and Ishidate, 1963). This framework is laid down early in embryonic life, thus establishing the mechanism responsible for governing the gland's shape. It has been suggested that each PAS-positive cell, with its surrounding lymphocytes, forms a spatially distinct subunit governing lymphocyte proliferation limited in activity by the number of cells present and the cell density (Franklin, Hinsull and Bellamy, 1978).

4.4.2 *Cell death and cell loss in the thymus*

The foregoing discussion presupposes that thymic homeostasis operates via the control of cell proliferation. However, the importance of cell loss in thymus autonomy is now generally accepted, but the mechanism of this loss is still the subject of considerable controversy. Early workers suggested that thymic lymphocytes were lost through pyknotic degeneration, and reported nearly equal percentages of pyknotic and mitotic cells (Nakamura and Metcalf, 1961). However, compared with the reports of other workers, the number of pyknotic nuclei reported by Nakamura and Metcalf are excessively high (Sainte-Marie and Peng, 1971; Hinsull, Bellamy and Franklin, 1977). Their calculations were dependent on the time course for the destruction of pyknotic nuclei. If the time required for pyknotic cell destruction exceeded the time taken for mitosis (Metcalf, 1964), the majority of thymic lymphocytes must be lost through migration. If pyknotic degeneration is a rapid process, comparable with the duration of mitosis (Matsuyama, Wiadrowski and Metcalf, 1966), relatively few thymocytes would migrate. This latter proposal remains unsubstantiated, as does the associated theory of cell death by thymocyte 'explosion'. In contrast with the lack of satisfactory evidence in favour of the majority of thymic lymphocytes dying *in situ*, there are substantial data in support of the alternative concept of emigration. Numerous radioautographic studies demonstrated that thymocytes are produced in the cortex and move to the medulla, from where they emigrate via the blood (Bryant, 1972; Borum, 1973; Joel, Chanana and Cronkite, 1974). In support of this latter point, Sainte-Marie and Peng (1971) demonstrated a marked increase in the blood lymphocyte content as blood flows through the thymus, and they concluded that the majority of thymocytes emigrate from the thymus. In contrast, experiments with labelled thymocytes indicated that only 16 per cent of the thymocytes emigrated (Larsson, 1967; Joel, Chanana, Cottier, Cronkite and Laissue, 1977). However, Sainte-Marie and Peng (1974) have stated that the small thymocytes, formed in the cortex, only become motile and capable of migration on reaching the medulla. Hence an injection of [³H] thymidine would not favour the labelling of fully mature motile thymocytes. Therefore Sainte-Marie and Peng (1974) destroyed the majority of the cortical thymocytes with corticosterone and labelled the remaining small thymocytes with [³H] cytidine prior to transfer. Using this technique it was found that between 80 and 100 per cent of cells emigrated, the majority being found in the spleen and lymph nodes.

Despite this conclusive evidence for thymocyte emigration, the observation of apoptotic bodies in the thymus (Abraham *et al.*, 1969) means that the role of cell death *in situ* in the thymus still requires investigation.

The lack of satisfactory evidence for the role of cell death in thymus homeostasis is partially due to the lack of distinct criteria, other than pyknosis, by which to characterize dead and dying cells at the light microscope level. Hinsull *et al.* (1977) suggested that the use of Trypan Blue *in vivo* offers a useful approach for assessing cellular death within the intact thymus. Using this technique it has been shown that at 6 weeks of age, when thymus mass is still increasing, the mitotic index is higher than the Trypan Blue index, whereas at 8 weeks of age, when the thymus may be considered to be in a steady state, no significant difference is apparent between the two indices (Fig. 4.1). At 14 weeks the rapid involution of the thymus is paralleled by a rise in the ratio of dying to mitotic cells. This ratio declines at 20 and 29 weeks of age, when the rate of involution also declines, but the number of Trypan Blue-positive cells is still higher than the mitotic index. Unfortunately the Trypan Blue technique yields no information on the duration of cell death. In accordance with the foregoing results Bowen and Lewis (1980) have reported an increase in acid phosphatase-positive cells from 1.27 per cent in the thymus glands from 8-week-old animals to 2.4 per cent in glands from animals at 42 weeks of age. In addition, Claesson (1970) and Claesson and Hartmann (1976), using a nigrosin dye-exclusion test, found that, during the first months of life, only 3–5 per cent of thymocytes were dying, whereas at 12 months of age 20 per cent were dying. The fact that the latter experiments were carried out *in vitro* may account for the high percentages of dying cells reported compared with the previously described findings.

The thymus has a very distinctive bimodal cell density distribution pattern, the mode of the lower densities representing the distribution of medullary areas and the mode in the higher density areas representing the distribution of cortical areas (Fig. 4.2). Both mitotic index and the Trypan Blue index are related to cell density.

Mitotic figures in 6-week-old animals are most frequent at the mean cell density. The non-viable cells are distributed in a similar way at this stage (Fig. 4.2). In older animals an increasing number of non-viable cells are found at the lower densities, resulting in an inverse relationship between mitotic and non-viable cell distribution. This inverse relationship suggests that the lymphocytes divide in the cortex, but move to the medulla to die. This movement of cells has been studied using pulse-labelling by Sainte-Marie and Leblond (1964) and Borum (1973). Labelled cells have been shown to migrate through the cortex to the medullary venules.

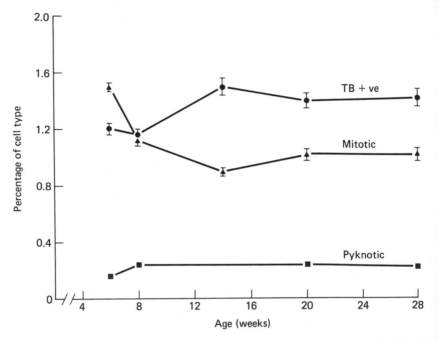

Fig. 4.1 Percentage of Trypan Blue-positive (TB + ve) mitotic and pyknotic cells in the thymus gland of rats with increasing age.

The specific distribution patterns of mitotic and non-viable cells in relation to cell density supports the idea of a density-dependent control mechanism of cell death and mitosis. However, the age differences in the distribution of the two cell types, together with the varying relationship between mitotic index and cell density during recovery from cortisol-induced involution (Bellamy and Hinsull, 1977), indicates a control mechanism which may be influenced by extrathymic factors.

We may conclude that cell death is a significant parameter in the maintenance of cell populations and that a very accurate homeostatic mechanism involving the regulation of cell death is necessary to maintain healthy organs which contain mitotic cells. When this system breaks down, communication between the cells is greatly modified and the cells assume the properties of disorderly growth, arrangement and behaviour; this happens in age involution and during the development of neoplasia.

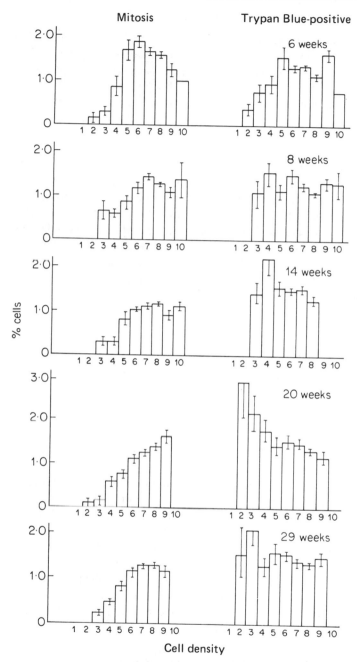

Fig. 4.2 Changes with age in percentage mitotic and percentage Trypan Blue-positive cells in relation to cell density per unit area.

4.5 Homeostasis in malignant tissue

Cell kineticists have long been interested in how the balance between cell proliferation and cell loss, encountered in this neoplastic situation, compares with that observed in healthy tissue. It has been estimated that between 40 per cent and 95 per cent of tumour cells are lost, according to the type of tumour studied (Cooper, 1973). Loss of tumour cells takes place by exfoliation, migration or death.

The relative contribution of these three modes is dependent on the type and location of the tumour. For instance tumours situated adjacent to the peritoneal or thoracic cavities are more susceptible to loss by exfoliation than a tumour enclosed within an organ. Similarly, tumours growing into major blood vessels will lose cells into the circulation, whereas tumours located in tissues which are intimately associated with the lymphatic system, such as the mammary gland, will suffer loss via migration into the lymphatic ducts. Both of these situations, particularly the latter, are liable to result in the formation of secondary growths.

The site of tumour establishment influences the tumour cell population in other ways. For instance, exposure to enzymes, urine, bacteria or viruses may inhibit tumour cell proliferation, shorten cell lifespan or directly cause cell death (Cooper, 1973).

4.5.1 *Tumour necrosis*

Folkman and Hochberg (1973) have suggested that tumour growth consists of two phases, avascular and vascular. During the avascular phase the tumours survive by simple diffusion, but become dormant at a diameter beyond 1 to 2 mm. Further tumour growth is only possible when the tumour is penetrated by blood capillaries. Since tumours are incapable of generating their own capillary supply, vessels must be elicited from the host tissue. Folkman (1974) has reported that the capillary supply to tumours is stimulated by the tumour angiogenesis factor (TAF), a chemical signal which diffuses into the host tissue causing proliferation of host capillaries and small venules. This rate of proliferation is said to be a function of the local concentration of TAF which, in turn, is proportional to the rate of tumour cellular proliferation (Folkman, Merler, Abernathy, and Williams, 1971).

With time, the peak tumour cell density and rate of proliferation shifts towards the periphery of the tumour, with decline in blood vessel surface area at the tumour centre (Liotta, Saidel, and Kleinerman, 1977). As the nutrient and oxygen availability falls below some critical minimal level, tumour cell death occurs (Tannock 1968, 1970; Tannock and Hayashi, 1972). This, together with constriction of the capillaries by tumour cells and thrombotic accidents, may produce the large areas of necrosis which are the most obvious forms of cell death in many tumours.

4.5.2 *Immunological cytolysis and differentiation*

Spontaneous regression of tumours suggests that immunological cytolysis may contribute to tumour cell death. There is an increasing amount of evidence to suggest that sensitized lymphocytes kill tumour cells and that the serum contains cytotoxic antibodies. However, the immunological elimination of tumour cells appears to be a weak reaction which is easily overcome by a rapidly growing tumour and it remains uncertain what contribution immunological attack makes to total tumour cell death (Siegel and Cohen, 1978; Mikulski and Muggia, 1978; Castro, 1978; Fidler, Gersten and Kripke, 1979).

Blood supply and immunological reaction are thus two major extrinsic causes of cell death, but perhaps more intriguing is the mechanism of intrinsic cell death. The differentiation which takes place in certain tumours in itself provides a mechanism for cell loss. For instance, differentiated tumours of the skin resemble their normal counterparts to the extent that they lose cells by keratinization. In undifferentiated tumours Lala (1972) has shown that there is both a progressive transition of dividing cells into a resting state and an increased rate of cell loss with age. Lala and Patt (1968) also found that non-proliferative cells retain the ability to recycle with certain stimuli, such as retransplantation or partial removal of the ascites. It is possible that this resting state may be a result of some deficiency in the larger tumours which may, in turn, limit the lifespan of their cells. Alternatively the cell may die as the result of innate senescence. What still remains uncertain is whether this senescence is the result of an accumulation of genetic errors or the failure of some biochemical pathway.

It is well established that removal of a primary tumour frequently results in increased proliferative activity in the secondaries (Rockwell and Kallman, 1972). Radiation therapy has little effect on the non-proliferating hypoxic cell fraction of tumours, but these cells are known to re-enter the cell cycle and re-establish tumour growth when the proliferative cells have been destroyed by radiation. The long intervals between the detection of a primary tumour and the appearance of secondaries may be explained by prolonged dormancy of cells in a privileged site (Willis, 1948). In this context (Willis, 1948) found that small tumours remained dormant for one year when transplanted into the anterior chamber of the eye, but grew normally when retransplanted into muscle.

Taken together, this varied evidence points to three possibilities. Firstly, the fact that tumour cells respond to changes in the mass of the tumour suggests autoregulation. Secondly, the tumour cell population is responsive to the micro-environment provided by the host. Thirdly, the tumour might create its own self-stimulatory micro-environment.

4.5.3 *Autoregulation in tumours*

With regard to autoregulation in tumours a number of investigators have postulated that tumours arise owing to the total breakdown in the normal tissue autoregulatory system. Wheldon (1975) proposed that tumour cells are incapable of producing the usual mitotic inhibitors, thus supporting the earlier theory of Osgood (1957) that early cell death in tumours results in the lack of differentiation of inhibitor-secreting cells. Subsequently many workers have shown that tumour cells are capable of producing a factor which inhibits mitosis in the equivalent tissue type, although only low concentrations of this factor are found in malignant compared with normal tissue (Rytomaa and Kiviniemi, 1968; Bichel, 1973; Kuo and Yoo, 1977; Makai and Gergely, 1980). This may be a result of inactivation of the inhibitor in the tumour (Wheldon, Gray, Kirk and Orr, 1970). It has been suggested that malignant cells fail to respond to inhibitory factors owing to a change either in the cell-specific carrier or receptor, resulting in non-recognition of the cell by its tissue-specific inhibitor (Riley, 1969; Rothbath, Maier, Schopf and Werner, 1977).

Similarly, in Walker 256 carcinoma as in the thymus, the highest rate of mitosis occurs at the mean cell density, but at higher densities cellular proliferation is inhibited (Bellamy and Hinsull, 1978b). According to the model postulated for the thymus this pattern could result from a balance between internally produced inhibitors and stimulators. In particular, high cell densities are thought to be associated with high inhibitor concentrations. This idea fits the data for the Walker tumour in so far as the mean mitotic index is reduced at the same time as cell density is increased. Alternatively, in tumours the excessive rate of cell death may prevent the population from attaining this limiting density, and so contribute to the failure of mitotic autoregulation to stabilize at a fixed cell density. In this way early cell senescence and death in tumours may contribute to the breakdown of the tissue's autoregulatory system.

4.5.4 *Microenvironment of tumours*

In old hosts where experimental tumours grow more slowly, the density pattern of mitotic tumour cells is closer to that observed in normal tissues. Thus division and death of tumour cells is influenced by the host's microenvironment. It is also well established that certain tumours have a predilection for establishment in specific tissues. For instance reticulum cell sarcoma accumulates in spleen and lymph nodes (Parks, 1974), Cloudman melanoma in lung (Kinsey, 1960) and plasma cytomas in bone marrow (Potter, Fahey and Pilgrim, 1957). It has been suggested that nutritional factors characteristic to the site of arrestment (Zeidman, 1961), as well as intrinsic factors peculiar to the particular tumour line, dictate the tissue of establishment.

Once the tumour cells have established in a site of preference, their behaviour still appears to be influenced by the tissue in which they have lodged. Bellamy and Hinsull (1978a) have shown that the rate of cellular proliferation of Walker 256 carcinoma lodged in the lungs or lymph nodes is significantly higher than that observed in a subcutaneous tumour. These results support the earlier work of Simpson-Herren, Sanford and Holmquist (1974) who showed that the cell cycle and S phase are shorter in Lewis lung carcinoma metastases than in the primary tumour. Trope (1975) has also found a difference between primary and secondary growths located in various tissues in so far as tumours residing in the lungs show a greater sensitivity to drugs.

As well as the overall differences in mitotic rate, it has also been shown that the central mass of the lung tumour has a higher rate of cellular proliferation than in the peripheral mass (Bellamy and Hinsull, 1978a). In Walker carcinoma growing beneath the kidney capsule there is a gradation in the mitotic rate, with the lowest levels of mitosis observed adjacent to host kidney cells (Hinsull and Bellamy, unpublished data). In addition Broyn (1975) has shown a greater rate of tumour growth in association with ulceràtion of the gastric mucosa and the consequent release of a stimulating wound hormone. It appears possible, therefore, that the host tissue may influence tumour growth via the normal chalone system and the differences in the effectiveness of host inhibitors may be involved in different site patterns of tumour growth.

Despite this evidence for the influence of host tissue on tumour establishment and growth there are still many areas which need investigating, particularly concerning events during the initial stages of establishment. The initial basic cellular reaction to an experimental cancer inoculum involves the establishment of a localized population of host cells from the lymphomyeloid complex which interact with but do not kill the invading tumour cells. Once the tumour has established a substantial cell population and has formed connective tissue barriers between itself and the adjacent host organs it is likely that it will maintain its own extracellular fluid in a unique steady state. Indirect evidence for this difference comes from work showing that the extracellular fluid of tumours has a slightly lower pH than the host's tissues when measured *in vivo* using radiotelemetry (H. R. Colson, Hinsull and Bellamy, unpublished data). Working on the assumption that tumours could maintain themselves at a different redox potential, compared with the host, an attempt was made to influence development of experimental tumours by manipulating the electron availability at the site of implantation, by using a simple metalplastic implant (Bellamy, Hinsull, Watson and Blache, 1979), consisting of a loop of platinum wire, 0.5 cm diameter sealed with silicone adhesive to 8 cm of medical grade silicone rubber tubing which was in turn sealed to a thin platinum disc of 1 cm diameter. Eight days after inoculation of tumour cells into the centre of the ring, it was found that the implant had a profound inhibitory influence on the proliferative behaviour of the tumour. Where tumour growth did occur, distinctive patterns of tumour cell distribution and cellular proliferation were observed, suggesting that the implants established a field effect on the tumour tissue. Similar implants, incorporating 1 or 10 μA batteries, also inhibited tumour growth. However, a 1 μA current did not significantly increase the degree of inhibition compared with the passive implant, and a 10 μA current resulted in an adverse host reaction (Hinsull, Bellamy, Franklin and Watson, 1980). The authors have postulated that this inhibitory effect is the result of the prevention of the establishment of a microenvironment by the tumour. For the simple implant to work there would have to be a different steady state redox potential between tumour and host tissue. The question of how this difference is generated and becomes necessary to promote the development of the tumour remains unresolved. However, the possibility of changing the tumour's microenvironment in such a way that growth is inhibited opens a number of avenues for investigating both possible tumour therapy and tumour cell kinetics.

4.5.5 *Cancer chemotherapy*

One attractive approach to tumour therapy, based on the theory that growth of tumours requires the generation of a unique redox potential, is to investigate the action of oxidizing agents on experimental tumours. Indeed, many years ago it was reported that Methylene Blue added directly to animal tumours caused them to regress.

The search for new drugs effective against cancer usually proceeds on the basis that the best way to control cancer is to prevent division of the cancerous cells. Although drugs in current clinical use inhibit cell division and interact with DNA, other solutions are possible.

An example would be the mode of action of platinum compounds. The different redox state of the experimental tumours might promote the formation of platinum-based electrode products in the vicinity of the tumour. Platinum ammonium chloride, the parent compound of the drug series, is very effective against experimental rat tumours, not only in preventing the small implanted fragments from developing, but also in causing regression of large solid tumours into which the untreated fragments grow. On the basis of the mitotic index, platinum ammonium chloride inhibits cell division and from the rapid rate at which large tumours regress, it also causes massive cell death. Are the two phenomena connected and is the primary site of action in DNA?

Indirect evidence for platinum drugs interacting with the cell nucleus is that generally G1-phase cells are more sensitive to platinum ammonium chloride than mid-S-phase cells. This is taken to mean that replication of a whole genome immediately after platinum damage would be expected to result in more damage at the end of the cell cycle compared with mid-S-phase when only part of the DNA would be replicated on the damaged template.

If the two are connected, regression and the interaction with DNA should be closely correlated, but, this correlation has not been tested directly. In an experiment comparing the sensitivity of cell cultures to platinum compounds it was found that the loss of viability of HeLa and Chinese hamster cells did not match the interaction of *cis*-platinum II diamine dichloride with DNA. Although the HeLa cells were 3.3 times more sensitive to the killing action of a given dose of *cis*-platinum II, on the basis of the different slopes for the relationship between DNA binding and viability the HeLa cells were only 1.5 times as sensitive (van den Berg, Fraval and Roberts, 1977).

Work on the distribution of platinum drugs within the body and within cells of tissue cultures indicated that the compounds have a general affinity for proteins and interact with all organelles and the cell membrane. A time lapse investigation of the cytotoxic action of *cis*-platinum II on sarcoma 180 ascites cells showed that the initial response during the first 15 min of exposure indicated a peripheral cellular reaction which caused the cell contents to concentrate in a distinct band around the nucleus. Between 30 and 45 min there was a very large increase in cell volume, indicating a failure of osmotic regulation. This time sequence suggests that there is an inital effect on the cell membranes and microfilaments, completely altering the flow of information within the cytoplasm (Aggarwal, Ofosus and Waku, 1977). It also appears from similar work on sarcoma 180 implanted into mice that *cis*-platinum II alters the volume-regulatory mechanisms without affecting DNA replication, resulting in the formation of giant cells each containing many nuclei with a common envelope (Sodhi, 1977). Further evidence for cytoplasmic interactions comes from studies on the staining reactions of platinum–Pyrimidine Blue complexes on sarcoma 180 ascites cells observed under the electron microscope (Aggarwal, 1977). It appears therefore that, if we are ever to have the ability to readjust a malfunctioning tissue autoregulatory system, we must find a technique which not only prevents cell division but also promotes cell death.

References

Abraham, R., Morris, M. and Hendy, R. (1969), Lysosomal changes in epithelial cells of the mouse thymus after hydrocortisone treatment. *Histochemie,* **17**, 295–311.

Aggarwal, S. K. (1977), Fine structural studies using platinum coordination complexes. *Clin. Hematol. Oncol.,* **7**, 760–796.

Aggarwal, S. K., Ofosus, G. A. and Waku, Y. (1977), Cytotoxic effects of platinum-pyrimidine complexes on the mammalian cells *in vitro*: A fine structural study. *Clin. Hematol. Oncol.,* **7**, 547–561.

Bartelmez, G. W. (1957), The phases of the menstrual cycle and their interpretation in terms of the pregnancy cycle. *Am. J. Obstet. Gynecol.,* **74**, 931–955.

Bellamy, D. and Hinsull, S. M. (1975), On the role of the reticular epithelial complex in transplantation of the thymus. *Differentiation,* **3**, 115–121.

Bellamy, D. and Hinsull, S. M. (1977), Density-dependent cell division after cortisol treatment of rat thymus in relation to age involution. *Virch. Arch. B. Cell Pathol.,* **24**, 251–261.

Bellamy, D. and Hinsull, S. M. (1978a), Density-dependent mitosis in the Walker 256 carcinoma and the influence of host age on growth. *Eur. J. Cancer,* **14**, 747–751.

Bellamy, D. and Hinsull, S. M. (1978b), Influence of lodgement site on the proliferation of metastases of Walker 256 carcinoma in the rat. *Br. J. Cancer,* **37**, 81–85.

Bellamy, D., Hinsull, S. M. and Phillips, J. G. (1976), Factors controlling growth and age involution of the rat thymus. *Age Ageing,* **5**, 12–19.

Bellamy, D., Hinsull, S., Watson, B. and Blache, L. A. (1979), Inhibition of the development of Walker 256 carcinoma with a simple metal–plastic implant. *Eur. J. Cancer,* **15**, 223–232.

Bichel, P. (1973), Self-limitation of ascites tumour growth – a possible chalone regulation. *Nat. Cancer Inst. Monogr.,* **38**, 359–367.

Borum, K. (1973), Cell kinetics in mouse thymus studied by simultaneous use of ^{3}H-thymidine and colchicine. *Cell Tissue Kinet.,* **6**, 545–552.

Bowen, I. D. and Lewis, G. H. J. (1980), Acid phosphatase activity and cell death in mouse thymus. *Histochemistry,* **65**, 173–179.

Brandes, D. and Groth, D. P. (1963), Functional ultrastructure of the rat prostatic epithelium. *Nat. Cancer Inst. Monogr.,* **12**, 47–62.

Broyn, T. (1975), The interaction between Walker tumour cells and mucosa cells in the lamina propria of gastric mucosa in rats. *Virch. Arch. B. Cell Pathol.,* **19**, 27–36.

Bryant, B. J. (1972), Renewal and fate in the mammalian thymus: mechanisms and inferences of thymokinetics. *Eur. J. Immunol.,* **2**, 38–45.

Bullough, W. S. (1952), The energy relations of mitotic activity. *Biol. Rev. Cambridge Philos. Soc.*, **27**, 133–168.

Bullough, W. S. (1972), The control of epidermal thickness. *Br. J. Dermatol.*, **87**, 187–199.

Castro, J. (1978), *Immunological Aspects of Cancer*. MTP Press Ltd., Lancaster.

Claesson, M. H. (1970), An autoradiographic study of the normal decay of lymphoid cells in the mouse thymus. *Acta Pathol. Microbiol. Scand.*, **78**, 556–564.

Claesson, M. H. and Hartmann, M. R. (1976), Cytodynamics in the thymus of young adult mice: quantitative study on the loss of thymic blast cells and non-proliferative small lymphocytes. *Cell Tissue Kinet.*, **9**, 273–291.

Cooper, E. H. (1973), The biology of cell death in tumours. *Cell Tissue Kinet.*, **6**, 87–95.

Cowdry, E. V. (1950), *Textbook of Histology*. Lea and Feibiger, Philadelphia.

Dukor, P., Miller, J. F., House, A. P. and Allman, V. (1965), Regeneration of thymus grafts. I. Histological and cytological aspects. *Transplantation*, **3**, 639–665.

Farbman, A. I. (1968), Electron microscope study of palate fusion in mouse embryos. *Dev. Biol.*, **18**, 9.

Fidler, I. J., Gersten, D. M. and Kripke, M. L. (1979), Influence of immune status on the metastasis of three murine fibrosarcomas of different immunogenicities. *Cancer Res.*, **39**, 3816–3821.

Folkman, J. (1974), Tumour angiogenesis factor. *Cancer Res.*, **34**, 2109–2113.

Folkman, J. and Hochberg, M. (1973), Self-regulation of growth in three dimensions. *J. Exp. Med.*, **138**, 745–753.

Folkman, J., Merler, E., Abernathy, C. and Williams, G. (1971), Isolation of a factor responsible for tumour angiogenesis. *J. Exp. Med.*, **133**, 275–288.

Franklin, A., Hinsull, S. and Bellamy, D. (1978), On the relationship between PAS positive cells and lymphocyte proliferation in the cortisol treated rat thymus. *Biol. Cell.*, **33**, 137–144.

Helminen, H. J. and Ericsson, J. L. E. (1968), Studies on mammary gland involution. *J. Ulstrastruct. Res.*, **25**, 193–252.

Helminen, H. J. and Ericsson, J. L. E. (1971), Ultrastructural studies on prostatic involution in the rat. Mechanisms of autophagy in epithelial cells with special reference to the rough surfaced endoplasmic reticulum. *J. Ultrastruct. Res.*, **36**, 708–724.

Helminen, J. J. and Ericsson, J. L. E. (1972), Ultrastructural studies on prostatic involution in the rat. Evidence for focal irreversible damage to epithelium and heterophagic digestion in macrophages. *J. Ultrastruct. Res.*, **39**, 443–455.

Hinsull, S. M. and Bellamy, D. (1974), Development and involution of thymus grafts in rats with reference to age and sex. *Differentiation*, **2**, 299–305.

Hinsull, S. M., Bellamy, D. and Franklin, A. (1977), A quantitative histological assessment of cellular death in relation to mitosis, in rat thymus during growth and age involution. *Age Ageing*, **6**, 77–84.

Hinsull, S. M., Bellamy, D., Franklin, A. and Watson, B. W. (1980), The inhibitory influence of a metal–plastic implant on cellular proliferation patterns in an experimental tumour compared with normal tissue. *Eur. J. Cancer*, **16**, 159–166.

Hopwood, D. and Levison, D. A. (1976), Atrophy and apoptosis in the cyclical human endometrium. *J. Pathol.*, **119**, 159–166.

Houck, J. C. (1976), *Chalones*. North-Holland Publishing Co., Amsterdam.

Ishidate, M. and Metcalf, D. (1963), The pattern of lymphopoiesis in the mouse thymus after cortisone administration or adrenalectomy. *Austr. J. Exp. Biol.*, **41**, 637–649.

Joel, D. D., Chanana, A. D., Cottier, H., Cronkite, E. P. and Laissue, J. A. (1977), Fate of thymocytes: Studies with ^{125}I-iododeoxyuridine and ^{3}H-thymidine in mice. *Cell Tissue Kinet.*, **10**, 57–69.

Joel, D. D., Chanana, A. D. and Cronkite, E. P. (1974), Thymus cell migration. *Ser. Haematol.*, **7**, 414–471.

Kerr, J. F. R. (1971), Shrinkage necrosis: a distinct mode of cellular death. *J. Pathol.*, **105**, 13–20.

Kerr, J. F. R. (1973), Some lysosome function in liver cells reacting to sublethal injury. In *Lysosomes in Biology and Pathology*, vol. 3 (ed. J. T. Dingle), North-Holland Publishing Co., Amsterdam, pp. 365–394.

Kerr, J. F. R. and Searle, J. (1973), Deletion of cells by apoptosis during castration – induced involution of the rat prostate. *Virch. Arch. Abt. Z. Zellpathol.*, **13**, 87–102.

Kerr, J. F. R., Wyllie, A. H. and Currie, A. R. (1972), Apoptosis: a basic biological phenomenon with wide ranging implications in tissue kinetics. *Br. J. Cancer*, **26**, 239–257.

Kinsey, D. L. (1960), An experimental study of preferential metastasis. *Cancer*, **13**, 674–676.

Kuo, C. Y. and Yoo, T. J. (1977), *In vitro* inhibition of tritiated thymidine uptake in Morris hepatoma cells by normal rat liver extract: A possible liver chalone. *J. Natl. Cancer Inst.*, **59**, 1691–1695.

Lala, P. K. (1972), Evaluation of the mode of cell death in Ehrlich ascites tumor. *Cancer*, **29**, 261–266.

Lala, P. K. and Patt, H. M. (1968), A characterization of the boundary between the cycling and resting states in the ascites tumor cells. *Cell Tissue Kinet.*, **1**, 137–146.

Lamerton, L. F. (1969), Cell population kinetics in relation to homeostasis. In *Ciba Foundation Symposium: Homeostatic Regulators* (eds. G. E. W. Wolstenholme and J. Knight), J. and A. Churchill Ltd., London, pp. 5–28.

Larsson, B. (1967), Export and import of ^{3}H-thymidine labelled lymphocytes in the thymus of normal and steroid treated guinea-pigs. *Acta Pathol. Microbiol. Scand.*, **70**, 390–416.

Liotta, L. A., Saidel, G. M. and Kleinerman, J. (1977), Diffusion model of tumour vascularization and growth. *Bull. Math. Biol.*, **39**, 117–128.

Lockshin, R. A. and Beaulaton, J. (1974), Programmed cell death. *Life Sci.*, **15**, 1549–1565.

Makai, G. S. and Gergely, H. (1980), Effect of Ehrlich ascites cell chalone on nascent DNA synthesis in isolated nuclei. *Cell Tissue Kinet.*, **13**, 65–73.

Matsuyama, M., Wiadrowski, N. and Metcalf, D. (1966), Autoradiographic analysis of lymphopoiesis and lymphatic migration in mice bearing multiple thymus grafts. *J. Exp. Med.*, **123**, 559–576.

Mescher, A. L. and Tassava, R. A. (1975), Denervation effects on DNA replication and mitosis during the initiation of limb regeneration in adult newts. *Dev. Biol.* **44**, 187–197.

Metcalf, D. (1964), The thymus and lymphopoiesis. In *The Thymus and Immuno biology* (eds. R. A. Good and A. E. Gabrielson), Harper and Row, New York pp. 150–179.

Metcalf, D. and Ishidate, M. (1963), PAS-positive reticulum cells in the thymus cortex of high and low leukemia strains of mice. *Austr. J. Exp. Biol. Med.* **40**, 57–71.

Mikulski, S. M. and Muggia, F. M. (1978), The suppressor mechanisms and their significance in tumor immunology. *Cancer Immunol. Immunother.*, **4**, 139–142.

Moller, G. (1969), Regulatory mechanisms in antibody synthesis. In *Ciba Foundation Symposium: Homeostatic Regulators* (eds. G. E. W. Wolstenholme and J. Knight), J. and A. Churchill Ltd., London, pp. 197–221.

Nakamura, K. and Metcalf, D. (1961), Quantitative cytological studies on thymic lymphoid cells in normal preleukaemic and leukaemic mice. *Br. J. Cancer*, 15, 306–315.

Obearade, M., Chen, J., Ove, P. and Lansing, A. I. (1974), Functional regeneration of liver of old rats after partial-hepatectomy. *Exp. Gerontol.*, 9, 181–190.

Osgood, E. E. (1957), A unifying concept of the etiology of the leukemias, lymphomas and cancers. *J. Natl. Cancer Inst.*, 18, 155–166.

Parks, R. C. (1974), Organ-specific metastasis of a transplantable reticulum cell sarcoma. *J. Natl. Cancer Inst.*, 52, 971–973.

Perris, A. D., Weiss, L. A. and Whitfield, J. F. (1970), Parathyroidectomy and the induction of thymic atrophy in normal, adrenalectomized and orchidectomized rats. *J. Cell. Physiol.*, 76, 141–149.

Potter, M., Fahey, J. L. and Pilgrim, H. I. (1957), Abnormal serum protein and bone destruction in transmissible mouse plasma cell neoplasm. *Proc. Soc. Exp. Biol. Med.*, 94, 9327–333.

Rahes H. M. (1976), Kinetics of hepatocellular proliferation after partial resection of the liver. *Prog. Liver Dis.*, 5, 85–99.

Riley, P. A. (1969), Heuristic model of mitotic autoregulation in cell populations. *Nature (London)*, 223, 1382–1383.

Rockwell, S. and Kallman, R. I. (1972), Growth and cell population kinetics of single and multiple KHT sarcomas. *Cell Tissue Kinet.*, 5, 449–457.

Rothbath, K., Maier, A., Schopf, E. and Werner, D. (1977), Inhibition of DNA synthesis by a factor from ascites tumor cells. *Eur. J. Cancer*, 13, 1195–1196.

Rytomaa, T. and Kiviniemi, K. (1968), Control of cell production in rat chololeukemia by means of the granulocytic chalone. *Nature (London)*, 220, 136–138.

Sainte-Marie, G. and Leblond, C. P. (1964), Thymus cell population dynamics. In *The Thymus in Immunobiology* (eds. R. A. Good and A. E. Gabrielson), Harper and Row, New York, pp. 207–226.

Sainte-Marie, G. and Peng, F. S. (1971), Emigration of thymocytes from the thymus. A review and study of the problem. *Rev. Can. Biol.*, 30, 51–78.

Sainte-Marie, G. and Peng, F. S. (1974), Distribution of transferred mature thymocytes of the rat. *Rev. Can. Biol.*, 33, 61–65.

Shirley, D. G. (1976), Development and compensatory renal growth in the guinea pig. *Biol. Neonate*, 30, 169–180.

Siegel, B. V. and Cohen, S. (1978), Tumor immunity. *Fed. Proc. Fed. Am. Soc. Exp. Biol.*, 37, 2212–2214.

Simpson-Herren, L., Sanford, A. H. and Holmquist, J. P. (1974), Cell population kinetics of transplanted and metastic Lewis lung carcinoma. *Cell Tissue Kinet.*, 7, 349–361.

Simard, A., Corneille, L., Deschamps, Y. and Verly, W. G. (1974), Inhibition of cell proliferation in the livers of hepatectomized rats by a rabbit hepatic chalone. *Proc. Natl. Acad. Sci. U.S.A.*, 71, 1763.

Sodhi, A. (1977), Origin of giant cells in the regressing sarcoma 180 after *cis*dichlorodiamine platinum (II) treatment. *J. Clin. Oncol. Hematol.*, 7, 569–579.

Speraff, L. and Van de Wiele, R. (1971), Regulation of the human menstrual cycle. *Am. J. Obstet. Gynecol.*, 109, 234–247.

Tannock, I. F. (1968), The relation between cell proliferation and the vascular system in a transplanted mouse mammary tumor. *Br. J. Cancer*, 22, 258–273.

Tannock, I. F. (1970), Population kinetics of carcinoma cells, endothelial cells and fibroblasts in a transplanted mouse mammary tumor. *Cancer Res.*, **30**, 2470–2476.
Tannock, I. F. and Hayashi, S. (1972), The proliferation of capillary endothelial cells. *Cancer Res.*, **32**, 77–82.
Thornton, C. S. (1968), Amphibian limb regeneration. *Adv. Morphol.*, **7**, 205–249.
Tinsley, J. C., Moore, C. V., Dubach, R., Minnich, V. and Grinstein, M. (1949), The role of oxygen in the regulation of erythropoiesis. *J. Clin. Invest.*, **28**, 1544–1556.
Trope, C. (1975), Different sensitivity to cytostatic drugs of primary tumor and metastasis of the Lewis carcinoma. *Neoplasma*, **22**, 171–180.
Tsirel'mkov, H. I. (1976), Effect of injury to the maternal liver on reactive changes in the liver of the young. *Exp. Biol. Med.*, **82**, 1395–1398.
van den Berg, H. W., Fraval, H. N. A. and Roberts, J. J. (1977), Repair of DNA damaged by neutral platinum complexes. *J. Clin. Hematol. Oncol.*, **7**, 349–373.
Verly, W. G. (1976), The control of liver growth. In *Chalones* (ed. J. C. Houck), North-Holland Publishing Co., Amsterdam, pp. 401–427.
Vernon, C. A., Banks, B. E. C., Banthorpe, D. V., Berry, A. R., Davies-Hiff, S., Lamont, D. M., Pearce, F. L. and Kenning, K. A. (1969), Nerve growth and epithelial growth factors. In *Ciba Foundation Symposium: Homeostatic Regulators* (eds. G. E. W. Wolstenholme and J. Knight), J. and A. Churchill Ltd., London, pp. 57–74.
Weaver, J. A. (1955), Changes induced in the thymus and lymph nodes of the rat by the administration of cortisone and sex hormones and by other procedures. *J. Pathol. Bacteriol.*, **69**, 133–148.
Wheldon, T. E. (1975), Mitotic autoregulation of normal and abnormal cells. *J. Theor. Biol.*, **53**, 421–433.
Wheldon, T. E., Gray, W. M., Kirk, J. and Orr, J. S. (1970), Mitotic autoregulation of populations of normal and malignant cells. *Nature (London)*, **226**, 547.
Willis, R. A. (1948), *The Pathology of Tumours*. Butterworth, London.
Wyllie, A. H., Kerr, J. F. R. and Currie, A. R. (1973a), Cell death in the normal neonatal adrenal cortex. *J. Pathol.*, **111**, 255–261.
Wyllie, A. H., Kerr, J. F. R., Macaskill, I. A. M. and Currie, A. R. (1973b), Adrenocortical cell deletion: The role of ACTH. *J. Pathol.*, **111**, 85–94.
Zeidman, I. (1961), The fate of circulating tumour cells. I. Passage of cells through capillaries. *Cancer Res.*, **21**, 38–39.

5 Cell senescence and death in plants

P. B. GAHAN

5.1 Introduction

In spite of the wide-spread occurrence of cell death in plants, the phenomenon is a topic rarely discussed in textbooks of plant anatomy or physiology. It is implicated extensively in the death of whole plants, e.g. monocarpic plants which reproduce once only and die, or polycarpic plants which reproduce either a few or many times before dying, and with the death of parts of polycarpic plants. This latter point is seen in the case of flower, leaf and fruit fall, bark-shedding or with the loss of the shoot after reproduction in some herbaceous plants. In addition, cell death is involved in both morphogenesis and a number of normal physiological events in cryptogams and phanerogams (Table 5.1). It is the aim of this chapter to consider the possible mechanisms leading to cell death, the involvement of cell death in the various, normal events in plants, and the ways in which cell death might be determined in various cell populations. Many of the examples quoted will be from flowering plants, since it is in this domain that most work has been performed. Space will not permit an extensive treatment of the literature and, instead, as many and varied examples as possible will be used to illustrate the points under discussion.

Table 5.1 Some examples of senescence and cell death in plants

(i)	Maturation of fibres; vessels; tracheids; sclereids; sieve tubes; root cap cells; phellem [3,4]
(ii)	Formation of air spaces; fenestra in leaves; food reserves [1,3,4]
(iii)	Emergence of aerial and lateral roots causing mechanical destruction of cortical and epidermal cells [3,4]
(iv)	Aging of pith and wood parenchyma cells; root hairs [8]
(v)	Mobilization of food reserves [1,7]
(vi)	Prefertilization; tapetum; pollen tubes; style cells; oogonia [2-6]
(vii)	Postfertilization: flower parts; fruit ripening; leaves; seed capsule dehiscence; spore and gamete dispersal; gametophytes [2-6]

[1] Bewley and Black (1978) [2] Bower (1935) [3] Esau (1974) [4] Fahn (1974)
[5] Fritsch (1935) [6] Gwynne-Vaughan and Barnes (1951) [7] Matile (1975)
[8] Molisch (1938)

5.1.1 *Terminology*

Two terms which will be used extensively are 'cell death' and 'cell senescence'. The latter is used to cover the series of changes which occur with the passing of time, and which result in the decreased survival capacity of the individual cell. The former is more difficult to define, but will be taken to mean those events forming the boundary between the end of senescence and the moment when the cell exhibits no further functional control. This boundary is difficult to pin-point, as will be discussed below.

5.1.2 *Mechanisms leading to cell death*

Cell death would appear to be initiated in one of four basic ways, namely, through (a) a pathological lesion, (b) mechanical damage, (c) physiological insult or (d) senescence. A pathological lesion can be a primary factor in cell death, either through toxins or other substances such as enzymes released by the pathogen. Necrotrophic plant pathogens kill host cells in advance of invasion. Mechanical damage and physiological insult can result in cell death either directly because the cells are irreversibly damaged or indirectly because minor damage permits the entry of a pathogen which kills the cell.

Whilst these two mechanisms have their place in the system, perhaps the more generally interesting possibility is that of cell senescence, since, in addition to leading to the death of the whole or parts of the plant, this mechanism can be seen in so many of the morphogenetic and physiological events listed in Table 5.1. The little work which has been performed in this area indicates that plant cells have a finite lifespan, the length of which can vary enormously from cell type to cell type within a given organism, or for the same cell type between different species. The limited available data indicate that flower parts, and hence their components, can have very short existences. If longevity is measured as the time from the opening of the flower to the withering or fall of the calyx and/or stamens, then the lifetime can be as short as 3 h, e.g. *Hibiscus trionium*, though at the opposite end of the spectrum is the flower of *Phalaenopsis shilleriana* with a lifespan of 90 to 135 days (Molisch, 1938). The problem of the mechanisms of cell death in flower parts will be given a more detailed consideration later.

Cell types in the main axis have also been studied, and it is considered that pith cells can vary in longevity from a few months in *Sambucus nigra* to about 40 years in *Fagus sylvatica*, *Tilia parvifolia* and *Sorbus Aucuparia* (Massopust, 1906). Wood parenchyma cells may also be long-lived, Schorler (1883) recording a range of 21 to 86 years for a variety of tree species. Root hair cells are short-lived, however, 55 h (McElgunn and Harrison, 1969) or less (Holden, 1975) being recorded for barley roots.

Again, very little information is available concerning the capacity for division of plant cells. Whilst many animal cell studies have shown limited division capacity for a number of cell types, as evidenced by population doublings measured *in vitro* (reviewed in Finch and Hayflick, 1977), no such experiments have been made for plants. The loss of mitotic capacity seen in primary root meristems of monocarpic plants may be indicative of this, or may merely be a marker of other changes in the plant leading to a loss of mitotic capacity as a visible secondary event (Gahan and Hurst, 1976; Mozaffari and Gahan, 1978).

5.1.3 *Cell senescence and possible mechanisms*

Of the studies made on cell senescence, the majority by far have concerned animal cells. A number of possible causal mechanisms have been suggested, of which some may be pertinent to the plant cell. However, in the case of plant cell senescence it is, perhaps, important to distinguish between cell senescence involved in morphogenetic processes, such as xylogenesis and phelloderm formation, and those involving organs which the main axis will shed, e.g. fruits and leaves. The reason for such a separation in treatment is because, in organs which are shed, cells are seen to senesce either because they are cut off from the parent plant by abscission as in the case of fruit, or because the system is cannibalized by the main axis prior to abscission, e.g. leaves (Woolhouse, 1974). In the latter case it is difficult to determine at which point there is a deterioration of leaf performance due to cell senescence, and at which point there is a switch-on of the process of removal of useful material from the cells to the main axis for either transport to the developing seed/fruit or storage in a part of the main axis. This process can be summarized into the following major steps:

Phase I, maximum functioning of the leaf as a site of, e.g., photosynthesis;
Phase II, possible aging of leaf cells and concomitant decline in performance;
Phase III, organized removal of leaf contents via main axis; and
Phase VI, abscission layer formation and leaf fall.

Thus the death of the cells, if not occurring in Phase II, will be precipitated in Phase III and finalized in Phase IV. The signalling mechanisms by which the transition from Phase II to Phase III occurs are not clear, though frequently, this transition coincides with a point in time after fertilization; it is not just a mechanism for transporting food supplies to the developing embryo and seed, since the events occur also in male plants which do not undergo such processes. Unfortunately, most studies have been performed upon senescing leaves which have been removed from the plant and so investigating possible control mechanisms has not proved easy (see reviews by Woolhouse, 1974, 1978; Butler and Simon, 1971). It would appear to be easier to review events in the main axis in the light of theories advanced to explain cell senescence. Of the many theories available (Curtis, 1966; Strehler, 1977; Finch and Hayflick, 1977) the following appear to be the most interesting from a plant cell point of view.

The somatic mutation theory of aging has been under discussion for many years and implies that there is an instability of the genetic structures which leads eventually to senescence (Curtis, 1966).

The error catastrophe theory of aging was first suggested by Medvedev (1963) and finally formulated by Orgel (1963, 1973). This theory required that macromolecules not be synthesized with complete accuracy and that a proportion of faulty molecules will themselves be responsible for further errors. A sufficiently large feedback of errors will result in an accumulated lethal damage in either the proteins or nucleic acids. The third theory proposes that aging processes are programmed, and stems from the idea that senescence is an extension of the developmental process. These possibilities will be further discussed in the light of the subsequent data.

5.1.4 *Cell senescence and death*

Whether cells die as the result of mechanical damage, infection or senescence, a number of clear-cut features are observed in the dead cells in all cases, including irreversible plasmolysis, disruption of cellular membranes with a concomitant loss of compartmentalization, loss of respiratory activity, changes in the subcellular distribution of enzymes such as hydrolases, and the inability to control the transport of molecules such as dye molecules. These features indicate that the cell is dead, returning us to the question raised in defining the moment of cell death. There is no satisfactory answer to this question for either plant or animal cells as there are no clear-cut reference points to indicate the passing from the late stages of senescence to the state of cell death. Matile (1975) has proposed that 'senescence ends with the abolition of compartmentalisation', and he further cites Berjak and Villiers (1972d) in that there is 'the chaotic appearance of the cytoplasm and the diffused state of acid phosphatase [becomes] evident when senescence has proceeded beyond a certain point' which we may take as cell death. It is of interest to take this as a starting point and to consider its implication in the light of available evidence to see whether such a criterion is suitable for signalling the end-point of senescence. If so, it may then be possible to interpret the various explanations as to how the senescent state arises. It may be, of course, that the loss of compartmentalization is merely a demonstration that senescence is over in the way that wrinkled skin indicates that man has aged, without stating 'when' or 'how'.

5.2 Examination of senescent and dying cells

The methods available for the study of the late stages of cell senescence and death are not always very suitable. Thus biochemical studies are useful if the system concerns single cells, but unfortunately most of the interesting problems involve tissues, or whole plants. Here, the increasing fragility of the membrane may result in misleading information through the rupturing of membranes with the rather traumatic biochemical methods available. Biochemistry is not, however, ruled out, and indeed, some useful data have been gained with this approach. However, biochemical results must relate primarily to a tissue rather than to individual cells, and hence may need to be interpreted in the light of data available from other methods. One such alternative is that of ultrastructural studies which will tell us something about the membrane changes and reorganization of the dying cells. A second alternative is quantitative cytochemistry, which has the advantage of combining biochemistry with the ultrastructure and has the bonus of allowing measurements to be made on individual cells in the tissue and so high-lighting variations which may be lost in the average values obtained by biochemical tests.

5.2.1 *Vital dye studies*

Early workers produced a number of interesting observations on cell death by the use of vital dyes, which are still valuable indicators in modern studies. The dyes have been employed in botany since at least 1886 (Baker, 1958; Aterman, 1979), and are used in low concentrations at which even toxic molecules such as Janus green B and neutral red have little effect upon cell viability for a number of hours. Three basic approaches have been employed, and two of these exploit dyes to determine only whether cells are alive or dead. In doing so, they indicate the intactness of the membrane system and the capacity of the cells to either actively transport and accumulate dye molecules in vacuoles or actively exclude the dye molecules from the cells. Thus, the additional information will concern the compartmentalization within the cell.

The first of these methods, involving uptake of the dye and accumulation in vacuoles, employs dyes including neutral red, neutral violet, cresyl blue or acridine orange (Guilliermond, 1941; Conn, 1969). The second method involving exclusion of dye by viable cells is based upon the animal cell method using trypan blue, and uses a derivative, Evans blue (Gaff and Okong'o-Ogola, 1971). In both cases dead cells are totally stained by the dyes. The third method (and second approach) uses dyes which interact chemically with a system in the living cell and yield information on respiratory activity, redox capacity and compartmentalization in the cell. Thus, Janus green B (Michaelis, 1900) normally is in an oxidized form with a blue–green colour. However, on entering the living cell, the dye is reduced to its colourless form, being re-oxidized by the terminal oxidase system in the mitochondria (Lazarow and Cooperstein, 1953). Thus, in a living cell, only the mitochondria will appear blue whilst a dead cell will stain completely. Fluorescein diacetate is another such compound, used extensively for testing the viability of isolated protoplasts. On entering the cell, the molecule is cleaved by esterase activity releasing the fluorescent fluorescein part of the molecule which can remain in the cell (it is polar) or diffuse out again. Owing to the high esterase activity, the living cell will be seen to contain a high concentration of fluorescein.

A further test of viability exploited by seed workers concerns the use of tetra-zolium salts to check whether or not the embryo in a seed is viable and hence will germinate (Lakon, 1942; Roberts, 1972; Berjak and Villiers, 1972a). This test relies upon the ability of the living cells in the embryo to reduce the colourless tetra-zolium salt to its coloured formazan. This approach has been extended to the use of reactions for glucose 6-phosphate dehydrogenase, succinate dehydrogenase and cytochrome *c* oxidase (Thomas, Esterly and Ruddat, 1977).

5.2.2 *Ultrastructural studies*

This approach is useful in that investigations can be made of cytological alterations which occur *in vivo*, especially with respect to the integrity of membranes, degree of compartmentalization, and controlled or uncontrolled degradation of the cytoplasm, for example, if autophagosomes are present, and if in great numbers. Many systems involving cell senescence and death have been examined in this way, including the fate of the root cap cells, xylogenesis, phloem development in the main axis, and evolution of leaves and flower parts as shed appendages.

(a) *Root cap* The root cap has at least two main functions, namely the geotropic response of the core cells, and lubrication for the root tip penetrating the soil, provided by the peripheral cells. The peripheral cells are sloughed off during this process, but not because of it, since sloughing occurs even in water culture or moist air (Juniper, Gilchrist and Robins, 1977). As cells move from being core cells to peripheral cells there are changes in ultrastructure initially paralleling the functional change, but ultimately resulting in degenerative changes. The cells would still seem to be alive when lost from the root cap. Such changes have been reviewed by Juniper *et al.* (1977) and include in the final stages, heavy vacuolation, loss of amyloplasts, disorganization of the endoplasmic reticulum, fracturing of the plasmodesmata, degeneration of nuclei and severe degradation of the cell walls such that at the moment of being sloughed off only the protoplast surrounded by wall vestiges remains. The root cap has formed the basis of one of the most complete ultrastructural studies on senescent and dying cells. Berjak and Villiers (1970, 1972a, 1972b, 1972c, 1972d) worked with seeds of *Zea mays* either unaged or aged in an accelerated fashion by storage at 40°C in 14% moisture for up to 34 days. Under such conditions, the mean viability period was about 18 days and viability was totally lost by 30 days. Both sets of seeds were allowed to imbibe for 12, 24 or 48 h at 25°C and prepared for electron microscopy. Whilst attention was directed generally to the radicle tip, most emphasis was placed upon changes in the root cap in which Berjak believes aging to be genetically programmed. They found three types of damaged embryos, namely: Type I with reversible damage (abnormalities seen at 12h of imbibition had disappeared from embryos by 24 or 48h); Type II with irreversible damage, and hence probably non-viable embryos; and Type III with irreversible damage, and certainly non-viable embryos. The general ultrastructural changes observed may be summarized as follows:

(a) a tendency to increasing nuclear lobing with the chromatin more susceptible
    to staining with potassium permanganate;
(b) swelling and distortion of the inner and outer membranes of the plastids,
    probably the first cytoplasmic organelles to show damage;
(c) disorientation of stacks and loss of the dictyosomes;

(d) initial distortion of the profiles of the endoplasmic reticulum, with increase in lumen, subsequent thinning, and final appearance as short distended profiles;

(e) disaggregation of polysomes (which are present in unaged embryos) 12 h after they commence imbibition; Type III embryos show only monosomes;

(f) vacuolar-like spaces in the mitochondria, which are swollen and disorganized;

(g) damaged membranes of phytolysosomes/vacuoles, and presence of auto-phagosomes.

Abu-Shakra and Ching (1967) also showed mitochondria to be damaged in seeds of *Glycine max* which had been stored for 3 years. The cristae were found to be inflated, the matrix coagulated and the outer membrane frequently missing. To this list may be added the well-known chromosome damage which can be seen on germination. The degree of spontaneous chromosome aberrations in main roots increases with increasing age of the embryo (Nawashin and Gerassimova, 1936; Roberts, Abdalla and Owen, 1967; Abdalla and Roberts, 1968).

(b) *Tapetal cells* Tapetal cells also are secretory cells which are physiologically important to pollen development and which are seen to degenerate after having served their function (Fahn, 1974). Precocious tapetal cell degeneration is equally important because of its relevance to cytoplasmic male sterility (Artschwager, 1947). Two types of tapetum occur, the secretory tapetum, the cells of which remain in the position in which they are formed, and the amoeboid tapetum in which cells penetrate between pollen mother cells and the developing pollen grains where they fuse together to form a tapetal periplasmodium. Degeneration can occur prior to microspore mitosis, e.g. *Beta vulgaris*, or shortly before anther dehiscence, e.g. *Helleborus, Tradescantia*. Many studies have been performed upon the changes occurring (Vasil, 1967), though ultrastructural studies have been more limited (Heslop-Harrison, 1963; Echlin and Godwin, 1968; Marquardt, Barth and Von Rahden, 1968; Mepham and Lane, 1969; Hoefert, 1971; Steer, 1977a, 1977b; Dickinson and Bell, 1976a, 1976b). The changes in the secretory tapetal cells of angiosperms start with the dissolution of the inner tangential walls, followed by the anticlinal walls, at which point the tapetal cells extend into the anther locule, ultimately being separated from it only by the plasmalemmae. The latter then rupture, releasing the cytoplasm into the locule. The chromatin clumps and the nuclear membrane breaks down. Subsequent degeneration of nuclear material and cytoplasmic organelles proceeds in the locule. In the case of *Pinus banksiana* (Dickinson and Bell, 1976b) the nuclear chromatin becomes highly condensed and the nucleus divides into two or more daughter nuclei much smaller than the original nucleus. Subsequently, there is a loss of dictyosomes, by which time the tapetal cells are no longer viable, according to the earlier definition. The cytoplasms then fragment.

(c) *Xylem and phloem* A detailed electron microscope study has also been made on the degenerative changes initiated during the differentiation of xylem (Esau, Cheadle and Risley, 1963; Hepler and Newcombe, 1963; Whaley, Kephart and Mollenhauer, 1964; Wooding and Northcote, 1964; Cronshaw and Bouck, 1965; Esau, Cheadle and Gill, 1966; Butler, 1967; O'Brien and Thimann, 1967; O'Brien, 1970; Srivastava and Singh, 1972). These observations may be considered in three stages: Stage I, concerning changes between typically meristematic cells and the initiation of the deposition of the secondary wall; Stage II, during which the secondary wall is formed; and Stage III, in which occurs autolysis of the protoplast and the primary cell wall. The last is a necessary step to achieve the tube system through which solutions can flow. Thus, in addition to the loss of the nucleus, there is a disappearance of the dictyosomes, plastids and microtubules, the last organelles to be lost being the mitochondria. All of these steps are concomitant with the breakdown of the vacuolar membrane (or tonoplast) and the plasmalemma, and dissolution of the non-lignified parts of the primary cell wall.

This dissolution of the tonoplast at the moment of cell lysis is parallelled in the cotyledon leaves of flax, where this situation is induced 6 h after cotyledons pretreated with paraquat are illuminated (Harris and Dodge, 1972). In general, the pattern of events is similar to that found by other workers, and is reviewed by Butler and Simon (1971). In some ways, the maturation of the phloem presents a situation closer to the problem of defining the moment of cell death than does differentiating xylem. The maturation of angiosperm sieve tube elements yields final stages of differentiation in which there is selective autophagy resulting in disorganization and/or loss of all the cellular components except the plasmalemma, mitochondria, plastids, some endoplasmic reticulum, and the slime or P-protein (if normally present).

One of the first structures to disappear is the tonoplast, and yet, unlike the situation in the xylem, root cap cells and cotyledons, the remaining cytoplasmic organelles appear to function. Similar events have been recorded for gymnosperms, except that no P-proteins have been noted (reviewed by Evert, 1977; Butler and Simon, 1971).

(d) *Green tissues* Ultrastructural changes occurring in green tissues have been well-reviewed by Butler and Simon (1971). General features emerge from studies made on detached leaves of *Elodea canadensis*, wheat, kidney bean and brussel sprout, or on leaves on plants of *Nicotiana rustica, Lemna gibba, Elodea, Xanthium* and *Perilla*, on detached or attached cotyledons of *Cucumiss sativus*, or the pigmented outer layers of pears. The results may be summarized in that, as the leaf passes from the maximally active state of Phase I to Phase II (see Section 5.1.3), there are striking changes in the chloroplasts with a re-orientation of the grana and stroma lamellae, loss of stroma density, and a decrease in the ribosome content. Grana then lose their compactness, the thylakoids swell and lipid droplets appear in the chloroplasts. Eventually the choroplast matrices and thylakoids disappear leaving the envelope membrane which is last to go (Woolhouse, 1978). The general sequence varies, but the pattern is basically that changes occur in the endoplasmic reticulum together with a loss of ribosomes. Often the free monosomes disappear before the polysomes, the last to be lost being the membrane-bound ribosomes.

Concomitant with ribosome loss are the changes in the chloroplasts which may precede those of the mitochondria, the nucleus remaining intact until a very late stage. In some cases, a number of changes occur before the tonoplast appears to fragment, though this situation is not typical, and the last structure to break down is often the plasmalemma. Interestingly mitochrondrial and chloroplast debris often remain at the end of this sequence of events.

(e) *Storage tissue* Storage tissues have been investigated as well, and though little ultrastructural information is available on cell death in the starchy endosperm of cereals, changes in cotyledons upon germination have been considered in *Phaseolus vulgaris* (Opik, 1965, 1966), *Pisum sativum* (Bain and Mercer, 1966), peanuts (Cherry, 1963) and *Vicia faba* (Briarty, Coult and Boulter, 1970). Removal of the stored reserves is followed by a degeneration of the cytoplasm of the storage cells with a loss of free ribosomes and endoplasmic reticulum, and a breakdown of the tonoplast. Any chloroplasts formed from pro-plastids and amyloplasts degenerate along with the mitochondria. The rate of degeneration varies according to the cell types examined.

(f) *Air spaces* In many plants, spaces form through the breakdown of cells, either to create large air spaces in tissues or to give the patterns of fenestration seen in leaves, e.g. *Monstera*. The formation of air spaces was studied in tomato roots (Morriset, 1968) and three stages were reported. In the first, all cytoplasmic organelles were present including many ribosomes; but incipient damage was identified in the second stage by a total loss of ribosomes, though other organelles seemed unimpaired, and the third stage showed cells largely devoid of organelles, with mitochondria having a dense matrix and being swollen. The subsequent image was of a dead cell in which the plasmalemma had broken down.

Degeneration of pollen tubes has also been studied (Larson, 1965; Jensen and Fisher, 1970), though curiously in this tissue the Golgi apparatus appears to remain stable as the other organelles degenerate.

5.3 Biochemical and cytochemical consideration

In assessing the meaning of biochemical and cytochemical studies on cell senescence and death in terms of the morphological images, it is important to remember the two basic patterns emerging from the ultrastructural studies, namely, either the destruction of cell organelles in conjunction with the loss of integrity of the tonoplast, or the loss of integrity of the tonoplast and continued functioning of selected organelles. Most of the studies fall into two groups, those concerned with acid hydrolases and others linked to lipid, nucleic acid and protein turnover. It is of interest to commence with the latter and more varied group, which includes studies related to respiratory activity, plastid function and membrane repair. Protein synthesis is of interest because it is well established that protein levels are maintained through an equilibrium between rates of protein synthesis and protein degradation. Thus, an increased rate of protein synthesis accompanied by an equally increased rate of protein degradation will maintain a *status quo* in the cell whereas an increased rate of protein synthesis but a stationary rate of protein degradation will yield an increase in total proteins. Most studies made on plants and virtually all studies on senescent and dying plant cells concern protein synthesis and exclude protein degradation so that it is difficult to draw significant conclusions from such studies in terms of protein turnover. Protein-turnover rates are difficult to measure when using isotopically labelled precursors (reviewed by Huffaker and Peterson, 1974), although a way to overcome some of the problems was devised by Trewavas (1972), who estimated radioactivities of an exogenously supplied amino acid in a distinct amino acid pool, aminoacyl-transfer RNA, i.e. in the immediate protein precursor pool. Similarly, an improvement was offered by Humphrey and Davies (1975) using ^3H_2O and a ^3H exchange method, but like Trewavas, they worked with *Lemna* and it is difficult to envisage these methods being applied directly to many non-aquatic plants. The double-label method (Arias, Doyle and Schimke, 1969) does not yet seem to have been tried with plant tissues. Thus, most approaches have measured protein synthesis, and in the case of the third pair of leaves of *Perilla frutescens*, appreciable protein synthesis occurs up to the moment of abscission (Woolhouse, 1967). However, the above study also showed that there was a reduction in the protein (and chlorophyll) content of these leaves from the time of completion of leaf expansion to the time of abscission. Whilst the actual rate of protein degradation could not be measured, it would appear that the rate of protein synthesis was lower than that of degradation and hence there was a net loss of protein from the organs.

These biochemical observations relate to the observed ultrastructural changes when the leaf passes through the phase of leaf maturity. The reorientation of the grana and stroma lamellae of the chloroplasts is accompanied by a decrease in the number of chloroplast ribosomes, events which parallel the loss of capacity of the chloroplasts for synthesis of proteins (Woolhouse, 1974).

A similar type of study has been performed upon the ephemeral flowers of *Ipomoea tricolor* (Matile and Winkenbach, 1971). Here the demonstration was of an increase in protein content and rate of synthesis until anthesis, when a drop occurred in total corolla protein, lasting until abscission. It is of interest to note that levels of protease in the corolla did not change during this period, and it is concluded that proteolysis exceeded protein synthesis during senescence (Matile, 1975). Both of these studies indicate the same pattern of events, though interpretation in terms of individual cell death is difficult, since the biochemical determinations were made on whole tissues and each component cell could have been in a slightly different physiological state from that of its immediate neighbours. In contrast, the studies on aging root cap cells (Berjak and Villiers, 1972a, 1972b, 1972c) have attempted to present such evidence at a cellular level. Analysis of radioautographs from unaged and aged seeds of *Zea mays* labelled with [^{3}H] thymidine has indicated an incorporation into DNA in only the dividing cells of the root cap of unaged seeds, whereas aged seeds show increasing levels of incorporation during a 4 h feeding period in most cells, together with an incorporation into nuclei of differentiated and senescent cells. It is not clear if this incorporation is a net synthesis of DNA or simply an enhanced repair situation. A study of nuclear DNA in similar cells probed by terminal deoxynucleotidyltransferase (Barlow, 1976) indicated that strand breaks did not accumulate as cells passed from the proliferative to a degenerative state. A similar response to [^{3}H] thymidine labelling in aging collenchyma cells was observed in *Lycopersicon esculentum* (Hurst, Gahan and Snellen, 1974), which did not appear to be explainable in terms of either replication or repair (Gahan, 1976). It was thought to represent metabolic DNA (Pelc, 1972) in which extra copies of a particular gene are produced at the moment when there is pressure to produce a large amount of a given mRNA in a short time. Extra gene copies would permit this, and also allow protection of the master copy, which would not be used for the manufacture of mRNA. Radioautographic analysis of [^{3}H] leucine and [^{3}H] uridine incorporation into the root cap cells showed a sharp increase in incorporation of both precursors in all zones of the root cap of the aged seeds, but most noticeably in the mature region. In the case of the senescing cells, there was an increase in [^{3}H] uridine incorporation, though not as great as in the mature cells, and a large increase in the incorporation of [^{3}H] leucine. In some material, it was noted that acid phosphatase activity was confined within lysosome-like structures but in the outermost, highly disorganized, root cap cells, the enzyme was present throughout the cytoplasm. Similar studies have also been made concerning nucleic acids in both *Perilla* and *Ipomoea* systems. Thus Woolhouse (1978) showed that chloroplast RNA polymerase becomes inactive at the completion of leaf expansion and hence the synthesis of chloroplast RNA stops. In consequence, as RNA structures are worn out and lost, they cannot be replaced, giving a concomitant decline in the overall RNA content of the chloroplasts, with its obvious impact on protein synthesis.

Ipomoea corollas also show a decline in DNA and RNA content (Matile and Winkenbach, 1971) during the period between anthesis and abscission, but in contrast with the situation with protein loss, there is a striking increase in the levels of ribonuclease and deoxyribonuclease. In the case of ribonuclease, this increase has been shown to be synthesized *de novo* in the fading corolla (Baumgartner, Kende and Matile, 1975). Such increased ribonuclease activity is also reported for senescing leaves (Dove, 1973) and Sacher and Davies (1974), using ^{2}H$_2$O, were able to show synthesis of this enzyme *de novo* in sections from *Rhoea*. Such studies invoke hydrolases in the reduction of the amount of RNA present or indicate the loss of the RNA-forming capacity through loss of enzymes.

However, in the case of the senescence and death of xylem and phloem, the situation is rather more drastic. During phloem sieve-element maturation in angiosperms the nucleus may disappear completely by chromatolysis, or remnants of the chromatin and nuclear envelope may persist for a long time, depending upon the species (reviewed in Evert, 1977). Studies by Zee and Chambers (1969) on primary roots of *Pisum sativum* showed that on feeding with [^{3}H] uridine, young cells incorporated the precursor into RNA in the nucleus, nucleolus and cytoplasm.

As maturation progressed, labelling failed to occur first in the nucleolus, then the nucleus and finally the cytoplasm, indicating a switch-off of RNA synthesis. All ribosomes have disappeared from the sieve-tube member.

In gymnosperms, nuclear degeneration is by pyknosis. Frequently the nuclei persist as electron-dense masses which fail to incorporate [^{3}H] uridine, indicating a similar situation for RNA synthesis to that found in angiosperms (Evert, 1977). Developing xylem elements also lose their nuclei on maturation. These observations lead to the question of the ability of these cells to maintain protein synthesis in the absence of genes. Thus, when the nucleus is lost, either ribosomes, long-lived mRNA and tRNAs are present in high enough concentration, or the cells cannot survive for a long period after this irreversible loss. The latter would seem more likely for the xylem than for the phloem, in which companion cells are in close contact with the sieve elements and could even provide a number of their basic needs (Esau, 1974), though clearly not all. For example, no ribosomes are found in mature sieve-elements, so that protein synthesis could not occur *in situ*. The question arises as to whether or not in these cells, the loss of the nucleus is the basic signal for cell death, because, with all genes gone, only a limited series of events is feasible. It has already been shown that a decline in the rate of protein synthesis could lead to a decrease in the total protein levels if degradation continued at the same rate — as seems possible in *Ipomoea*. Thus an impairment in protein synthesis would eventually lead to a serious breakdown in the repair of membranes and in the replacement of certain enzymes of short half-life, such as glycerophosphate dehydrogenase (74.4 h — Kuehl and Sumsion, 1970) and acetyl-CoA carboxylase (59 h — Nakanishi and Numa, 1970). The situation in other systems is not quite so clear-cut, though observations upon the starchy endosperm cells of cereals might suggest a pattern similar to that for *Perilla* leaves and *Ipomoea* corollas. Thus studies on waxy single cross WF9 X Bear 38 maize (Rabson, Mans and Novelli, 1961) involved capacities for incorporation of amino acids by an unwashed particle system containing the polysome-rich fraction from maize kernel at different stages of development. Maximum incorporation occurred at 25 days after pollination, followed by a decline to almost zero by 50 days. The initial increase probably represented an increased synthesis of enzymes, since about 80 per cent of protein-synthesizing capacity will reside in the endosperm by 25–30 days after pollination. By 50 days, there will be so much starch and protein stored in the endosperm cells that the cells might have been damaged irreversibly and hence no further protein synthesis will occur (Bewley and Black, 1978). Work on an inbred line WF9 (Ingle, Beitz and Hageman, 1965) showed protein synthesis in the endosperm to be biphasic, the first peak being reached at about 25 days followed by a lag and a second burst at day 37. At the time of the first peak, the DNA content of the endosperm cells starts to fall, as does the RNA content which reaches about a third of its 27-day value by day 45, and drops to almost zero in the mature grain. The DNA content falls to a third of its original level in the mature grain. The RNA content increases when non-storage proteins are being synthesized and falls when storage proteins are being made. It is indicated that endosperm nucleases increase at about the time of the RNA and DNA decrease. It appears that the protein synthesis occurring during the latter stages of maturation employs RNA made during the early stages and that after being used it is degraded. The data also indicate the possibility that by about day 25, the repair systems of the endosperm cells no longer function adequately and the cells degenerate, having only the capacity to synthesize storage material.

5.4 Possible interpretations of the biochemical cytochemical and ultrastructural studies

5.4.1 *Respiratory control*

The biochemical and ultrastructural changes lead to the possibility of severely impaired protein synthesis in cell senescence. This impairment could in turn explain how membranes might degenerate, i.e. through an inadequate repair mechanism. Leaky or incompetent membranes could lead to:

(a) loss of compartmentalization;
(b) loss of capacity for reversible plasmolysis;
(c) increase of mitochondrial activity and hence respiratory activity or uncoupling; and
(d) dispersal of soluble enzymes such as hydrolases throughout the cell.

A sharp increase in respiratory activity is known to occur in animal cells just prior to cell death. Coupled with this increase is a leak of glutamate dehydrogenase from mitochondria and a decrease in mitochondrial malate dehydrogenase when cells are damaged. Again neither of these markers appears to have been used in plant studies. The measurement of respiratory activity does not seem to have been exploited to any great extent as a marker of cell damage and death. However, the modified mitochondria seen in dark-grown axes of old germinating soya-bean seeds showed an increased oxygen consumption of 110–140% when compared with mitochondria from new seeds, and this increase seemed to result from an uncoupling of oxidative phosphorylation in the old-seed mitochondria, as evidenced by a fall of 30–60% in P/O ratio as compared with the new-seed mitochondria (Abu-Shakra and Ching, 1967). As an extension of this argument, Abdul-Baki (1969) showed that the utilization, as opposed to uptake, of externally supplied glucose allowed a very early detection of seed deterioration, much before more classic markers. This result implied changed mitochondrial activity as an early step, perhaps explaining the high succinate dehydrogenase activity observed cytochemically in differentiating xylem tracheids and vessels in *Vicia faba* roots (Gahan, unpublished work). Delouche (1969) actually lists the degradation of cellular membranes and subsequent loss of the control of permeability and impairment of energy-yielding mechanisms as very early stages in the sequence of events during seed ageing. It is difficult to be sure if these are typical of all senescence situations, and, indeed, Eilam (1965) has demonstrated that permeability changes are independent of the respiratory system of some tissues. The impairment of energy-yielding mechanisms may be true of the climacteric rise in respiration in fruit-ripening, which might also be interpreted as a sharp rise in respiratory activity immediately prior to cell death (Biale, 1964). However, as discussed by Woolhouse (1978), the climacteric rise is noticeably absent from some fruits such as grapes, and so this latter suggestion may not be correct, especially as the mitochondria isolated from fruit at this stage of ripening appear to be fully coupled, i.e. electron transport is fully coupled to the production of ATP. Against this idea is the fact that the pathway is resistant to cyanide (Woolhouse, 1978), which could imply that the electrons are passing via an alternative cyanide-resistant pathway, as has been demonstrated for a number of other tissues, and in essence yielding a situation similar to that of uncoupling (Palmer, 1976). In this case, cyanide-resistance may be indicative of a changed metabolism with a smaller need for ATP and a possible deterioration of the cells.

5.4.2 *Dispersal of hydrolases*

The dispersal of hydrolases throughout the cell raises the important point as to the significance of this event in terms of cell death, a topic which has been the subject of debate since de Duve (1963) launched the concept of the 'suicide' particle. The presence of a lysosome-like structure in a broad range of phanerogams and crypto-gams has now been established, leading to the concept of a lysosomal vacuolar system in plant cells (Berjak, 1972; Gahan, 1973; Matile, 1975; Pitt, 1975). In particular, the large central vacuole present in many cells appears to behave as a large lysosome, since it contains many hydrolases (in Matile, 1978; Nishimura and Beevers, 1978; Boller and Kende, 1979) and materials, including cytoplasmic structures, would seem to pass to it for digestion (Coulomb, 1972).

As early as 1966, Gahan and Maple found that loss of the protoplast from the dying xylem cell during element differentiation in *Vicia faba* was accompanied by an increase in, and an apparent release of, acid β-glycerophosphatase activity. Similar results were also obtained *in vivo* with naphthol AS-BI phosphate and *p*-nitrophenyl phosphate as substrates (Beneš and Opatrná, 1964; Gahan, 1978). These findings have been confirmed on a number of occasions in *Pisum sativum* (Sexton and Sutcliffe, 1969; Stewart and Pitt, 1977), *Lepidium sativum* (Berjak and Lawton, 1973), *Phaseolus lunatus* (Thomas *et al.*, 1977), *Phaseolus aureus* (Stewart and Pitt, 1977), *Zea mays, Hordeum vulgare, Phaseolus vulgaris, Helianthus annuus* (Shaykh and Roberts, 1974) and *P. vulgaris* (Charvat and Esau, 1975), and a variety of bryophytes (Hébant, 1973). However, it has also been considered that such cytochemical images might also be interpreted as being due to an increased production of enzyme sites throughout the cell (Matile, 1969; Stewart and Pitt, 1977), raising also the question as to whether or not a net synthesis of hydrolase activity is necessary in such a situation. Whilst some workers have indicated that a net synthesis of some hydrolases can be measured during senescence and death in some tissues (see in Matile, 1975; Pitt, 1975), it is also clear that the biological half-life of some hydrolase molecules is sufficiently long so as not to require a further synthesis (Dawson and Gahan, 1979).

How may these differences be resolved? The increase in hydrolase activity has been measured biochemically, whilst the distribution of hydrolase activity tends to be observed cytochemically. The range of reliable cytochemical reactions for hydrolases is restricted, and, in consequence a majority of workers have limited their observations to the behaviour of acid phosphatases with β-glycerophosphate, p-nitrophenyl phosphate or naphthol AS-BI (AS-MX or AS-TR) phosphate as substrates (see Chapter 13).

It has been shown that p-nitrophenyl phosphate will indicate the enzymes demonstrated with β-glycerophosphate or naphthol AS-BI phosphate as substrates. Moreover, it acts as a substrate for glucose-6-phosphatase and K-activated acyl phosphatase (Gahan, Dawson and Fielding, 1978), nucleotide pyrophosphatase (Kole, Sierakowska and Shugar, 1976; Sierakowska, Gahan and Dawson, 1978; Gahan, Sierakowska and Dawson 1979), 5'(3')-ribonucleotide phosphohydrolase (Polya and Ashton, 1973) and nucleotide phosphotransferase (Brawerman and Chargaff, 1955). Most of these enzymes are either partly or totally membrane-bound, and hence it is difficult to envisage their being released from lysosomes to diffuse throughout and digest the cytoplasm. It is possible that fragments of membranes together with their phosphatases are distributed throughout the cytoplasm. Alternatively, there may be sufficient membrane alteration at an early stage of damage which results in an activation of all of the membrane-bound phosphatases. Equally, there could be the formation of many autophagosomes associated with the breakdown of the cytoplasm in some circumstances (Berjak and Lawton, 1972; Srivastava and Singh, 1972). Alteration of membranes may be partly involved as shown by the increased speed of response to the cytochemical test (Gahan and Maple, 1966). Clearly, more information and methods are required before we can try to draw too many conclusions concerning the involvement of lysosomes in cell death. The enzymes studied may be good markers of early membrane changes, but they may not always indicate a general release of acid phosphatases.

Other hydrolases and enzymes have been studied, including esterases (Thomas *et al.*, 1977; McLean, 1969; Gahan, 1972), leucine aminopeptidase, β-glucosidase, β-galactosidase, glucosaminidase (Thomas *et al.*, 1977) and respiratory and pentose phosphate pathway enzymes (Beneš, Lojda and Hořavka, 1961; Gahan, unpublished data; Thomas *et al.*, 1977). That most of these enzymes show increased activity during cell senescence might be interpreted as indicating that membrane changes occur during the senescence process. These changes may occur prior to the breakdown of the tonoplast, i.e. compartmentalization is lost the moment that membrane changes occur, and hence the cell may be dying and irreversibly damaged before there is a release of hydrolases. In consequence, it seems more likely that the frequently observed leak of hydrolases is a *post-mortem* event and is not linked to the process as a primary event (Barton, 1966). Even when hydrolases are released, it is not clear if the environment is at the correct pH for them to work efficiently, although it might be argued that the vacuolar sap may be present in sufficient volume to provide the necessary environment. However, in some instances, such as xylem-element differentiation, the flow of fluid through the cells may be sufficient to remove the hydrolases before they can have time to act. The picture of the involvement of lysosomes in cell death is far from clear, and may well prove to be misleading.

5.5 Mechanisms of cell senescence and death revisited

If the increase and changed distribution of hydrolases during cell senescence and death are secondary rather than primary events, what alternatives are there? It is perhaps, now worth reconsidering the three hypotheses concerning cell senescence proposed earlier in this chapter, in the light of the type of evidence presented so far.

5.5.1 *The somatic mutation theory*

Berjak and Villiers (1972c) have cited their observations in aged seed, together with those of Abdalla and Roberts (1968), as being in possible support of the somatic mutation theory (Curtis, 1966). Whilst Curtis has provided some interesting experimental evidence indicating such mutations as a major factor in aging, there are a number of aspects which may be open to question. Thus, for example, in the case of his radiation studies on mice he was able to produce instantly in young mice with 60–70R of neutron equivalent the entire lifetime dose of accumulated mutations. The dosage is inconsistent with the theory, since such a small radiation exposure does not decrease life expectancy (Strehler, 1977). As far as plants are concerned, Woolhouse (1974) argues against the somatic mutation theory by pointing out that senescence of monocarpic plants can be delayed by the removal of young flowers as soon as they form; or capitation of a tobacco plant, leaving behind just the yellowed basal leaves, will result in their turning green again with a concomitant synthesis of new proteins and nucleic acids; and, in some cases, rooted cuttings taken from old trees will develop into vigorous young trees. It might be argued also that the observed chromosome aberrations in aged seeds are not specifically in accord with the somatic mutation theory, since whilst the cells containing such chromosomes are postmitotic, they are kept under conditions differing considerably from postmitotic cell populations in mammals. In addition, aqueous extracts from either old or new seeds are themselves capable of inducing the high levels of chromosome damage in new seeds (Mota, 1952). The theory does not currently find great favour among zoologists (Strehler, 1977).

5.5.2 *Error catastrophe theory*

Since its inception, this theory has provided a strong debate which has yet to be resolved. There are a number of interesting pieces of experimental plant evidence which can be taken as supporting the theory, including the early work on fungi (Holliday, 1969). Some of the evidence presented in this chapter could be interpreted as being in support of this hypothesis. Thus, in the studies of root cap cells in aged seeds, the rates of synthesis of RNA and proteins are higher in embryos which have been deliberately aged up to 14 weeks as compared with normal unaged embryos (Berjak and Villiers, 1972a). On ageing the embryos for longer periods, there is a decline in the rates of synthesis of both RNA and protein, which is very marked at 18 days of aging. This situation could be explained in terms of an accumulation of error proteins requiring an increased rate of protein synthesis in order to produce adequate levels of 'good' protein. In doing so, more faulty protein is produced, which might include enzyme molecules related to RNA and/or protein synthesis, so leading to the observed decline in both RNA and protein synthesis. The theory could also be invoked to explain the radioautographic observations of Gahan and Hurst (1976, 1977), in which, on aging, the root apical meristems of *Zea mays* were shown to have a decreased nuclear labelling index ($[^3H]$ thymidine), a lengthening of the mean cell cycle time (due to the appearance of G_1 and a lengthening of G_2), and a reduced capacity for incorporation into mitochondrial and plastid DNA. Error proteins could lead to a situation found in aging fibroblasts by Petes, Farber, Tarrant and Holliday (1974) and by Linn, Kairis and Holliday (1976) in which there appears to be faulty DNA polymerase molecules, thus leading to reduced rate of $[^3H]$ thymidine incorporation.

Again, there are many arguments against the error catastrophe theory, including the fact that the incorporation of abnormal amino acids into protein does not necessarily lead to a shortening of the lifespan of the organisms concerned (Maynard-Smith, 1973, Popp, Bailiff, Hirsch and Conrad, 1976; Strehler, 1977).

Many of the situations of cell senescence and death reviewed above relate to predictable events at different stages in the lives of plants. One repeated observation is the loss of control over protein synthesis leading to a reduction in protein content of the cells or tissues under study. This observation has led to an interpretation in support of a programmed cell senescence (Butler and Simon, 1971). Berjak and Lawton (1972) went further in suggesting the possibility of groups of genes existing to control the senescence processes. Such a case has been argued for Werner's Syndrome in man (Fulder, 1977), where it is possible to demonstrate an autosomal recessive mode of inheritance of the disorder, from which it is suggested that a single or a closely linked group of genes might be responsible. However, in the case of plants there is no such genetic evidence other than a repeatable pattern of events, and this may not be so different from the repeating patterns seen in, e.g. embryogenesis and differentiation, a genetically determined series of events only in terms of those genes allowed to express through repression/depression by virtue of their being present in cells which are in defined tissue positions and hence in preprogrammed zones. Consequently, the fact that a group of genes could be involved does not mean *de facto* that they are genes for senescence rather than merely groups of normally involved genes switched into patterns of activity which will be detrimental to the cell. This may be programmed by the environment of a given cell. In the case of organs such as leaves, this switching may be dictated, at least in part, by the environment of the whole plant. Thus, in a sense, there is programming switched on through the interplay of the cell and its environment, and subsequently of nucleus and cytoplasm, but death is not programmed in the sense of a genetic finality.

Unfortunately, too little is known about the detailed changes in senescence, structurally, hormonally or metabolically, for any precise discussion of these events (Roberts, 1976; Woolhouse, 1978), and hence it is premature to consider if senscence is generally programmed in the sense of the theory, i.e. as a continuation of differentiation.

5.5.3 *Pathological and mechanical damage*

Cell death induced by either of these events has not been studied as extensively as might have been expected. Frequently, tests have been employed merely to determine if the pathological lesion has or has not damaged or killed the host cells. Ultrastructural studies of host cells infected by either viruses or fungi seem to indicate a series of degenerative changes in the host cells which parallel those observed in senescing non-infected cells (reviewed in Butler and Simon, 1971). Such studies have not been made in any great detail on mechanically damaged cells.

The pattern of events leading to cell death by infection due to the effects of toxins or enzymes should not necessarily be expected to be similar in all details to those events outlined above for senescence. Much work on the mechanism of the induction of cell death has concentrated on differences in response between species which are susceptible or resistant to a given pathogen. In susceptible strains, emphasis appears to have been placed on the interaction of the toxin with the plasmalemma and not with other membrane or synthetic systems. In fact, the resistance of a host plant to infection may depend upon the extreme susceptibility of some of its individual invaded cells, which die as the result of the hypersensitive reaction, so leading to the blocking of further invasion of the host plant. From the information available on biochemical and cytochemical studies (Pitt, 1975), there is no reason to believe that the conclusions offered above for events such as lysosomal involvement in cell death for normally senescing cells should be changed for infection.

5.5.4 *The timing and markers of the onset of cell death*

Matile's (1975) suggestion of the loss of compartmentalization was considered initially as an indicator of the moment of cell death. In view of the evidence presented, it would seem that a series of events such as the loss of control of synthesis of proteins and lipids could lead to the loss of compartmentalization. Hence loss of compartmentalization can signal that the cell death has occurred, and earlier events must be considered if the transition between senescence and death is to be observed. Some useful markers of these situations may be viewed in two ways.

(a) *Loss of compartmentalization* Loss of compartmentalization can be seen in a variety of situations already discussed, and including ultrastructural studies of membrane integrity, biochemical studies of membrane fragility and mobility of enzymes between subcellular fractions, cytochemical observations with vital dyes, changed localization of enzymes, increased availability to staining of membrane phospholipids, and the loss of the capacity for reversible plasmolysis.

(b) *Events leading to loss of compartmentalization* Events which precede loss of compartmentalization will include ultrastructural changes such as those related to the nucleus, nucleolus, ribosomes, endoplasmic reticulum and Golgi apparatus. Additional biochemical and cytochemical markers might occur through a sharp increase in mitochondrial (respiratory) activity which may possibly become cyanide-resistant, through increased activity of membrane-bound enzymes such as acid phosphatases when measured with naphthol AS-BI phosphate or *p*-nitrophenyl phosphate as substrates, through an increased rate of protein synthesis together with or without a similar increase in protein-degradation rates, and followed by a decline in protein-synthetic capacity, and through a decrease in the rates of lipid synthesis. A specific sequence, if any, remains to be elucidated.

Acknowledgements

I wish to thank the Science Research Council for partial financial assistance, and Dr J. B. Heale for critical reading of the manuscript.

References

Abdalla, F.H. and Roberts, E.H. (1968), Effects of temperature and oxygen on the induction of chromosome damage in seeds of barley, broad beans and peas during storage. *Ann. Bot.*, **32**, 119–136.

Abdul-Baki, A.A. (1969), Relation of glucose metabolism to germinability and vigour in barley and wheat seeds. *Comp. Sci.*, **9**, 732–737.

Abu-Shakra, S.S. and Ching, T.M. (1967), Mitochondrial activity in germinating new and old soybean seed. *Crop Sci.*, **7**, 115–117.

Arias, I.M., Doyle, D. and Schimke, R.T. (1969), Studies on the synthesis and degradation of proteins of the endoplasmic reticulum of rat liver. *J. Biol. Chem.*, **244**, 3303–3315.

Artschwager, E.F. (1947), Pollen degeneration in male-sterile sugar beets, with special reference to the tapetal plasmodium. *J. Agric. Res.*, **75**, 191–197.

Aterman, K. (1979), The development of the concept of lysosomes. A historical survey, with particular reference to the liver. *Histochem. J.*, **11**, 503–541.

Bain, J.M. and Mercer, F.V. (1966), Subcellular organization of the cotyledons in germinating seeds and seedlings of *Pisum sativum* L. *Austr. J. Biol. Sci.*, **19**, 69–84.

Baker, J.R. (1958), *Principles of Biological Microtechnique*. Methuen, London.

Barlow, P.W. (1976), The integrity and organization of nuclear DNA in cells of the root cap of *Zea mays* probed by terminal deoxynucleotidyl transferase and microdensitometry. *Z. Pflanzenphysiol.*, **80**, 271–278.

Barton, R. (1966), Fine structure of mesophyll cells in senescing leaves of *Phaseolus*. *Planta*, **71**, 314–325.

Baumgartner, B., Kende, H. and Matile, Ph. (1975), RNAase in senescing Japanese morning glory. Purification and demonstration of *de novo* synthesis. *Plant Physiol.*, **55**, 734–737.

Beneš, K., Lojda, Z. and Hořavka, B. (1961), A contribution to the histochemical demonstration of some hydrolytic and oxidative enzymes in plants. *Histochemie*, **2**, 313–321.

Beneš, K. and Opatrná, J. (1964), Localization of acid phosphatase in the differentiating root meristem. *Biol. Plant. (Praha)*, **6**, 8–16.

Berjak, P. (1972), Lysosomal compartmentalization: Ultrastructural aspects of the origin, development and function of vacuoles in root cells of *Lepidium sativum*. *Ann. Bot.*, **36**, 73–81.

Berjak, P. and Lawton, J.R. (1973), Prostelar autolysis: a further example of a programmed senescence. *New Phytol.*, **72**, 625–637.

Berjak, P. and Villiers, T.A. (1970), Ageing in plant embryos. I. The establishment of the sequence of development and senescence in the root cap during germination. *New Phytol.*, **69**, 929–938.

Berjak, P. and Villiers, T. A. (1972a), Ageing in plant embryos. II. Age induced damage and its repair during early germination. *New Phytol.*, 71, 135–144.

Berjak, P. and Villiers, T. A. (1972b), Ageing in plant embryos. III. Acceleration of senescence following artificial ageing treatment. *New Phytol.*, 71, 513–518.

Berjak, P. and Villiers, T. A. (1972c), Ageing in plant embryos. IV. Loss of regulatory control in aged embryos. *New Phytol.*, 71, 1069–1074.

Berjak, P. and Villiers, T. A. (1972d), Ageing in plant embryos. V. Lysis of the cytoplasm in non-viable embryos. *New Phytol.*, 71, 1075–1079.

Bewley, J. D. and Black, M. (1978), *Physiology & Biochemistry of Seeds. 1. Development, germination and growth.* Springer-Verlag, Berlin.

Biale, J. B. (1964), Growth, maturation and senescence in fruits. *Science*, 146, 880–888.

Boller, T. and Kende, H. (1979), Hydrolytic enzymes in the central vacuole of plant cells. *Plant Physiol.*, 63, 1123–1132.

Bower, F. O. (1935), *Primitive land plants.* Macmillan and Co. Ltd., London.

Brawerman, G. and Chargaff, E. (1955), Distribution and significance of the nucleoside phosphotransferases. *Biochim. Biophys. Acta*, 16, 524–532.

Briarty, L. G., Coult, D. A. and Boulter, D. (1970), Protein bodies of germinating seeds of *Vicia faba. J. Exp. Bot.*, 21, 513–524.

Butler, R. D. (1967), The fine structure of senescing cotyledons of cucumber. *J. Exp. Bot.*, 18, 535–543.

Butler, R. D. and Simon, E. W. (1971), Ultrastructural aspects of senescence in plants. *Adv. Gerontol. Res.*, 3, 73–129.

Charvat, I. and Esau, K. (1975), An ultrastructural study of acid phosphatase localization in *Phaseolus vulgaris* xylem by the use of an azo-dye method. *J. Cell. Sci.*, 19, 543–561.

Cherry, J. H. (1963), Nucleic acid, mitochondria, and enzyme changes in cotyledons of peanut seeds during germination. *Plant Physiol.*, 38, 440–446.

Conn, H. J. (1969), *Biological Stains*, 8th edn. (ed. R. D. Lillie), Williams and Wilkins Co., Baltimore.

Coulomb, C. (1972), Processus lytiques dans les vacuoles des cellules radiculaires méristématiques de la Scorscinère (*Scorzonera hispanica*). Ph. D. Thesis, University of Marseilles.

Cronshaw, J. and Bouck, G. B. (1965), The fine structure of differentiating xylem elements. *J. Cell Biol.*, 24, 415–431.

Curtis, H. J. (1966), *Biological Mechanisms of Ageing.* Ch. C. Thomas, Springfield, Illinois.

Dawson, A. L. and Gahan, P. B. (1979), Longevity of hydrolase molecules during primary xylem differentiation. *Ann. Bot.*, 43, 251–254.

de Duve, C. (1963), In *Lysosomes* (eds. A. V. S. de Reuck, and M. P. Cameron), Churchill, London, pp. 1–35.

Delouche, J. C. (1969), Planting seed quality. Journal paper No. 1721. Mississippi Agric. Exp. Stn., Mississippi State Univ. Proc. Beltwide Cotton Production Mechanization Conf., New Orleans, pp. 16–18.

Dickinson, H. G. and Bell, P. R. (1976a), Development of the tapetum in *Pinus banksiana* preceding sporogenesis. *Ann. Bot.*, 40, 103–113.

Dickinson, H. G. and Bell, P. R. (1976b), The changes in the tapetum of *Pinus banksiana* accompanying formation and maturation of the pollen. *Ann. Bot.*, 40, 1101–1109.

Dove, L. D. (1973), Ribonucleases in vascular plants: cellular distribution and changes during development. *Phytochemistry*, 12, 2561–2570.

Echlin, P. and Godwin, H. (1968), The ultrastructure and ontogeny of pollen in *Helleborus foetidus. J. Cell Sci.,* 3, 161–174.

Eilam, Y. (1965), Permeability changes in senescing tissue. *J. Exp. Bot.,* 16, 614–627.

Esau, K. (1974), *Plant Anatomy,* 2nd edn., John Wiley and Sons Ltd., New York.

Esau, K., Cheadle, V. I. and Gill, R. H. (1966), Cytology of differentiating tracheary elements. I. Organelles and membrane systems. *Am. J. Bot.,* 53, 756–764.

Esau, K., Cheadle, V. I. and Risley, E. B. (1963), A view of ultrastructure of *cucurbita* xylem. *Bot Gaz.,* 124, 311–316.

Evert, R. F. (1977), Phloem structure and histochemistry. *Ann. Rev. Plant Physiol.,* 28, 199–222.

Fahn, A. (1974), *Plant Anatomy,* 2nd edn., Pergamon Press, Oxford.

Finch, C. E. and Hayflick, L. (1977), *Handbook of the Biology of Aging.* Van Nostrand Reinhold Co., New York.

Fritsch, F. E. (1935), *The structure and reproduction of the algae.* Vols. I and II, Cambridge University Press, Cambridge.

Fulder, S. (1977), A pathological race through life. *New Sci.,* 21, 122–3.

Gaff, D. F. and Okong'o-Ogola, O. (1971), The use of non-permeating pigments for testing the survival of cells. *J. Exp. Bot.,* 22, 756–758.

Gahan, P. B. (1972), The role of acid hydrolases in the differentiation of plant cells. In *Cell Differentiation* (eds. R. Harris, P. Allin and D. Viza), Munksgaard, Copenhagen, pp. 223–224.

Gahan, P. B. (1973), Plant Lysosomes. In *Lysosomes in Biology and Pathology,* vol. 3 (ed. J. T. Dingle), North-Holland, Amsterdam, London, pp. 69–88.

Gahan, P. B. (1976), DNA turnover in nuclei of cells from higher plants. *Riv. Istochim. Norm. Pat.,* 20, 108–109.

Gahan, P. B. (1978), A re-interpretation of the cytochemical evidence for acid phosphatase activity during cell death in xylem differentiation. *Ann. Bot.,* 42, 755–757.

Gahan, P. B., Dawson, A. L. and Fielding, J. (1978), Paranitrophenyl phosphate as a substrate for some acid phosphatases in roots of *Vicia faba. Ann. Bot.,* 42, 1413–1420.

Gahan, P. B. and Hurst, P. R. (1976), Effects of ageing on the cell cycle in *Zea mays. Ann. Bot.,* 40, 887–890.

Gahan, P. B. and Hurst, P. R. (1977), Reduced incorporation of ^{3}H-thymidine into cytoplasmic DNA in ageing cell populations *in vivo. Exp. Gerontol.,* 12, 13–15.

Gahan, P. B. and Maple, A. J. (1966), The behaviour of lysosome-like particles during cell differentiation. *J. Exp. Bot.,* 17, 151–155.

Gahan, P. B., Sierakowska, H. and Dawson, A. L. (1979), Nucleotide pyrophosphatase activity in dry and germinated seeds of *Triticum* and its relationship to general acid phosphatase. *Planta,* 145, 159–166.

Guilliermond, A. (1941), *The cytoplasm of the plant cell.* Chronica Botanica Co, Mass.

Gwynne-Vaughan, H. C. I. and Barnes, B. (1951), *The structure and development of the fungi.* 2nd edn., Cambridge University Press, Cambridge.

Harris, N. and Dodge, A. D. (1972), The effect of paraquat on flax cotyledon leaves: changes in fine structure. *Planta,* 104, 201–209.

Hébant, C. (1973), Acid phosphomonoesterase activities (β-glycerophosphatase and naphthol AS-MX phosphatase) in conducting tissues of bryophytes. *Protoplasma,* 77, 231–241.

Hepler, P. K. and Newcombe, E. H. (1963), The fine structure of young tracheary xylem elements arising by redifferentiation of parenchyma in wounded *Coleus* stem. *J. Exp. Bot.,* **14**, 496–503.

Heslop-Harrison, J. (1963), Ultrastructural aspects of differentiation in sporogenous tissue. In *Cell Differentiation* (ed. G. E. Fogg), *Symp. Soc. Exp. Biol.,* **17**, 315–339.

Hoefert, L. L. (1971), Ultrastructure of tapetal cell ontogeny in *Beta. Protoplasma,* **73**, 397–406.

Holden, J. (1975), Use of nuclear staining to assess rates of cell death in cortices of cereal roots. *Soil Biol. Biochem.,* **7**, 333–334.

Holliday, R. (1969), Errors in protein synthesis and clonal senescence in fungi. *Nature (London),* **221**, 1224–1228.

Huffaker, R. C. and Peterson, L. W. (1974), Protein turnover in plants and possible means of its regulation. *Ann. Rev. Plant Physiol.,* **25**, 363–392.

Humphrey, T. J. and Davies, D. D. (1975), A new method for the measurement of protein turnover. *Biochem. J.,* **148**, 119–127.

Hurst, P. R., Gahan, P. B. and Snellen, J. W. (1974), Turnover of labelled DNA in differentiated collenchyma. *Differentiation,* **1**, 261–266.

Ingle, J., Beitz, D. and Hageman, R. H. (1965), Changes in composition during development and maturation of maize seeds. *Plant Physiol.,* **40**, 835–839.

Jensen, W. A. and Fisher, D. E. (1970), Cotton embryogenesis: The pollen tube in the stigma and style. *Protoplasma,* **69**, 215–235.

Juniper, B. E., Gilchrist, A. J. and Robins, R. J. (1977), Some features of secretory systems in plants. *Histochem. J.,* **9**, 659–680.

Kole, R., Sierakowska, H. and Shugar, D. (1976), Novel activity of potato nucleotide pyrophosphatase. *Biochim. Biophys. Acta,* **438**, 540–550.

Kuehl, L. and Sumsion, E. N. (1970), Turnover of several glycolytic enzymes in rat liver. *J. Biol. Chem.,* **245**, 6616–6623.

Lakon, G. (1942), Topographischer Nachweis de Keimfahigkeit des Mais durch Tetrazoliumsalze. *Ber. Dtsch. Bot. Ges.,* **60**, 434.

Larson, D. A. (1965), Fine structural changes in the cytoplasm of germinating pollen. *Am. J. Bot.,* **52**, 139–154.

Lazarow, A. and Cooperstein, S. J. (1953), Studies on the enzymatic basis for the Janus green B staining reaction. *J. Histochem. Cytochem.,* **1**, 234–241.

Linn, S., Kairis, M. and Holliday, R. (1976), Decreased fidelity of DNA polymerase activity isolated from aging human fibroblasts. *Proc. Natl. Acad. Sci. U.S.A.,* **73**, 2818–2822.

McElgunn, J. D. and Harrison, C. M. (1969), Formation, elongation and longevity of barley root hairs. *Agron. J.,* **61**, 79–81.

McLean, J. (1969), A histochemical study of acid hydrolases in dividing and differentiating plant tissues. Ph. D. Thesis, University of London.

Marquardt, H., Barth, O. and von Rahden, U. (1968), Zytophotometrische und electronen mikroscopische Beobachtfungen über die Tapetumzellen in den Antheren von *Paeonia tenuifolia. Protoplasma,* **65**, 407–421.

Massopust, B. (1906), Über die Lebensdauer des Markes im Stamme und einige Fälle von Auflösung des Kalkoxalates in demselben. *Sitzber. Verbiondung 'Lotos' II,* **26**, 186–201.

Matile, Ph. (1969), Plant Lysosomes. In *Lysosomes in Biology and Pathology* (eds. J. T. Dingle and H. B. Fell), North-Holland, Amsterdam, pp. 406–430.

Matile, Ph. (1975), The Lytic Compartment of Plant Cells. *Cell Biol. Monogr.,* **1**, Springer-Verlag, New York.

Matile, Ph. (1978), Biochemistry and function of vacuoles. *Ann. Rev. Plant Physiol.,* **29**, 193–213.

Matile, Ph. and Winkenbach, F. (1971), Function of lysosomes and lysosomal enzymes in the senescing corolla of the morning glory (*Ipomoea tricolor*). *J. Exp. Bot.*, **22**, 759–771.

Maynard-Smith, J. (1973), Cells and ageing. In *Cell Biology in Medicine* (ed. E. Bittar), John Wiley and Sons, New York, pp. 681–695.

Medvedev, Z. A. (1963), *Protein Biosynthesis and Problems of Heredity, Development and Ageing*, Oliver and Boyd, Edinburgh.

Mepham, R. and Lane, G. (1969), Formation and development of the tapetal periplasmodium in *Tradescantia bracteata*. *Protoplasma*, **68**, 175–192.

Michaelis, L. (1900), Die vitale Farbung, eine Darstellungsmethode der Zellgranula. *Arch. Mikrosk. Anat.*, **55**, 558.

Molisch, H. (1938), *The Longevity of Plants*, Translated E. H. Fulling, New York.

Morriset, C. (1968), Observations ultrastructurales sur les espaces intercellulaires dans les meristemes radiculaires de germinations de Tomate (*Lycopersicon esculentum* L. Solanée). *C. R. Hebd. Séances Acad. Sci. Ser. D*, **267**, 845–848.

Mota, M. (1952), The action of seed extracts on chromosomes. *Argu. Patol.*, **24**, 336–357.

Mozaffari, F. D. S. and Gahan, P. B. (1978), Chromosome aberrations and ageing root meristems. *Ann. Bot.*, **42**, 1161–1170.

Nakanishi, S. and Numa, S. (1970), Purification of rat liver acetyl coenzyme A carboxylase and immunochemical studies on its synthesis and degradation. *Eur. J. Biochem.*, **16**, 161–173.

Nawashin, N. and Gerassimova, H. (1936), Natur and Ursachem der Mutationen. III. *Cytologia*, **7**, 437–465.

Nishimura, M. and Beevers, H. (1978), Hydrolases in vacuoles from castor bean endosperm. *Plant Physiol.*, **62**, 44–48.

O'Brien, T. P. (1970), Further observations on hydrolysis of the cell wall in xylem. *Protoplasma*, **69**, 1–14.

O'Brien, T. P. and Thimann, K. (1967), Observations on the fine structure of the oat coleoptile. III. Correlated light and electron microscopy of the vascular tissues. *Protoplasma*, **63**, 443–478.

Opik, H. (1965), Respiration rate, mitochondrial activity and mitochondrial fine structure in the cotyledons of *Phaseolus vulgaris* L. during germination. *J. Exp. Bot.*, **16**, 667–682.

Opik, H. (1966), Changes in cell fine structure in the cotyledons of *Phaseolus vulgaris* L. during germination. *J. Exp. Bot.*, **17**, 427–439.

Orgel, L. E. (1963), The maintenance of the accuracy of protein synthesis and its relevance to ageing. *Proc. Natl. Acad. Sci. U.S.A.*, **49**, 517–521.

Orgel, L. E. (1973), Ageing of clones of mammalian cells. *Nature (London)*, **243**, 441–445.

Palmer, J. (1976), The organization and regulation of electron transport in plant mitochondria. *Ann. Rev. Plant Physiol.*, **27**, 133–157.

Pelc, S. R. (1972), Metabolic DNA. *Int. Rev. Cytol.*, **32**, 327–348.

Petes, T. D., Farber, R. A., Tarrant, G. M. and Holliday, R. (1974), Altered rate of DNA duplication in ageing human fibroblasts in cultures. *Nature (London)*, **251**, 434–436.

Pitt, D. (1975), *Lysosomes and Cell Function*, Longman, London.

Polya, G. M. and Ashton, A. R. (1973), Inhibition of wheat seedling 5′ (3′)-ribonucleotide phosphohydrolase by adenosine 3′, 5′–cyclic monophosphate. *Plant Sci. Lett.*, **1**, 349–357.

Popp, R. A., Bailiff, E. G., Hirsch, G. P. and Conrad, R. A. (1976), Errors in human haemoglobin as a function of age. *Interdiscip. Top. Gerontol.*, **9**, 209–218.

Rabson, R., Mans, R. J. and Novelli, G. D. (1961), Changes in cell-free amino acid incorporating activity during maturation of maize kernels. *Arch. Biochem. Biophys.*, **93**, 555–562.

Roberts, E. H. (1972), *Viability of seeds*, Chapman and Hall Ltd., London.

Roberts, E. H., Abdalla, F. H. and Owen, R. J. (1967), Nuclear damage and the ageing of seeds with a model for seed survival curves. *Symp. Soc. Exp. Biol.*, **21**, 65–100.

Sacher, J. A. and Davies, D. D. (1974), Demonstrations of *de novo* synthesis of RNAase in *Rhoeo* leaf sections by deuterium oxide labelling. *Plant Cell Physiol.*, **15**, 157–161.

Schorler, B. (1883), Untersuchungen über die Zellkerne in den starke führenden Zellen der Hölzer. Diss. Univ. Jena, pp. 1–29.

Sexton, R. and Sutcliffe, J. F. (1969), The distribution of β-glycerophosphatase in young roots of *Pisum sativum* L. *Ann. Bot.*, **33**, 407–419.

Shaykh, M. M. and Roberts, L. W. (1974), A histochemical study of phosphatases in root apical meristems. *Ann. Bot.*, **38**, 165–174.

Sierakowska, H., Gahan, P. B. and Dawson, A. L. (1978), The cytochemical localization of nucleotide pyrophosphatase activity in plant tissues, using naphthyl esters of thymidine-5-phosphate. *Histochem. J.*, **10**, 679–693.

Srivastava, L. M. and Singh, A. P. (1972), Certain aspects of xylem differentiation in corn. *Can. J. Bot.*, **50**, 1795–1804.

Steer, M. W. (1977a), Differentiation in the tapetum in *Avena*. I. The cell surface. *J. Cell Sci.*, **25**, 125–138.

Steer, M. W. (1977b), Differentiation in the tapetum in *Avena*. II. The endoplasmic reticulum and Golgi apparatus. *J. Cell Sci.*, **26**, 71–86.

Stewart, P. and Pitt, D. (1977), Ultrastructural localization of acid phosphatase activity in root tips by the *p*-(acetoxymercuric) aniline diazotate reagent and a comparison with a Gomori procedure. *J. Cell Sci.*, **26**, 19–29.

Strehler, B. L. (1977), *Time, Cells and Ageing*, 2nd edn., Academic Press, New York.

Thomas, C. V., Esterly, J. R. and Ruddat, M. (1977), Light microscopical histochemistry of hydrolases in the root tip of Lima beans, *Phaseolus lunatus* L., in relation to cell differentiation and wound regeneration. *Protoplasma*, **91**, 229–242.

Trewavas, A. (1972), Determination of the rates of protein synthesis and degradation in *Lemna minor*. *Plant Physiol.*, **49**, 40–46.

Vasil, I. K. (1967), Physiology and cytology of anther development. *Biol. Rev. Cambridge Philos. Soc.*, **42**, 327–373.

Whaley, W. G., Kephart, J. E. and Mollenhauer, H. H. (1964), The dynamics of cytoplasmic membranes during development. *Symp. Soc. Dev. Growth*, **22**, 135–173.

Wooding, F. B. P. and Northcote, D. H. (1964), The development of the secondary wall of the xylem in *Acer pseudoplanatus*. *J. Cell. Biol.*, **23**, 327–337.

Woolhouse, H. W. (1967), The nature of senescence in plants. *Symp. Soc. Exp. Biol.*, **21**, 179–213.

Woolhouse, H. W. (1974), Longevity and the senescence of plants. *Sci. Prog. Oxf.*, **61**, 123–147.

Woolhouse, H. W. (1978), Senescence processes in the life cycle of flowering plants. *Biosciences*, **28**, 25–31.

Zee, S. Y. and Chambers, T. C. (1969), Development of the secondary phloem of the primary root of *Pisum sativum*. *Austr. J. Bot.*, **22**, 257–259.

6 The tissue kinetics of cell loss

N. A. WRIGHT

6.1 Introduction

The preceding chapters in this book have largely been concerned with the morphological description of the death of cells, with its mechanisms, and the role of cell death in maintaining tissue homoeostasis; in this chapter, however, we approach the phenomenon as a problem in cell population kinetics. Cell-renewal systems, such as the intestine and bone marrow, maintain a constant population size; hence cell loss must equal cell production. In malignant disease, cells are constantly being lost from the population by necrosis, migration to form metastases, exfoliation from surfaces and also by tumour differentiation. These phenomena immediately place us in the sphere of quantification; the aim of tissue kinetics is to define each cell population in terms of its size, its flux and its time, and in the present context, we need to express cell loss in terms of the number of cells involved, the rate at which they are being lost (flux) and the time they take to be lost to the population. Studies on the mechanism of cell loss, though naturally important, do not come within the scope of the present treatment, although we shall be concerned with the various proliferative states from which cells can be lost.

This chapter then is about the measurement of cell loss, about its importance as a kinetic parameter and its effects upon other kinetic parameters, notably those of proliferation. This, however, is a difficult task; cell kineticists have been largely concerned with the measurement of cell production and the factors controlling it, rather than cell loss. Although cell death is a well-recognized phenomenon in foetal tissues in organogenesis (see Glücksmann, 1951), it is only comparatively recently that cell deletion has been considered as a kinetic event in cell-renewal systems or expanding cell populations (Wyllie, Kerr and Currie, 1973a), and few authors have attempted to measure rates of cell loss in normal tissues. This lack is probably a reflection of the comparatively recent observation that cell death, almost exclusively by apoptosis, does occur within the proliferative compartments of a renewal system such as the small intestine (Murray and Rumsey, 1976).

On the other hand, the presence of large areas of frank necrosis in both human (Refsum and Berdal, 1967) and experimental tumours (Steel, 1967, 1968) rapidly led investigators to an early analysis of the role of cell loss, determining not only the pattern of tumour growth in the unperturbed state (Steel, 1968, 1977), but

171

also the response of the tumour to irradiation (Denekamp, 1972) and cytotoxic chemotherapy (Bagshawe, 1968), and our knowledge of cell loss in tumours is correspondingly greater.

These comments apply to the actual removal of a cell, but there is a body of opinion which regards the loss of reproductive capacity, rather than physical space, as the only important parameter of cell loss (Bagshawe, 1968), and in some situations this argument is reasonable. The ability of a tumour to regenerate after irradiation, or of a tumour to regrow after chemotherapy, depends upon the survival of cells with reproductive capacity, now usually called 'clonogenic cells'. Thus when we are considering cell loss as a factor affecting the ability of a tissue to mount a regenerative response, there is little point in measuring the overall cell loss; it becomes important to concentrate solely upon these clonogenic cells, and assess their survival characteristics in isolation. We have now begun to discuss the kinetic organization of tissues in relationship to cell loss phenomena, and, before proceeding to a more formal statement of definitions, to a survey of techniques and to some illustrative examples, it would be well to look briefly at the kinetic structure of the populations we shall be considering and the important events in the life cycle of proliferating cells.

6.2 The cell cycle

Many dynamic phenomena which occur in cell populations are age-dependent; that is to say, a migration or death process only occurs in cells after the effluxion of a certain amount of time, and here time is measured since the occurrence of a particular event; for example, cells which are born in the zona glomerulosa of the mammalian adrenal cortex are thought to migrate inwards to die in the zona reticularis (Wyllie *et al.*, 1973a; Wright and Voncina, 1977), and consequently the deletion of an adrenal cortical cell is dependent upon its age. Many other loss processes are age-dependent in that they occur at a specific point in the cell cycle, in which case age is measured from cell birth at the previous mitosis. The concept of the cell cycle is central to our discussion, and a useful diagrammatic representation is shown in Fig. 6.1. Cycling cells pass through a defined series of events the period of DNA synthesis usually occurs towards the end of the cell cycle, and is termed the S phase. The G_1 phase occurs before S, and G_2 after S and before mitosis. In these phases the cells prepare for DNA synthesis and mitosis respectively and the time taken for cells to complete this cycle is called, not unreasonably, the cell cycle time. In most cell populations *in vivo*, cells are continually leaving this cycle to perform other life-cycle events, and these avenues are also depicted in Fig. 6.1; cells leave the cycle to lose reproductive capacity and differentiate, and in renewal systems such as the epidermis or bone marrow to die in the normal process of keratinization and exfoliation, or sequestration in the spleen respectively. Thus the normal process of cell death is age-dependent. Cells can also leave the cycle to enter a putative resting state, sometimes called G_0, from which they can re-enter the proliferative cycle if conditions are appropriate. By no means all workers are

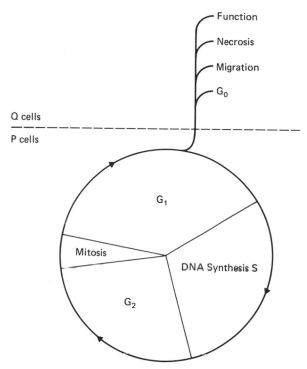

Fig. 6.1 A scheme showing the cell division cycle, indicating how a population can be partitioned into proliferating and non-proliferating cells, and
where most cells in the resting G_0 phase may re-enter the cell cycle.

agreed on the concept of the G_0 phase (Smith and Martin, 1973), although populations with large numbers of proliferatively resting cells which are subsequently
stimulated into the cell cycle do show biochemical differences compared with
continuously renewing cells (Baserga, 1971; Alison and Wright, 1979b). However,
from the point of view of the present discussion, it is important to appreciate that
most populations, and this includes even rapidly proliferating experimental tumours, contain a ratio of actively cycling cells to differentiating, resting or dying
non-cycling cells; this ratio is called the growth fraction, and the cell types are
termed P and Q cells respectively (Fig. 6.1). It is important to appreciate that
both P and Q cells have an age structure, noted from birth at mitosis (for P cells)
and from production by a decycling event in the case of Q cells; this age structure
is usually termed the age distribution, of which more will be said when considering
age-dependent loss processes (see Section 6.4). In the main, cells are considered
to leave the cell cycle, to differentiate and die, or enter the G_0 phase, sometime
before the onset of DNA synthesis, in late G_1 phase (Fig. 6.1), although in some
tissues, notably epidermis and liver, there is evidence for decycling events occurring
in G_2 phase (Gelfant, 1977).

6.3 The organization of cell populations

The kinetic organization of a cell population can be approached at several levels, and for our present purpose two approaches are important. Fig. 6.2 shows a simple diagram of a cell population from simplistic ideas of kinetic behaviour. The diagram can be considered to represent the situation in both renewal systems and tumours. All tissues contain a stem cell, which here can be simply defined as those cells which have the ability to produce a large family of descendants, and which, in normal cell populations, are relatively few in number compared with the total proliferative cellularity. As well as maintaining its own numbers, the stem cell compartment feeds the proliferative compartment, where the cell production is amplified by a number of transit divisions. Cells then lose the capacity for division, and may, in normal tissues, enter a maturing or differentiation pathway, which leads to death by a normal deletion process. In tumours, however, cells may enter a Q cell state, which may also lead to death by necrosis as a result of inadequate oxygenation (see Section 6.6). Applied to renewal systems such as the small intestinal mucosa, stem cells are housed in the crypt base (Cheng and Leblond, 1974; Wright, 1977) and feed the proliferative compartment in the lower two-thirds of the crypt. The transition between P and Q cells occurs in the upper portion of the crypt, which thereafter contains differentiating cells preparing for function in the villus epithelium, and eventual death in the intestinal lumen. In the epidermis there is no such spatial organization; both stem cell division and amplification divisions occur in the basal layer, and, indeed, in the mouse epidermis, cells lose reproductive capacity whilst still in the basal layer, and spend some time on the basal layer waiting to migrate as postmitotic maturing cells (Potten, 1977). Following migration from the basal layer, differentiation continues until death occurs in the granular layer. In tumours there may or may not be a spatial organization, largely dependent upon the arrangement of blood supply. In the mammary tumour studied by Tannock (1968, 1970), where cells only survive in close juxtaposition to the penetrating vessels, stem cells and proliferating cells were grouped around blood vessels, while decycled non-proliferating Q cells were found at a distance from the vessel, and hypoxic cell death ensued largely from the hypoxic Q compartment. In other tumours, which show central vascularization, such as some lymphomas, there is no regional necrosis, and consequently no spatial organization. Thus the kinetic organization in these several tissues exemplifies the age-dependent nature of dynamic cellular processes, including that of cell death.

The other approach relevant to our discussion is a classification of cell types in a population in relation to their operational behaviour. We noted above that only certain cells are capable of prolonged self-replication, the stem cells; other proliferative cells, though capable of a few transit divisions, are in fact doomed to leave the population and die. Stem cells which maintain the proliferative population in normal growth circumstances are usually called functional (Cairnie, 1976) or effective (Steel, 1977) stem cells. However, in abnormal growth conditions, such as after death of proliferative cells induced by irradiation, other cells may be induced to act as stem cells; these are referred to as potential stem cells (Cairnie,

1976). For example, in the small intestine, only a few cells in the crypt are effective stem cells (Cheng and Leblond, 1974), but measurement of clonogenic cell number, by radiobiological techniques which necessitate the killing of large numbers of proliferative cells, indicate that the potential stem cell number may approach the proliferative cellularity, which is about 150 cells in the mouse crypt (Withers, 1976; Cairnie, 1976; Hendry and Potten, 1974). These considerations become important when we are considering cell death in the context of the ability to mount a regenerative response.

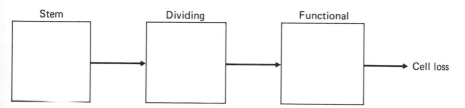

Fig. 6.2 A simple diagram of a three-compartment cell kinetic model, showing the ways a renewing population can be classified according to the kinetic behaviour of the component sub-population.

6.4 The measurement of the kinetics of cell loss

In this chapter I have been careful to use the term cell death only when the death process itself is inferred. This is because kinetic methods which purport to measure cell death in fact only measure cell loss; it is very difficult to distinguish between the loss of cells by death processes and loss by migration from the tissue or compartment. For example, in tumours, cells are frequently lost by metastatic processes involving blood and lymphatic channels, and in lymphoid tissues such as thymus and spleen, cell loss by migration becomes a major flux parameter (Claesson and Hartmann, 1976). It is important to remember this caveat in the ensuing discussion.

6.4.1 Theoretical aspects of the kinetics of cell loss

Much of the basic work towards an understanding of the kinetics and measurement of cell loss is due to Steel (1966, 1968, 1977), who started with the simple premise that the growth rate of a cell population, as reflected in the doubling time (or time required for the number of cells in a population to double), was seldom equal to the birth rate as reflected by the cell cycle time. The reasons for this inequality are evident but important: first, the existence of a growth fraction of less than unity, with numbers of non-productive differentiating or resting cells, and second, the presence of cell loss. Thus a potentially useful method for the measurement of cell loss becomes apparent. Consider an exponentially growing population with no cell loss: the increase in population size is a function of the birth rate exclusively, and the birth rate (k_B) will equal the growth rate (k_G). Extending the argument, if cell loss is occurring from the population, then k_G will be less than k_B and:

$$k_L = k_B - k_G \tag{6.1}$$

where k_L is the cell loss rate, expressed, as are all these rate parameters, in cells/cell per h or cells/1000 cells per h. The values k_G and k_B can be measured for various tissues (see Section 6.4.4) or can be calculated from other measurements, and, of course, have special relevance for tumour cell populations, but can be measured for normal tissues (see Section 6.5.3). In the case of the birth rate, again in exponential growth conditions:

$$k_B = \ln 2/T_c \tag{6.2}$$

where T_c is the cell cycle time, and if the growth fraction (I_p) is less than unity:

$$k_B = \ln(1 + I_p)/T_c \tag{6.3}$$

Similarly, the growth rate k_G can be calculated from measurements of the doubling time (t_D) of the tissue, and:

$$k_G = \ln 2/t_D \tag{6.4}$$

So the cell loss rate can be calculated from substitution in equation 6.1.

Steel (1968) has popularized a parameter called the potential doubling time (t_{PD}), which is a measure of the growth potential of a population; thus t_{PD} is equal to the population doubling time when there is no cell loss, and is calculated from the birth rate using the relationship:

$$t_{PD} = \ln 2/k_B \tag{6.5}$$

but can also be calculated from the duration of DNA synthesis (t_S) and the flash thymidine-labelling index (I_S), thus:

$$t_{PD} = \lambda t_S/I_S \tag{6.6}$$

where λ is a correction factor which depends upon the age distribution and the position of S in the cell cycle, and in most real cell populations takes a value from about 0.7 to 1.0 (see Section 6.4.5).

We can now rewrite equation 6.1; from equations 6.3 and 6.4:

$$k_L = \ln(1+I_p)/T_c - \ln 2/t_D \tag{6.7}$$

and from equation 6.5:

$$k_L = \ln 2/t_{PD} - \ln 2/t_D \tag{6.8}$$

Steel (1968) introduced a cell loss factor ϕ, which expresses the rate of cell loss as compared with the rate of cell birth, and assesses to what extent cell loss is affecting the proliferative potential of a population:

$$\phi = k_L/k_B \tag{6.9}$$

or from equation 6.8:

$$\phi = 1 - t_{PD}/t_D \tag{6.10}$$

Thus ϕ is the ratio of cell loss to cell birth; when there is no cell loss from populations, then ϕ is 0, and potential doubling time equals the actual measured doubling time. The other limit is where all cells are lost and $\phi = 1$; that is to say, each time a cell is born, one is lost, and the population growth rate is 0. This situation of course occurs in normal renewal tissues such as the gut, but is also approached in tumours towards the end of their growth period, where doubling time is extremely long.

6.4.2 *Age-dependent death processes*

Cooper (Cooper, 1973; Cooper, Bedford and Kenny, 1975) has classified the several avenues of cell death as (*a*) intrinsic mechanisms, such as cell aging (for example, death of neurons in the central nervous system) or differentiation-induced cell loss exemplified by the loss of cells which accompanies formation of epithelial 'pearls' by keratinization in well-differentiated squamous carcinomas (Pierce and Wallace, 1971), (*b*) physical deplacement, which includes cell migration, (cf. loss of cells from the zona glomerulosa by inward migration into the zona fasciculata) and exfoliation, as seen in the loss of small intestinal cells from the surfaces of the villi and (*c*) extrinsic mechanisms of special relevance to tumour cell populations, such as death from inadequate oxygenation and nutrition, owing to the relative failure of the tumour vasculature. In order to analyse the mechanism by which cells are lost, it is important to know where, in its life history, the deletion process occurs, and this is one of the contributions that tissue kinetics can reasonably be expected to make. It will be appreciated that many of the above processes are conceivably age-dependent, in that they usually (or invariably) occur at a particular age. In normal tissues, for example, cells usually migrate or differentiate in the G_1 phase of the cell cycle; we saw above that intestinal cells begin differentiation after a final mitosis in the proliferative compartment, and, similarly, there is good evidence that the migration process from the basal layer of the epidermis is not random, but occurs from the oldest G_1-phase cell nearest a mitosis (Iversen, Bjerknes and Devik, 1968).

It is not generally appreciated that death processes in tumours are also age- and cell-cycle-dependent. In the mouse mammary tumour studied by Tannock (1968, 1970), cords of tumour cells are seen growing around a central vessel, while the cords are separated from each other by areas of overt necrosis. Tannock measured kinetic parameters in three regions: close to the central vessel, an intermediate zone, and the region close to the peripheral necrotic area; the cell cycle time remained constant in all zones, but the growth fraction was found to fall progressively from 100% in the perivascular zone to 50% in the peripheral area. Cells appeared to be born mainly in perivascular regions, and migrate peripherally, losing proliferative ability and becoming hypoxic as they move, and finally dying. Because of the large numbers of decycled Q cells in peripheral areas, a major avenue of cell loss here must be from the Q cell compartment. Further, because the proportion of Q cells increases with distance from the vessel, a reasonable

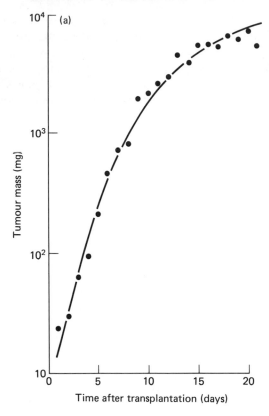

Fig. 6.3 The growth curves for (a) the Balb/c sarcoma, and (b) the rat adrenal gland.

suggestion is that the Q cells also have an age structure (Steel, 1968), and death may be random within the compartment, or occur at a fixed age (Steel, 1977). In the same context, death at mitosis is also considered to be an important avenue of cell death in tumours; many malignant tumours contain morphologically abnormal mitoses, and show numerous chromosome abberations which can reasonably be expected to lead to death of that cell (or its progeny) in mitosis (Cooper *et al.*, 1975). We may conclude that age-dependent processes are important in cell deletion phenomena, and proceed to an examination of some methods which are available for the estimation of cell loss.

6.4.3 *Methods for the estimation of loss parameters*

The theoretical treatment given above is clearly most applicable to cell populations which are growing, as a tacit assumption is the existence of a steady state of exponential growth. Such growth conditions are naturally found in the early growth phase of experimental tumours, where most work using this approach has been

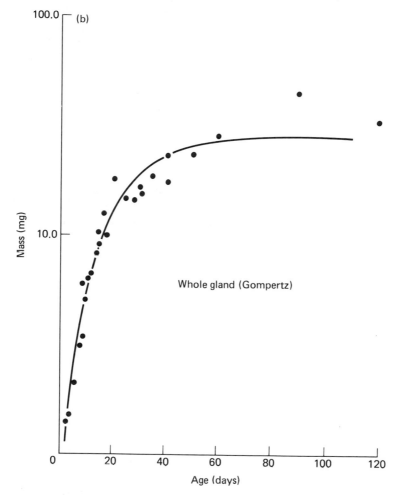

Fig. 6.3 Continued

carried out; however, the method has also found application in the analysis of cell loss in normal tissues, albeit growing, such as the neonatal adrenal cortex (Wright and Voncina, 1977; and see Section 6.5.3).

The first parameter to be measured in this procedure is the specific growth rate, k_G, or the doubling time t_D. Theoretically these are found by measuring the increase in the numbers of cells with the passage of time; this measurement is rarely possible *in vivo*, and measurements are usually made of overall tissue growth as reflected in its volume, assessed by linear dimensions or weight. This requires the necessary assumption that the rate of change in the tissue volume is an accurate index of the rate of change of cell number. A typical growth curve for a rapidly growing experimental tumour is shown in Fig. 6.3(a), and for a normal tissue (the rat adrenal gland) in Fig. 6.3(b). Note that the graph is plotted on a semilogarithmic scale, with the result that the straight line component of the curve reflects a state of exponential growth. On this part of the curve, i.e. between 3 and 7 days after transplantation in the tumour, and 1 and 20 days in the case of the adrenal cortex, the growth curve can be found from the slope of the line; after this, growth is no longer truly exponential, but as an approximation, exponential growth is assumed as over a brief period, and the growth rate at any instant is found from the slope of a tangent drawn to the growth curve. Alternatively, one can assume exponential growth over one doubling time (t_D) and calculate the growth rate from equation 6.5. The growth rate is usually expressed as cells/cell per h.

An alternative approach, which removes the element of subjectivity from drawing a tangent to the curve, is to fit the data by some suitable function, such as the Gompertz function, which is particularly useful in describing the growth of many tissues and organs, because exponential growth is allowed initially, but later the function is self-retarding and fits the slowing in growth rate so often observed in these tissues. The Gompertz function has the general form:

$$V_t = V_oA/\alpha(1-e^{-\alpha t}) \tag{6.11}$$

where V_o and V_t are the volumes at zero time and time t, α is a constant, and A the initial specific growth rate; simplifying and differentiating this equation allows the specific growth rate to be calculated at any time (see for example Wright and Voncina, 1977). Having then obtained an estimate of the growth rate or doubling time, we must then measure the birth rate or potential doubling time, to calculate k_L from equation 6.1 or 6.8. The simplest method is by a metaphase arrest or stathmokinetic technique, which relies on the accumulation of arrested metaphases after injection of a suitable agent such as vincristine or colchicine (Fig. 6.4). If the rate of accumulation of metaphases is linear, the birth rate is the slope of the accumulation line, in cells/cell per h, and the potential doubling time is then simply calculated from equation 6.5.

Alternatively, t_{PD} can be calculated from the measurements of labelling index and the duration of DNA synthesis (equation 6.6). Here we need to derive a value for λ. In its simplest form, and assuming that the duration of the G_2 phase is known and is a small fraction of the cell cycle, λ can be calculated from:

$$\lambda = t_D/t_S(exp[ln2/t_D(t_S+t_{G_2})]-exp[ln2/t_D.t_{G_2}]) \tag{6.12}$$

as shown by Steel (1968, 1977), and representative curves for λ, for a variety of values of $t_S+t_{G_2}$ are shown in Fig. 6.5). The minimum value is ln2 or 0.693, and the maximum value is 2ln2 or 1.38, which only occurs when the S phase is close to the beginning of the cycle. Populations which share the same doubling time, and where S is similarly sited in the cell cycle, have similar values, and $(t_S+2t_{G_2})$ is a constant (Steel, 1977); λ is then conveniently obtained from Fig. 6.5. The derived values are then substituted into the appropriate equations, and k_L and ϕ are calculated. The behaviour of the various methods are compared when specific examples of cell loss from various tissues are discussed (see Sections 6.5.3 and 6.6.1).

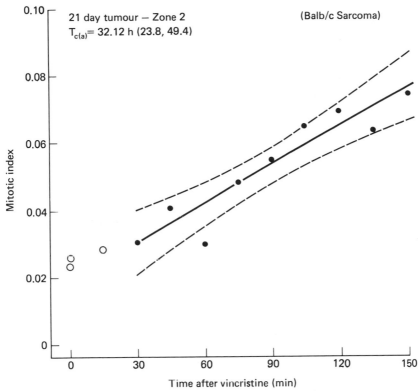

Fig. 6.4 A typical metaphase-arrest curve. Vincristine is injected at time 0, and readings are made of the increase in metaphase or mitotic index with time thereafter. The slope of the fitted line represents the birth rate; the method of fitting the line depends on the appropriate age distribution of the population concerned; in some tissues the age distribution can be calculated or measured, and use of an inappropriate age distribution can result in considerable errors (Duffill, Appleton, Dyson, Shuster and Wright, 1977).

6.4.4 *The assessment of age-dependent loss in tumour cell populations*

We have seen that cells can die within the cell cycle, for example at mitosis, or can die from the Q cell compartment, having left the cell cycle. There are methods which allow an approach to the avenue of cell death in tumour cell populations, which involve the computer analysis of continuous thymidine-labelling curves. This method (see Fig. 6.6) depends on making tritiated thymidine continuously available to the cell populations by intravenous infusion or by repeated injections; if the latter method is used, it is essential to ensure that injections are given at a time less than the median duration of the S phase itself, to ensure that cells do not pass through S without becoming labelled. The kinetics of such labelling is complex, because of the several different cell kinetic compartments which are present in a population; in Fig. 6.2 there are proliferating cells and non-proliferating cells which

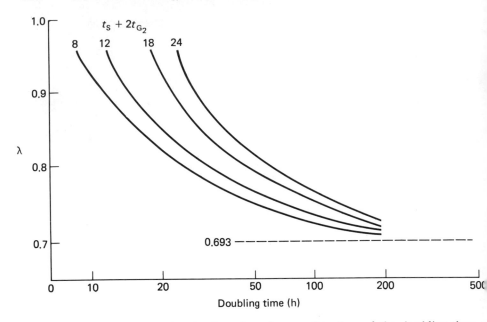

Fig. 6.5 A diagram showing how λ varies as a function of the doubling time t_D, for a small selection of $t_S + 2t_{G_2}$ (redrawn after Steel, 1977). The dotted line represents the lower limit for λ.

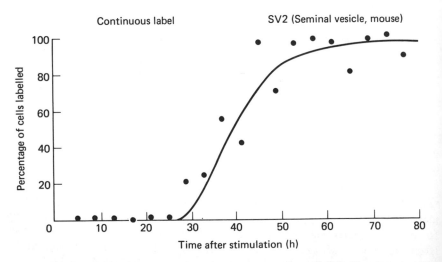

Fig. 6.6 A diagrammatic representation of a continuous labelling curve with [3]HdThd, which is made continuously available to the population. The plateau value, which, incidentally, is not a good estimator of the growth fraction (Steel, 1977) is reached in a time equivalent to $t_{G_2} + t_M + t_{G_1}$.

will be eventually lost but are fed from the proliferating compartment; if a resting G_0 phase exists, this adds to its complexity. The time taken to achieve a plateau value is related to the distribution of transit times through G_1 and G_2, but both this time and the actual level of the plateau value are also dependent upon the age distribution of the non-proliferating cells which arise from the dividing cells, and on the age at which non-proliferating cells are lost from the population (Steel and Hanes, 1971; Hartmann, Dombernowsky and Bichel, 1976; Steel, 1977); on the basis of the reasonable proposal that cells can be deleted at, for example, mitosis, or directly from the Q compartment, it is possible to formulate certain models for cell populations which imply a particular mode of cell loss; having proposed a cell loss mechanism, the age distribution of P and Q cells can then be drawn. Steel (1977) has analysed several such populations, and Fig. 6.7 shows an example of one of these: an exponentially growing cell population, where there is a fixed probability for the production of Q cells, and where Q cells are lost at a fixed age T_L; P and Q cells will then have the age distribution shown in Fig. 6.7. Table 6.1 summarizes Steel's analysis of age distributions in various populations which are subject to cell loss. Using these simple models one can then construct theoretical continuous labelling curves. For example, in Fig. 6.7, every cell which enters DNA synthesis will become labelled, and after division, the labelled progeny will either re-enter the cell cycle to be labelled again in the next S period, or will enter the non-proliferating compartment of Q cells as a labelled cell; the proposed mode of loss of the Q cells, or indeed of P cells where such cells are lost (see Table 6.1), will predict the format of the continuous labelling curve if that mode of cell loss is appropriate. Moreover, assuming that the proposed mechanisms of cell loss are the only kinetic alternatives available, some idea of the relative contribution of each cell loss pathway can be assessed. A worked example for an experimental tumour is given in Section 6.6.1.

6.4.5 *Pitfalls in the assessment of cell loss from growth curves*

There are two main problems which beset the appreciation of cell loss kinetics from growth curves. These both involve the difficulty in measuring the actual growth rate, or doubling time, from growth curves. From a growth curve we do not actually measure the growth rate of cells; we extrapolate from volume or weight to express k_G in cells/cell per h; we are essentially assuming that there is no change in the volume of cells over the observational period. There could, of course, be a change in the cell volume distribution, and other extraneous factors such as formation of extracellular constituents, such as collagen, and, in tumours, infiltration with leucocytes, haemorrhage and cyst formation can be important causes of an over-estimate of growth rate, with a consequent underestimate of cell loss rate. Many tumours also show large areas of overt necrosis, which are in a state of continuous flux, with cell death processes expanding the area at its periphery, and resorption of dead material tending to decrease the necrotic component. Unless special morphometric methods are undertaken to exclude the necrotic component, and

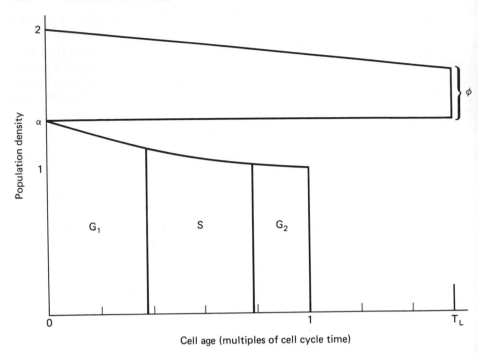

Fig. 6.7 An example of the age distribution of P and Q cells in a population from which non-proliferating cells are lost at a fixed age, T_L (redrawn from Steel, 1977).

Table 6.1. Types of cell populations which are suitable for analysis by modelling continuous thymidine-labelling curves (from Steel, 1977).

A	Uniform population of proliferating cells only, with no cell loss.
B	As A with no cell loss, but with a fixed possibility for production of non-proliferative cells (Q cells).
C	As B, but where non-proliferative Q cells are lost at a fixed age, T_L.
D	As B, where cells are lost at mitosis.
E	As B with random cell loss from both P and Q cell compartments.

therefore to measure the growth characteristics of the apparently viable fraction only (Smith, Thomas and Riches, 1974), then it is necessary further to assume that there is no change in the size of the necrotic portion with time, in which case the growth rate or doubling time in the whole tumour will reflect that in the viable portion as well.

The other problem concerns the mode of growth. In our above analysis, it is a tacit assumption that growth and age distribution are both exponential. In the later stages of both organ and tumour growth, owing possibly to decreases in growth fraction and increases in cell loss rate or cell cycle time, growth rate slows, and we can no longer assume an exponential age distribution. While we can to some extent compensate for this on the growth curve by fitting a self-retarding curve such as that given by the Gompertz function, we still have to make an assumption about the age distribution when we calculate the birth rate from the metaphase arrest method or from labelling data. This also applies if we fit continuous labelling curves using the cell loss models depicted in Table 6.1. In renewal systems such as the normal epidermis, the age distribution can be calculated from observations (Duffill, Appleton, Dyson, Shuster and Wright, 1977; Appleton, Wright and Dyson, 1977), but in perturbed kinetic situations such as arise in recovery from irradiation, or even when the growth rate slows for whatever reason, we have a non-exponential population which is then difficult to analyse.

6.4.6 *Estimation of cell loss rates from measurements of the turnover of labelled thymidine or thymidine analogues*

It might be thought that, if cells could be labelled with specific DNA precursors, and radioactivity retained and shared among subsequent progeny, measurements of the loss of such label would reflect cell loss rate. Steel (1966) labelled transplantable tumour cells, growing in rats, with tritiated thymidine and followed the total amount of bound tritium in the tumour. After a short period during which the level remained constant, over the next 7 days the tritium level in the tumour rose; this rise was due to label released from the normal death processes of labelled cells in renewal systems, such as the bone marrow and gut, and its subsequent reutilization by proliferating tumour cells. Fortunately Feinendegen, Bond and Hughes (1966), comparing the reutilization of tritiated thymidine with that of the thymidine analogue ^{131}I-labelled 5-iododeoxyuridine (^{131}IdUrd), reported minimal reutilization of ^{131}IdUrd, and later, Porschen and Feinendegen (1969) showed that reutilization levels of only about 5% were shown by the related compound ^{125}IdUrd; the method found extensive use for the measurement of cell loss from leukaemic and ascites tumours, by labelling cells either *in vivo* or *in vitro*, injecting them into suitable hosts and following the loss of radioactivity by wholebody counting (Hofer, 1969). A typical ^{125}IdUrd decay curve is shown in Fig. 6.8; assuming that all the label is in DNA and that there is no reutilization, the slope is equal to k_L, the cell loss rate, and:

$$k_L = \ln2/T_{1/2} \tag{6.13}$$

where $T_{1/2}$ is equivalent to the potential doubling time, t_{PD}. Now from equations 6.8 and 6.10:

$$k_L = \ln2/t_{PD} - \ln2/t_D$$

and since

$$\phi = 1 - t_{PD}/t_D$$
$$\phi = t_D(T_{1/2} + t_D) \tag{6.14}$$

(Begg, 1977).

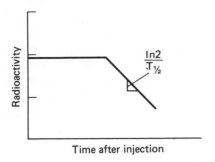

Fig. 6.8 A diagrammatic representation of the loss of incorporated ^{125}IdUrd from a tumour cell population subject to cell loss.

Hofer (Hofer, Prensky and Hughes, 1969; Hofer, 1969), reported that after intraperitoneal injections of labelled L1210 leukaemic cells, the half-life ($T_{1/2}$) for the loss of radioactivity was 5 days, and administration of cytotoxic agents increased the rate of loss of radioactivity. Methotrexate, BCNU (bischloroethyl nitrosourea) and cyclophosphamide were found to increase the cell death rate and increase the lifespan of the tumour-bearing mice, but when levels of cell survival were very low, quantitative assessment of the cell death rate became difficult; furthermore these chemotherapeutic agents appeared to decrease the rate of excretion of ^{125}I liberated from dead tumour cells.

Porschen and Feinendegen (1969), Dethlefsen (1969, 1970, 1971) and Begg and Fowler (1974) have approached the more difficult problem of assessment of cell loss in solid tumours. Dethlefsen also found that, in normal mouse tissues, reutilisation of ^{125}IdUrd was minimal, but in mammary carcinoma detectable reutilization of ^{125}IdUrd occurred, and with increased cell death induced by cytoxic agents, reutilization was considerable. Nevertheless, the method gave a $T_{1/2}$ of 122 ± 23(SD)h, which apparently compared reasonably with a cell loss rate of 10% per day, equivalent to a $T_{1/2}$ of 158h, derived from orthodox measurements (Mendelsohn and Dethlefsen, 1968; and see Sections 6.4.2 and 6.4.3).

Porschen and Feinendegen (1969) advocated the use of external counting in the solid tumours, although Dethlefsen (1971) found it necessary to extract the DNA before counting ^{125}I, with loss of the facility to monitor cell loss in the same tumour continuously. This was because of the problem of variable non-incorporated ^{125}IdUrd content in the acid-soluble pool of the tumour, and in its overlying skin. However, Begg and Fowler (1974) and Begg (1977) maintained that useful information on cell loss in solid tumours can be obtained by external counting; more valuably, both Begg (1977) and Dethlefsen, Sorensen and Snively (1977) have compared ^{125}IdUrd and growth rate/radioautographic methods in several different mouse solid tumours. The results are summarized in Table 6.2 and it can be seen that Begg, who corrected for an increasing area of necrosis (see Section 6.4.3), concluded that the methods compared well in some tumours, but in others, radioactivity did rise with time, possibly owing to the influx of labelled inflammatory

Table 6.2. A comparison of cell loss factor (ϕ) in several experimental tumours, measured by ^{125}IdUrd and ^{3}HdThd (after Begg, 1977)

	^{125}IdUrd		^{3}HdThd
	in situ	Excisions	
C_3H mammary carcinoma	68–77	41–55	67–78
CBA carcinoma NT	20–49	20–40	53–60
	27[+]		
CBA sarcoma S	39–49	36	30
CBA sarcoma F	*	*	26
	23[+]		
WHT carcinoma MT	*	*	54
	33[+]		

[*] Rising tumour radioactivity.
[+] Obtained by whole-body counting.

cells, and reutilization was a problem. Nevertheless, Begg (1977) concluded that the non-invasive ^{125}IdUrd method was useful, particularly with low rates of cell death, although ϕ appeared to be underestimated in high-cell-loss states. On the other hand, Dethlefsen *et al.* (1977) were less optimistic; they compared cell loss kinetics, measured using the two methods, in three mouse mammary tumours *in vivo*. The radioactive decay curves were analysed using three models: (1) exponential cell loss from both P and Q compartments; (2) as (1) but introducing a delay compartment, D, fed from the Q compartment, consisting of degenerate cells and debris which delays loss of radioactivity from the tumour; and (3) in which the delay was constant, and in Q itself; after traversing the delay, removal was then exponential. The third model was considered to give the best fit to the decay curves, but Dethlefsen *et al.* (1977) considered that ^{3}HdThd and ^{125}IdUrd results are not generally interchangeable, and, since ^{125}IdUrd results indicate a long delay before death, suggested that the method measures net flow from incorporation to exit, whereas ^{3}HdThd techniques measure the flux of cells between compartments, but not viable to degenerate or necrotic compartments. It is therefore possible that the methods are measuring different things; perhaps the best approach would be to perform both measurements, and to interpret the results in the light of the experimental conditions, and the necessary enforced assumptions (see Section 6.4.3).

6.4.7 *The assessment of clonogenic cell loss*

We have seen that cell populations contain stem cells, which are capable of giving rise to a large family of descendants in normal circumstances, or can be induced to do so by some set of abnormal circumstances. From the viewpoint of population maintenance, death of such stem cells is the over-riding consideration. Indeed, Roper and Drewinko (1976), in comparing several methods of determining growth-induced lethality on cells growing *in vitro* (including the doubling time, labelling index, [3]HdThd incorporation into DNA, dye exclusion and release of [51]chromium) concluded that colony-forming ability was the only dose-dependent index of cell lethality, and that other estimators of cell death grossly over- or under-estimated the amount of induced death. The main problem in this sphere is the recognition of stem cells, and to be in a position to measure them; it is probably true to say that stem cells as such cannot be recognized, or their number measured directly, but only their performance in an operational or 'clonogenic assay' can be appreciated (Steel, 1977). Clonogenic assays necessitate subjecting cells to a test environment in which their growth capacity is measured in a (usually) distinctly abnormal situation; it is hoped, but by no means certain, that clonogenic capacity in an assay reflects stem cell complement in a tissue. It is as well to note this distinction, and perhaps to confine ourselves to talking of clonogenic cells, recognizing that these are the cells which can form numbers of descendents in our assay system.

(a) *Measuring clonogenic cell loss in normal tissues* In the bone marrow it is possible to identify several sorts of cells with different clonogenic properties; the earliest form is the pluripotent stem cell, which is recognized and quantified by its ability to form macroscopic spleen colonies, containing erythroid, myeloid and megakaryocytic cells, when injected into irradiated recipient mice (Till and McCulloch, 1961), and usually assessed at 10 days after injection. The proliferative status of these colony-forming units (CFU-S) is assessed by incorporating a lethal amount of [3]HdThd ([3]HdThd suicide), which kills a number of CFU-S equivalent to those in the S phase (or the labelling index), and leads to a reduction in the number of spleen colonies on subsequent injection. Later in the developmental series, committed progenitor cells with colony-forming abilities are detected by their ability to grow in culture, and include the CFU-C cells in granulopoiesis (Bradley and Metcalfe, 1966) and CFU-E cells in erythropoiesis (Stephenson, Axelrad, McLeod and Shreeve, 1971); numerous publications have now assessed the responses and sensitivity of clonogenic marrow cells to irradiation and cytotoxic drugs (see, for example, Lajtha, Pozzi, Schofield and Fox, 1969).

The assessment of clonogenic capacity in epithelial tissues is more difficult, and investigations to date depend on the response of the tissue to a series of graded doses of irradiation, and estimating survival by some method which allows detection of colony-forming ability; for example, in the small intestine, it is customary to measure the survival of cryptogenic cells by measuring crypt survival in the Withers microcolony assay (Withers and Elkind, 1969), in which cross-sections of the intestine are studied at 3.5 days after irradiation, and the number of regenerating colonies per circumference are noted. A typical survival curve is shown in Fig. 6.9; after an amount of irradiation which has no effect on crypt survival (D_Q or quasi-threshold dose) there is an exponential decrease in survival, and the radiation dose which produces 37 per cent survival levels is the D_O, which for small intestinal clonogenic cells is about 150 rads. Calculations about the number of clonogenic cells per crypt can be made from this curve (Withers and Elkind, 1969), or, possibly better, from similar curves produced by splitting the dose of irradiation

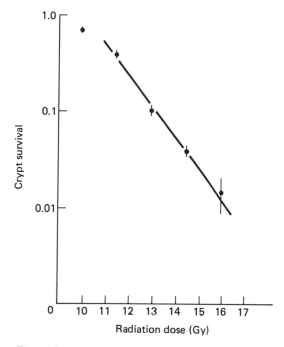

Fig. 6.9 A typical whole-crypt-survival curve for mice in air, using graded doses of irradiation.

(Hendry and Potten, 1974). The measurement of clonogenic crypt cell number (and their death and survival characteristics) are discussed below (see Section 6.5.4). Similar irradiation techniques for measuring clonogenic cell survival in epidermis are available (Potten and Hendry, 1973; Al-Barwari and Potten, 1976).

(b) *Measuring clonogenic cell loss in tumours* The ability to measure the amount of induced cell death in clonogenic tumour cells is quite critical, since it is upon the survival of such cells that regrowth potential, and consequent clinical response to therapy, depends so heavily. There are several methods for assaying clonogenic tumour cells, all of which were developed using experimental transplantable tumours, but, more recently, methods have been adapted to include human tumours. They all depend upon having a reliable method for the preparation of a viable cell suspension, which is of course no problem in leukaemias or ascites, but can be difficult in some solid tumours. They are, severally: (1) the spleen colony assay (described for bone marrow above) has also been adapted to experimental leukaemia and lymphomas (Bruce and Van Der Gaag, 1963; Bruce and Meeker, 1964), (2) the lung colony assay, which is usually applicable to transplantable tumours of non-lymphoid origin that colonize the lung after intravenous injection (Hill and Bush, 1969), and consists of injecting a known number of viable tumour cells into the tail vein of a mouse and counting the subsequent tumour nodules which appear in the lung; (3) the limiting-dilution assay, in which the number of cells from an unperturbed tumour required to produce 50 per cent positive takes in recipient animals is compared with the number of cells from a treated tumour required to produce the same proportion of takes; the ratio of the two values gives an estimate of the fraction of clonogenic tumour cells killed by the treatment; (4) colony assays *in vitro*, which involve the direct inoculation of a known number of viable cells into suitable media *in vitro*, with subsequent evaluation of the number of colonies formed. This has been described above in respect of later committed haemopoietic precursors, but the method has also been used to study leukaemic cells (Brown and Carbone, 1971); at the same time it was possible to develop similar methods for solid tumours which have been specially adapted to grow *in vitro* (Reinhold, 1966; Barendsen and Boerse, 1969), but, more recently, it has proved possible to clone experimental tumours directly without such prior adaptation *in vitro*; for example, Courtenay (1976) reported the use of clonogenic assays of the Lewis lung carcinoma and B16 melanoma by direct cloning in soft agar. Moreover, Courtenay, Smith, Peckham and Steel (1976) were able to clone human tumour cells from xenografts growing in immune deprived mice, and Salmon (1978) using feeder layers, has described the cloning of cells taken directly from human tumours.

6.5 Some examples involving the measurement of cell loss kinetics in normal tissues

6.5.1 *Differentiation-induced cell loss*

Methods have been evolved for measuring naturally occurring cell loss rates in renewing tissues, both as a result of the steady-state requirement and during increased cell loss as a result of an induced curtailment of the lifespan of functional cells, or an increase in the cell production rate. The measurement of red cell lifespan by tagging cells with radioactive chromium is well known, and in abnormal states such as haemolytic anaemias, such survival studies can be of use in determining the nature of the disease state (Dacie, 1967). Cell loss measurements in epithelial tissues are not so widely appreciated; Croft, Loehry and Creamer (1968) measured the amount of DNA in exfoliated cells in perfusates of the human small intestine, ensuring by electron microscopy that the exfoliated cells were mainly epithelial in origin. Croft *et al.* (1968) were able to show an increased rate of mucosal cell loss in patients with the avillous mucosa of coeliac disease; the steady-state requirement was balanced by Wright, Watson, Morley, Appleton and Marks (1973a, 1973b), who showed that the enlarged crypt systems found in the mucosa in coeliac disease produce greatly increased numbers of cells. Recently Wright, Watson, Appleton and Marks (1979) showed that increased cell production rates also occurred in mucosal biopsy specimens with less morphological evidence of mucosal damage, and presumably increased rates of cell loss are also found in these mucosae.

There have been considerable technical problems in measuring the rate of cell loss in epidermal tissues, and possibly the most satisfactory approach has been that of Marks (1979), who used an apparatus which scrubbed a measured portion of human epidermis, and counted the cells resulting from the manoeuvre; this cell loss measurement, or desquamation rate, was found to correlate with the transit time through the stratum corneum measured at multiple sites, indicating its reliability.

6.5.2 Cell loss parameters in the adrenal cortex

Assessment of cell loss kinetics from growth rate measurements is usually confined to tumours, but, theoretically at least, the methods could be adaptable to growing organs, and, indeed, this has been carried out in the case of the adrenal cortex (Wright, Voncina and Morley, 1973; Wright and Voncina, 1977). There have been two conflicting theories of adrenocortical cytogenesis: the migration theory, which holds that cells are produced mainly or exclusively in the outer zona glomerulosa and migrate inwards to die in the inner zona reticularis; the competing concept was the zonal theory, which proposed that each zone was responsible for the maintenance of its own structural integrity. Although numerous authors have described degenerating cells in the zona reticularis, the problem was approached on a quantitative basis by Wyllie and his colleagues (Wyllie, Kerr, Macaskill and Currie, 1973; Wyllie, Kerr and Currie, 1973), who described apoptotic bodies located in the inner part of the adrenal cortex of the adult rat, largely confined to zona reticularis and the deepest levels of the zona fasciculata; in foetal glands the process was somewhat more widely distributed, but again was most frequently evident in the deep cortical areas. Suppression of endogenous ACTH (adrenocorticotrophin), by steroids in adults and by decapitation of foetuses, produced increased apoptosis. They further showed that the decrease in adrenocortical weight which occurs in the early neonatal period is also associated with increased apoptosis; this decrease in weight is thought to be caused by a decrease in endogenous ACTH (Hiroshige and Sato, 1970). Exogenous ACTH effectively prevented this increased apoptosis.

Apoptosis is localized to the inner zones, the cells of which, on the basis of a migration theory, would be expected to be chronologically the oldest. Withdrawal of ACTH leads to apoptosis only in the older cells, while administration of ACTH extends their lifespan. Wyllie *et al.* (1973a) went on to suggest that cell deletion played an important part in maintaining adrenocortical cell population size; the action of ACTH on structural homoeostasis could be mediated by its action on cell deletion, possibly associated with an increase in the cell production rate. Fig. 6.10(a) shows the growth curve for the adrenal cortex, and Figs. 6.10(b) and 6.10(c) similar curves for the zona glomerulosa and the zona reticularis respectively; these curves were established by calculating the fractional weight of each component from point-counting serially sectioned glands and from the overall gland weight; the curves are those given by the Gompertz function (Wright and Voncina, 1977), and from these, specific growth rates can be calculated for each zone. The birth rates were calculated from the equation:

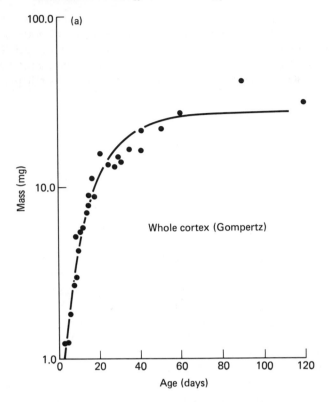

Fig. 6.10 Growth curves for the several zones of the adrenal cortex of the postnatal rat: (a) whole cortex; (b) zona glomerulosa; (c) zona reticularis. Measured by the method described in the text, and fitted by the Gompertz function.

$$k_B = I_S/t_S \qquad\qquad (6.15)$$

and the several kinetic parameters are summarized in Table 6.3. Note that the time chosen for comparison of growth and birth rates is 14 days, where growth is exponential, and, since t_S is short in comparison with T_C, and occurs near the end of the cell cycle (Wright, 1971), this equation will hold for exponential growth.

Table 6.3 shows that the zona glomerulosa is losing 7 cells/1000 cells per h, while the zona reticularis has a negative cell loss rate, in effect a *cell gain rate* of nearly 5 cells/1000 cells per h. The measurements of the cell loss factor, calculated as detailed above (Section 6.4.1), indicate that, in the zona glomerulosa, for every 100 cells born, 57 migrate into the inner zones. In the zona reticularis on the other hand, for every 100 cells born by cell division, nearly 2000 cells enter it by migration from without. Thus the zona glomerulosa would appear to be feeding cells inwards, which are received by the zona reticularis, and despite the fact that cells are dying in this zone by apoptosis, these migrating cells are mainly responsible for the

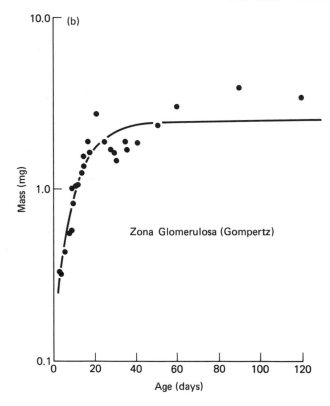

Fig. 6.10 continued.

Table 6.3. Cell loss kinetics in the adrenal cortex of the male rat at 14 days of age (Wright and Voncina, 1977).

	k_B (cells/1000 cells per h)	k_L (cells/1000 cells per h)	ϕ
Zona glomerulosa	6.9	3.9	0.57
Zona reticularis	0.25	−4.8[+]	−19.7[+]

* Calculated from $\phi = k_L/k_B$.
[+] The possession of a negative value here indicated a net cell accretion rate.

postnatal growth of the zona reticularis.

We have to repeat our usual caveat about the equivalence of converting growth rates in cells, but we may conclude that our measurement of the kinetics of cell loss in the adrenal cortex, coupled with the observation that apoptosis is confined to the inner part of the cortex, supports the migration theory of adrenocortical cytogenesis.

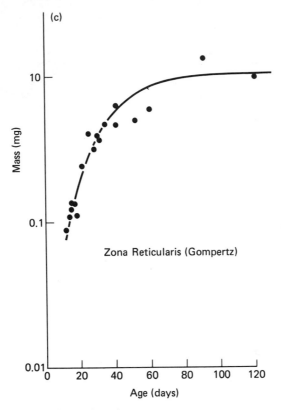

Fig. 6.10 continued.

6.5.3 *Cell loss in the human epidermis*

The increased cell production rate which accompanies psoriasis in the human epidermis is now well known (Weinstein and Frost, 1967; Duffill, Wright and Shuster, 1976). Goodwin, Hamilton and Fry (1973) noted that the flash thymidine-labelling index of both normal and psoriatic epidermis was much higher than the mitotic index; naturally some disparity would be expected, because the duration of DNA synthesis is longer than the duration of mitosis, but Goodwin *et al.* considered the difference to be much larger than that expected from time considerations alone. Ralfs, Dawber, Ryan and Wright (1981) measured the rates of entry of cells into mitosis and DNA synthesis (Table 6.4); it can be seen that the rate of entry into DNA synthesis is 5 to 6 times the rate of entry into mitosis. Since these measurements are fluxes, they cannot be explained by time differentials of DNA synthesis and mitosis, nor can they be explained by an age-distribution effect. Ralfs *et al.* considered that the probable cause of this disparity was the death of numbers of cells between DNA synthesis and mitosis. Subsequent examination of the epidermis in psoriasis does show considerable numbers of apoptotic bodies. Olson and Everett (1975) have also remarked on the presence of epidermal apoptotic bodies in u.v.-light-damaged epidermis, although it is not clear what the proliferative status of the epidermis was at the time of study. It would appear that cell deletion processes are active in rapidly proliferating epidermis, and could be responsible for the disparity between fluxes into S and M in psoriasis; any possible mechanism is, at the moment, obscure, and would repay further study.

Table 6.4. Fluxes into mitosis and DNA synthesis in psoriatic epidermis.

Patient	Flux into M* (cells/1000 cells per h)	Flux into S[+] (cells/1000 cells per h)
1	7.7	43
2	6.5	44
3	9.1	40
4	8.0	57

* Measured by a metaphase-arrest method with vincristine.
[+] Measured by a double-labelling method with [3]HdThd.

6.5.4 Cell death in the intestinal mucosa

The intestinal mucosa, and notably that of the small intestine, has been extensively used as a model system in which to examine the effects of numerous cytocidal manoeuvres, both irradiation (for example Withers and Elkind, 1969) and cytotoxic drugs (for example Al-Dewachi, Wright, Appleton and Watson, 1977). However, it has only recently been noted that evidence of cell death, in the form of apoptotic bodies, is present even on the normal unperturbed mucosa; Cheng and Leblond (1974) reported experiments in which the nuclear remnants of basal crypt cells, killed by the apparent extreme sensitivity to irradiation in the form of β-rays from incorporated [3]HdThd, were phagocytosed by neighbouring crypt cells; Cheng and Leblond concluded that phagocytosis of debris resulting from cell death was a normal pathway for removal of dead cells; moreover, Murray and Rumsey (1976) and Elmes (1977) reported the occurrence of occasional apoptotic bodies in the crypt of normal rats, while Potten (1977) counted about 0.2 apoptotic bodies/crypt section in intact mice. Thus cell death does appear to be a normal phenomenon in the small intestine, although in perturbed states, such as in starvation (Murray and Rumsey, 1976; Elmes, 1977), in the atrophic mucosa of Thiry-Vella loops (Murray and Rumsey, 1976) and in zinc deficiency (Elmes, 1977), apoptosis is increased. After irradiation, Potten (1977) reported that even very low doses of X-irradiation resulted in an increased apoptotic yield, and furthermore, at this low radiation dose, apoptotic bodies were concentrated in the crypt base, which observation supports the concept that basal crypt cells are extremely radiosensitive.

This conclusion is singularly interesting, since the crypt base has been proposed as the site of origin of both functional crypt stem cells (Cheng and Leblond, 1974) and potential crypt stem cells (Hendry and Potten, 1974; Wright, 1977). Potten

(1977) has speculated that this function could be related to the possibility that epithelial stem cells contain a mechanism by which new, damaged DNA strands are selectively distributed to the committed cells, which would ensure stability of the stem cell DNA. Whatever the mechanism, it would appear that basal crypt cells are remarkably radiosensitive.

Thus death of cells in the proliferative compartment of crypts is a normal feature, and is also increased in pathological states. But perhaps more important for the maintenance of intestinal homoeostasis is the concept of *reproductive death*, which in this context means those cells which are able to repopulate the crypt after extensive death of crypt cells, in effect, the *cryptogenic cells*, and numerous studies have now assessed the survival characteristics of these cells to irradiation (for example, Hendry and Potten, 1974) and cytotoxic drugs (Moore, 1979).

In Section 6.4.6 the measurement of cryptogenic cell number was discussed; although there is some controversy about the best method of measurement, perhaps the best estimate for the mouse crypt is 84 ± 31 cryptogenic cells per crypt (Potten and Hendry, 1975), and these cells have a D_0 of about 1.4Gy (1Gy = 100 rads); these are the cells which repopulate the crypt after death of proliferative cells, and may be localized to the crypt base; certainly, after proliferative cell death induced by the DNA poison hydroxyurea, the basal crypt cells are the first to respond, increasing the flux of cells into the depleted proliferative compartment by a rapid decrease in their cell cycle time (Wright, 1977).

There is further evidence that cryptogenic and proliferative cells form separate subpopulations (Hendry and Potten, 1974); in the mouse crypt, after 9 Gy of gamma rays, regeneration characteristics for cryptogenic and proliferative cells differ: cryptogenic cells were assayed by a split-dose survival curve, and proliferative cells were recognized by radioautography. Cryptogenic cells were the first to increase in number (within 24h) whereas proliferative cell number remained low for 1.5 to 2 days after irradiation. Thus it does appear that the two populations are distinct, and detailed radioautographic analysis of regeneration patterns after proliferative cell death does indicate that cryptogenic cells are housed in the crypt base (Hendry and Potten, 1974; Al-Dewachi *et al.*, 1977; Wright, 1977). Cheng and Leblond (1974) maintain that the crypt base also contains the functional stem cells, so a reasonable hypothesis would be that the two populations are synonymous. However, since there are only about 20-30 non-Paneth basal crypt cells (the functional stem cells of Cheng and Leblond), the potential stem cell population must also contain some committed cells. Further work must establish the exact relationship between these cells.

6.5.5 *Cell death in the prostate complex*

Although there has been little work on cell loss and its importance in the normal prostate, after castration there is a marked reduction in cellularity in rodent accessory sex organs. Organ weight declines rapidly following castration, with a loss of DNA (reflecting cell loss) of 40 per cent in the mouse prostate (Kochakian and Harrison, 1962), 50 per cent in the mouse coagulating gland (Alison, Appleton, Morley and Wright, 1974a) and 65–70 per cent in the mouse seminal vesicle (Alison, Appleton, Morley and Wright, 1974b), all measured at 14 days after castration. That this reduction in DNA content reflected cell loss is confirmed by Lesser and Bruchovsky's (1973) finding of an 85 per cent reduction in the number of nuclei isolated from the rat prostate 14 days after castration. Finally Kerr and Searle (1973) have shown, by thorough ultrastructural studies, that castration-induced involution of the rat prostate is characterized by the formation of typical apoptotic bodies. Alison (Alison *et al.*, 1974a, 1974b; Alison, Wright, Morley and Appleton, 1976; Alison and Wright, 1979a) has analysed the induced proliferative response which occurs in the prostatic complex after injection of androgens, by detailed radioautographic studies; the size of the cell deficit does appear to determine the extent of the proliferative response, in that animals stimulated at increasing times after castration (and thus with an increased cell deficit) also showed a proportional increase in the magnitude of the proliferative response. As to the proliferative status of those cells lost, it does appear that there is no preferential loss of P or Q cells after castration, i.e. there is no apparent stable stem cell cohort, with consequent selective loss of non-proliferative Q cells only; Alison and Wright (1979a and b) showed that there was a considerable Q→P cell transition during castration-induced regression in mouse accessory sex glands.

Alison and Wright (1979b) have analysed an intriguingly selective insensitivity of androgen-stimulated regenerating prostate tissue to the necrogenic effects of the poisons of DNA synthesis hydroxyurea (HU) and cytosine arabinoside (AraC). Whilst a single injection of either agent brought about a rapid inhibition of DNA synthesis (as evidenced by [3]HdThd labelling), and whilst both agents had a lethal effect upon the majority of S-phase cells in the intestinal crypts, only a very small proportion of S-phase cells in the accessory sex glands showed evidence of necrosis. The mechanism of this differential lethal effect is uncertain, but could be due to intrinsic differences between continuously cycling cells and cells re-entering the cycle from a G_0 compartment.

6.5.6 *The kinetics of cell death in lymphoid tissue*

Claesson and Hartmann (1976) have presented an interesting analysis of cell loss and migration in the mouse thymus; by a combination of stathmokinetic and radioautographic methods, they considered that about 17% of cells which show up as [3]HdThd-labelled blasts in the S phase will not reach mitosis, i.e. will die between S and M. This conclusion was, in effect, reached by comparing the flux into S with that into M (and in essence similar to the method summarized in Table 6.4). However, this cell loss rate was confirmed by a computer analysis of continuous labelling curves (Sections 6.4.4 and 6.6.1).

Thus, in both normal thymus and in psoriatic epidermis, a major cell loss pathway appears between the S and M phases, and death in G_2 may be an important avenue of cell loss.

6.5.7 Assessment of cell loss in vitro

Many morphological methods are suitable for application *in vitro*, including such techniques as Trypan Blue exclusion, although, as we have seen, these have been extensively criticized by Roper and Drewinko (1976), who consider that clonogenic assays are the only worthwhile assay methods in this context. The only rigorous studies concerning the measurement of the kinetics of cell death are by Jagers and Norrby (Jagers, 1970; Norrby 1970; Jagers and Norrby, 1974); the method is based on the theory of branching processes, and makes several assumptions: (i) a steady state of exponential growth; (ii) a growth fraction of unity; (iii) equal probability of death from all cells. Consequently, because at least the first two of these requirements are only found *in vitro*, the method cannot be adapted for use *in vivo*; furthermore, the calculation of cell loss from this method required a computation of the distribution function of t_m, the duration of mitosis, which is only possible by time-lapse cinematography. Norrby (1970) has applied this method to monolayer cultures.

6.6 The kinetics of cell loss in tumours

The theoretical and practical aspects of cell loss assessment have been discussed in Section 6.4 *et seq.* and here it is proposed to consider some examples of where cell kinetic methods have been able to quantify cell loss and indicate possible avenues along which deletion processes occur.

6.6.1 Cell loss in experimental tumours

Fig. 6.3(a) shows the growth curve for a rapidly growing transplantable sarcoma, growing in Balb/c mice; the curve is representative of many such tumours, and can be satisfactorily described by the Gompertz function (see Section 6.4.4). Measurements of k_L and of ϕ can be made, as described in Section 6.4.4, by comparing the birth rate with the measured growth rate. The results of these calculations are given in Table 6.5, which shows results for 7, 14 and 21 days after transplantation; k_L and ϕ are calculated by three methods, each of which, however, use k_G values from the growth curve shown in Fig. 6.3(a): (i) by calculating the birth rate from the duration of S (t_S) and the flash labelling index (measuring λ from Fig. 6.5); t_S was measured by the fraction-of-labelled-mitoses method; (ii) by measuring the birth rate directly by a metaphase-arrest method with vincristine; (iii) by analysing continuous thymidine-labelling curves by a computer model, which proposes that cells can die at mitosis, from the non-proliferative Q cell compartment or randomly from any phase.* The computer analysis allows an assessment of the relative importance of each loss pathway. An illustrative continuous labelling curve is shown in Fig. 6.11, and the computed cell loss pathways are shown in Table 6.6.

Table 6.5 shows that each method agrees that there is an increase in the cell

*This analysis was carried out by Niels Hartmann, of the Finsen Institute, Copenhagen.

Table 6.5. ϕ values in Balb/c sarcoma measured by three methods at 7, 14 and 21 days after transplantation.

	7 days	14 days	21 days
Fraction labelled mitoses	0.43	0.78	0.93
Vincristine	0.35	0.65	0.99
Continuous labelling and computer modelling*	0.19	0.81	0.93

*Analysed by Niels R. Hartmann.

Table 6.6. Cell loss pathways for the Balb/c sarcoma at 7, 14 and 21 days after transplantation, analysed by continuous thymidine labelling.*

	ϕ_M^+	ϕ_Q^+	ϕ_R^+	ϕ
7 days	0.00	0.19	0.00	0.19
14 days	0.51	0.00	0.30	0.81
21 days	0.55	0.00	0.38	0.93

* Analysed by Niels R. Hartmann.
+ These are the fractional ϕ values due to mitotic, Q cell and random cell loss respectively.

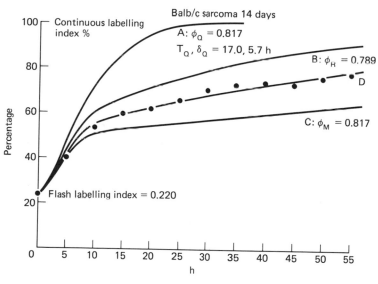

Fig. 6.11 An example of a continuous [3]HdThd-labelling curve for the Balb/c sarcoma, at 14 days after transplantation. The curves are for the models shown, and the experimental points are also indicated (analysed by N. R. Hartmann).

loss rate as tumour growth proceeds, and, because the birth rate shows a small decrease with time, the cell loss factor ϕ also increases over the growth period, until at 21 days most cells which are produced at mitosis are dying, and the growth curve shows that there is little effective growth at this time. The increase in k_L is largely responsible for the curtailment of growth (Aherne, Camplejohn and Wright, 1977). It should be noted from Table 6.5 that k_L and ϕ, as measured by the [3]HdThd and vincristine methods, show good agreement, emphasizing that reasonable results are obtainable by the simple and more convenient stathmokinetic method. However, the computer model, although indicating the same trend in k_L and ϕ, does show some difference in absolute values compared with the other two methods.

Table 6.6 shows that, although at 7 days cell loss is shown as occurring exclusively from the Q cell compartment, at 14 and 21 days after transplantation random cell loss and loss at mitosis appear as major loss pathways, and at 21 days, most cell loss seems to be happening at mitosis. This is the converse of the conclusion of Hermens (1973); studying a transplantable rhabdomyosarcoma, Hermens reported that death of cells in mitosis was an important source of cell loss in the early stages of growth, but later, as the tumour enlarged, other modes of loss, particularly interphase death, predominated. However, measurement of nuclear DNA content in the Balb/c sarcoma showed that there was a large increase in polyploid and aneuploid nuclei with time after transplantation, and that this was accompanied by an increase in morphologically abnormal mitoses, which observations would support an increased loss of cells at mitosis. As noted above (Section 6.4.4), this method is particularly sensitive to the effects of non-exponential growth, and of course, after 7 days growth is no longer truly exponential, as judged by the growth curve (Fig. 6.3a). This limits the usefulness of this technique in the assessment of cell loss after perturbation of tumours by irradiation or cytotoxic chemotherapy.

Lala (1972) reported an interesting and detailed analysis of cell loss from the Ehrlich ascites tumour; Lala noted that, in tumour cells at 7 days after transplantation, the maximum continuous labelling index after 13h of thymidine labelling (equivalent to the cell cycle time) was 65 per cent. Thus, if all proliferating (P) cells are labelled, any further increase in the continuous labelling index must reflect the transformation of labelled P cells to non-proliferative labelled Q cells, and this occurred at a rate of 2.4 per cent of cells/h. Now, if the growth fraction remains constant, the fractional loss from the Q cell compartment will equal the overall cell loss rate multiplied by the fraction of Q cells, which equalled 0.83 per cent/h; this represents the contribution of the non-proliferating cells to cell death. Consequently, since the P→Q transformation occurs at 2.4 per cent/h, there must be preferential loss from the Q cell compartment (if the growth fraction remains constant). Furthermore, if this loss occurred in an age-specific manner (i.e. with a constant T_L, see Fig. 6.7), so-called 'pipe-line cell loss', Lala considered that the rate of increase of the continuous labelling index (i.e. the P→Q transformation rate) should equal the maximum cell loss rate, and both were found to equal 2.4 per cent/h. Nevertheless, it was not possible to exclude a small amount of random cell loss, or some death at mitosis, as fragmented mitoses were evident.

Thus kinetic methods provide an insight into the avenues of cell death in tumours, and can also provide reliable estimates of the cell death rate; however, in interpreting these results, as in any cell kinetic method, it is well to remember the assumptions upon which the results are based. Useful tabular data for cell loss parameters are given for a large range of experimental and human tumours by Cooper *et al.* (1975) and Steel (1977).

6.6.2 *Loss of clonogenic cells in tumours*

Although overall cell loss rates give a valuable insight into the mode of tumour growth, the regrowth potential of any tumour depends on its content of clonogenic cells; the several methods that are available are discussed in Section 6.4.7. A detailed discussion of clonogenic cell survival in tumours would be out of place here, but the subject has been excellently reviewed by Steel (1977). The main value of such studies has been in assessing the mode of action and possible optimizing of therapy with cytotoxic chemotherapeutic agents. A valuable application was reported by Bruce, Meeker and Valeriote (1966), who compared the survival response of normal bone marrow with that of a transplantable mouse lymphoma, and, on the basis of these studies, have proposed a widely used classification of such agents into (i) proliferation-independent, (ii) cell-cycle-phase-specific and (iii) proliferation-dependent agents. While a detailed discussion is out of place here, we might note that the clonogenic cell survival curve for each class of agent does differ; for example, S-phase-specific drugs, such as 5-fluorouracil and cytosine arabinoside, tend to give survival curves that fall to a plateau, indicating that cells in some part of the cell cycle are escaping.

The clonogenic cell content of experimental tumours varies; thus L1210 leukaemia has a large clonogenic fraction, many of which are proliferating, and, while most clonogenic PC-5 plasmacytoma cells are also proliferating, they constitute only a small fraction of the total cell population (Steel, 1977); such considerations are of direct importance in the prediction of tumour response to treatment.

6.7 Tissue responses to cell loss

In kinetic terms, it is difficult to consider the problem of induced cell loss and death in isolation, without some mention of its effect on the remaining proliferating cells. Thus we have seen that regeneration characteristics of any tissue will depend largely on the nature of the remaining cells, and their clonogenic capacity. Studies on the effect of induced cell loss or death on cell proliferation are important in our understanding of growth control mechanisms; for example the proliferative response of epidermis after removal of cells from the horny layer may involve ejection of cells from a G_0 compartment, a shortening of the cell cycle time or both (Fowler and Denekamp, 1976), but whatever the kinetic mechanism, such studies bring us into the realm of the control of tissue mass, a field in which the role of cell-line-specific growth control agents such as chalones has been emphasized (Houck, 1976). However, the role of cell loss in maintaining tissue homoeostasis is considered elsewhere in this volume (see Chapter 4).

6.8 Conclusions

In this chapter we have been largely concerned with the measurement of the rate of cell loss and death in tissues; we have not given much thought to the mechanism of cell loss beyond classifying likely candidates. Moreover, many of the techniques described involve numerous assumptions, many of which are difficult to substantiate. Nevertheless, the methods described do give considerable insight into the dynamics of tissue populations, and further work may enable a more stringent quantitative analysis.

Acknowledgements

The personal work described herein was carried out with the support of the Cancer Research Campaign and the Wellcome Trust. I am grateful to Neils R. Hartmann for the computer analysis of the continuous labelling curves, and to Mrs Barbara Gill for expert secretarial assistance.

References

Aherne, W. A., Camplejohn, R. S. and Wright, N. A. (1977), *Cell Population Kinetics*, Edward Arnold, London.

Al-Barwari, S. E. and Potten, C. S. (1976), Regeneration and dose-response characteristics of irradiated dorsal epidermal cells. *Int. J. Radiat. Biol.*, **30**, 201–216.

Al-Dewachi, H. S., Wright, N. A., Appleton, D. R. and Watson, A. J. (1977), The effect of a single injection of hydroxyurea on cell population kinetics in the rat jejunal crypt. *Cell Tissue Kinet.*, **10**, 203–213.

Alison, M. R., Appleton, D. R., Morley, A. R. and Wright, N. A. (1974a), Cell population growth in the castrate mouse prostate complex; experimental verification of computer simulation. *Cell Tissue Kinet.*, **7**, 425–431.

Alison, M. R., Appleton, D. R., Morley, A. R. and Wright, N. A. (1974b), Influence of cell loss upon the cytokinetics of induced DNA synthesis in the accessory sex organs of castrated mice. *J. Endocrinol.*, **63**, 18.

Alison, M. R. and Wright, N. A. (1979a), Testosterone induced cell proliferation in the accessory sex glands of mice at various times after castration. *Cell Tissue Kinet.*, **12**, 461–475.

Alison, M. R. and Wright, N. A. (1979b), Differential lethal effects of both cytosine arabinoside and hydroxyurea on jejunal crypt cells and the testosterone stimulated accessory sex glands. *Cell Tissue Kinet.*, **12**, 477–491.

Alison, M. R., Wright, N. A., Morley, A. R. and Appleton, D. R. (1976), Cell proliferation in the prostate complex of the castrate mouse. *J. Microsc.*, **106**, 221–237.

Appleton, D. R., Wright, N. A. and Dyson, P. (1977), Age distribution in stratified squamous epithelium. *J. Theor. Biol.*, **68**, 769–779.

Bagshawe, K. D. (1968), Tumour growth and antimitotic action. The role of spontaneous cell losses. *Br. J. Cancer*, **22**, 698–713.

Barendsen, G. W. and Boerse, J. J. (1969), Experimental therapy of a rat rhabdomyosarcoma with 15MeV neutrons and 300KV X-rays. 1. Effect of single exposures. *Eur. J. Cancer*, **5**, 373–391.

Baserga, R. (1971), *The Cell Cycle and Cancer*, Marcel Dekker, New York.

Begg, A. C. (1977), Cell loss from several types of solid murine tumours; comparisons of ^{125}I-iododeoxyuridine and tritiated thymidine methods. *Cell Tissue Kinet.*, **10**, 409–427.

Begg, A. C. and Fowler, J. F. (1974), A rapid method for the evaluation of tumour cell loss. *R. B. E. Br. J. Radiol.*, **47**, 154–160.

Bradley, T. R. and Metcalfe, D (1966), The growth of mouse bone marrow cells *in vitro*. *Aust. J. Exp. Biol. Med. Sci.*, **44**, 287–300.

Brown, C. H. and Carbone, P. P. (1971), *In vitro* growth of normal and leukaemic human bone marrow. *J. Natl. Cancer Inst.*, **46**, 989–1000.

Bruce, W. R. and Meeker, B. E. (1964), Dissemination and growth of transplanted isologous mouse lymphoma cells. *J. Natl. Cancer Inst.*, **32**, 1145-1159.

Bruce, W. R., Meeker, B. E. and Valeriote, F. A. (1966), Comparison of the response of normal haemopoietic and transplanted lymphoma colony-forming cells to chemotherapeutic agents administered *in vivo*. *J. Natl. Cancer Inst.*, **37**, 233-245.

Bruce, W. R. and Van der Gaag, H. (1963), A quantitative assay for the number of murine lymphoma cells capable of proliferation *in vivo*. *Nature (London)*, **199**, 79-80.

Cairnie, A. B. (1976), Homeostasis in the small intestine. In *Stem Cells of Renewing Cell Populations* (eds. A. B. Cairnie, P. K. Lala and D. G. Osmond), Academic Press, New York and London, pp. 67-77.

Cheng, H. and Leblond, C. P. (1974), Origin, differentiation and renewal of the four main epithelial types in the mouse small intestine. V. Unitarian theory of the origin of the four epithelial types. *Am. J. Anat.*, **141**, 537-562.

Claesson, M. H. and Hartmann, N. R. (1976), Cytodynamics of the thymus of young adult mice; a quantitative study on the loss of thymic blast cells and non-proliferative small thymocytes. *Cell Tissue Kinet.*, **9**, 273-291.

Cooper, E. H. (1973), The biology of cell death in tumours. *Cell Tissue Kinet.*, **6**, 87-95.

Cooper, E. H., Bedford, A. J. and Kenny, T. E. (1975), Cell death in normal and malignant tissues. *Adv. Cancer Res.*, **21**, 59-120.

Courtenay, V. D. (1976), An *in vitro* colony assay in soft agar for the Lewis lung tumour and the B16 melanoma taken directly from the mouse. *Br. J. Cancer*, **34**, 39-45.

Courtenay, V. D., Smith, I., Peckham, M. J. and Steel, G. G. (1976), The *in vitro* and *in vivo* radiosensitivity of human tumour cells obtained from a pancreatic carcinoma xenograft. *Nature (London)*, **263**, 771-772.

Croft, D. N., Loehry, C. A. and Creamer, B. (1968), Small bowel cell loss and weight loss in the coeliac syndrome. *Lancet*, **2**, 68-70.

Dacie, J. C. (1967), *The Haemolytic Anaemias*, Churchill, London.

Denekamp, J. (1972), The relationship between the cell loss factor and the immediate response to irradiation in animal tumours. *Eur. J. Cancer*, **8**, 335-340.

Dethlefsen, L. A. (1969), Comparisons of tumour radioactivity after the administration of either [125]I- or ^{3}H-labelled 5-iodo-2'-deoxyuridine. *Cancer Res.*, **29**, 1717-1720.

Dethlefsen, L. A. (1970), Reutilisation of ^{131}I-5-iodo-2'-deoxyuridine as compared to ^{3}H-thymidine in mouse duodenum and mammary tumour. *J. Natl. Cancer Inst.*, **40**, 389-395.

Dethlefsen, L. A. (1971), An evaluation of radioiodine labelled 5-iodo-2'-deoxyuridine as a tracer for measuring cell loss from solid tumours. *Cell Tissue Kinet.*, **4**, 123-138.

Dethlefsen, L. A., Sorensen, J. and Snively, J. (1977), Cell loss from 3 established lives of the C_3H mouse mammary tumour: a comparison of the [125]IUdR and autoradiographic methods. *Cell Tissue Kinet.*, **10**, 447-459.

Duffill, M. B., Appleton, D. R., Dyson, P., Shuster, S. and Wright, N. A. (1977), The measurement of the cell cycle time in squamous epithelium using the metaphase arrest method with vincristine. *Br. J. Dermatol.*, **96**, 493-502.

Duffill, M. B., Wright, N. A. and Shuster, S. (1976), Cell population kinetics in psoriasis measured by three independent techniques. *Br. J. Dermatol.*, **94**, 355-362.

Elmes, M. E. (1977), Apoptosis in the small intestine of zinc deficient and fasted rats. *J. Pathol.*, **123**, 219-223.

Feinendegen, L. E., Bond, V. P. and Hughes, W. L. (1966), Physiological thymidine neutralisation in rat bone marrow. *Proc. Soc. Exp. Biol. Med.*, **122**, 448–455.

Fowler, J. and Denekamp, J. (1976), Regulation of epidermal stem cells. In *Stem Cells of Renewing Cell Population* (eds. A. B. Cairnie, P. K. Lala and D. Osmond), Academic Press, New York, pp. 117–134.

Gelfant, S. (1977), The cell cycle in psoriasis – a reappraisal. *Br. J. Dermatol.*, **95**, 577–590.

Glücksmann, A. (1951), Cell deaths in normal vertebrate ontogeny. *Biol. Rev. Cambridge Philos. Soc.*, **26**, 59–86.

Goodwin, P. G., Hamilton, S. and Fry, L. (1973), Comparison between DNA synthesis and mitosis in involved and uninvolved psoriatic epidermis and normal epidermis. *Br. J. Dermatol.*, **89**, 619–623.

Hartmann, N. R., Dombernowsky, P. and Bichel, P. (1976), Resting cells in L1210 and JB1 ascites tumours with special emphasis on recycling. *Cancer Treatment Rep.*, **60**, 1861–1870.

Hendry, J. H. and Potten, C. S. (1974), Cryptogenic and proliferative cells in intestinal epithelium. *Int. J. Radiat. Biol.*, **25**, 583–588.

Hermens, A. L. (1973), Variation in the cell kinetics and growth rate in an experimental tumour during natural growth and after irradiation. Thesis, University of Amsterdam; Radiobiological Institute, TNO, Risjerick, Netherlands.

Hill, R. P. and Bush, R. S. (1969), A lung colony assay to determine the radiosensitivity of cells of a solid tumour. *Int. J. Radiat. Biol.*, **15**, 435–444.

Hiroshige, T. and Sato, T. (1970), Circadian rhythm and stress changes in hypothalamic content of corticotrophin releasing activity during postnatal development in the rat. *Endocrinology*, **86**, 1184–1186.

Hofer, K. G. (1969), Tumour cell death *in vivo* after administration of chemotherapeutic agents. *Cancer Chemother. Rep.*, **53**, 273–281.

Hofer, K. G., Prensky, W. and Hughes, W. L. (1969), Death and metastatic distribution of tumour cells in mice monitored with [125]IUdR. *J. Natl. Cancer Inst.*, **43**, 763–773.

Houck, J. C. (ed.) (1976), *Chalones*, North-Holland, Amsterdam.

Iversen, O., Bjerknes, R. and Devik, F. (1968), Kinetics of cell renewal, cell migration and cell loss in the hairless mouse dorsal epidermis. *Cell Tissue Kinet.*, **1**, 365–379.

Jagers, P. (1970), The composition of branching cell populations; a mathematical result and its application to determine the incidence of death in a cell population. *Mathemat. Biosci.*, **8**, 227–238.

Jagers, P. and Norrby, K. (1974), Estimations of the mean and variance of cycle times in cinematographically recorded cell populations during balanced exponential growth. *Cell Tissue Kinet.*, **7**, 201–211.

Kerr, J. F. R. and Searle, J. (1973), Deletion of cells by apoptosis during castration-induced involution of the rat prostate. *Virchows Arch. Abt. B, Cell Pathol.*, **13**, 87–92.

Kochakian, C. D. and Harrison, D. G. (1962), Regulation of nucleic acid synthesis by androgens. *Endocrinology*, **70**, 99–106.

Lajtha, L. J., Pozzi, L. V., Schofield, R. and Fox. M. (1969), Kinetic properties of haemopoietic stem cells. *Cell Tissue Kinet.*, **2**, 39–49.

Lala, P. K. (1972), Evaluation of the mode of cell death in Ehrlich ascites tumour. *Cancer*, **29**, 261–266.

Lesser, B. and Bruchovsky, N. (1973), The effects of testosterone 5 α-dihydrotestosterone and adenosine 3′:5′-monophosphate on cell proliferation and differentiation in rat prostate. *Biochim. Biophys. Acta*, **308**, 426–435.

Marks, R. (1979), Quantitative aspects of keratinisation. In *The Epidermis in Disease* (eds. R. Marks and E. Christophers), in the press.

Mendelsohn, M. L. and Dethlefsen, L. A. (1968), Cell proliferation and volumetric growth of fast line, slow line and spontaneous C_3H mammary tumours. *Proc. Am. Assoc. Cancer Res.*, 9, 47–55.

Moore, J. V. (1979), Ablation of murine jejunal crypts by alkylating agents. *Br. J. Cancer*, 39, 175–181.

Murray, B. and Rumsey, R. D. E. (1976), Cell deletion in the mucosa of the rat small intestine. *J. Physiol. (London)*, 263, 197–198P.

Norrby, K. (1970), Population kinetics of normal, transforming and neoplastic cell lines. *Acta Pathol. Microbiol. Scand.*, 78, Suppl. 214.

Olson, R. L. and Everett, M. A. (1975), Epidermal phagocytosis; cell deletion by phagocytosis. *J. Cut. Pathol.*, 2, 53–57.

Pierce, G. B. and Wallace, C. (1971), Differentiation of malignant to benign cells. *Cancer Res.*, 31, 127–134.

Porschen, W. and Feinendegen, L. E. (1969), *In vivo* Bestimmung der Zell verlustrate bei Experimental tumouren mit markiertear Joddeoxyuridin. *Strathlen-therapie*, 137, 718–723.

Potten, C. S. (1977), Epithelial proliferative subpopulations. In *Stem Cells and Tissue Homeostasis* (eds. B. I. Lord, C. S. Potten and R. J. Cole), Cambridge University Press, Cambridge, pp. 317–334.

Potten, C. S. and Hendry, J. H. (1973), Clonogenic cells and stem cells in epidermis. *Int. J. Radiat. Biol.*, 27, 413–424.

Potten, C. S. and Hendry, J. H. (1975), Differential regeneration of intestinal proliferative cells and cryptogenic cells after irradiation. *Int. J. Radiat. Biol.*, 27, 413–424.

Ralfs, I., Dawber, R., Ryan, T. and Wright, N. A. (1980), Studies on cell proliferation in psoriatic epidermis, using the metaphase arrest method with vincristine. *Br. J. Derm*, (in press).

Refsum, S. B. and Berdal, P. (1967), Cell loss in malignant tumours in Man. *Eur. J. Cancer*, 3, 235–236.

Reinhold, H. S. (1966), Quantitative evaluation of the radiosensitivity of the cells of a transplantable rhabdomyosarcoma of the rat. *Eur. J. Cancer*, 2, 33–42.

Roper, P. R. and Drewinko, B. (1976), Comparisons of *in vitro* methods to determine drug-induced cell lethality. *Cancer Res.*, 36, 2182–2188.

Salmon, S. E. (1978), Bioassay of human tumour stem cells. A new approach to the evaluation and treatment of cancer. *Ariz. Med.*, 35, 109–111.

Smith, J. A. and Martin, L. (1973), Do cells cycle? *Proc. Natl. Acad. Sci. U.S.A.*, 70, 1263–1267.

Smith, R. S., Thomas, D. B. and Riches, A. C. (1974), Cell production in tumour isografts measured using vincristine and colcemid. *Cell Tissue Kinet.*, 7, 529–536.

Steel, G. G. (1966), Delayed uptake by tumours of tritium from thymidine. *Nature (London)*, 210, 806–808.

Steel, G. G. (1967), Cell loss as a factor in the growth rate of human tumours. *Eur. J. Cancer*, 3, 381–383.

Steel, G. G. (1968), Cell loss from experimental tumours. *Cell Tissue Kinet.*, 1, 193–207.

Steel, G. G. (1977), *Growth kinetics of tumours*, Oxford University Press, Oxford.

Steel, G. G. and Hanes, S. (1971), The technique of labelled mitoses; analysis by automatic curve fitting. *Cell Tissue Kinet.*, 4, 93–105.

Stephenson, J., Axelrad, A., McLeod, D. and Shreeve, N. M. (1971), Induction of colonies of haemoglobin-synthesising cells by erythropoietin *in vitro. Proc. Natl. Acad. Sci. U.S.A.,* **68**, 1542–1546.

Tannock, I. (1968), The relationship between cell proliferation and the vascular system in a transplanted mouse mammary tumour. *Br. J. Cancer,* **22**, 258–273.

Tannock, I. (1970), Population kinetics of carcinoma cells, capillary endothelial cells, and fibroblasts in a transplanted mouse mammary tumour. *Cancer Res.,* **30**, 2470–2476.

Till, J. E. and McCulloch, E. A. (1961), A direct measurement of the radiation sensitivity of normal mouse bone marrow. *Radiat. Res.,* **14**, 213–222.

Weinstein, G. D. and Frost, P. (1967), Abnormal cell proliferation in psoriasis. *J. Invest. Dermatol.,* **50**, 254–259.

Withers, H. R. (1976), Colony forming units in the intestine. In *Stem Cells of Renewing Cell Populations* (eds. A. B. Cairnie, P. K. Lala and D. G. Osmond), Academic Press, New York and London, pp. 33–40.

Withers, H. R. and Elkind, M. M. (1969), Radiosensitivity and fractionation response of crypt cells of mouse jejunum. *Radiat. Res.,* **38**, 598–613.

Wright, N. A. (1971), Cell proliferation in the prepubertal male rat adrenal cortex; an autoradiographic study. *J. Endocrinol.,* **49**, 599–609.

Wright, N. A. (1977), The cell population kinetics of repopulating cells in the intestine. In *Stem Cells and Tissue Homeostasis* (eds. B. I. Lord, C. S. Potten and R. J. Cole), Cambridge University Press, Cambridge, pp. 335–358.

Wright, N. A. and Voncina, D. (1977), Postnatal growth of the rat adrenal cortex. *J. Anat.,* **123**, 147–156.

Wright, N. A., Voncina, D. and Morley, A. R. (1973), An attempt to demonstrate cell migration from the zona glomerulosa in the prepubertal male rat adrenal cortex. *J. Endocrinol.,* **59**, 451–459.

Wright, N. A., Watson, A. J., Appleton, D. R. and Marks, J. (1979), Crypt cell kinetics in convoluted mucosa. *J. Clin. Pathol.,* in the press.

Wright, N. A., Watson, A. J., Morley, A. R., Appleton, D. R. and Marks, J. (1973a), Cell kinetics in the flat (avillous) mucosa of the human small intestine. *Gut,* **14**, 701–710.

Wright, N. A., Watson, A. J., Morley, A. R., Appleton, D. R., Marks, J. and Douglas, A. (1973b), The cell cycle time in the flat (avillous) mucosa of the human small intestine. *Gut,* **14**, 603–606.

Wyllie, A. H., Kerr, J. F. R. and Currie, A. R. (1973a), Cell death in the normal neonatal rat adrenal cortex. *J. Pathol.,* **111**, 255–281.

Wyllie, A. H., Kerr, J. F. R., Macaskill, I. A. M. and Currie, A. R. (1973b), Adrenocortical cell deletion: the role of ACTH. *J. Pathol.,* **11**, 85–94.

7 Cell death and the disease process. The role of calcium

BENJAMIN F. TRUMP, IRENE K. BEREZESKY AND
ALVARO R. OSORNIO-VARGAS

7.1 Introduction

The cellular processes involved in the death of the cell are of vital importance to biology and medicine: in biology because they occur as a normal part of the economy of every organism even during development and, of course, in its ultimate demise, and in medicine because cell death forms an important part of virtually every disease. We have been studying the events associated with cellular death for over 20 years and, in this chapter, will briefly review not only the knowledge that we have gained but also correlate it with data obtained from other laboratories. In addition, we will review our current working hypothesis concerning ion redistribution, particularly calcium, and their effects which lead to cell death.

We regard cell death as an irreversible reaction to an injury. An injury is defined as any perturbation that alters the 'normal' homoeostasis of cells. The severity of injurious agents ranges from those that kill the cell virtually instantaneously, such as high temperature, chemicals such as osmium tetroxide, and mechanical disruption, to those that kill the cell more slowly, resulting in a progression with time through a series of degenerative stages ultimately resulting in irreversible cell injury or cell death. With the latter type of injury, the cell continues to show reactions even after its death. These are predominantly degradative and convert the dead cell to debris which approaches physical chemical equilibrium with the environment. Many of the acute instantaneous injuries do not occur in necrosis, as they simultaneously inactivate the enzymes and processes responsible for necrosis. This is, of course, why such agents are employed as fixatives for morphological study. There is still another type of injury which, even though prolonged, does not result in cell death. These sublethal, often chronic, injuries will not be considered in this review.

Following most types of acute lethal injury, such as total ischaemia or damage to the cell membrane with complement or certain chemical toxins, the cell undergoes a series of reactions, first reversible then irreversible, the two being separated by the 'point of no return' or 'point of cell death'. The first part of the chapter will review these events, which, for purposes of convenience, we have designated as stages. The latter part of the chapter will consider the possible mechanisms involved, especially those concerning ion redistributions and, in particular, ionized calcium.

7.2 Stages of cell injury

Although we have published the stages of cell injury several times through the years (Trump and Mergner, 1974; Trump, Laiho, Mergner and Arstila, 1974a; Trump and Arstila, 1975) and most recently in some detail (Trump, Berezesky, Laiho, Osornio, Mergner and Smith, 1980), we will review them again here in order to provide the reader with sufficient detail to facilitate easy comprehension of our concepts concerning cell death.

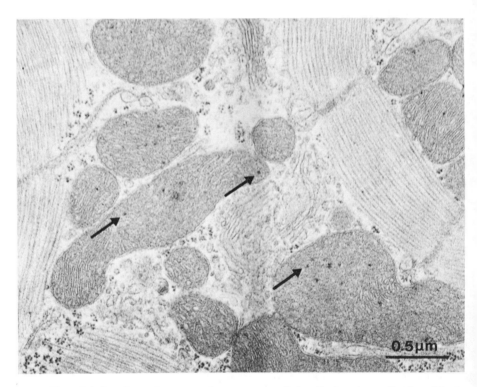

Fig. 7.1 Portion of a normal rat myocardial cell showing mitochondria in orthodox conformation. Note the presence of normal mitochondrial granules (arrows). *Stage 1.*

Stage 1 is a normal cell with normal appearance and relationships of organelles (Fig. 7.1). Following lethal injury, progression through stages 2-7 occurs. In the first stage, an injury such as ischaemia results in a sudden decrease in the oxygen tension within the extracellular fluid and the cell. Because of the rapid fall in oxygen tension, the cell seems to 'attempt' to maintain its homoeostasis by using very rapid control mechanisms. Such control mechanisms involve changes in ion concentrations as well as in various co-enzymes and regulating enzyme activities. As the oxygen concentration falls, mitochondrial phosphorylation decreases rapidly and contraction of the inner mitochondrial compartment begins. The drop in cellular ATP is believed to stimulate the activity of phosphofructokinase (PFK) in many cells, and this results in an increased rate of anaerobic glycolysis, assuming that an ample supply of glycogen is present. This accelerated glycolysis leads to the accumulation of lactate which, together with the increased inorganic phosphate from ATP hydrolysis, soon reduces the intracellular pH. Since the circulation has ceased, metabolites such as lactate and hydrogen ions also tend to accumulate in the extracellular microenvironment, depressing the extra- as well as the intracellular pH. This decreased pH appears to be protective in that it has a 'stabilizing' effect on the cell membrane (Penttila and Trump, 1974, 1975a, 1975b, 1975c; Penttila, Glaumann and Trump, 1976). The change in intracellular pH is probably reflected by clumping of nuclear chromatin (*Stage 1a*), which occurs very rapidly but is clearly reversible. This could also result from loss of bound K^+ which occurs very quickly. Clumping of nuclear chromatin is known to be associated with decreased nuclear RNA synthesis. However, this is of no immediate consequence to the cell, since if circulation is not restored, the cell will die and undergo necrosis long before the decreased RNA synthesis has any effect. In addition to nuclear chromatin clumping and reduction in glycogen, cells in *Stage 1a* also show the disappearance of mitochondrial matrical granules, as seen after glutaraldehyde/ osmium fixation and Epon embedment. Meanwhile, the cell also begins to show the effects of the reduced ATP concentration. Among the consequences of the fall in ATP concentration is a decrease in the activity of the ion pumps at the cell membrane that leads to movements of Na^+, K^+, Ca^{2+}, and Mg^{2+} down their concentration gradients. Using X-ray microanalysis, we have seen such ion shifts as early as 5 min after ischaemia (Osornio, Berezesky, Mergner and Trump, 1981).

One of the earliest ultrastructural changes which can be seen in cells after a variety of acute lethal injuries is enlargement in the volume of the endoplasmic reticulum (ER) that is reflected in electron micrographs as dilation of the cisternae (*Stage 2*). This initial change can be correlated with increased sodium and decreased potassium content of the cell and often with increased water content, although studies by Laiho and Trump (1974a, 1974b) indicate that significant dilation of the ER can sometimes occur with decreased total cell volume. The implied mechanism is intracellular redistribution of ions and water. This stage is clearly reversible and occurs before the 'point of no return'. Also in *Stage 2*, cell surface changes in the form of protrusions or 'blebs' begin to appear. These alterations in cell shape appear to reflect acute changes in the microtubules and microfilaments whose normal function is to maintain cell shape. Recent data have implied that lack of control of cell Ca^{2+} levels during this phase, probably through effects on the cytoskeleton, may result in these changes (Trump, Berezesky, Mergner and Phelps, 1978a; Trump, Penttila and Berezesky, 1979b, 1979c). In addition, alterations in the membrane-associated Mg^{2+}-dependent ATPase can be seen (Mergner, Smith and Trump, 1972; Mergner, Chang, Marzella, Kahng and Trump, 1979). As the pH continues to drop in the cytosol, it is probable that inhibition of PFK occurs, with a gradual decrease in the rate of glycolysis, tending to stabilize the glycogen levels at low but significantly greater than zero values. The fall in cellular ATP concentration continues and ADP rises. This decreased ATP concentration begins to be reflected by decreased functions of the various energy-requiring systems such as protein synthesis, ion pumps at the cell surface, and most probably ion accumulation by mitochondria which begin to undergo shrinkage of the inner compartment.

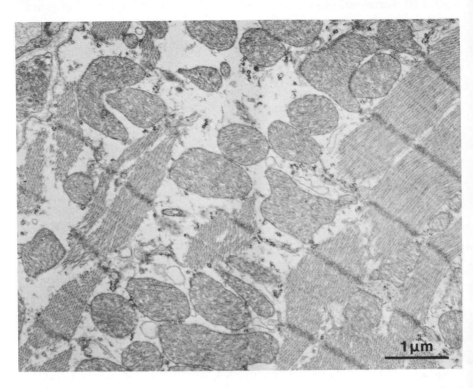

Fig. 7.2 Portion of a rat myocardial cell after 5 min coronary occlusion. Note condensation of the inner mitochondrial compartments and expansion of the intracristal spaces. This change represents a short and transient stage in the ischaemic rat heart. *Stage 3*.

In *Stage 3*, the mitochondria appear dense, showing marked condensation of the inner compartment and enlargement of the space between the inner and outer membranes of the envelope and the intracristal space (Fig. 7.2). This is associated with dilation of the ER and the cell sap, both of which correlate with the gradual increases in Na^+ and water contents as well as with decreases in K^+ and Mg^{2+}. Loss of K^+ from mitochondrial inner compartments could result in shrinkage of that structure, while movements into the ER could explain the dilation of such cisternae. Decreased K^+ content is also a factor which is associated with inhibited protein synthesis, although at this stage the membrane-bound polysomes appear unaltered and maintain both their ordered arrangements and their membrane attachments.

The increased water content at this state is often reflected by the presence of blebs at the cell surface, although the latter may also involve modulations of the cytoskeleton. Mitochondrial function is inhibited because of the continued lack of oxygen and, as a consequence, ATP and now ADP levels are low (Kahng, Berezesky and Trump, 1978). If cells in this stage are restored to normal conditions, the mitochondrial inner compartments re-expand, the ER cisternae contract, and the cells extrude Na^+ and Ca^{2+} along with water and Cl^-. Therefore, although the mitochondria are temporarily and dramatically changed in form, they retain their ability to respire and phosphorylate as well as the ability to regain their normal internal compartment volume following restitution of a normal microenvironment. At this stage, protein synthesis is severely inhibited and both free and bound polysomes are present as monosomes, some of which maintain their membrane attachments. The cell sap is swollen because of the increased water content and the lysosomes are clear and often swollen, although significant leaks in the lysosomal membranes apparently do not occur. At present, it is not clear whether the release of lysosomal hydrolases to the cytosol plays a role in this or later stages.

In *Stage 4*, the transition across the 'point of no return' begins. Here, the mitochondria begin to show high amplitude swelling which seems to indicate initially reversible and later irreversible loss of the inner membrane function. As a result, some mitochondrial profiles show high amplitude inner compartment swelling, while others remain condensed. In some cells, such as renal tubular epithelium, where two or more inner compartments are present in a given mitochondrion, one compartment may be condensed while others are swollen (Glaumann and Trump, 1975; Glaumann, Glaumann, Berezesky and Trump, 1975; Trump, Strum and Bulger, 1974b; Kahng *et al.*, 1978). The remaining organelles resemble those seen in Stage 3. This series of membrane-permeability changes involving both the mitochondria and the cell membrane probably reflect basic chemical as well as structural modifications in the molecular architecture. The mitochondrial swelling is associated with increased content of Na^+ and Ca^{2+} in the inner compartments, although the precise ionic composition of the mitochondria probably depends upon that of the cytoplasm. Among the factors which may result in this structural damage to the mitochondrial inner membrane is the continued Mg^{2+} deficiency and the destruction of phospholipids and release of fatty acids. Mitochondrial membranes possess endogenous phospholipases which can be activated by factors such as increased levels of Ca^{2+} known to be present at this time (Goracci, Porcellati and Woelk, 1978).

In *Stage 4a*, in addition to the alterations mentioned above, all mitochondria are swollen with inner compartments expanded (Fig. 7.3). In *Stage 4b*, not only are all mitochondria swollen but some also contain tiny dense aggregates (Fig. 7.4). This stage was recently described by Glaumann, Glaumann, Berezesky and Trump (1977a) and Glaumann, Glaumann and Trump (1977b) in proximal tubules from the pars convoluta of rat kidney following 30 and 60 min of ischaemia.

In *Stage 5*, all mitochondria exhibit massive swelling and show marked increases in membrane permeability associated with loss of proteins including matrical

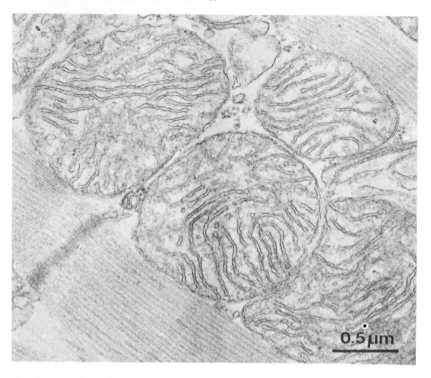

Fig. 7.3 Portion of a myocardial cell from the dog following 15 min *in vivo* occlusion showing mitochondrial swelling. *Stage 4a* (Courtesy of Trump *et al.*, 1976c).

enzymes and cofactors to the cell sap and ultimately through the cell membrane to the extracellular space. Cell sap enzymes are probably lost before mitochondrial or other organelle enzymes and the sum of adenosine nucleotides is only a small fraction of normal (Kahng *et al.*, 1978). Also at this stage, prominent flocculent densities appear in the inner compartment presumably the result of denaturation of matrix proteins (Fig. 7.5). Cells are swollen and the membranes in the cytocavitary network are fragmented. Distortion of intracellular membrane systems with whorl-like configurations often can be seen at the cell surfaces. Chromatin begins to undergo enzymic attack and in later stages is completely dissolved. Lysosomes begin to swell as evidenced by increased size, clarification of the matrix and condensation of the contents. The lysosomal swelling may result in hydrolase leak. In *Stage 5a*, calcifications of the inner mitochondrial compartments can be seen, consisting of amorphous or crystalline deposits along the inner membrane (Fig. 7.6).

In *Stage 6*, the changes are mainly characterized by a rapid acceleration in the rate of digestion of intracellular constituents as measured by marked increases in free fatty acids, free amino acids and inorganic phosphorous, and decreases in phospholipids, triglycerides, protein, DNA and RNA. This is reflected morphologically

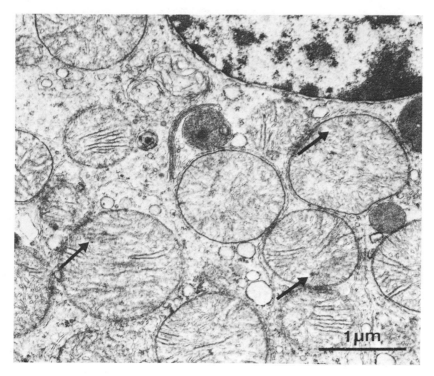

Fig. 7.4 Portion of a rat proximal convoluted tubule cell following 60 min *in vivo* ischaemia. Mitochondria are swollen and some contain tiny dense aggregates (arrows). *Stage 4b*. (Courtesy of Trump *et al.*, 1980.)

by beginning karyolysis and changes in the staining pattern of the cytoplasm. There is evidence that, at this stage, the lysosomal contents can escape into the cytosol and, indeed, the disappearance of the macromolecular components of the cell, as well as the persistence of lysosomal hydrolases as compared with many other enzymes, strongly indicates that lysosomal leakage is followed by digestion. Other membrane systems show dramatic rearrangements, including vesiculation, wrappings, formation of tubular forms in the inner compartment of the mitochondria, disappearance of ribosomes from the ER membrane surface, alterations in the nucleolus, disappearance of microbody matrices and frequent visualization of breaks in the plasma membrane contour.

In *Stage 7*, in spite of the fact that the cell is virtually completely degraded, new structures in the cytoplasm occur as large dense inclusions composed of dense homogeneous osmiophilic material with lamellar regions in lacunae within bodies or at the margins of the bodies. These periodic regions resemble stacks of membranes, although the fine structure within the bodies has not been discerned. However, they probably correspond to the myelin of Virchow, so-named because of the resemblance to the bodies which Virchow described following autolysis in

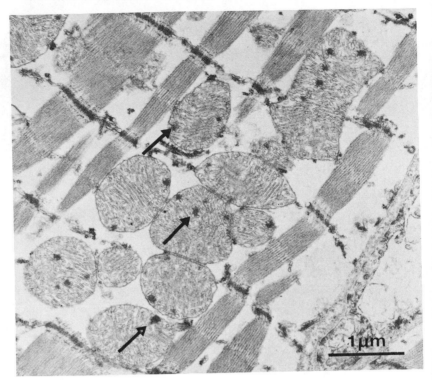

Fig. 7.5 Portion of a rat myocardial cell after 30 min coronary occlusion. All mitochondria are swollen and contain large flocculent densities (arrows), the hallmark of irreversibility. The cytoplasm is less electron dense and the sarcomeres are relaxed. *Stage 5.*

various cell types especially in the CNS. In general, at this stage, most enzymic activity has approached zero. Unless other sources of enzymes such as those produced by micro-organisms are present, the cells are stabilized for long periods of time in this state. However, equilibration once again occurs during this phase when the extracellular fluid pH approaches physiological levels. Greater and greater Ca^{2+} binding by these cells is observed, possibly related to the change in pH as well as to changes in reactive protein sites. How much of the Ca^{2+} binding is mitochondrial is not known. At this stage, it is almost certainly not energy-dependent. A diagrammatic summary illustrating mitochondrial profiles following cell injury is shown in Fig. 7.7.

7.2.1 *Comments on the stages*

The rate of progression through these stages varies dramatically in different cell types and different types of injury. We do not yet know all of the reasons for this

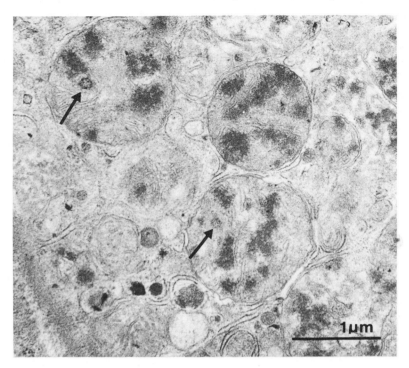

Fig. 7.6 Portion of rat proximal convoluted tubule cell following 120 min *in vivo* ischaemia followed by 24h reflow. Mitochondria are swollen and contain large flocculent densities and occasional calcifications (arrows). *Stage 5c.* (Courtesy of Trump *et al.*, 1980.)

variation. For example, Stage 3 is extremely variable. In hypoxic Ehrlich ascites tumour cells (EATC; Trump and Laiho, 1975) or renal cortical slices at 0-4°C (Kahng *et al.*, 1978; Trump *et al.*, 1974b), Stage 3 is prominent as it also is with acute cell membrane damage by organic mercurials (Penttila and Trump, 1975a,c) or complement (Hawkins, Ericsson, Biberfeld and Trump, 1972). On the other hand, this stage is absent or transient in acute total ischaemia of rat kidney (Glaumann and Trump, 1975; Glaumann *et al.*, 1975), dog (Trump, Berezesky, Collan, Kahng and Mergner, 1976b; Trump, Mergner, Kahng and Saladino, 1976c) or rat myocardium (Osornio-Vargas, unpublished work), human or rat pancreas (Jones, Garcia, Mergner, Pendergrass, Valigorsky and Trump, 1975) and human or hamster bronchus (Barrett, McDowell, Harris and Trump, 1977). Both smooth and skeletal muscle also represent special cases possibly based on their remarkable capacity for anaerobic glycolysis (Paul, Bauer and Prease, 1979). We are currently investigating the latter for its possible relationship to hypertension and athero-sclerosis.

In addition, recent studies in our laboratories have necessitated modifications of Stages 4 and 5. In proximal tubules from the pars convoluta of rat kidney following 30 and 60 min of ischaemia, all mitochondria not only appeared swollen (indicative of Stage 4a) but some contained tiny dense aggregates (Glaumann *et al.*, 1977a). These alterations were classified into Stage 4b. Since these changes were compatible with the apparently complete recovery seen in the cortex, it was inferred that this stage was also a reversible one. The dense aggregates are believed to represent reversible denaturation of matrix proteins in the mitochondria; it is well known that protein denaturation is initially reversible (Lehninger, 1970). Moreover, Majno, Lagattuta and Thompson (1960) reported reversibility of protein denaturation prior to the 'point of no return' in light-scattering studies of tissue slices following ischaemic injury.

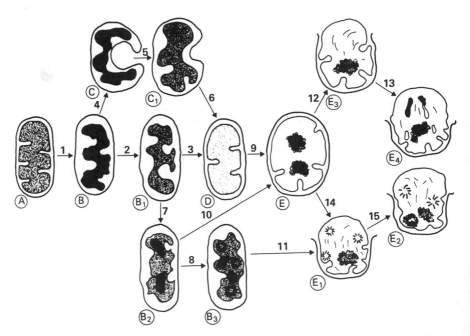

Fig. 7.7 A diagrammatic representation of mitochondrial profiles following cell injury. A, Orthodox mitochondria (*Stage 1*); B, condensed (*Stage 3*); B_1, transitional form of condensed; B_2 and B_3, condensed mitochondria with flocculent densities or calcifications (*Stage 3c*); C and C_1, ring-formed condensed mitochondria; D, slightly swollen (*Stage 4a*); E, E_1, E_2, E_3 and E_4, various forms of highly swollen mitochondria with flocculent densities (*Stage 5*) or calcifications (*Stage 5c*). (Courtesy of Laiho and Trump, 1975.)

Stages 3c, 4c and 5c represent altered cells in which there is calcification of mitochondria in addition to the other changes mentioned. This modification does not occur with total ischaemia or chemicals which inhibit respiration or oxidative phosphorylation but does appear after ischaemia followed by reflow (Glaumann *et al.*, 1977a, 1977b) or after lethal injury with membrane-damaging agents such as complement and antibody (Hawkins *et al.*, 1972), amphotericin B (Saladino,

Bentley and Trump, 1969), penetrating or non-penetrating mercurials (Gritzka and Trump, 1968; Sahaphong and Trump, 1971), inhibition of the $Na^+ + K^+$ ATPase with ouabain (Ginn, Shelburne and Trump, 1968) (Fig. 7.8), and direct mechanical damage (Trump *et al.*, 1974b).

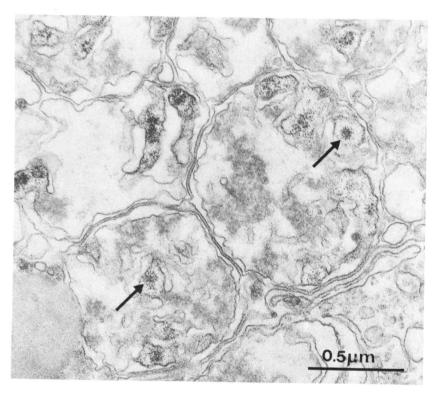

0.5μm

Fig. 7.8 Portion of a tubule cell from an isolated flounder nephron treated with 10^{-5} M-ouabain for 4 h. Note the massive enlargement of the mito-chondrial matrical compartments, interruptions in the continuity of the basilar infoldings, and calcium deposits (arrows).

Different cell types do not reach the irreversible Stage 5 at the same time interval. For example, totally ischaemic rat renal proximal tubule cells require approximately 1-2 h (Glaumann and Trump, 1975; Glaumann *et al.*, 1975) while hamster and human bronchial epithelia require over 3 h (Barrett *et al.*, 1977). The ambient temperature of the cells during the injurious process also has obvious relevance to the rate of change through these stages. Kidney cells at 37°C progress to Stage 5 in 1-1.5 h while at 4°C, this same progression requires over 48 h (Kahng *et al.*, 1978). Similarly, at 0-4°C, human bronchus and mammary epithelia survive for over 1 week (Barrett *et al.*, 1977).

Another stage modification is that of mitochondrial membrane fusions. In acute myocardial infarction in the rat, and to some extent also in acute ischaemia of the rat kidney, the mitochondrial inner membranes exhibit early fusions, resulting in the formation of intracristal or intermembrane helical arrays with a periodicity of 15 nm (Osornio, Berezesky, Mergner and Trump, 1980). These fusions typically appear in large numbers during early time intervals following occlusion and tend to decline and disappear in the later intervals (Fig. 7.9). Although the nature of these inner membrane fusions is not presently known, we speculate that they represent the effect of a calcium-induced phospholipase action, with the release of membrane-fusing agents such as fatty acids or lysophosphatides, or some direct action of calcium itself. As the time after injury increases, the inner membrane fusions decrease in frequency. This could mean that completion of the fusion causes a return to a trilaminar membrane, as was recently described in secretory vesicles (Sun, Day and Ho, 1978), or that the fusion precedes membrane disruptions, as we have suggested (Osornio-Vargas, Berezesky and Trump, unpublished work).

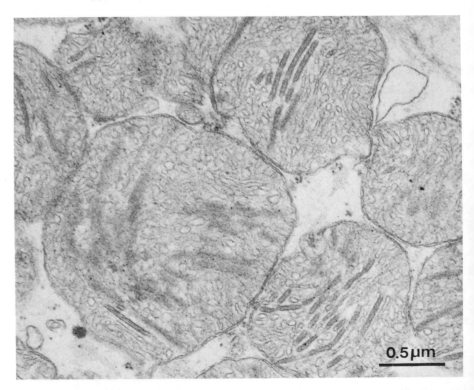

Fig. 7.9 Mitochondria from a rat myocardial cell in the central portion of a 60 min infarct illustrating abundant fusions of the cristae. This alteration could represent cristae modifications secondary to phospholipase activation.

In muscle, intermembrane mitochondrial inclusions have been observed in several diverse conditions such as ischaemia (Hanzlikova and Schiaffino, 1977; Karpati, Carpenter, Melmed and Eisen, 1974), post-mortem aging (Cheah and Cheah, 1977;

Saito, Smigel and Fleisher, 1974) and anoxia (Schiaffino, Severin and Hanzlikova, 1979). However, the exact nature of these inclusions and the factors responsible for their formation remain to be determined. Recently, Hanzlikova and Schiaffino (1977) reported that they were protein in nature, as shown by enzyme-digestion experiments, and suggested that they may result from the polymerization of enzymes such as creatine kinase present in the mitochondrial intermembrane space. However, the stain affinity of these structures also suggests the presence of lipids.

7.3 Mechanisms of progression

It is evident from consideration of these stages that one of the characteristic types of change that occurs is change in the volume of intracellular compartments and redistribution of the state of dispersion of their contents. This originally led to explorations of the theory that regulation of cell ions and water could be important factors in the process and might even be crucial to the onset of irreversible damage. Indeed, studies in a variety of model systems as well as in experimental and human disease have revealed that such shifts commonly do occur. For example, early nuclear chromatin clumping could well be the result of increased hydrogen ion especially with factors that inhibit mitochondrial phosphorylation and lead to accumulation of lactated phosphate. Likewise, the dilation of the ER and the swelling of the cell itself could be the result of increased accumulation of sodium, and water passively following. During late stages of necrosis, magnesium levels in the cell tend to fall, but this does not seem to occur until after the 'point of no return'. Recently, we have been intrigued by the distinct possibility that control of ionized calcium concentration could represent a key factor in inducing the 'point of no return'. This is discussed more extensively below in the section relating to our current hypothesis.

One set of observations stands out in all of our studies, namely that two major types of agent lead to acute cell death followed by necrosis. These are: (1) agents that inhibit or uncouple mitochondrial oxidative phosphorylation and (2) agents that affect the permeability or transport system of the cell membrane. Examples of the first type of agent include anoxia, ischaemia and treatment with metabolic inhibitors including potassium cyanide, carbon monoxide, dinitrophenol and antimycin. Examples of the second type of agent include complement and antibody interacting with the cell membrane, mechanical damage to the cell membrane, ionophorous antibiotics such as amphotericin and inhibitors of Na^+-K^+ ATPase such as the cardiac glycosides. This leads us to speculate that agents of unknown mechanisms which result in acute cell death probably act through one or both of these mechanisms. Future experiments will obviously be necessary to confirm or refute this hypothesis. In any case, many if not most of the causes of cell death in human disease can be directly attributable to one or both of these factors. Although, through the years, there have been numerous hypotheses imputing the cause of cell death to one or another organelle malfunction, definity of support cannot be deduced for any but these two. It seems indisputable, however, that total inhibition of nuclear function would result in losses of cell viability, but treatment of many cells with inhibitors of DNA synthesis or even direct removal by micro-manipulator of the nucleus does not result in cell death. On the other hand, some systems, such as dividing cell populations, are inhibited. Also, in the case of the haematopoietic system or epithelial linings of the gastrointestinal tract where constant cell renewal is essential, viability is certainly affected by inhibition of DNA, RNA and protein synthesis. On the contrary, other interphase cells can continue their viability, even with virtually 100 per cent inhibition of protein synthesis. It is true that such conditions could result in a more chronically progressive and slower cell death, and, indeed, such may be the case. At the same time, it would be our prediction that, once again, such inhibition of transportation and translation would ultimately result in killing of the cell through failure to maintain the barrier function of the plasmalemma and the energetic function of the mitochondrion.

What then of the lysosomes? Ever since De Duve's original papers on the subject in which he proposed the 'suicide bag' concept of lysosomes (De Duve and Wattiaux, 1966), it has been speculated that triggering of lysosomal enzyme release into the cytoplasm might be a common denominator of many if not all types of acute cell death. This hypothesis has been difficult to either confirm or refute (Hawkins *et al.*, 1972; Wildenthal, 1978; Decker and Wildenthal, 1980). In many common types of acute lethal injury, such as anoxia or ischaemia, it seems that leaky lysosomes do not occur until after the 'point of no return' (Trump and Ginn, 1969). On the other hand, in certain systems, including uptake of uric crystals or silicon, evidence can be produced that is compatible with, although not proving that, the suicide bag hypothesis is correct. At the same time, we know that in many types of injury, lysosomal enzymes are released into the extracellular space and seemingly have injurious effects on neighbouring tissue cells, as in the acute inflammatory process. The mechanism here is probably once again the effect of these hydrolases on the cell membranes of the affected neighbouring cells and proceeds through pathways not unlike those in our models of cell membrane damage.

7.4 The role of ion shifts in cell injury

Numerous studies over the last few years have strongly implicated physiologically active ions such as Na^+, Mg^{2+}, P^{2-}, Cl^-, K^+, and Ca^{2+} as playing an extremely important role in both normal and abnormal modifications of a variety of cell processes. It has become increasingly evident that the critical control of ion concentrations is not only responsible for normal cell functions such as maintenance of cell shape, cell transport, cell junctions, cell-cell communication, energy production, protein and nucleic acid synthesis, and control of the cell cycle but also that any modifications of such concentrations are very important in the pathophysiology of many disease states, including acute renal failure, myocardial infarction, stroke, shock, metaplasia, regeneration and malignant transformation. The recent development of

X-ray microanalysis (Russ, 1978) has now made possible the measurement and quantification of these diffusible ions (Trump, Berezesky, Chang and Bulger, 1976a; Trump, Berezesky, Pendergrass, Chang, Bulger and Mergner, 1978b; Trump, Berezesky, Chang, Pendergrass and Mergner, 1979a; Trump *et al.*, 1980) and thus will allow further elucidation of their quality and quantity in various cell compartments (Fig. 7.10). As a result, correlation with altered cell function will be possible and will provide a greatly improved basis for the development of a more refined understanding of the pathophysiology of many important diseases.

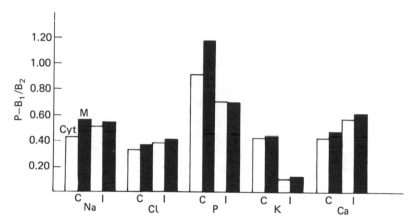

Fig. 7.10 Bar graph illustrating X-ray microanalysis peak to background ratios for Na, Cl, P, K, and Ca where P is the peak of interest, B_1 is the background under the peak, and B_2 is the continuum count between 5.50 to 6.50 keV after the continuum from the support film has been subtracted. Measurements were obtained over ultrathin frozen-dried cryosections of quench-frozen control rat kidney cortex (C) and over cortex following 1 h in vivo ischaemia (I). Mitochondria (M). Cytoplasm (Cyt). (Courtesy of Trump et al., 1979a.)

The various effects of ion shifts and/or movements from their normal cell compartments following cell injury have been briefly discussed under the section on stages. Although all ion shifts can be deleterious to the cell, we will confine our comments in the remainder of this chapter to the role of calcium, since considerable evidence has been accumulated which indicates that abnormal distributions and elevations of this ion in the cytosol can initiate catabolic processes directly leading the cell beyond the 'point of no return' or cell death.

7.5 Calcium and cell injury

Calcium is an extremely important intracellular electrolyte whose unique role as activator and regulator of many diverse cellular activities is just beginning to be realized. Its distribution across the cell membrane is not in equilibrium; under normal conditions, free intracellular calcium approximates to 10^{-7} M, while extracellular calcium is 10^{-3} M (Baker, 1972; Blaustein, 1974; Kretsinger, 1976). Control of calcium levels, especially that portion which is ionized and which exists extracellularly in only a small fraction of the total, has recently received much attention. However, total cell calcium may have little relationship to that which is physiologically active. In squid axon, of 50 μmol/kg, only 0.06 per cent is in the ionized form, although ionized calcium is higher in concentration in the extracellular compartment (Brinley, 1978). Also, the distribution of intracellular calcium is uneven, with almost 90 per cent being localized in mitochondria (Brinley, 1978). As for calcium levels in various intracellular compartments, further studies need to be performed before our knowledge can be considered complete. However, using X-ray microanalysis both we (Trump *et al.*, 1978b, 1979a) and Somlyo's group (Somlyo, Somlyo and Shuman, 1979) have failed to find very great differences in total concentrations between mitochondria and cytosol in normal cells (see Fig. 7.10). This does not necessarily mean that large gradients could not exist for ionized calcium but rather that adequate methods for measurement at the subcellular level have not been completely understood or fully developed. As should be evident from the foregoing, calcium levels may well be the common thread that could explain the progression of cells through the reversible and irreversible stages of cell injury and finally to cell death. We believe therefore that knowledge of the activities of this important cation may well be pivotal in the understanding of these events.

The relationship between tissue calcification and necrosis as a terminal stage of cell injury has been recognized by the pathologist for many years (Fig. 7.11). However, it was only recently that we learned that total cellular calcium levels are directly related to cell death. Accumulated evidence from our laboratory over the past 20 years (Gritzka and Trump, 1968; Saladino *et al.*, 1969; Croker, Saladino and Trump, 1970; Sahaphong and Trump, 1971; Laiho and Trump, 1974b; Trump *et al.*, 1978a) has given support to the hypothesis that the initial rise in calcium concentration in the cytosol which occurs following cell injury may be detrimental to the cell. The concept that such calcium accumulation might be primary to cell death as well as a late and therefore secondary phenomenon has only recently been added. Investigations which support this concept of dramatic increases in calcium following injury include those of Hearse (1977) on ischaemic myocardium, Shen and Jennings (1972) on reflow studies, Kloner, Ganote, Whalen and Jennings (1974) on the transient ischaemic model, and Goring and Spieckermann (1978) during reduced perfusion. On the basis of the importance of the well-regulated role of calcium (especially in its ionized form) in normal cellular function (Cheung, 1980), one can easily deduce that any modifications in these calcium levels will exert a deleterious effect on the cell. As has been previously shown, variations in cytosolic ionized calcium in normal cells occurs within relatively narrow limits (Carafoli, 1979), with organelles such as mitochondria and endoplasmic reticulum involved in its control.

Although the possibility that calcium might not act in its free ionic form but rather in the presence of a binding protein has been known for some time (Meyer,

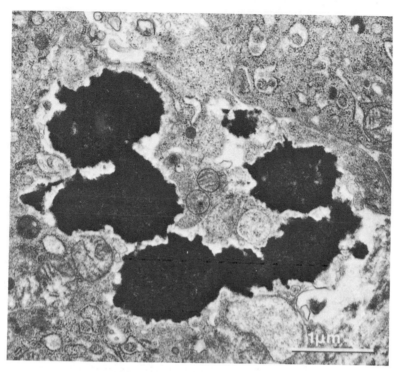

Fig. 7.11 An area from the central portion of a 7-day rat myocardial infarct exhibiting large dense calcium deposits surrounded by cellular debris and fibroblasts. This picture represents the traditional association between calcium and cell death. (Courtesy of Trump *et al.*, 1980.)

Fischer and Krebs, 1964; Cheung, 1970; Kakiuchi, Yamazaki and Nakajima, 1970; Wolff and Siegel, 1972), it was not until recently that this same calcium-binding protein (calmodulin) was found to act as a calcium regulator in a variety of cellular enzyme systems and in most types of cell motility (Rasmussen and Goodman, 1977; Rubin, 1974; Dedman, Brinkley and Means, 1979; Cheung, 1980; Means and Dedman, 1980). This protein, apparently found in all nucleated cells, appears to play a major role in normal and abnormal cellular regulation because of its role as an intracellular intermediary for calcium. Apart from the already mentioned more or less productive activities, calcium, through calmodulin, may be involved in destructive cell processes as well, since calcium can activate the seriously destructive cellular phospholipases. Until relatively recently, however, the manner in which calcium could modify or control such seemingly diverse functions defied analysis. With the discovery of calmodulin, we have, for the first time, possible insight into the mechanisms.

In addition to the above-mentioned support for the hypothesis that high calcium levels are detrimental, excellent correlation has been obtained between cellular

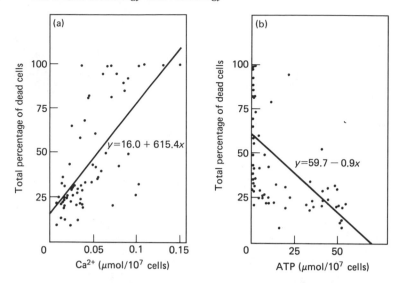

Fig. 7.12 (a) Scatter plot with regression line showing a good positive corre-lation between Ca^{2+} content of cells and total percentage of dead cells. Cellular injury was produced by an inhibitor of function of the plasma membrane [PCMBS (P-choromercuribenzene sulfonic acid)] or by inhibitors of respiration (antimycin A), glycolysis (iodoacetic acid), oxidative phos-phorylation (2, 4-dinitrophenol) or by combinations of the last three. ($r = 0.68$, $P < 0.001$, $N = 62$.) (b) Scatter plot with regression line showing a moderate correlation between ATP content and the total percentage of dead cells. Cellular injury was produced as in Fig. 7.12(a) ($r = 0.58$, $P < 0.001$, $N = 72$.) (Courtesy of Laiho and Trump, 1974b.)

calcium levels and loss of viability in EATC injured by agents which inhibit energy metabolism and inflict direct cell membrane damage (Laiho, Shelburne and Trump, 1971) (Fig. 7.12). Although it is still not known exactly in what manner calcium levels are modified following cell injury, there are at least three possible mechan-isms. These include: (1) increased permeability or decreased extrusion at the plasma membrane; (2) increased efflux or decreased accumulation by mitochondria; and (3) increased efflux or decreased accumulation by the ER. However, whether such modifications take place in the ionized or total calcium remains to be elucidated. Following ischaemic or anoxic injury, decrease in ATP production results in the arrest of the plasma-membrane ion-transport mechanisms, including calcium ex-trusion and cessation of the Na^+-K^+ pump. This, in turn, leads to subsequent calcium accumulation with the resultant Na^+ increase further interfering with mitochondrial calcium accumulation, at least in excitable cells (Crompton, Moser, Ludi and Carafoli, 1978).

We have previously reported, for example, that in toad bladders, following amphotericin treatment, sucrose Ringer's buffer promoted much more mitochon-drial calcification than did sodium Ringer's buffer (Saladino *et al.*, 1969). As a consequence of the loss of volume control which results from the cessation of the Na^+-K^+ pump, cellular swelling may accentuate the calcium control problem by cell membrane 'stretching', with increased permeability and increased calcium influx.

Progression of the damage as the injury persists is accompanied by sequential and degenerative changes in the inner mitochondrial membrane. Although these alterations cannot presently be defined at the molecular level, such changes are associated with impairment of the mitochondrial calcium-accumulation ability (Trump, Croker and Mergner, 1971; Mergner, Smith, Sahaphong and Trump, 1977) and may correlate with alterations in mitochondrial phospholipids, as seen following ischaemia (Smith, Collan, Kahng and Trump, 1980).

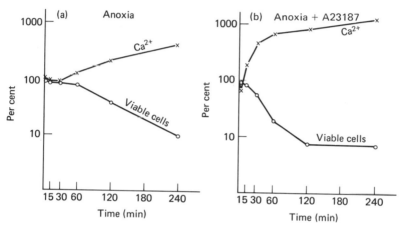

Fig. 7.13 (a) Graph illustrating percentage of viable cells and Ca^{2+} uptake in EATC following 0–240 min of anoxia. (Courtesy of Trump *et al.*, 1980.) (b) Graph illustrating percentage of viable cells and Ca^{2+} uptake in anoxic EATC following 0–240 min treatment with 10 μM-ionophore A23187. (Courtesy of Trump *et al.*, 1980.)

The use of calcium ionophores, such as A23187, has contributed significantly to our understanding of the role of calcium in the disease process. Recently, we reported the synergistic effect of this compound on the progression of cell death in EATC following anoxia (Trump, Jones, Berezesky, Phelps and Laiho, 1978c; Laiho, Berezesky and Trump, unpublished work) (Fig. 7.13). Similar results were obtained by Schanne, Kane, Young and Farber (1979), who found that rat hepatocytes exposed to several membrane-active toxins were killed more rapidly in the presence of calcium rather than in its absence. Studies on liver and kidney have suggested that ionophorous compounds are not only responsible for cellular calcium influx from the extracellular fluid but that they also can induce intracellular redistributions of calcium (Borle and Studer, 1978). Publicover, Duncan and Smith (1978), studying ultrastructural damage by A23187 in mammalian muscle, postulated that the action of this ionophore resulted in the release of calcium from the sarcoplasmic reticulum followed by its uptake in mitochondria which then, after 40 min, released the calcium. These findings support the observations by Wrogemann and Pena (1976), who reported that cellular necrosis in various muscle disease was a consequence of an increased net influx of cellular calcium over-loading, which, in turn, caused calcium overloading of the mitochondria.

Zimmerman, Daems, Hulsmann, Snijder, Wisse and Durrer (1967) described marked cell damage when myocardial tissue was exposed to calcium after short periods of perfusion with calcium-free medium. This phenomenon, known as the 'calcium paradox', has provided evidence for the calcium-damaging effect in the heart and kidney. Recently, Bulkley, Nunnally and Hollis (1978) demonstrated correlation between morphological irreversible changes and decreases in creatine phosphate and ATP in hearts subjected to the 'calcium paradox'. On the other hand, Ashraf, White and Bloor (1978) have observed a delay in the initiation of ultrastructural and enzymic changes after calcium-free perfusion. Although the phenomenon is not completely understood, it appears that the lack of calcium modifies the membrane lipid–protein relationship, altering its permeability and then facilitating the entry of calcium. Also, morphological studies of tissues exposed to a calcium-free medium have demonstrated alterations in cell shape and cellular junctions (Bulger and Trump, 1969).

7.5.1 *Mechanisms*

Sufficient evidence appears to be at hand to relate elevated cytosolic calcium levels, caused by modification of cell membrane permeability to this cation, to cell death. As was mentioned earlier, such modifications are common in human disease, occurring not only with non-specific mechanical damage, as in violent trauma, but also with various other specific and non-specific perturbations, including those caused by certain heavy metals, antigen–antibody reactions which activate complement, ionophorous antibiotics, and agents of unknown mechanisms including endotoxin. When calcium levels exceed the cell's capability to modulate them through the mitochondria, endoplasmic reticulum and cytoplasmic proteins such as calmodulin, severe consequences can be expected. Some of these adverse effects include activation of calcium-dependent membrane-bound enzymes such as phospholipase, excessive mitochondrial calcium accumulation with crystal deposition and loss of function, and alterations in the cytoskeleton. Of these, the effect which has the most consequence to the cell in the disease process is the perpetuation of plasma membrane and mitochondrial membrane damage by membrane-bound phospholipases, a hypothesis which we have postulated for the past few years (Trump *et al.*, 1976b; Smith *et al.*, 1980).

7.5.2 *Phospholipases*

It is well-known that phospholipases play an important role in cell lipid metabolism by carrying on a major role in phospholipid-membrane turnover (Gan-Elepano and Mead, 1978). This process is performed by membrane-bound phospholipases in diverse cellular membrane systems; some phospholipases require the presence of calcium as a cofactor and present a constant and rapid fatty acid release and/or replacement, which is useful as an immediate precursor pool. An example of this is in prostaglandin synthesis where the activated phospholipases serve as a source of fatty acids, releasing free arachidonate, the basic structure in prostaglandins. Soluble acidic calcium-inhibited phospholipases A1 and A2 have been localized in lysosomes, and membrane-bound alkaline calcium-activated phospholipases A1 and A2 have been demonstrated in the plasma membrane, ER and Golgi apparatus. The mitochondria also contain membrane-bound phospholipases, but only phospholipase A2 has been found in the inner and outer membranes. Different agents are capable of stimulating or inhibiting phospholipases, and some of these have been used to induce or protect against cell injury. For example, calcium and the ionophore A23187 are among the inductive group, while local anaesthetics are in the second group (Kunze, Nahas, Traynor and Wurl, 1976). Goracci *et al.* (1978) have reviewed the properties and localization of these enzymes in the heart and liver.

Cell injury is known to result in release of fatty acids. This release has been found under such diverse conditions as in both tissue and perfusate following myocardial ischaemia (Lochner, Kotze, Benade and Gevers, 1978), in aged mitochondria (Parce, Cunningham and Waite, 1978), in ischaemic rat liver (Chien, Abrams, Serroni, Martin and Farber, 1978) and in response to membrane peroxidation and vitamin E deficiency (Pappu, Fatterpaker and Sreenivasan, 1978). Fatty acids and lysophosphatides also have a deleterious effect on membranes and can be considered as sustaining agents of cell injury (French, Norum, Ihrig and Todoroff, 1971). Their removal with albumin has demonstrated a diminution of the noxious effect. We have recently observed that one major source of fatty acids is mitochondrial phospholipids, which decrease early following acute ischaemic injury in the rat kidney (Smith *et al.*, 1980). These studies also demonstrated that cardiolipin is decreased at a greater rate than other mitochondrial phospholipids. It is known that cardiolipin is the major phospholipid component of the mitochondrial inner membrane and that it can be correlated with the maintenance of the mitochondrial Mg^{2+}-ATPase, NADH oxidase and cytochrome oxidase function (Luzikov, 1973). It is probable then that the degradation of mitochondrial phospholipids contributes significantly to the loss of mitochondrial function (Smith *et al.*, 1980).

Although the release of fatty acids after cell injury has several possible explanations, it is our present hypothesis that membrane-bound phospholipases are the responsible factor. Acidic lysosomal phospholipases could also provide a possible explanation, but some of our experiments can be interpreted as excluding any role for lysosomal enzymes. For example, we have observed that a reduction in extracellular pH has a protective effect against anoxic injury, cell membrane damage and hyperthermia (Penttila and Trump, 1974; Penttila *et al.*, 1976), and that lysosomal enzyme release is a late effect which occurs after the 'point of no return' (Hawkins *et al.*, 1972). There is currently no explanation for the protective effect of acidosis, but the results are highly reproducible and inconsistent with any hypothesis that suggests a primary role for release of lysosomal hydrolases.

Other authors have also studied mitochondria, lysosomes and sarcolemma as sources of fatty acids after ischaemia and anoxia. Vasdev, Kako and Biro (1979) reported a decrease in mitochondrial phospholipids and an increase in the lysosomal fraction after 3 h of ischaemia in the dog. Owens, Pang and Weglicki (1979) observed fatty acid release from the sarcolemma in the presence of calcium without detection of lysophospholipase activity. Both groups suggested a major role for membrane-bound phospholipids.

On the other hand, externally added phospholipases mimic the sequence of mitochondrial dysfunction in ischaemia (Luzikov, 1973). Seifulla, Onishchenko and Artamonov (1979) suggested a relationship between the content of intra-mitochondrial calcium, the concentration of fatty acids and the degree of passive swelling following mitochondrial damage by phospholipases.

7.5.3 Cell shape modifications

Probably one of the earliest reactions to injury is modification in cell shape. In the acutely ischaemic kidney, for example, the apical cell membrane of the proximal tubular cell blebs cut into the lumen (Fig. 7.14). Similar changes occur in anoxic EATC (Trump *et al.*, 1979b, 1979c) (Fig. 7.15) and in isolated flounder kidney tubules when generation of ATP is affected (Sahaphong and Trump, 1971). These blebs can be released into the extracellular space as membrane-bound vesicles with no consequences to isolated cells; however, in the case of the kidney, the 'pinching off' effect can obstruct the tubular lumens according to the acute renal-failure obstruction theory (Donohoe, Venkatachalam, Bernard and Levinsky, 1978).

Although cell-shape modifications after injury can be related to fluid redistri-bution, studies in our laboratory strongly suggest that they can be induced by alterations in the cytoskeleton (Trump, Phelps, Shamsuddin and Harris, 1979d). Changes similar to those observed following ischaemia or anoxia have been induced by the use of vinblastine, cytochalasin and local anaesthetics. On the basis of the effects of ionized calcium on microtubules (Schliwa, 1976), it is our present hy-pothesis that the high cellular calcium levels following injury initiate and/or maintain cell-shape modifications.

Fig. 7.14 (a) SEM (Scanning electron micrograph) of a portion of a control rat kidney proximal tubule. The microvillus border (MB) is seen projecting into a patent lumen. (b) SEM of a portion of a rat kidney proximal tubule following 15 min *in vivo* ischaemia. Note distortion of the microvillus border forming large blebs (B) which protrude into the lumen. Some lateral membrane infoldings appear greatly swollen. (Courtesy of Trump *et al.*, 1976b.)

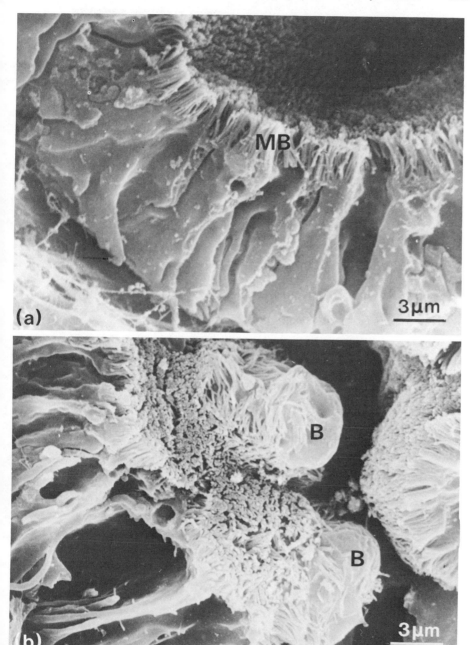

(a)

(b)

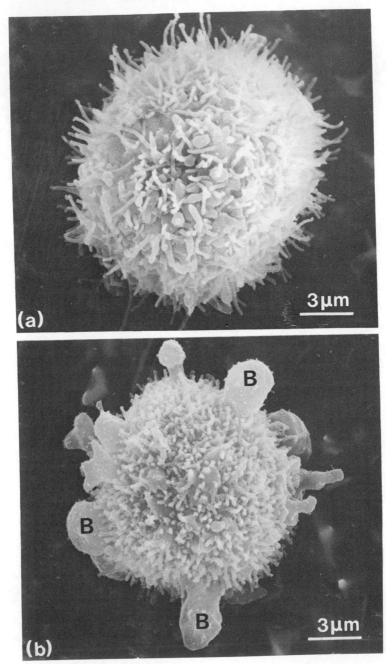

7.6 Hypothesis

Earlier in this chapter we divided the cellular events following injury into several morphological stages in order to provide a better understanding of the series of changes which lead the cell to the 'point of no return'. In order to clarify our hypothesis further, a conceptualization of the effects of ischaemia on ion distribution and the cellular events which are triggered is shown in Fig. 7.16. Following total interruption of the blood supply, acute oxygen and substrate deficiencies occur quickly within the cell, leading to cessation of mitochondrial oxidative phosphorylation. This, in turn, results in a rapid decrease in ATP levels and de-energization of mitochondria, causing the cell to progress through its series of stages. Because of the reduction in ATP levels, a rapid inhibition of calcium-translocation systems occurs in the cell membrane, in the ER and in the mitochondria, leading to increased cytosolic calcium as a result of its influx from the extracellular space and release from both the ER and the mitochondria. In some tissues, this is often compounded by simultaneous stoppage of the Na^+-K^+-translocation systems at the plasmalemma, with an increase in Na^+ which displaces calcium even further from the mitochondria. These increased levels of cytosolic calcium have many important effects, as was discussed earlier, including cell-shape modifications due to alterations in the cytoskeleton and phospholipase activation causing membrane damage. This results in increased membrane permeability, further increasing intracellular calcium and resulting in organelle swelling. Mitochondrial calcium uptake and mitochondrial calcification, however, do not occur in the energy-depleted state, but do readily occur following reflow, when there is an oxygen and energy source to support active uptake. Additional damage leads to irreversible changes in the mitochondrial inner membrane, resulting in the formation of flocculent densities. In contrast with ischaemia, there are other types of lethal injury that are caused by agents that primarily attack the cell membrane, such as heavy metals, complement, endotoxin and ionophores. When this injury occurs, there is a primary increase in calcium entry into the cell, either through specific diffusion sites or by non-specific increases in membrane permeability resulting from alterations in protein and/or lipid. As the calcium-buffering capacity of the mitochondria, ER and cytosol is exceeded, an increase in cytosolic ionized calcium occurs, which, we believe, eventually triggers the phospholipase-induced membrane damage. These effects can be extremely deleterious to the cell, causing rapid cell-volume changes, cell-shape modifications and early cell lysis (Fig. 7.17). The progression of the cell from normal to Stage 5 or necrosis is much more rapid with these plasma-membrane-damaging agents than following ischaemia. We believe this is because the primary attack is directed against the cell membrane

---

Fig. 7.15 (a) SEM of a preincubated control EATC. Note the many delicate closely packed microvilli. (Courtesy of Trump *et al.*, 1979b.) (b) SEM of an EATC following 1 h of anoxia. Although numerous microvilli still remain, some areas of the cell surface contain large elongated blebs (B). (Courtesy of Trump *et al.*, 1979c.)

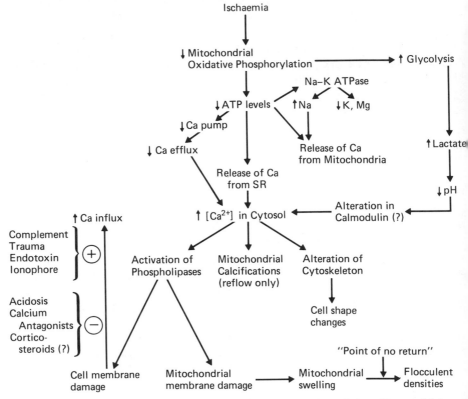

Fig. 7.16 A diagram illustrating a conceptualization of the effects of injury on ion distributions and the cellular events which are triggered. (Courtesy of Trump *et al.*, 1980.)

itself and that the mitochondrial membrane is not involved, at least not initially. Calcium is then translocated into the mitochondria, and, since phosphate is present and increased as ATP hydrolysis occurs, it is precipitated as calcium phosphate. Initially, this begins as annular deposits which do not resemble normal matrix granules or flocculent densities. At later stages, needle-shaped crystals occur, which, in some cases, represent calcium hydroxyapatite (Fig. 7.18). The mitochondria continue to progress through the stages, and Stage 5 is reached, in which flocculent densities are associated with the mitochondrial calcifications. Calcium precipitates also occur in condensed mitochondria (Stage 3a). Agents that reduce cell swelling, such as mannitol, or those that modify calcium permeability, such as Verapamil and methylprednisolone, may modify this process. However, extracellular acidosis delays the response and modifies cell killing.

7.7 Summary

As we stated earlier, the understanding of the events following various types of cell

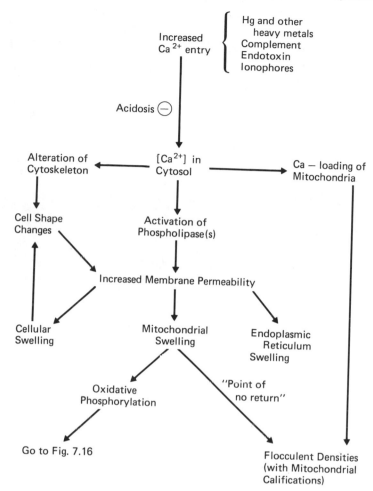

Fig. 7.17 A diagram illustrating a conceptualization of the effects of increased Ca^{2+} entry into the cell and the events which are triggered. (Courtesy of Trump *et al.*, 1980.)

njury and the mechanisms involved in leading the cell beyond the 'point of no return' to cell death is of extreme importance to the development of our knowledge concerning the pathogenesis, treatment and prevention of human diseases. Through our involvement for many years in the study of these processes one consistent link has emerged: namely, that many of these cellular changes are initiated and modified by primary and secondary effects of ion redistribution. In particular, we have been concerned with the role of calcium, both ionized and total, since considerable recent evidence has accumulated indicating that abnormal levels and distributions of this ion can, indeed, initiate catabolic processes that directly lead the cell to

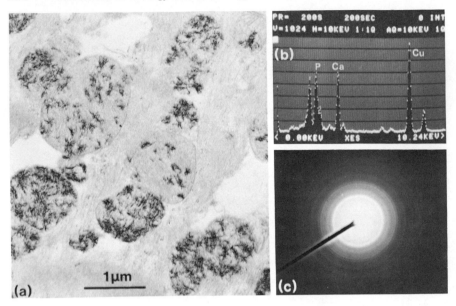

Fig. 7.18 (a) Mitochondria from the ischaemic region of a 60 min rat myocardial infarct containing numerous needle-shaped inclusions. (b) Typical X-ray spectrum obtained from analysis of these inclusions. Note the significant Ca and P peaks. Cu is from the support grid. (c) Electron diffraction pattern obtained over an inclusion revealing the characteristic pattern of hydroxyapatite.

death. No matter what type of injury the cell undergoes, calcium accumulation, by impaired energy metabolism and/or plasma membrane alterations, occurs. We have briefly covered the ramifications to the cell of this increased cytosolic calcium level and, as can be seen, certain points continue to need clarification. Although it is now known that phospholipases are calmodulin-modulated enzymes and that calcium concentrations are controlled by this protein, nevertheless, there is currently no knowledge of how calmodulin is altered during cell injury. Also, experimentation needs to be directed toward methods of measuring ionized versus total calcium levels as a function of time following cell injury. With the recent development of X-ray microanalysis, we can now quantify total cell calcium and organelle-localized calcium, but several points need much more work. Since the importance of determining the event(s) responsible for cell death is directly related to the potential capability of their manipulation, knowledge of these event(s) could well lead to the development and/or modification of pharmacological interventions. At the present time, some promising results have been obtained by manipulation of cell injury with drugs, such as calcium antagonists and corticosteroids (Nayler, Fassold and Yepez, 1978), and also with acidosis experiments (Studer and Borle, 1979). Since the movement of calcium between extra- and intra-cellular compartments and between various intracellular compartments appears to be the key to determining cellular reactions to injury, only a complete understanding of the distribution of this cation will allow us possibly to control and/or prevent cell death.

References

Ashraf, M., White, F. and Bloor, C. M. (1978), Ultrastructural influences of reperfusing dog myocardium with calcium-free blood after coronary artery occlusion. *Am. J. Pathol.*, **90**, 423–434.

Baker, P. F (1972), Transport and metabolism of calcium ions in nerve. *Prog. Biophys. Mol. Biol.*, **24**, 177–223.

Barrett, L. A., McDowell, E. M., Harris, C. C. and Trump, B. F. (1977), Studies on the pathogenesis of ischemic cell injury. XV. Reversal of ischemic cell injury in hamster trachea and human bronchus by explant culture. *Beitr. Pathol. Bd.*, **161**, 109–121.

Blaustein, M. P. (1974), The interrelationship between Na and Ca fluxes across the cell membranes. *Rev. Physiol. Biochem. Pharmacol.*, **70**, 33–82.

Borle, A. B. and Studer, R. (1978), Effects of calcium ionophores on the transport and distribution of calcium in isolated cells in liver and kidney slices. *J. Membr. Biol.*, **38**, 51–72.

Brinley, F. J. Jr. (1978), Calcium-buffering in squid axons. *Annu. Rev. Biophys. Bioeng.*, **7**, 363–392.

Bulger, R. E. and Trump, B. F. (1969), Ca^{2+} and K^+ ion effects on ultrastructure of isolated flounder kidney tubules. *J. Ultrastruct. Res.*, **28**, 301–319.

Bulkley, B., Nunnally, R. L. and Hollis, D. P. (1978), 'Calcium paradox' and the effect of varied temperature on its development: A phosphorous nuclear magnetic resonance and morphological study. *Lab. Invest.*, **39**, 133–140.

Carafoli, E. (1979), The calcium cycle of mitochondria. *FEBS Lett.*, **104**, 1–6.

Cheah, K. S. and Cheah, A. M. (1977), Inclusions in aged mitochondria. *J. Bionerg. Biomembr.*, **9**, 105–115.

Cheung, W. Y. (1970), Cyclic 3′, 5′-nucleotide phosphodiesterases: Demonstration of an activator. *Biochem. Biophys. Res. Commun.*, **33**, 533–538.

Cheung, W. Y. (1980), Calmodulin plays pivotal role in cellular regulation. *Science*, **207**, 19–27.

Chien, K. R., Abrams, J., Serroni, A., Martin, J. T. and Farber, J. L. (1978), Accelerated phospholipid degradation and associated membrane dysfunction in irreversible ischemia liver cell injury. *J. Biol. Chem.*, **253**, 4809–4817.

Croker, B. P., Saladino, A. J. and Trump, B. F. (1970), Ion movements in cell injury: Relationships between energy metabolism and the pathogenesis of lethal injury in toad bladder. *Am. J. Pathol.*, **59**, 247–278.

Crompton, M., Moser, R., Ludi, H. and Carafoli, E. (1978), The interrelations between the transport of sodium and calcium in mitochondria of various mammalian tissues. *Eur. J. Biochem.*, **82**, 25–31.

Decker, R. S. and Wildenthal, K. (1980), Lysosomal alterations in hypoxic and reoxygenated hearts. I. Ulstructural and cytochemical changes. *Am. J. Pathol.*, **98**, 425–444.

Dedman, J. R., Brinkley, B. R. and Means, A. R. (1979), Regulation of micro-filaments and microtubules by calcium and cyclic AMP. *Adv. Cyclic Nucleo-tide Res.*, **11**, 131–174.

De Duve, C. and Wattiaux, R. (1966), Function of lysosomes. *Ann. Rev. Physiol.*, **28**, 435–492.

Donohoe, J. F., Venkatachalam, M. A., Bernard, D. B. and Levinsky, N. G. (1978), Tubular leakage and obstruction after renal ischemia: Structural–functional correlations. *Kidney Int.*, **13**, 208–222.

French, S. W., Norum, H. L., Ihrig, T. J. and Todoroff, T. (1971), Effect of phospholipid hydrolysis on the structural and functional integrity of mito-chrondria. *Lab. Invest.*, **25**, 427–434.

Gan-Elepano, M. and Mead, J. F. (1978), The function of phospholipase A in the metabolism of membrane lipids. *Biochem. Biophys. Res. Commun.*, **83**, 247–251.

Ginn, F. L., Shelburne, J. D. and Trump, B. F. (1968), Disorders of cell volume regulation. I. Effects of inhibition of plasma membrane adenosine triphos-phatase with ouabain. *Am. J. Pathol.*, **53**, 1041–1071.

Glaumann, B., Glaumann, H., Berezesky, I. K. and Trump, B. F. (1975), Studies on the pathogenesis of ischemic cell injury. II. Morphological changes of the pars convoluta (P_1 and P_2) of the proximal tubule of the rat kidney made ischemic *in vivo*. *Virch. Arch. B Cell Pathol.*, **19**, 281–302.

Glaumann, B., Glaumann, H., Berezesky, I. K. and Trump, B. F. (1977a), Studies on cellular recovery from injury. II. Ultrastructural studies on the recovery of the pars convoluta of the proximal tubule of the rat kidney from temporary ischemia. *Virch. Arch. B Cell Pathol.*, **24**, 1–18.

Glaumann, B., Glaumann, H. and Trump, B. F. (1977b), Studies of cellular recovery from injury. III. Ultrastructural studies of the recovery of the pars recta of the proximal tubule (P_3 segment) of the rat kidney from temporary ischemia. *Virch. Arch. B Cell Pathol.*, **25**, 281–308.

Glaumann, B. and Trump, B. F. (1975), Studies on the pathogenesis of ischemic cell injury. III. Morphological changes of the proximal pars recta tubules (P_3) of the rat kidney made ischemic *in vivo*. *Virch. Arch. B Cell Pathol.*, **19**, 303–323.

Goracci, G., Porcellati, G. and Woelk, H. (1978), Subcellular localization and distribution of phospholipases A in liver and brain tissue. *Adv. Prostaglandin Thromboxane Res.*, **3**, 55–67.

Goring, G. G. and Spieckermann, P. G. (1978), Ca^{2+} uptake and release phenomena from cardiac mitochondria under normal and ischemic conditions. *Basic Res. Cardiol.*, **73**, 126–132.

Gritzka, T. L. and Trump, B. F. (1968), Renal tubular lesions caused by mercuric chloride. Electron microscopic observations: Degeneration of the pars recta. *Am. J. Pathol.*, **52**, 1225–1277.

Hanzlikova, V. and Schiaffino, S. (1977), Mitochondrial changes in ischemic skeletal muscle. *J. Ultrastruct. Res.*, **60**, 121–133.

Hawkins, H. K., Ericsson, J. L. E., Biberfeld, P. and Trump, B. F. (1972), Lysosome and phagosome stability in lethal cell injury. *Am. J. Pathol.*, **68**, 255–287.

Hearse, D. J. (1977), Reperfusion of the ischemic myocardium. *J. Mol. Cell Cardiol.*, **9**, 605–616.

Jones, R. T., Garcia, J. H., Mergner, W. J., Pendergrass, R. E., Valigorsky, J. M. and Trump, B. F. (1975), Effects of shock on the pancreatic acinar cell. Cellular and subcellular effects in humans. *Arch. Pathol.*, **99**, 634–644.

Kahng, M. W., Berezesky, I. K and Trump, B. F. (1978), Metabolic and ultrastructural response of rat kidney cortex to *in vitro* ischemia. *Exp. Mol. Pathol.*, **29**, 183–198.

Kakiuchi, S., Yamazaki, R. and Nakajima, H. (1970), Properties of a heat stable phosphodiesterase activating factor isolated from brain extract. *Proc. Jpn. Acad.*, **46**, 587–591.

Karpati, G., Carpenter, S., Melmed, C. and Eisen, A. A. (1974), Experimental ischemic myopathy. *J. Neurol. Sci.*, **23**, 129–161.

Kloner, R. A., Ganote, C. E., Whalen, D. A. and Jennings, R. B. (1974), Effect of transient period of ischemia on myocardial cells. II. Fine structure during the first few minutes of reflow. *Am. J. Pathol.*, **74**, 399–422.

Kretsinger, R. H. (1976), Evolution and function of calcium-binding proteins. *Int. Rev. Cytol.*, **46**, 323–393.

Kunze, H., Nahas, N., Traynor, J. R. and Wurl, M. (1976), Effects of local anaesthetics on phospholipases. *Biochim. Biophys. Acta*, **441**, 93–102.

Laiho, K. U., Berezesky, I. K. and Trump, B. F. (1980), Studies on the modification of the cellular response to injury. VII. Effect of the ionophore A23187 on ATP and ion content of Ehrlich ascites tumour cells following anoxia. (In preparation)

Laiho, K. U., Shelburne, J. D. and Trump, B. F. (1971), Observations on cell volume, ultrastructure, mitochondrial conformation and vital-dye uptake in Ehrlich ascites tumor cells: Effects of inhibiting energy production and function of the plasma membrane. *Am. J. Pathol.*, **65**, 203–230.

Laiho, K. U. and Trump, B. F. (1974a), Relationship of ionic, water, and cell volume changes in cellular injury of Ehrlich ascites tumor cells. *Lab. Invest.*, **31**, 207–215.

Laiho, K. U. and Trump, B. F. (1974b), The relationship between cell viability and changes in mitochondrial ultrastructure, cellular ATP, ion and water content following injury of Ehrlich ascites tumor cells. *Virch. Arch. Abt. B Zellpathol.*, **15**, 267–277.

Laiho, K. U. and Trump, B. F. (1975), Mitochondrial changes, ion and water shifts in the cellular injury of Ehrlich ascites tumor cells. *Beitr. Pathol. Bd.*, **155**, 237–247.

Lehninger, A. (1970), Proteins and their biological functions. In *Biochemistry* (ed. A. Lehninger), Worth Publishers, Inc., New York, p. 59.

Lochner, A., Kotze, J. C. N., Benade, A. J. and Gevers, W. (1978), Mitochondrial oxidative phosphorylation in low flow hypoxia: Role of free fatty acids. *J. Mol. Cell Cardiol.*, **10**, 857–875.

Luzikov, V. N. (1973), Stabilization of the enzymatic systems of the inner mitochondrial membrane and related problems. *Sub-Cell. Biochem.*, **2**, 1–31.

Majno, G. Lagattuta, M. and Thompson, T. (1960), Cellular death and necrosis: Chemical, physical and morphological changes in rat liver. *Virch. Arch. Pathol. Anat.*, **333**, 421–465.

Means, A. R. and Dedman, J. R. (1980), Calmodulin – An intracellular calcium receptor. *Nature (London)*, **285**, 73–77.

Mergner, W. J., Chang, S., H., Marzella, L., Kahng, M. W. and Trump, B. F. (1979), Studies on the pathogenesis of ischemic cell injury. VIII. ATPase of rat kidney mitochondria. *Lab. Invest.*, **40**, 686–694.

Mergner, W. J., Smith, M. W., Sahaphong, S. and Trump, B. F. (1977), Studies on the pathogensis of ischemic cell injury. VI. Accumulation of calcium by isolated mitochondria of ischemic rat kidney cortex. *Virch. Arch. B. Cell Pathol.*, **26**, 1–16.

Mergner, W. J., Smith, M. W. and Trump, B. F. (1972), Structural and functional effects of the negative stains silicotungstic acid, phosphotungstic acid, and ammonium molybdate on rat kidney mitochondria. *Lab. Invest.*, **27**, 372–383.

Meyer, W. L., Fischer, W. H. and Krebs, E. G. (1964), Activation of skeletal muscle phosphorylase b kinase by Ca^{2+}. *Biochemistry*, **3**, 1033–1039.

Nayler, W. G., Fassold, E. and Yepez, C. (1978), Pharmacological protection of mitochondrial function in hypoxic heart muscle: Effect of verapamil, propranolol, and methylprednisolone. *Cardiovasc. Res.*, **12**, 152–161.

Osornio, A. R., Berezesky, I. K., Mergner, W. J. and Trump, B. F. (1980), Mitochondrial membrane fusions in experimental myocardial infarction. *Fed. Proc. Fed. Am. Soc. Exp. Biol.*, **39**, 634.

Osornio-Vargas, A. R., Berezesky, I. K. and Trump, B. F. (1980b), Mitochondrial membrane fusions: An early ultrastructural change in experimental rat myocardial infarction. (In preparation)

Owens, K., Pang, D. C. and Weglicki, W. B. (1979), Production of lysophospholipids and free fatty acids by a sarcolemmal fraction from canine myocardium. *Biochem. Biophys. Res. Commun.*, **89**, 368–373.

Pappu, A. S., Fatterpaker, P. and Sreenivasan, A. (1978), Phospholipase A of rat liver mitochondria in vitamin E deficiency. *Biochem. J.*, **172**, 349–352.

Parce, J. W., Cunningham, C. C. and Waite, M. (1978), Mitochondrial phospholipase A activity and mitochondrial aging. *Biochemistry*, **17**, 1634–1639.

Paul, R. J., Bauer, M. and Prease, W. (1979), Vascular smooth muscle: Aerobic glycolysis linked to sodium and potassium transport processes. *Science*, **206**, 1414–1416.

Penttila, A., Glaumann, H. and Trump, B. F. (1976), Studies on the modification of the cellular response to injury. IV. Protective effect of extracellular acidosis against anoxia, thermal, and *p*-chloromercuribenzene sulfonic acid treatment of isolated rat liver cells. *Life Sci.*, **18**, 1419–1930.

Penttila, A. and Trump, B. F. (1974), Extracellular acidosis protects Ehrlich ascites tumor cells and rat renal cortex against anoxic injury. *Science*, **135**, 277–278.

Penttila, A. and Trump, B. F. (1975a), Studies on modifications of the cellular response to injury. I. Protective effect of acidosis on *p*-chloromercuribenzene sulfonic acid-induced injury of Ehrlich ascites tumor cells. *Lab. Invest.*, **32**, 690–695.

Penttila, A. and Trump, B. F. (1975b), Studies on the modification of the cellular response to injury. II. Electron microscopic studies on the protective effect of acidosis on anoxic injury of Ehrlich ascites tumor cells. *Virch. Arch. B Cell Pathol.*, **18**, 1–16.

Penttila, A. and Trump, B. F. (1975c), Studies on the modification of the cellular response to injury. III. Electron microscopic studies on the protective effect of acidosis on *p*-chloromercuribenzene sulfonic acid-(PCMBS) induced injury of Ehrlich ascites tumor cells. *Virch. Arch. B Cell Pathol.*, **18**, 17–34.

Publicover, S. J., Duncan, C. J. and Smith, J. L. (1978), The use of A23187 to demonstrate the role of intracellular calcium in causing ultrastructural damage in mammalian muscle. *J. Neuropathol. Exp. Neurol.*, **37**, 554–557.

Rasmussen, H. and Goodman, D. B. (1977), Relationships between calcium and cyclic nucleotides in cell activation. *Physiol. Rev.*, **57**, 421–509.

Rubin, R. P. (1974), Calcium and the secretory process. Plenum, New York.

Russ, J. C. (1978), Electron probe X-ray microanalysis-principles. In *Electron Probe Microanalysis in Biology*, (ed. D. A. Erasmus), Chapman and Hall, London, pp. 5–36.

Sahaphong, S. and Trump, B. F. (1971), Studies of cellular injury in isolated kidney tubules of the flounder. V. Effects of inhibiting sulfhydryl groups of plasma membrane with the organic mercurials PCMB (parachloromercuribenzoate and PCMBS (parachloromercuribenzenesulfonate). *Am. J. Pathol.*, **63**, 277–298.

Saito, A., Smigel, M. and Fleisher, S. (1974), Membrane junctions in the intermembrane space of mitochondria from mammalian tissues. *J. Cell Biol.*, **60**, 653–663.

Saladino, A. J., Bentley, P. J. and Trump, B. F. (1969), Ion movements in cell injury. Effect of amphotericin B on the ultrastructure and function of the epithelial cells of the toad bladder. *Am. J. Pathol.*, **54**, 421–466.

Schanne, F. A. X., Kane, A. B., Young, E. E. and Farber, J. L. (1979), Calcium dependence of toxic cell death: A final common pathway. *Science*, **206**, 700–702.

Schiaffino, S., Severin, E. and Hanzlikova, V. (1979), Intermembrane inclusions induced by anoxia in heart and skeletal muscle mitochondria. *Virch. Arch. B Cell Pathol.*, **31**, 169–179.

Schliwa, M. (1976), The role of divalent cations in the regulation of microtubule assembly. *In vivo* studies on microtubules of the heliozan axopodium using the ionophore A23187. *J. Cell Biol.*, **70**, 527–540.

Seifulla, R. D., Onishchenko, N. H. and Artamonov, S. D. (1979), Stabilizing effect of the antioxidants alpha-tocopherol and sodium selenite following phospholipase damage in mitochondrial membranes in a model of anoxia. *Farmakol. Toksikol.*, **42**, 157–163.

Shen, A. C. and Jennings, R. B. (1972), Myocardial calcium and magensium in acute ischemic injury. *Am. J. Pathol.*, **67**, 417–440.

Smith, M. W., Collan, Y., Kahng, M. W. and Trump, B. F. (1980), Changes of fatty acids and phospholipids in ischemic rat kidney. *Biochim. Biophys. Acta* (in press).

Somlyo, A. P., Somlyo, A. V. and Shuman, H. (1979), Electron probe analysis of vascular smooth muscle. Composition of mitochondria, nuclei and cytoplasm. *J. Cell Biol.*, **81**, 316–335.

Studer, R. K. and Borle, A. B. (1979), Effect of pH on the calcium metabolism of isolated rat kidney cells. *J. Membr. Biol.*, **48**, 325–341.

Sun, S. T., Day, E. P. and Ho, J. T. (1978), Temperature dependence of calcium-induced fusion of sonicated phosphatidylserine vesicles. *Proc. Natl. Acad. Sci. U.S.A.*, **75**, 4325–4328.

Trump, B. F. and Arstila, A. U. (1975), Cell membranes and disease processes. In *Pathobiology of Cell Membranes*, Vol. I (eds. B. F. Trump and A. U. Arstial), Academic Press, New York, pp. 1–52.

Trump, B. F., Berezesky, I. K., Chang, S. H. and Bulger, R. E. (1976a), Detection of ion shifts in proximal tubule cells of the rat kidney using X-ray microanalysis. *Virch. Arch. B Cell Pathol.*, **22**, 111–120.

Trump, B. F., Berezesky, I. K., Chang, S. H., Pendergrass, R. E. and Mergner, W. J. (1979a), The role of ion shifts in cell injury. In *Scanning Electron Microscopy/III*, SEM Inc., AMF O'Hare, Illinois 60666, pp. 1–14.

Trump, B. F., Berezesky, I. K., Collan, Y., Kahng, M. W. and Mergner, W. J. (1976b), Recent studies on the pathophysiology of ischemic cell injury. *Beitr. Pathol. Bd.*, **158**, 363–388.

Trump, B. F., Berezesky, I. K., Laiho, K. U., Osornio, A. R., Mergner, W. J. and Smith, M. W. (1980), The role of calcium in cell injury. A review. In *Scanning Electron Microscopy*, SEM Inc., AMF O'Hare, Illinois 60666 (in press).

Trump, B. F., Berezesky, I. K., Mergner, W. J. and Phelps, P. C. (1978a), The role of calcium in cell injury. *Micron*, **9**, 5–6.

Trump, B. F., Berezesky, I. K., Pendergrass, R. E., Chang, S. H., Bulger, R. E., and Mergner, W. J. (1978b), X-ray microanalysis of diffusible elements in scanning electron microscopy of biological thin sections. Studies of pathologically altered cells. In *Scanning Electron Microscopy/II*, SEM Inc., AMF O'Hare, Illinois 60666, pp. 1027–1039.

Trump, B. F., Croker, B. P., Jr. and Mergner, W. J. (1971), The role of energy metabolism, ion, and water shifts in the pathogenesis of cell injury. In *Cell Membranes: Biological and Pathological Aspects* (eds. G. W. Richter and D. G. Scarpelli), Williams and Wilkins Co., Baltimore, pp. 84–128.

Trump, B. F. and Ginn, F. L. (1969), The pathogenesis of subcellular reaction to lethal injury. In *Methods and Achievements in Experimental Pathology*, vol. IV (eds. E. Bajusz and G. Jasmin), Karger, Basel, pp. 1–29.

Trump, B. F., Jones, R. T., Berezesky, I. K., Phelps, P. C. and Laiho, K. U. (1978c), Role of calcium in cell injury. *Fed. Proc. Fed. Am. Soc. Exp. Biol.*, **37**, 403.

Trump, B. F. and Laiho, K. U. (1975), Studies of cellular recovery from injury. I. Recovery from anoxia in Ehrlich ascites tumor cells. *Lab. Invest.*, **33**, 706–711.

Trump, B. F., Laiho, K. U., Mergner, W. J. and Arstila, A. U. (1974a), Studies on the subcellular pathophysiology of acute lethal cell injury. *Beitr. Pathol. Bd.*, **152**, 243–271.

Trump, B. F. and Mergner, W. J. (1974), Cell injury. In *The Inflammatory Process*, vol. 1, 2nd Ed. (eds. B. W. Zweifach, L. Grant and R. T. McClusky), Academic Press, New York, pp. 115–257.

Trump, B. F., Mergner, W. J., Kahng, M. W. and Saladino, A. J. (1976c), Studies on the subcellular pathophysiology of ischemia. *Circulation*, **53**, *Suppl.* I, 17–26.

Trump, B. F., Penttila, A. and Berezesky, I. K. (1979b), Studies on cell surface conformation following injury. I. Scanning and transmission electron microscopy of cell surface changes following *p*-chloromercuribenzene sulfonic acid (PCMBS)-induced injury of Ehrlich ascites tumor cells. *Virch. Arch. B Cell Pathol.*, **29**, 281–296.

Trump, B. F., Penttila, A. and Berezesky, I. K. (1979c), Studies on cell surface conformation following injury. II. Scanning and transmission electron microscopy of cell surface changes following anoxic injury in Ehrlich ascites tumor cells. *Virch. Arch. B Cell Pathol.*, **29**, 297–307.

Trump, B. F., Phelps, P. C., Shamsuddin, A. M. and Harris, C. C. (1979d), Cell surface changes in premalignant and malignant lesions of the colon. *Lab. Invest.*, **40**, 289.

Trump, B. F., Strum, J. M and Bulger, R. E. (1974b), Studies on the pathogenesis of ischemic cell injury. I. Relation between ion and water shifts and cell ultrastructure in rat kidney slices during swelling at 0–4°C. *Virch. Arch. B. Cell Pathol.* **16**, 1–34.

Vasdev, S. C., Kako, K. J. and Biro, G. P. (1979), Phospholipid composition of cardiac mitochondria and lysosomes in experimental myocardial ischemia in the dog. *J. Mol. Cell. Cardiol.*, **11**, 1195–1200.

Wildenthal, K. (1978), Lysosomal alterations in ischemic myocardium: Result or cause of myocellular damage. *J. Mol. Cell. Cardiol.*, **10**, 595–603.

Wolff, D. J. and Siegel, F. (1972), Purification of a calcium-binding phosphoprotein from pig brain. *J. Biol. Chem.*, **247**, 4180–4185.

Wrogemann, K. and Pena, S. D. J. (1976), Mitochondrial calcium overload: A general mechanism for cell necrosis in muscle diseases. *Lancet*, i, 672–674.

Zimmerman, A. N. E., Daems, W., Hulsmann, W. C., Snijder, J., Wisse, E. and Durrer, D. (1967), Morphological changes of heart muscle caused by successive perfusion with calcium-free and calcium-containing solutions (calcium paradox). *Cardiovasc. Res.*, **1**, 201–209.

8 Cell death *in vitro*

LEONARD HAYFLICK

8.1 Introduction

The death of cells cultured in laboratory glassware, after various periods of time, has claimed the attention of cell culturists ever since the techniques were first developed at the turn of this century. During the subsequent fifty years it was commonly believed that the inability to propagate vertebrate cells *in vitro* indefinitely was attributable to an incomplete understanding of optimum culture conditions.

The explanations that were most frequently offered for the death of cultured cells after several weeks or months of propagation included (1) inadequate knowledge of nutritional requirements, (2) improperly prepared glass or plastic substrates, (3) cell damage from subcultivation procedures or (4) a myriad of other unknown technical limitations. One or more of these reasons was often invoked to explain the death of cultured cells because of the implication that if cells could divide several times *in vitro*, then they must have the capacity to divide indefinitely. This notion was supported further by the conviction that most cells *in vivo* have the capacity to divide for periods of time in excess of their capacity to do so *in vitro*. Hence, the lack of optimum *in vitro* culture conditions was thought to be the cause of the failures.

In the late 1940s and early 1950s a number of cultured cell populations were discovered that did appear to have the remarkable property of immortality. That is, they could be serially propagated in laboratory glassware for what appeared to be indefinite periods of time. Indeed, many of these cell populations continue to flourish in laboratories throughout the world even to this day. These cells double in numbers, during the exponential phase of growth, in about 24 h. The best-known human immortal cell line prototype is the HeLa cell, derived from a human cervical carcinoma in 1952. One of the first known continuously propagable non-human cell lines was the L cell derived from mouse mesenchyme in 1943. There soon followed descriptions of several hundred more immortal cell lines derived from a number of vertebrates including humans, and from a variety of their normal and cancerous tissues. Each of these cell lines seemed to arise spontaneously from the initial cultured cell population because they acquired recognizable morphological changes that distinguished them from their precursors.

243

The phenomenon was called transformation. Further studies showed that these transformed immortal cells differed significantly from normal cell populations. The most important differences found were changes in karyology, biochemical abnormalities, and, in most cases, the formation of tumours when these cells were inoculated into laboratory animals.

Until the early 1960s, immortal cell populations were only known to arise spontaneously. At that time the remarkable observation was made that the addition of certain viruses to cultured cells could produce a purposeful transformation. The SV_{40} virus, found in some monkeys, is capable of transforming many human cell types and the Epstein–Barr virus can transform human lymphocytes. The polyoma virus can transform cultured mouse and other mammalian cells but not those from man. Other transforming viruses are known. Collectively, these viruses are called oncogenic viruses and have the capacity to transform cells *in vitro* to cells with one or more abnormal properties in addition to the acquisition of the property of immortality.

The widespread use of spontaneously transformed cell lines for a variety of research purposes in laboratories throughout the world is subject to the criticism that the cells are not characteristic of any cell type found in normal human or animal tissue. Many experimental data generated from the use of such cell populations cannot be extrapolated to apply to cells that characterize the animal species from which they were originally descended. Consequently, the use of 'immortal' cell lines is questionable, since such cells undoubtedly represent laboratory artefacts whose behaviour may be unrelated to that of cells found *in vivo*. The use of transformed cells is widespread because they are generally believed to be easier to culture than untransformed or normal cells, but that is not the case.

Twenty years ago, Moorhead and I found that cultured normal human fibroblasts underwent a finite number of serial subcultivations or cell population doublings and then died (Hayflick and Moorhead, 1961). Spontaneous transformations did not occur. For cells derived from human embryonic tissue, even under the most favorable culture conditions, cell death was inevitable after 50 ± 10 population doublings. This number of doublings was reached after about 6 months of serial subcultivations. We divided the culture history of normal cells into three phases: Phase I represented the initial culture, Phase II, the period of serial transfer and exponential cell increase, and Phase III, the death of the cells. Untransformed normal cell populations with a finite capacity to replicate are called cell strains.

Our studies also showed that Phase III was not due to the then prevailing view which involved inadequate nutrition or other failures in technique as explanations for the death of cultured cells. We demonstrated that the death of cultured normal or untransformed cells is an inherent property of the cells themselves (Hayflick and Moorhead, 1961; Hayflick, 1965). That observation has now been confirmed in hundreds of laboratories in which variations in medium composition and cultural conditions have been as numerous as the laboratories in which the studies were done. We interpreted the Phase III phenomenon to be aging at the cell level.

8.2 Cell aging and death *in vitro*

The notion that aging occurs in animals that reach a fixed size after maturity is beyond dispute, but is the inevitability of the aging and death of individual cells composing that organism predetermined? A superficial consideration of this thought may provoke some incredulity, since it is intuitively obvious that a dead or aging organism must consist of dead or aging cells. Nevertheless, whatever causes age changes and death in the whole organism undoubtedly does not produce similar changes, and at the same rate, in each cell composing that organism. If the rates of aging vary among organs, tissues, and their constituent cells, then the root causes of aging may occur as a consequence of decrements in some few cell types where the rate is fastest and the effects greatest. Are normal somatic cells predestined to undergo irreversible functional decrements that presage aging in the whole organism?

There are at least two ways in which this question has been put to the test. First, vertebrate cells have been grown and studied in cell culture, and second, eukaryotic cells containing specific markers, allowing them to be distinguished from host cells, have been serially transplanted in isogenic laboratory animals. The goal of such studies has been directed toward answering this fundamental question: Can vertebrate cells, functioning and replicating under ideal conditions, escape from the inevitability of aging and death, which is universally characteristic of the whole animals from which they were derived?

Among the studies undertaken in cell culture, one investigation stands out as the classic response to this intriguing question. In the early part of this century, Alexis Carrel, a noted cell culturist, described experiments purporting to show that the fibroblast cells derived from chick heart tissue could be cultured *seriatum* indefinitely. The culture was voluntarily terminated after 34 years (Ebeling, 1913; Parker, 1961). The importance of this experiment to gerontologists was the implication that if cells released from *in vivo* control could divide and function normally for periods in excess of the lifespan of a species, either the type of cells cultured play no part in the aging process or aging is the result of changes occurring at the supracellular level. That is, aging would be the result of decrements that occur only in organized tissue or whole organs as a result of the physiological interactions between those organized cell hierarchies. The inference would be that aging, *per se*, is not the result of events occurring at the cellular level.

In the years that followed Carrel's observations, support for his experimental results seemed to be forthcoming from the many laboratories in which it was observed that cultured cell populations, derived from many tissues of a variety of animal species and from man, had the striking ability to replicate apparently indefinitely. However, these transformed cell populations, as indicated previously, are abnormal.

Of central importance to the question of aging is whether the cell populations studied *in vitro* are composed of normal or abnormal cells. Clearly the aging of animals occurs in normal cell populations. If we are to equate the behaviour of normal cells *in vivo* to similar cells *in vitro*, then the latter must be shown to be normal as well. For this reason the immortal or transformed cell lines described above, of which the HeLa and L cell populations are prototypes, must be excluded from consideration. They are composed of cells that are abnormal in one or more important properties. Normal cells, on the other hand, have a limited capacity to divide, and, even under the best culture conditions, ultimately die (Hayflick and Moorhead, 1961; Hayflick, 1965).

Since normal diploid cell strains have a limited doubling potential *in vitro*, studies on any single strain would be severely curtailed were it not possible to preserve these cells at subzero temperatures for apparently indefinite periods of time. The reconstitution of frozen human foetal diploid cell strains has revealed that, regardless of the doubling level reached by the population at the time it is preserved, the total number of doublings that can be expected is about 50 when those made before and after preservation are summed (Hayflick and Moorhead, 1961; Hayflick, 1965). Storage of human diploid cell strains merely arrests the cells at a particular population doubling level but does not influence the total number of expected doublings.

We have reconstituted a total of 130 ampules of our human diploid cell strain WI-38, which was placed in liquid nitrogen storage 19 years ago. Since 1962 one ampule was reconstituted approximately each month, and all have yielded cell populations that have undergone 50 ± 10 cumulative population doublings. This represents the longest period of time that viable normal human cells have been arrested at subzero temperatures.

Since normal human embryo fibroblasts are able to undergo only a fixed number of reproductive cycles *in vitro*, we postulated that this observation might be interpreted to be aging at the cellular level. Although we and others were sceptical of this interpretation at first, subsequent experimental data have tended to support the validity of this notion. However, before considering the newer developments in this field, it is necessary to reconsider Carrel's experiment, in which a presumptive normal chicken cell population was cultured for 34 years and then voluntarily terminated. The cells cultured by Carrel are presumed to have been normal simply because abnormal immortal cell lines have never been known to arise from chick tissue.

In the years following Carrel's observation, and even in recent times when more sophisticated cell culture techniques have been used, no one has been able to culture chick fibroblasts serially in a state of continuous rapid proliferation beyond one year (Hayflick, 1977a). Thus, there is serious doubt that the original interpretation of Carrel's work can be accepted because it has never been confirmed. We have proposed one explanation for Carrel's finding in which the method of preparation of chick embryo extract, used as a source of nutrients for his cultures and prepared daily under conditions of low-speed centrifugation, allowed for the survival and introduction of new viable fibroblasts into the so-called 'immortal' culture at each feeding (Hayflick, 1970, 1972).

One possible exception to the lack of confirmation of Carrel's study is the finding that chick embryo muscle explants have survived successive transplants *in vitro* up to 44 months on a collagen substrate but not on commonly used glass surfaces (Gey, Svotelis, Foard and Bang, 1974). However, because of the difficulty in determining population doublings in explant cultures, it is probable that this longevous culture represents a considerable amount of cell maintenance time and not necessarily an unusually large number of population doublings. Maintenance of slowly dividing cell populations can occur over much longer periods of time than will cultures of similar cells allowed to proliferate at their maximum rate. We and others have found this to be so when culturing normal human and chick cells in which cell replication was slowed by diminishing serum concentrations or reducing the incubation temperature (Hay and Strehler, 1967; Dell'Orco, Mertens and Kruse, 1973; Goldstein and Singal, 1974). The essential point is that, although such cultures may survive for longer periods of time, the maximum number of population doublings achieved is comparable with that for similar populations cultured under conditions allowing a maximum rate of replication.

8.3 Donor age versus cell doubling potential

Because cultured normal human cells derived from embryonic tissue have a finite proliferative capacity of about 50 population doublings and because this may represent cellular aging, it is important to determine the proliferative capacity of normal cells derived from human adults of varying ages. Our first report of such studies did, indeed, show a diminished proliferative capacity for cultured normal human adult fibroblasts in which 14 to 29 doublings occurred in cells derived from the lungs of eight adult donors (Hayflick, 1965). This compared with a range of 35 to 63 doublings found in cells cultured from the lungs of 13 human embryos. Although it was clear that fibroblasts cultured from human adults as a group had a greatly diminished proliferative capacity *in vitro* compared with cells derived from embryos, we were unable to detect a direct correspondence between donor age and population doubling potential. A precise relationship, however, does exist between donor age and population doubling potential and our inability to detect it can be attributed to the lack of precision by which population doublings were measured at that time.

Subsequent to our studies, reports by Martin, Sprague and Epstein (1970) and LeGuilly, Simon, Lenoir and Bourel, (1973) not only confirmed the principle we observed but extended it significantly. The former investigators cultured fibroblasts derived from biopsies taken from the upper arm of human donors ranging from foetal to 90 years of age. They found the regression coefficient, from the first to the ninth decade, to be -0.20 population doublings per year of life with a standard deviation of 0.05 and a correlation coefficient of -0.50. The regression coefficient was significantly different from zero ($P<0.01$). LeGuilly *et al.* (1973), studying human liver cells, obtained results similar to these. Schneider and Mitsui (1976) and Goldstein, Moerman, Soeldner, Gleason and Barnett (1978) also studied fibroblasts from human skin biopsies, and Bierman (1978) examined human arterial smooth muscle cells. All of these studies made with cells derived from four different kinds of human tissues tend to support the original contention that the number of population doublings achievable by cultured normal human cells is inversely proportional to donor age. Nevertheless, these findings must be tempered with observations that the relationship may be clouded when different tissue sites are compared (Schneider, Mitsui, Aw and Shorr, 1977) or where the physiological state of the donor is abnormal. For example, strains derived from diabetics have been found to undergo fewer doublings *in vitro* than their normal age-matched counterparts (Goldstein *et al.*, 1978).

Comparative properties of normal human cell strains derived from old versus prenatal donors are beginning to yield useful data. Table 8.1 lists those variables that have been compared and for which differences have been found.

Table 8.1 Properties of cultured normal adult human cells that distinguish them from similar cells from foetal, embryonic or newborn donors

Fewer population doublings	Hayflick, (1965)
	Martin *et al.* (1970)
	LeGuilly *et al.* (1973)
	Schneider and Mitsui (1976)
Easier induction of chromosome 21-directed anti-viral gene(s)	Tan *et al.* (1975)
Acyl turnover and release of radioisotope (Foetal cells retain phospholipid acyl groups)	Rosenthal and Geyer (1976)
More gradual decrease in proportion of [^{3}H] dThd-labelled cells	Vincent and Huang (1976)
Increased amount of insulin bound to specific receptor sites	Rosenbloom *et al.* (1976)
Greater response to hydrocortisone stimulation of growth rate	Rowe *et al.* (1977)
Decrease in rate of fibroblast migration	Schneider and Mitsui (1976)
Fewer cell numbers at confluency	Schneider and Mitsui (1976)
Slower cell population replication rate	Schneider and Mitsui (1976)
Greater activity of adenosine deaminase	Murphy *et al.* (1978)
Longer latent period for cell migration from explants	Waters and Walford (1970)
	Schneider and Chase (1976)
Fewer numbers of cells synthesizing DNA per day in early population doublings and longer period to reach saturation density	Macieira-Coelho (1970)

8.4 Species lifespan versus cell doubling potential

Although the data are fragmentary, it is interesting to explore the possibility that the population doubling potential of normal fibroblasts derived from the embryonic tissue of a variety of animal species might be proportional to the mean maximum lifespan for those species (Table 8.2). Some of the data need confirmation, and

Table 8.2 The finite lifetime of cultured normal embryonic human and animal fibroblasts

Species	Range of population doublings for cultured normal embryo fibroblasts	Mean maximum lifespan (years)	Reference
Galapagos tortoise	90–125*	175(?)	Goldstein (1974)
Man	40–60	110	Hayflick and Moorhead (1961)
Mink†	30–34	10	Porter (1974)‡
Chicken	12–35	30(?)	Hayflick (1977a)
Mouse	14–28	3.5	Rothfels *et al.* (1963) Todaro and Green (1963)

* Young donors, not embryos.
† Data from 20 embryos.
‡ Private communication, D. D. Porter, Pathology Department, University of California, Los Angeles.

many more exhaustive studies should be done. If such a correlation is ultimately found to exist, a significant finding will have been made. It would then be possible to determine the mean maximum lifespan for any animal species simply by assessing the number of population doublings of which their cultured normal fibroblasts are capable. Results would include (1) more precise figures for the longevity of particular animal species; (2) the determination of species lifespans without observing animals throughout their lifespan; and (3) predicting lifespans for species whose records for longevity are scant or non-existent. Of considerable interest is the recent finding that species lifespans are inversely correlated with the ability of their cultured normal cells to metabolize hydrocarbons (Moore and Schwartz, 1978; Schwartz and Moore, 1977).

8.5 The finite lifetime of normal cells transplanted *in vivo*

If the concept of the finite lifetime or senescence of cells grown *in vitro* is related to aging in the whole animal, then it is important to know whether normal cells, given the opportunity, can proliferate indefinitely *in vivo*. If all of the multitude of animal cell types were continually renewed, without loss of function or capacity for self-renewal, we would expect that the organs composed of such cells would function normally indefinitely and that their host would live on forever. Unhappily, however, renewal cell populations do not occur in most tissues and when they do, a proliferative finitude is often manifest. Although this important question has been discussed previously (Hayflick, 1966; Hayflick, 1970), much new information has become available in recent years.

It is apparent to all that most animal species have a specific lifespan and then die. The normal somatic cells composing their tissues obviously die as well. An important question then arises: Is it possible to circumvent the death of normal animal cells that results from the death of the 'host' by transferring marked cells to younger animals *seriatum*? If such experiments could be conducted, then we would have an *in vivo* counterpart of the *in vitro* experiments and would predict that normal cells transplanted serially to proper inbred hosts would, like their *in vitro* counterparts, ultimately age and die. Such experiments would largely rule out those objections to *in vitro* findings that are based on the artificiality of cell cultures. The question could be answered by serial orthotopic transplantation of normal somatic tissue to a new young inbred host each time the former host approaches old age.

Data from eight different laboratories in which mammary tissue (Daniel, deOme, Young, Blair and Faulkin, 1968), skin (Krohn, 1962) and haematopoietic cells (Ford, Micklem and Gray, 1959; Cudkowicz, Upton, Shearer and Hughes, 1964; Siminovitch, Till and McCulloch, 1964; Williamson and Askonas, 1972; Harrison, 1973, 1975; Hellman, Botnick, Hannon and Vigneulle, 1978) were employed, indicate that normal cells, serially transplanted to inbred hosts, do not survive indefinitely. Furthermore, the trauma of transplantation does not appear to influence the results (Krohn, 1962), and finally, in heterochronic transplants, survival time is related to the age of the grafted tissue (Krohn, 1962). It is well known that under similar conditions of tissue transplantation, cancer cell populations can be serially passed indefinitely (Stewart, Snell, Dunham and Schylen, 1959; Till, McCulloch and Simonovitch, 1964; Daniel, Aidells, Medina and Faulkin, 1975). The implications of this result may be that acquisition of potential for unlimited cell division or escape from senescent changes by mammalian cells *in vitro* or *in vivo* can only be achieved by somatic cells which have acquired some or all properties of cancer cells. Paradoxically this leads to the conclusion that in order for mammalian somatic cells to become biologically 'immortal' they first must be induced to an abnormal or neoplastic state either *in vivo* or *in vitro* at which time they can then be subcultivated or transplanted indefinitely. Krohn (1966) found some transplants capable of surviving several years beyond the maximum lifespan of the mouse, but they were ultimately lost. It is important to note that long survival time is not equivalent to proliferation time or rounds of division, hence these long surviving grafts consist of cells having a very slow rate of reproductive turnover. This is analogous to maintaining cell cultures at room temperature which also extends calendar time for cell survival but does not result in an increase in population doublings.

A series of experiments has been reported (Daniel and Young, 1971; Daniel *et al.*, 1968; Daniel, 1973) which clearly show that the birth rate of mouse mammary epithelial cells declines during *in vivo* serial transplantation. See Daniel (1977) for a review of this area. Mouse mammary epithelium was propagated in isogenic female hosts by periodic transplantation of tissue samples into the mammary fat pads from which the host gland was surgically removed. These workers repeatedly observed that, unlike cancerous and precancerous mammary tissue, normal mammary epithelium displays a characteristic decline in proliferative capacity with repeated transplantations (Daniel *et al.*, 1968). It has also been found that when transplants that are allowed to proliferate continuously are compared with transplants in which growth is restricted, the decline in cell proliferation is related to numbers of population doublings undergone rather than to the passage of metabolic time (Daniel and Young, 1971). This represents the first *in vivo* confirmation of our *in vitro* data (Hayflick and Moorhead, 1961; Hayflick, 1965). At that time we suggested that somatic cells have an intrinsic predetermined capacity for division under the most favourable environmental conditions. Others, however, (Hay and Strehler, 1967; McHale, Moulton and McHale, 1971) have invoked the passage of 'metabolic time' as the determinant of senescence *in vitro*. That metabolic time is not the governing factor but that population doublings are, has been recently demonstrated rigorously (Dell'Orco *et al.*, 1973; Harley and Goldstein, 1978). The *in vivo* finding that population doublings and not 'metabolic time' dictate cell senescence (Daniel and Young, 1971) is especially significant, as it results from studies on cells grown entirely *in situ* and thus circumvents arguments levelled at similar data obtained from the alleged 'artificial' conditions of *in vitro* cell culture. These investigators conclude that 'the ability of grafts from old donors to proliferate rapidly in young hosts suggests that the life span of mammary glands is influenced primarily by the number of cell divisions rather than by the passage of chronological or metabolic time' (Young, Medina, deOme and Daniel, 1971).

The study of single antibody-forming cell clones *in vivo* has shown that these cells are also capable of only a limited capacity to replicate after serial transfer *in vivo* (Williamson and Askonas, 1972; Williamson, 1972). Harrison (1972, 1973) reports that when marrow cell transplants from young and old normal donors are made to a genetically anaemic recipient mouse strain, the anaemia is cured. He further reports that such transplants to anaemic mice ultimately expire, exhibiting, once again, the finitude of normal cell proliferation *in vivo* (Harrison, 1975).

In connection with findings that bear upon proliferative capacity of cells *in vivo* as a function of age, several other intriguing reports bear mentioning here. The most informative of such studies are those bearing upon the major cell-renewal systems: the haematopoietic cells of the bone marrow, the epidermis, the lymphatic cells of the thymus, the spleen and lymph nodes, the sperm cells in the seminiferous tubules of the testes, and the epithelial cells lining the gastrointestinal tract and in the crypts of the epithelial lining of the small intestine. It has been found, for example, in the last system in the mouse that cell generation time increases with age (Lesher, Fry and Kohn, 1961a. 1961b; Thrasher and Greulich, 1965). The increase in generation time is not a linear increase from early youth to extreme old age (Lesher and Sacher, 1968). Generation time increases up to approximately 1 year of age and levels off during the second year, and between 675 and 825 days generation time increases again. Furthermore, the intestinal crypt of a 3-month-old mouse contains a population of about 132 dividing cells, while the proliferative compartment of a 27-month-old animal contains only 92 cells. These data support the finding of Thrasher and Greulich (1965), who also found, in older animals, a decrease in the number of crypt cells that are capable of dividing and synthesizing DNA. Lesher and Sacher (1968) conclude that 'because of the increase in generation time and the decrease in the proliferating cell population, production of new cells in old animals is approximately 50% less than the production in 100-day-old animals.'

Post and Hoffman (1964) find in rats that, following birth, there is an exponential decrease in the rate of body and liver growth as well as in the number of cells engaged in DNA synthesis and in mitosis. Furthermore, the size of the replicating pool decreases and there is lengthening of the replication time and of its component parts, DNA synthesis, G_2, mitosis and G_1.

Their data reveal that, as the rat reaches maturity (6 months), there is a marked decline in the percentage of hepatic cells and hepatocytes engaged in DNA synthesis and mitosis, after which a low level of each is maintained. Schneider and Fowlkes (1976) find an increased proportion of human diploid cells in the G_1 period as they approach Phase III. Suzuki and Withers (1978) make the significant finding that even mouse spermatogonial stem cells decrease exponentially with age under circumstances in which tissue transplantation is avoided.

Thus we can see a wide variety of examples of constraints *in vivo* on the proliferative capacity of replicating cell cohorts. It is likely that different proliferating cell systems may display variations on this general theme. On the basis of our *in vitro* data, the fibroblast may represent the upper limit of proliferative potential, although one should not exclude a higher limit being placed on such cell-renewal systems as haematopoietic cells or skin epithelium. The latter has been shown recently to have a division potential *in vitro* much greater than was formerly realized (Green, Kehinde and Thomas, 1979).

8.6 Population doublings *in vivo*

If normal human embryonic cells grown *in vitro* have a finite lifetime of 50 ± 10 doublings, how does one account for the lifelong multiplication of such cells as those found in bone marrow and intestinal and skin epithelium *in vivo*? Presumably these cell populations undergo far more than 50 population doublings during an individual's lifetime. There are several answers to this question, all of which are based on the quantity of cells produced by a primary cell population that undergoes the maximum 50 doublings. That number, as we reported some years ago, is 20 million metric tons of cells (Hayflick and Moorhead, 1961), certainly a quantity sufficient to account for all cells generated during the lifetime of an individual.

In a more elegant consideration of this problem, Kay (1965) argued: Cells can divide in several ways, the two extremes of which are (1) tangential division, in which a stem cell multiplies to produce another stem cell and one differentiated cell, and (2) logarithmic division, in which — if all mitoses are synchronous — a single cell divides yielding a doubled number of cells at each division (similar to our hypothetical model for the ultimate yield of 20 million tons of WI-38 if all cells remain after 50 population doublings). Kay pointed out that a maintained cell output in such rapidly dividing tissue as bone marrow could be maintained by an asynchronous division of cells within the logarithmic model. The essential factor is a variation in the rate of primitive stem cell division so as to produce a continuous release of mature differentiated cells. The advantage of this system, called 'clonal succession', compared with the tangential model would be a reduction in the number of cell generations required to yield a given population of mature cells, thus allowing a much closer adherence to the original genetic message (Kay, 1965).

Only 54 population doublings would be required for a single cell to produce, by the logarithmic pattern, the total output of human erythrocytes and leucocytes during 60 years of life. This figure is surprisingly close to the 50 ± 10 doubling limit that we found for normal human fibroblasts grown *in vitro*. The tangential system would require about 12 000 doublings. Thus the asynchronous logarithmic division model can account for all cells produced during an individual's lifetime, because it would contain several generations of dormant ancestral cells lingering, for example, at the 25th population doubling and which would be successively promoted to form clones of maturing stem cells.

8.7 Organ clocks

Is it possible that a limit on cell function and cell proliferation in some strategic organ could orchestrate the entire phenomenon of senescence and cell death? Burnet (1970a, 1970b) has speculated that, if this is so, the most likely organ is the thymus and its dependent tissues. Burnet reasons that aging is largely mediated by autoimmune processes that are influenced by progressive weakening of the function of immunological surveillance. He further argues that weakening of immunological surveillance may be related to weakness of the thymus-dependent immune system. He concludes that the thymus and its dependent tissues are subject to a proliferative limit similar to the Phase III phenomenon or senescence *in vitro* described by us for human cells. Whether the role played by the thymus and its dependent tissues as the pacemaker in senescence is important or not still remains to be established. However, the immunological theory of aging, being based on the occurrence of somatic mutations (Walford, 1969), has been questioned because the somatic mutation theory of aging is thought to be unlikely (Comfort, 1979; Strehler, 1977).

8.8 Clonal variation

Clonal variation in onset of Phase III occurs with cultured human diploid cell strains. Some isolated single embryonic cells give rise to progeny capable of about 50 doublings (Hayflick, 1965), and others yield colonies composed of various numbers of cells or no clones at all (Merz and Ross, 1969; Smith and Hayflick, 1974). However, the uncloned or wild embryonic cell population always undergoes about 50 population doublings. See also G. M. Martin (1977a, 1977b) for a discussion of clonal growth patterns.

Colony size determinations have been recently introduced as a potential measure of lifespan completed or *in vitro* age of cultured human diploid cells. It is reported that there is a highly significant linear correlation between the percentage of colonies of 16 or more cells and the number of population doublings remaining in the *in vitro* lifespan. Thus the colony size distribution can be used to predict the number of population doublings remaining in the *in vitro* lifespan of a human diploid cell strain without knowledge of its previous passage history (Smith, Pereira-Smith and Good, 1977).

Of equal importance in this context is the intriguing possibility that cultures of bacteria in which exchange of genetic information between individual organisms is prevented might also reveal a senescence phenomenon. It is commonly believed that bacteria and cultures of other unicellular organisms are immortal but most protozoa are not (Hayflick, 1975). They are immortal only in the sense that the human species is immortal over the eons, but individual members of the species age and die. Until the possibility for exchange of genetic information is prevented in bacterial cultures, individual organisms can only be assumed to age and die. In order to test this hypothesis, daughter micro-organisms would have to be kept isolated from other members of the culture as they replicate serially. The technical problems that would be encountered in constructing such an experiment are probably sufficient reason why this experiment has not yet been reported. Perhaps of interest in this connection are reports of the exchange of genetic material by mammalian cells *in vitro* (Barski, Sorieul and Cornefert, 1960; Ephrussi, Scaletta, Stenchever and Yoshida, 1964). Is it possible that immortal cell populations can only occur when genetic material is periodically exchanged to somehow recycle the clock that leads to age changes?

8.9 Irradiation, DNA repair and effects of visible light

One of the major theories of age changes involves decreased or less efficient mechanisms for cells to repair DNA damage as a function of age. Much work has been done on cultured cells in the past few years to elucidate this phenomenon.

An early study by Lima, Malaise and Macieira-Coelho (1972), using chick embryo fibroblasts, was one of the first to show the effects of ^{60}Co irradiation on cell longevity *in vitro*. Saturation densities were unaffected, but Phase III was initiated earlier in the irradiated cultures. The percentage of dividing cells was unchanged but the number of population doublings reached was less in the irradiated cultures. The reduction in the total number of cells produced was found to be directly proportional to the total number of rads received. Lima *et al.* (1972) concluded that even very small doses of irradiation may have long-term damaging effects on cultured normal cells.

Epstein, Williams and Little (1973a, 1973b, 1974) reported deficient DNA strand rejoining after X-irradiation of Phase III normal human cells but Regan and

Setlow (1973) and Clarkson and Painter (1974) could not confirm this. Goldstein (1971) and Painter, Clarkson and Young (1973) showed a decreased repair in u.v.-induced damage in Phase III human fibroblasts, but Mattern and Cerrutti (1975) found that isolated nuclei from Phase II WI-38 cells lost their capacity to excise damaged bases from α-irradiated DNA. In 1974 Hart and Setlow made the important observation that cultured skin fibroblasts from long-lived species have a greater capacity to repair u.v.-induced damage than do cells from shorter-lived species. In 1976 these workers (Hart and Setlow, 1976) also reported that the average amount of unscheduled DNA synthesis decreases as cultured human cells approach Phase III.

Bowman, Meek and Daniel (1976), working with WI-38 cells, showed that unscheduled DNA synthetic capacity was significantly reduced in Phase III cultures. This reduction was a result of decreased repair capacity which correlated well with a similar decrease in the proportion of dividing cells in the population.

Bradley, Erickson and Kohn (1976), using alkaline elution and alkaline sedimentation, found that the rate of rejoining of X-ray-induced DNA single-strand breaks was unchanged in Phase III WI-38 cells. They also found no increase in DNA cross-linking in Phase III WI-38 cells.

Macieira-Coelho, Diatloff, Billard, Bourgeois and Malaise (1977) showed that human diploid cells during late Phase III are more sensitive to low dose rate ionizing radiation than are cells in Phase II. In 1978, Macieira-Coelho, Diatloff, Billard, Fertil, Malaise and Fries compared fibroblasts from embryonic and adult origin. The response of adult cells irradiated during Phase II was found to be similar to that of embryonic cells irradiated during Phase III.

Duker and Teebor (1975) demonstrated that WI-38 cells contain an endonuclease activity that breaks u.v.-irradiated circular DNA. When compared with xeroderma pigmentosum fibroblasts (which cannot repair DNA), the endonuclease activity was similar. They concluded that there is no correlation between enzyme activity and rates of repair synthesis. This difference suggested that the endonuclease activity is not the u.v.-repair endonuclease.

Macieira-Coelho and Diatloff (1976) Macieira-Coelho, Diatloff and Malaise (1976) and Macieiro-Coelho *et al.* (1977) reported that ionizing radiation shortens the lifespan of chicken cells, accelerates spontaneous transformation in mouse cells and has an intermediate effect on normal human fibroblasts where no reduction in population doubling potential was seen to occur.

The diminished repair capacity of progeria cells was found to be enhanced by co-cultivation with a hamster cell line, a mid-Phase II diploid human cell strain or a second progeria strain [Brown, Epstein and Little (1976) – progeria is a rare disease in which the victims appear to age prematurely and die very young].

Smith and Hanawalt (1976) found no significant difference in the amount of repair replication performed, its dose response or the time course between growing and confluent WI-38 cells, early passage and Phase III cells, or normal WI-38 cells and SV40-transformed WI-38 cells.

Kantor, Mulkie and Hull (1978) reported that no accumulation of damage induced by u.v. affects lifespan, thymidine-labelling index, growth rates or confluent cell densities of human diploid cells. They further found no selection of a population with altered sensitivity. They conclude that genome hits may not be a prime reason for the Phase III phenomenon.

Pereira, Smith and Packer (1976), in studying the damaging effects of near-u.v. and visible light on WI-38 cells, reported that the cells were killed by light doses ranging from 2×10^3 to 10×10^3 W/m²h. There was found an inverse correlation between population doubling level and photosensitivity. The effect was not related to capacity for DNA synthesis and cell division. Flavins were clearly implicated as endogenous photosensitizers and several antioxidants were found to protect cells from light damage. Products of lipid peroxidation were detected in cell homogenates irradiated in the presence of riboflavin.

Parshad and Sanford (1977a, 1977b) have shown that short exposure (3 or 4 h) to fluorescent light (6 to 9 W/m) one to three times weekly enhances the proliferation rate of three human diploid cell strains. The effect is reversible and mediated through the culture medium. The enhancement was reflected by (1) an increase in population doublings, (2) cumulative increase in cell number, (3) duration of cell survival and (4) duration of growth. The foetal strains studied, including WI-38, showed about a 20 per cent increase in total population doublings, which was observed earlier by Litwin (1972). Cells of adult origin, in contrast, which had never been exposed to light of wavelength less than 500 nm during the first seven population doublings, showed a 70-98 per cent increase in total population doublings. The only other factor known to increase proliferation rate and extend lifespan is hydrocortisone (Macieira-Coelho, 1966; Cristofalo, 1972). The effect of intermittent exposure to fluorescent light is apparently the production of some mitogenic substances in the culture medium since the proliferative response of cells grown in preilluminated medium simulated that of cells directly exposed to light.

Paffenholz (1978), using mouse fibroblasts from three different inbred strains, made the interesting observation that unscheduled DNA synthesis following u.v. irradiation of these cells correlated with the different life expectations of the mouse strains. In each case, however, the ability to perform repair synthesis paralleled Phase III, although semiconservative DNA synthesis and proliferative potential of the cells remained unchanged until Phase III.

8.10 Cytogenetic studies

Schneider and his colleagues have made several interesting observations on sister chromatid exchanges (SCE) in cultured normal human diploid fibroblasts. Interest in this phenomenon is based on the possibility that at least one kind of SCE represents a DNA-repair process which is of great interest to many gerontological theoreticians. SCEs are associated with a variety of agents that are capable of inducing chromosomal aberrations, and it is for this reason that spontaneous SCEs are of interest. Their appearance may reflect certain kinds of DNA damage and subsequent repair.

Schneider and Monticone (1978) found that baseline SCEs in cultured normal human diploid cells remain relatively constant throughout their *in vitro* lifespan. However, a significant decline in SCEs was found to occur in middle Phase II and Phase III cells induced by mitomycin-C, ethyl methanesulphonate and *N*-acetoxyl-2-acetylaminofluorene. Cell concentration and replication rate did not alter the effect. Schneider and Kram (1978) also report a decline in SCE frequency in older donor cells at high mitomycin-C concentration and a higher frequency of chromosomal alterations. Studies done by Kram and Schneider (1978), using four strains of young and old mice perfused with mitomycin-C, found a reduced frequency of induced SCEs in AKR mice and in old rat and mouse bone marrow when compared with young cells. They conclude that there is a diminished response of old cells, both *in vitro* in human cells and *in vivo* in mouse and rat cells, to high levels of DNA damage induced by mitomycin-C, as manifested by both diminished SCE formation and increased chromosomal aberrations.

8.11 Error accumulation

Since the proposal was made that accumulation of errors in the protein-synthesizing machinery of cells could result in age changes, a substantial amount of experimental evidence has been accumulated that bears on this as an explanation for the Phase III phenomenon. Orgel (1973) has reviewed this area and especially those studies done by Holliday and his colleagues that seem to support the hypothesis (Holliday and Tarrant, 1972). Nevertheless, substantial data have been obtained which do not support the notion that the loss of cell-doubling capacity *in vitro* is due to an accumulation of errors in enzymes or other proteins (Ryan, Duda and Cristofalo, 1974; Holland, Koline and Doyle, 1973; Pitha, Adams and Pitha, 1974; Tomkins, Stanbridge and Hayflick, 1974; Wright and Hayflick, 1975a, 1975b; Goldstein and Csullog, 1977).

8.12 The proliferating pool

One of the first decrements shown to occur as cell populations approach and enter Phase III is the numbers of cells in the proliferating pool (Merz and Ross, 1969; Cristofalo and Sharf, 1973; Smith and Hayflick, 1974). More recently, using phase-contrast time-lapse cinemicrophotography, Absher, Absher and Barnes (1974) and Absher and Absher (1976) showed that constancy of interdivision time decreased and asynchronous division events occured in cells at later population doubling levels. Variations were also noted in clone size and generations per clone. Correlation coefficients for interdivision times of sister pairs are high in young clones and generally low in aged clones. A consistent division pattern at all population doubling levels is one of short average time for early and late generations of a clone and long average interdivision time for the middle range of generations of a clone. It was concluded that non-cycling cells are present in increased amounts during *in vitro* cell senescence; nevertheless, a cogent argument in opposition to this view has also been made (Macieira-Coelho, 1974). A mathematical model describing the variation in length of mitotic cycle as cells approach Phase III has been put forward by Good and Smith (1974).

8.13 Efforts to increase population doubling potential

Several reports have appeared in which additives to the usual cell culture media or environmental changes have resulted in an increase in population doubling potential. The first of these reports (Todaro and Green, 1963) claimed that the addition of serum albumin resulted in a 20 per cent increase in population doubling potential. However, there has been no confirmation of this result. Despite several attempts (Sakagami and Yamada, 1977) there has also been failure to confirm the claim that the addition of vitamin E can significantly prolong the population doubling potential of cultured normal human cells (Packer and Smith, 1974a, 1974b). In fact, this claim has been formally retracted by these authors (Packer and Smith, 1977). There is, nevertheless, ample evidence to support the finding that cortisone and hydrocortisone can extend *in vitro* lifespan by 30 to 40 per cent (Macieria-Coelho, 1966; Cristofalo, 1970). The earlier these substances are administered, the greater is the effect, and when hydrocortisone is removed from cells which have grown beyond the lifespan of controls, cell death occurs within two subcultivations. The mechanism by which these compounds produce this effect is unknown, but some recent results suggest that the events controlling cell proliferation are sensitive to hydrocortisone modulation during the G_1 and possibly the G_2 periods (Grove and Cristofalo, 1977).

Poiley, Schuman and Pienta (1978) claim to have obtained over 100 population doublings for an allegedly diploid human cell strain. However, this cell population is clearly abnormal as evidenced by the high levels of aneuploidy which far exceed generally acceptable criteria for normalcy. A list of factors reported to extend population doublings over that found using conventional procedures is given in Table 8.3. There are no reports describing normal cell populations which have an unlimited capacity for replication. Even those studies in which extensions of replicative capacity have been observed must be evaluated in the light of the numerous possibilities for miscalculating population doublings that, until recently, have received little attention (Hayflick, 1977b).

8.14 Phase III in cultured mouse fibroblasts

It is sometimes erroneously reported that cultured mouse fibroblasts do not exhibit the Phase III phenomenon because, unlike chicken or human fibroblasts, spontaneous transformation frequently occurs. This belief is spurious because the normal mouse fibroblasts, which do in fact exhibit a Phase III phenomenon (like normal fibroblasts from all other species), are simply over-run by the proliferating transformed cells which mask Phase III occurring in the untransformed component. The most recent substantiation of this fact emerges from the studies of Meek,

Table 8.3 Factors reported to increase lifespan (population doublings) of normal human diploid cells over that found using conventional procedures

Serum albumin concentration	Todaro and Green (1964)
Various glucocorticoids	Macieira-Coelho (1966)
Hydrocortisone	Cristofalo (1970)
	Grove and Cristofalo (1977)
Medium volume	Ryan *et al.* (1975)
10% oxygen	Packer and Fuehr (1977)
1% oxygen	Taylor *et al.* (1978)
	[No effect by partial pressures of O_2 of 6, 24, 26, 49, 50 or 137 mmHg reported by Balin *et al.* (1977, 1978)]
Vitamin E	Packer and Smith (1974a, 1974b)
	[Results subsequently unconfirmed, (Packer and Smith 1977; Sakagami and Yamada, 1977; Balin *et al.* 1977)]
Aneuploidy	Poiley *et al.* (1978)
	(Cells clearly are not normal, although are stated to be by authors)
Some foetal bovine serum batches	Schneider *et al.* (1978)
Intermittent exposure to fluorescent light	Parshad and Sanford (1977a, 1977b)

Bowman and Daniel (1977). Normal mouse embryo cells were clearly shown to display Phase III. A decreased rate in the percentage of cells able to incorporate [^{3}H] thymidine was also found as was first described in human cells. Subsequent to these events, however, continuously proliferating abnormal transformed cells did arise. Thus it is inaccurate to argue that normal mouse fibroblasts do not exhibit Phase III. Normal mouse embryo fibroblasts have been shown to undergo between 14 and 28 population doublings *in vitro* (Todaro and Green, 1963; Rothfels, Kupelwieser and Parker, 1963).

8.15 Phase III theories

Holliday, Huschtscha, Tarrant and Kirkwood (1977) have proposed a commitment theory of *in vitro* fibroblast aging in which the starting cell population is uncommitted and potentially immortal. An assumption is made that as the cells divide there is a given probability that fibroblasts will arise which are irreversibly committed to senescence and death. Ultimately all of the committed cells die. They argue that if the probability of commitment is high and the number of population doublings is sufficiently high, then the number of uncommitted cells progressively declines and they are lost. Thus all remaining cells are committed to senescence and the population becomes mortal with individual cells having different growth potentials. The basis of this proposal is dependent on computer analysis and

Holliday *et al.* (1977) conclude that the senescence of these cultures is an artefact of normal culturing procedures. There are several arguments that can be raised against this interpretation. Considering the tremendous number of cell cultures that have been established from WI-38, which is the most widely studied population, it would be expected that at least one immortal or uncommitted cell would have survived. This has never been experienced in the nineteen years during which WI-38 has been cultured in hundreds of laboratories. Also, if the hypothesis were tenable, one would expect that cloning populations at early population doubling levels would also yield a few immortal cell populations. This also has never been observed. Harley and Goldstein (1978) have provided important data and cogent arguments contrary to the commitment theory.

Finally, if potentially immortal cells become committed to senescence, it would be difficult to explain how the 500 or 600 immortal cell populations (HeLa, for example) circumvent the theory. Good (1975) has also proposed other theoretical concepts of the Phase III phenomenon which have been criticized by Macieira-Coelho (1977).

8.16 Can cell death be normal?

The death of cells and the destruction of tissues and organs is, indeed, a normal part of morphogenic or developmental sequences in animals. It is the common method of eliminating organs and tissues that are useful only in the larval or embryonic stages of many animals: for example, the pronephros and mesonephros of higher vertebrates, the tail and gills of tadpoles, larval insect organs and, in many cases, the thymus. The degeneration of cells is an important part of development and it is a widespread occurrence in mammalian cells (Saunders, 1966). During the development of vertebrate limbs, cell death and cell resorption model not only digits but also thigh and upper arm contours (Whitten, 1969). In the limbs of vertebrates the death clocks function on schedule, even when heterochronic tissue grafts are made (Saunders and Fallon, 1966). The following remarks by Saunders (1966) attest to the universality and normality of cell death *in vivo*.

> 'One confronts less than comfortably the notion that cellular death has a place in embryonic development; for why should the embryo, progressing towards an ever more improbable state squander in death those resources of energy and information which it has laboriously won from a less ordered environment? Nevertheless, abundant death, often cataclysmic in its onslaught, is a part of early development in many animals: it is the usual method of eliminating organs and tissues that are useful only during embryonic or larval life or that are but phylogenetic vestiges (phylogenetic death). For example, the pronephros and mesonephros of the higher vertebrates, the anuran tail and gills, and larval organs of holometablous insects; it plays a role in the differentiation of organs and tissues, as exemplified by the histogenesis and remodeling of cartilage and bone (histogenetic death); and it accompanies the formation of folds and the confluence of anlagen (morpho-

genetic death). It occurs frequently in the early blastoderm of the chick embryo, more or less at random, and dead cells are scattered throughout the tissues of the older embryo. Indeed, degeneration of cells and tissues is a very prominent part of development, and it is unfortunate that little has been done to analyze its significance in the processes with which it is associated, its embryonic control, the biochemical events in the onset and realization of necrosis, and the development of products of degeneration.'

Thus physiological decrements that lead to cell death are an intrinsic part of development. These same events also may be an intrinsic part of aging. If so, the underlying mechanism which induces them may be identical. To the casual observer the contemplation that normal human embryo cells grown *in vitro* will die after dividing vigorously for 50 population doublings may be difficult to accept. Yet the same logic that makes acceptable aging and death in whole animals as being universal and inevitable should apply equally when the same phenomenon is seen to occur in cultured cells derived from these same animals.

8.17 Dividing, slowly dividing and non-dividing cells

The fact that highly differentiated normal epithelial cell populations double only a few times *in vitro* has been offered as evidence for the inadequacy of cell culture conditions. Yet it is at least equally plausible to suggest that such cell populations are as incapable of many rounds of cell division *in vitro* as they are *in vivo* and for the same inherent reasons. Thus it is possible that the likelihood of finding *in vitro* conditions that would permit many rounds of division in highly differentiated cells is as unlikely as finding similar conditions that would permit the indefinite proliferation of normal fibroblasts. The search for conditions permitting extended proliferation of normal differentiated cells may be just as futile as the search has been for conditions that would allow normal fibroblasts to replicate indefinitely.

Since our suggestion that the Phase III phenomenon might be an expression of aging at the cell level, the point has been made that, because the major age-related changes in animals are regarded as being an expression of events that occur in non-dividing or slowly dividing highly differentiated cells, the Phase III phenomenon in cultured fibroblasts cannot be regarded as a useful subject for gerontological inquiry. We believe this to be spurious because it is predicated on the implied premise that cell division is not a function or that it is somehow quantitatively less of a function than, say that of neurons, kidney or liver cells. Our contention is that whatever the causes are that produce functional decrements in non- or slowly dividing cells are equally likely to cause that functional decrement recognized as cessation of proliferative ability. There is no justification for distinguishing between slow or non-dividing cells on the one hand and dividing cells on the other hand on the basis of the former cells being functional and the latter not. Cell division is, after all, a functional property whose likelihood for change with age is equally as probable as any other cell function.

We do not contend that aging in animals is the result of the loss of proliferative capacity in those cells capable of division. What is more likely is that the determinants that decrease cell function with time also affect replicative capacity. This determinant is likely to repose in the cell genome (Hayflick, 1976).

8.18 Aging or differentiation?

Although the idea is not new, the suggestion has been made again that the Phase III phenomenon may not be a manifestation of aging but of differentiation (Bell, Marek, Levinstone, Merrill, Sher, Young and Eden, 1978). Until an unambiguous and testable distinction can be made between those two terms, the notion that the Phase III phenomenon represents one and not the other will not be meaningful. The processes of differentiation and aging are so inextricably interwoven that it would be futile, if not impossible, to make any fundamental distinction. To say that aging is a constellation of physiological decrements that follow differentiation, is to say little that permits experimentation. These processes represent a continuum of cell changes that are probably governed by the same fundamental mechanism. Cell differentiation, cell aging and cell death may occur in embryos, newborn, adult or aging animals.

It is well documented that during embryogenesis, phenomena indistinguishable from cell aging occur. The death of cells and the destruction of tissues and organs is a normal part of developmental sequences in animals (Saunders, 1966).

The fact is that normal cells cultured *in vitro* or serially transplanted *in vivo* ultimately incur physiological decrements and die. Whether one regards this as aging or 'differentiation to death' is more likely to interest semanticists than to lead to meaningful experimentation.

8.19 Functional and biochemical changes that occur in cultured normal human cells

As previously indicated, the likelihood is remote that animals age because one or more important cell population loses its proliferative capacity. Normal cultured cells have a finite capacity for replication, and we have proposed that this finite limit is rarely, if ever, reached by cells *in vivo*. We would therefore suggest that many of the functional losses now known to occur in cultured cells prior to their loss of division capacity also occur in cells *in vivo* much before their replicative capacity is lost. It is these functional decrements which herald the loss of mitotic capability *in vitro* that we believe to be the causes of age changes *in vivo*. It is reasonable to suppose that not all functional decrements occur in cells at the same time or at the same rate. Loss of division capacity by mitotically active cells occurs well after the occurrence of many other decrements. We are now becoming aware of a plethora of functional changes taking place in normal human cells grown *in vitro* and expressed well before they lose their capacity to replicate. These changes have been reported by many laboratories. An exhaustive, but undoubtedly incomplete, listing of these increments and decrements may be found in Tables 8.4 and 8.5.

Table 8.4 Properties that increase as normal human diploid cells approach the end of their *in vitro* lifespan (Phase III)

Lipids	
Lipid content	Kritchevsky and Howard (1966)
Lipid synthesis	Chang (1962)
	Yuan and Chang (1969)
Requirement for delipidized serum protein for stimulation of lipid synthesis	Howard *et al.* (1976)
Carbohydrates	
Glucose utilization	Goldstein and Trieman (1974)
Glycogen content	Cristofalo (1970)
Proteins and Amino Acids	
Protein content	Cristofalo (1970)
	Wang *et al.* (1970)
	(Also reported not to change)
Proteins with increased proteolytic susceptibility (only at last PDL*)	Bradley *et al.* (1975)
Protein component P8	Stein and Burtner (1975)
Breakdown rate of proteins when treated with an amino acid analogue	Shakespeare and Buchanan (1976)
Albumin uptake	Berumen and Macieira-Coelho (1977)
Stimulatory activity of polyornithine on protein uptake	Berumen and Macieira-Coelho (1977)
Proportion of rapidly degradable proteins	Dean and Riley (1978)
Amino acid efflux	Goldstein *et al.* (1976)
Error level for mis-incorporation of methionine into histone H1	Buchanan and Stevens (1978)
Number of methionine residues	Buchanan and Stevens (1978)
Complexity of H1 polypeptide chains	Buchanan and Stevens (1978)
RNA	
RNA content	Cristofalo and Kritchevsky (1969)
	Hill *et al.* (1978)
	Schneider *et al.* (1973)
RNA turnover	Michl and Svoboda (1967)
Proportion of RNA and histone in chromatin	Srivastava (1973)
Enzymes	
Lysosomes and lysosomal enzymes	Cristofalo *et al.* (1967)
	Robbins *et al.* (1970)
	Wang *et al.* (1970)
	Milisauskas and Rose (1973)

* PDL, population doubling level.

Table 8.4 continued. . .

Heat lability and activity of glucose 6-phosphate dehydrogenase and 6-phosphogluconate dehydrogenase	Holliday (1972) Holliday and Tarrant (1972) Fulder and Holliday (1975) Duncan *et al.* (1977) Fulder and Tarrant (1975)
Activity of 'chromatin associated enzymes' (RNAase, DNAase, protease, nucleoside triphosphatase, NAD+ pyrophosphorylase)	Srivastava (1973) Ryan and Cristofalo (1972)
Esterase activity	Turk and Milo (1974)
Acid phosphatase band 3	Turk and Milo (1974)
Acid phosphatase	Cristofalo and Kabakjian (1975)
β-Glucuronidase activity	Turk and Milo (1974) Cristofalo and Kabakjian (1975)
Membrane-associated ATPase activity	Turk and Milo (1974)
pH 3.4 protease activity (decreased in middle population doublings)	Bosmann *et al.* (1976)
N-Acetyl-β-glucosaminidase	Sun *et al.* (1975)
5'-Nucleotidase	Sun *et al.* (1975)
Cytochrome oxidase	Sun *et al.* (1975)
Tyrosine aminotransferase activity	Fulder and Tarrant (1975)
Acid hydrolases: α-D-Mannosidase N-Acetyl-β-D-galactosaminidase N-Acetyl-β-D-glucosaminidase β-L-fucosidase β-D-galactosidase	Bosmann *et al.* (1976)
γ-Glutamyltransferase [(5-glutamyl)-peptide − amino acid 5-glutamyltransferase] activity	Takahashi *et al.* (1978)
Monoamine oxidase activity	Edelstein *et al.* (1978)

Cell Cycle

Prolongation of population doubling time	Hayflick and Moorhead (1961) Hayakawa (1969) Macieira-Coelho *et al.* (1966a) Bolton and Barranco (1976)
Heterogeneity in length of division cycle	Macieira-Coelho *et al.* (1966a)
Slow or non-replicating cells	Mitsui and Schneider (1976b)
Mitotic cycle	Macieira-Coelho (1977) Grove and Cristofalo (1977)
Cell longevity with certain fractions of cigarette smoke condensate	Litwin *et al.* (1978)

Morphology

Cell size and volume	Simons (1967) Simons and van den Broek (1970)

Table 8.4

Cell size and volume (cont'd...)	Cristofalo and Kritchevsky (1969)
	Macieira-Coelho and Ponten (1969)
	Bowman *et al.* (1975)
	Greenberg *et al.* (1977)
	Mitsui and Schneider (1976b)
	Schneider and Fowlkes (1976)
Number and size of lysosomes	Lipetz and Cristofalo (1972)
	Robbins *et al.* (1970)
	Brunk *et al.* (1973)
Number of residual bodies	Brunk *et al.* (1973)
Cytoplasmic microfibrils	Lipetz and Cristofalo (1972)
Endoplasmic reticulum, constricted and 'empty'	Lipetz and Cristofalo (1972)
Particulate intracellular fluorescence	Deamer and Gonzales (1974)
Nuclear size of the slow or non-replicating cell component	Mitsui and Schneider (1976a)
Mean nuclear area in cells increases progressively from donors aged 8, 40 and 84 years	Lee *et al.* (1978)
Cell sizes in both G1 and G2+M	Schneider and Fowlkes (1976)
Particles on freeze-fracture face E	Kelley and Skipper (1977)
Microvilli on cell surfaces, chromatin condensation and dense bodies	Johnson (1979)
Cells containing long thin dense mito-chondria and bizarre shapes, cells exhibiting filamentous degeneration	Johnson (1979)
Blebs and marginal ruffling	Wolosewick and Porter (1977)
Fraction of cells with one large nucleolus per nucleus	Bemiller and Lee (1978)
Mean nucleolar dry mass and area	Bemiller and Lee (1978)

Unclassified

Cyclic AMP level/mg protein	Goldstein and Haslam (1973)
Tolerance to sublethal radiation damage	Cox and Masson (1974)
Time needed to respond to proliferative stimulus by medium change	Hill (1977)
Reversion rate of herpes virus temperature-sensitive mutant E	Fulder (1977)
DNA repair in cells arrested by low serum concentration	Dell'Orco and Whittle (1978)
Rate of uridine transport	Polgar *et al.* (1978)
Cyclic AMP levels	Polgar *et al.* (1978)

Table 8.5 Properties that decrease as normal human diploid cells approach the end of their *in vitro* lifespan (Phase III)

Proteins	
Collagen synthesis	Macek *et al.* (1967)
	Houck *et al.* (1971)
Collagenolytic activity	Houck *et al.* (1971)
Rate of histone acetylation	Ryan and Cristofalo (1972)
Total protein and hydroxyproline	Shen and Strecker (1975)
Proteolytic capacity	Goldstein *et al.* (1976)
Incorporation of isotope into histone	Dell'Orco *et al.* (1978)

RNA	
Rate of RNA synthesis	Macieira-Coelho *et al.* (1966a)
Ribosomal RNA content	Levine *et al.* (1967)
	(May have been due to mycoplasma contamination)
RNA synthesizing activity of chromatin	Srivastava (1973)
	Ryan and Cristofalo (1972)
Chromatin template activity	Ryan and Cristofalo (1975)
(also reported not to change)	Stein and Burtner (1975)
Template activity of isolated nuclei (only in late Phase III)	Hill *et al.* (1978)
Incorporation of tritiated uridine into cellular RNA	Schneider *et al.* (1973)
Nucleolar synthesis of RNA	Bowman *et al.* (1976)

DNA	
DNA synthesis	Choe and Rose (1974)
	Macieira-Coelho *et al.* (1966a, 1966b)
Rate of DNA chain elongation	Petes *et al.* (1974)
Rate of DNA strand rejoining and repair rate	Epstein *et al.* (1974)
	Mattern and Cerutti (1975)
Induced sister chromatid exchanges	Schneider and Monticone (1978)
DNA polymerase activity	Linn *et al.* (1976)

Enzymes	
Lactate dehydrogenase isoenzyme pattern	Childs and Legator (1965)
Glycolytic enzymes	Wang *et al.* (1970)
	(also reported not to change)
Transaminases	Wang *et al.* (1970)
	(also reported not to change)
Alkaline phosphatase	Turk and Milo (1974)
Specific activity of lactate dehydrogenase	Lewis and Tarrant (1972)
Response to induction of ornithine decarboxylase	Fulder and Tarrant (1975)

Table 8.5 continued. . .

pH 7.8 neutral protease activity	Bosmann *et al.* (1976)
Prolyl hydroxylase activity	Chen *et al.* (1977)
Ornithine decarboxylase activity	Duffy and Kremzner (1977)
Ascorbate dependence of the prolyl hydroxylase system	Chen *et al.* (1977)
Glutamine synthetase specific activity and heat lability	Viceps-Madore and Cristofalo (1978)

Cell Cycle

Numbers of cells in proliferating pool	Cristofalo and Sharf (1973) Merz and Ross (1969) Smith and Hayflick (1974)
Cell saturation density	Macieira-Coelho (1970)
Population doubling potential as a function of donor age	Hayflick (1965) Martin *et al.* (1970) LeGuilly *et al.* (1973) Schneider and Mitsui (1976) Bierman (1978)
Incorporation of tritiated thymidine	Cristofalo and Sharf (1973) Vincent and Huang (1976) Hill *et al.* (1978)
Synchronous division, constancy of interdivision time and motility	Absher *et al.* (1974)
Growth potential	Hill (1976)
Percent of colonies exceeding a given colony size	Smith *et al.* (1977)
Cell longevity decreased by one fraction of cigarette smoke condensate	Litwin *et al.* (1978)

Morphology

Proportion of mitochondria with completely transverse cristae	Lipetz and Cristofalo (1972)
Distribution and development of specialized structures for intracellular contact and communication	Kelley (1976)
Particles on freeze-fracture face P	Kelley and Skipper (1977)
Number of mitochondria (also reported not to change)	Johnson (1979)

Synthesis, Incorporation and Stimulation

Mucopolysaccharide synthesis	Kurtz and Stidworthy (1969)
Pentose phosphate shunt	Cristofalo (1970)
Cyclic AMP level (molar values)	Goldstein and Haslam (1973)
Stimulation of growth with putrescine	Duffy and Kremzner (1977)
Radioactive uridine, thymidine, protein hydrolysate, acetate, oleic acid, cholesterol	Razin *et al.* (1977)

Table 8.5 continued. . .

Synthesis of glycosaminoglycans	Schachtschabel and Wever (1978)
Interferon production	Horoszewicz *et al.* (1978)
Number of cells responding to proliferative stimulus by medium change	Hill (1977)
Ability to synthesize proteins and amino acids	Razin *et al.* (1977)

Unclassified

HLA specificities (cloned cells)	Goldstein and Singal (1972)
Adherence to polymerizing fibrin and influence on fibrin retraction	Niewiarowski and Goldstein (1973)
Electrophoretic mobility (net negative cell surface charge)	Bosmann *et al.* (1976)
Lifespan after chronic exposure to elevated partial pressures of oxygen	Balin *et al.* (1977, 1978)
Reversion rate of herpes virus temperature-sensitive mutant G	Fulder (1977)

Table 8.6 Properties that do not change as normal human diploid cells approach the end of their *in vitro* lifespan (Phase III)

DNA and RNA

Soluble RNAase, soluble DNAase, soluble seryl-t-RNA synthetase, soluble and chromatin-associated DNA polymerase	Srivastava (1973)
Mean temperature of denaturation of DNA and chromatin	Comings and Vance (1971)
Histone/DNA ratio	Ryan and Cristofalo (1972)
Rate of DNA strand rejoining and ability to perform repair replication	Clarkson and Painter (1974)
Chromatin template activity	Hill (1975)
Nucleoplasmic synthesis of RNA	Bowman *et al.* (1976)
Heat lability in relative profiles of DNA, RNA and protein precursors (minor increase at last doubling)	Goldstein and Csullog (1977)
Level of unscheduled DNA synthesis in confluent cultures	Dell'Orco and Whittle (1978)

Enzymes

Respiratory enzymes	Hakami and Pious (1968) Wang *et al.* (1970)
Glutamate dehydrogenase	Wang *et al.* (1970)
Alkaline phosphatase	Cristofalo *et al.* (1967) Wang *et al.* (1970)

Table 8.6 continued. . .

Superoxide dismutase activity	Yamanaka and Deamer (1974)
Heat stability of glucose 6-phosphate dehydrogenase	Pendergrass *et al.* (1976) Kahn *et al.* (1977a)
Catalase activity	Sun *et al.* (1975)
Lysyl hydroxylase activity	Chen *et al.* (1977)
Enzymes of the 'γ-glutamyl cycle'	Takahashi *et al.* (1978)
Specific activity or thermostability of phosphoglucose isomerase	Shakespeare and Buchanan (1978)
Glucose 6-phosphate dehydrogenase alterations	Pendergrass *et al.* (1976)
N-Acetyl-α-D-galactosaminidase *N*-Acetyl-β-D-glucosaminidase-D-glucosidase Sulphite cytochrome *c* reductase	Houben and Remacle (1978)

Cell Cycle

S phase of cell cycle	Grove and Cristofalo (1977) Kapp and Klevecz (1976)
Generation time of selected mitotic cells	Kapp and Klevecz (1976)

Karyology

Diploidy (only changes in late Phase III)	Saksela and Moorhead (1963) Thompson and Holliday (1975) Miller *et al.* (1977) Benn (1976)
Prematurely condensed chromosomes	Yanishevsky and Carrano (1975)
Proportion of colchicine-induced polyploid cells	Thompson and Holliday (1978)

Cyclic AMP, GMP

Cyclic AMP concentration	Haslam and Goldstein (1974)
Cyclic AMP phosphodiesterase activity	Polgar *et al.* (1978)
Decrease in K_m of cyclic AMP phosphodiesterase with serum stimulation	Polgar *et al.* (1978)
Levels of cyclic AMP and cyclic GMP	Ahn *et al.* (1978)

Virology

Virus susceptibility	Hayflick and Moorhead (1961) Pitha *et al.* (1974)
Polio virus and herpes virus titre, mutation rate and protein chemistry	Holland *et al.* (1973) Tomkins *et al.* (1974)
Chromosome 21-directed anti-viral gene(s)	Tan *et al.* (1975)
Reversion rate of herpes virus temperature-sensitive mutant D	Fulder (1977)
Synthesis of vesicular stomatitis virus RNA	Danner *et al.* (1978)
Susceptibility to interferon	Horoszewicz *et al.* (1978)

Morphology

Numbers of mitochondria	Lipetz and Cristofalo (1972)
Number of intramembrane particles	Kelley and Skipper (1977)

Synthesis and Degradation

Glycolysis	Cristofalo and Kritchevsky (1966)
Mis-synthesized or post-translationally modified proteins	Kahn *et al.* (1977b)
Conversion of glutamate to proline or hydroxyproline	Shen and Strecker (1975)
Hydralazine inhibition of hydroxylation	Chen *et al.* (1977)
Increase in V_{max} of uridine transport with serum stimulation	Polgar *et al.* (1978)
Ability to degrade normal or analogue-containing proteins	Dean and Riley (1978)

Unclassified

Cell viability at sub-zero temperatures (19 years)	Hayflick (1965), Hayflick (unpublished observations)
Respiration	Cristofalo and Kritchevsky (1966)
HLA specificities (mass cultures)	Brautbar *et al.* (1972, 1973)
Irreversible adsorption/uptake of foreign macromolecules	Press and Pitha (1974)
Membrane fluidity	Polgar *et al.* (1978)
Phospholipid and neutral fat content	Polgar *et al.* (1978)
Effect of nicotine	Litwin *et al.* (1978)
Effect of polynuclear hydrocarbon carcinogens	Litwin *et al.* (1978)

Table 8.6 lists those variables that have not been observed to change as normal human cells approach Phase III. It is more likely that the changes, described in Tables 8.4 and 8.5, which herald the approach of loss of division capacity play the central role in the expression of aging and result in death of the individual animal well before replicative failure of its somatic cells occurs.

The measurement of loss of cell division potential is, after all, one of many cell functions that could be studied. If the manifestations of age changes are due to loss of cell function, other than loss of cell division, as we believe is more likely, then *in vitro* systems are all the more useful for gerontological research.

8.20 Immortal cells

Speculation on how immortal cell populations might arise was proposed several years ago (Hayflick, 1975) and has been elaborated upon more recently by

Bernstein (1979). Meiosis or fusion of adjacent somatic cells (or a cell and a virus) and subsequent nuclear reorganization may confer immortality on a cell lineage. Repair of DNA lesions may be a major function of meiosis and somatic cell recombination. In the latter rare circumstance, the immortal cell population may subsequently develop the full complement of cancer cell attributes.

Bernstein (1979) points out that a fertilized egg cell has none of the physiological or morphological decrements found in some old cells and that the entire developmental programme for that organism is reinitiated at the beginning. The lesions in germ line DNA may be removed in several ways. The extraperitoneal location of the testes in mammals has been reported to result in a twofold reduction of the spontaneous mutation rate in male gonadal tissue that would be caused if these tissues were maintained at body temperature (Baltz, Bingham and Drake, 1976). In bacteria and bacteriophage, major repair processes occur only when homologous chromosomes are allowed to interact. One efficient form of repair of many kinds of DNA lesions in bacteriophages is multiplicity reactivation where such interaction occurs.

If chromosome pairing promotes repair, then for cells undergoing meiosis, the greatest sensitivity to lethal agents would be expected to occur after the pairing of extended chromosomes is completed. In studies on 10 different organisms by 18 groups, the meiotic stages where radiation sensitivity was highest was found to occur during or after pachytene (Eriksson and Tavrin, 1965). Bernstein (1979) argues that since, in most organisms, pairing of extended chromosomes occurs prior to (Grell, 1978) and during zygotene (which just precedes pachytene), these results support the thesis that pairing of chromosomes promotes their repair. Martin (1977) has accounted for the aging and death of vegetatively grown ciliates or their rejuvenation by repair of DNA damage following induction of meiosis, autogamy or conjugation. Martin postulates that repair is a major function of meiosis. Bernstein (1977) postulates that there is a selective advantage for meiotic recombination. Recombination, in addition to being a mechanism for generating genetic diversity, may also play an important role as a DNA-repair process.

If DNA lesions, which accumulate in aging somatic cells (and in cultured cells) were to accumulate in the germ line, then fertilized eggs at each succeeding generation would be in an increasingly 'older' condition (Bernstein, 1979). Aged progeny would be expected to be produced. Since they are not, DNA lesions must be removed at some point in the lifecycle. Martin (1977) and Bernstein (1977) provide evidence in support of the view that meiosis is that point.

These authors have not considered a mechanism for the immortality of cultured cell lines and continuously propagable cancer cells *in vivo*. Normal somatic cells may, under rare circumstances, fuse, exchange genetic information, and in a process similar to meiosis, undergo repair of DNA lesions. The widespread occurrence of somatic cell fusion found *in vitro* may lend support for their notion. When it occurs, however, in normal somatic cells, an immortal cancer cell population results. Meiosis may repair DNA damage at each generation, resulting in the immortality and continuity of the germ plasm. The repair of DNA lesions in normal somatic cells may be brought about by somatic cell fusion mediated by viruses or chemical carcinogens which, among other changes, will also confer immortality.

Acknowledgements

This work was supported, in part, by the Glenn Foundation for Medical Research, Manhasset, New York, and Grant AG 00850, from the National Advisory Council on Aging, National Institute on Aging, National Institutes of Health, Bethesda, Maryland, USA.

References

Absher, P. M. and Absher, R. G. (1976), Clonal variation and aging of diploid fibroblasts. *Exp. Cell Res.*, **103**, 247–255.

Absher, P. M., Absher, R. G. and Barnes, W. D. (1974), Genealogy of clones of diploid fibroblasts. *Exp. Cell Res.*, **88**, 95–104.

Ahn, H. S., Horowitz, S. G., Eagle, H. and Makman, M. H. (1978), Effects of cell density and cell growth alterations on cyclic nucleotide levels in cultured human diploid fibroblasts. *Exp. Cell Res.*, **114**, 101–110.

Balin, A. K., Goodman, D. B. P., Rasmussen, H. and Cristofalo, V. J. (1977), The effect of oxygen and vitamin E on the lifespan of human diploid cells *in vitro*. *J. Cell Biol.*, **74**, 58–67.

Balin, A. K., Goodman, D. B. P., Rasmussen, H. and Cristofalo, V. J. (1978), Oxygen-sensitive stages of the cell cycle of human diploid cells. *J. Cell Biol.*, **78**, 390–400.

Baltz, R. H., Bingham, P. M. and Drake, J. W. (1976), Heat mutagenesis in bacteriophage T4; The transition pathway. *Proc. Natl. Acad. Sci. U.S.A.*, **73**, 1269–1273.

Barski, G., Sorieul, S. and Cornefert, F. (1960), Production dans des cultures *in vitro* de deux souches cellulaires en association de cellules de caractere 'hybride'. *C. R. Hebd. Séances Acad. Sci. Ser. D*, **251**, 1825.

Bell, E., Marek, L. F., Levinstone, D. S., Merrill, C., Sher, S., Young, I. T. and Eden M. (1978). Loss of division potential *in vitro*: Aging or differentiation? *Science*, **202**, 1158–1163.

Bemiller, P. M. and Lee, L-H. (1978), Nucleolar changes in senescing WI-38 cells. *Mech. Aging Dev.*, **8**, 417–427.

Benn, P. A. (1976), Specific chromosome aberrations in senescent fibroblast cell lines derived from human embryos. *Am. J. Hum. Genet.*, **28**, 465–473.

Bernstein, C. (1979), Why are babies young? Meiosis may prevent aging of the germ line. *Persp. Biol. Med.*, **22**, 539–544.

Bernstein, H. (1977), Germ line recombination may be primarily a manifestation of DNA repair processes. *J. Theor. Biol.*, **69**, 371–380.

Berumen, L. and Macieira-Coelho, A. (1977), Changes in albumin uptake during the lifespan of human fibroblasts *in vitro*. *Mech. Aging Dev.*, **6**, 165–172.

Bierman, E. L. (1978), The effect of donor age on the *in vitro* lifespan of cultured human arterial smooth-muscle cells. *In Vitro*, **14**, 951–955.

Bosmann, H. B., Gutheil, R. L., Jr. and Case, K. R. (1976), Loss of a critical neutral protease in ageing WI-38 cells. *Nature (London)*, **261**, 499–501.

Bowman, P. D., Meek, R. L. and Daniel, C. W. (1975), Ageing of human fibroblasts *in vitro*: Correlations between DNA synthetic ability and cell size. *Exp. Cell Res.*, **93**, 184–190.

Bowman, P. D., Meek, R. L. and Daniel, C. W. (1976), Decreased synthesis of nucleolar RNA in aging human cells *in vitro*. *Exp. Cell Res.*, **101**, 434–437.

Bradley, M. O., Dice, J. F., Hayflick, L. and Schimke, R. T. (1975), Protein alterations in aging WI-38 cells as determined by proteolytic susceptibility. *Exp. Cell Res.*, **96**, 103–112.

Bradley, M. O., Erickson, L. C. and Kohn, K. W. (1976), Normal DNA strand rejoining and absence of DNA crosslinking in progeroid and aging human cells. *Mutat. Res.*, **37**, 279–292.

Brautbar, C., Payne, R. and Hayflick, L. (1972), Fate of HL-A antigens in aging cultured human diploid cell strains. *Exp. Cell Res.*, **75**, 31–38.

Brautbar, C., Pellegrino, M. A., Ferrone, S., Reisfeld, R. A., Payne, R. and Hayflick, L. (1973), Fate of HL-A antigens in aging cultured human diploid cell strains. II. Quantitative absorption studies. *Exp. Cell Res.*, **78**, 367–375.

Brown, W. T., Epstein, J. and Little, J. B. (1976), Progeria cells are stimulated to repair DNA by co-cultivation with normal cells. *Exp. Cell Res.*, **97**, 291–296.

Brunk, U., Ericsson, J. L. E., Ponten, J. and Westermark, B. (1973), Residual bodies and 'aging' in cultured human glia cells. *Exp. Cell Res.*, **79**, 1–14.

Buchanan, J. H. and Stevens, A. (1978), Fidelity of histone synthesis in cultured human fibroblasts. *Mech. Aging Dev.*, **7**, 321–334.

Burnet, F. M. (1970a), *Immunological Surveillance*, pp. 224–257, Pergamon Press, New York.

Burnet, F. M. (1970b), An immunological approach to aging. *Lancet*, ii, 358–360.

Chang, R. S. (1962), Metabolic alterations with senescence of human cells. Some observations *in vitro*. *Arch. Intern. Med.*, **110**, 563–568.

Chen, K. H., Paz, M. A. and Gallop, P. (1977), Collagen prolyl hydroxylation in WI-38 fibroblast cultures: Action of hydralazine. *In Vitro*, **13**, 49–54.

Childs, V. A. and Legator, M. S. (1965), Lactic dehydrogenase isozymes in diploid and heteroploid cells. *Life Sci.*, **4**, 1643–1650.

Choe, B.-K. and Rose, N. R. (1974), Synthesis of DNA binding protein in WI-38 cells stimulated to synthesize DNA by medium replacement. *Exp. Cell Res.*, **83**, 261–270.

Clarkson, J. M. and Painter, R. B. (1974), Repair of X-ray damage in aging WI-38 cells. *Mutat. Res.*, **23**, 107–112.

Comfort, A. (1979), *The Biology of Senescence*, 3rd edn., Elsevier/North-Holland, New York.

Comings, D. E. and Vance, C. K. (1971), Thermal denaturation of DNA and chromatin of early and late passage human fibroblasts. *Gerontologia*, **17**, 116–121.

Cox, R. and Masson, W. K. (1974), Changes in radiosensitivity during the *in vitro* growth of diploid human fibroblasts. *Int. J. Radiat. Biol.*, **26** 193–196.

Cristofalo, V. J. (1970), Metabolic aspects of aging in diploid human cells. In *Aging in Cell and Tissue Culture* (eds. E. Holečková and V. J. Cristofalo), Plenum Press, New York, pp. 83–119.

Cristofalo, V. J. (1972), Animal cell cultures as a model for the study of aging. *Adv. Gerontol. Res.*, **4**, 45–79.

Cristofalo, V. J. and Kabakjian, J. (1975), Lysosomal enzymes and aging *in vitro*: Subcellular enzyme distribution and effect of hydrocortisone on cell lifespan. *Mech. Aging Dev.*, **4**, 19–28.

Cristofalo, V. J. and Kritchevsky, D. (1966), Respiration and glycolysis in human diploid cell strain WI-38. *J. Cell. Comp. Physiol.*, **67**, 125–132.

Cristofalo, V. J. and Kritchevsky, D. (1969), Cell size and nucleic acid content in the diploid human cell line WI-38 during aging. *Med. Exp.*, **19**, 313–320.

Cristofalo, V. J., Parris, N. and Kritchevsky, D. (1967), Enzyme activity during the growing and aging of human cells *in vitro. J. Cell. Physiol.*, **69**, 263–271.

Cristofalo, V. J. and Sharf, B. B. (1973), Cellular senescence and DNA synthesis. *Exp. Cell Res.*, **76**, 419–427.

Cudkowicz, G., Upton, A. C., Shearer, G. M. and Hughes, W. L. (1964), Lymphocyte content and proliferative capacity of serially transplanted mouse bone marrow. *Nature (London)*, **201**, 165–167.

Daniel, C. W. (1973), Finite growth span of mouse mammary gland serially propagated *in vivo. Experientia*, **29**, 1422–1424.

Daniel, C. W. (1977), Cell longevity: *In vivo*. In *Handbook of the Biology of Aging* (eds. C. Finch and L. Hayflick), VanNostrand Reinhold Co., New York, pp. 122–158.

Daniel, C. W., Aidells, B. D., Medina, D. and Faulkin, L. J., Jr. (1975), Unlimited division potential of precancerous mouse mammary cells after spontaneous or carcinogen-induced transformation. *Fed. Proc. Fed. Am. Soc. Exp. Biol.*, **34**, 64–67.

Daniel, C. W., deOme, K. B., Young, L. J. T., Blair, P. B. and Faulkin, L. J., Jr. (1968), The *in vivo* lifespan of normal and preneoplastic mouse mammary glands: A serial transplantation study. *Proc. Natl. Acad. Sci. U.S.A.*, **61**, 53–60.

Daniel, C. W. and Young, L. J. T. (1971), Influence of cell division on an aging process. *Exp. Cell Res.*, **65**, 27–32.

Danner, D. B., Schneider, E. L. and Pitha, J. (1978), Macromolecular synthesis in human diploid fibroblasts. *Exp. Cell Res.*, **114**, 63–67.

Deamer, D. W. and Gonzales, J. (1974), Autofluorescent structures in cultured WI-38 cells. *Arch. Biochem. Biophys.*, **165**, 421–426.

Dean, R. T. and Riley, P. A. (1978), The degradation of normal and analogue-containing proteins in MRC-5 fibroblasts. *Biochim. Biophys. Acta*, **539**, 230–237.

Dell'Orco, R. T., Guthrie, P. L. and Simpson, D. L. (1978), Age related alterations in the chromosomal proteins from human diploid fibroblasts. *Mech. Aging Dev.*, **8**, 435–444.

Dell'Orco, R. T., Mertens, J. G. and Kruse, P. F., Jr. (1973), Doubling potential, calendar time, and donor age of human diploid cells in culture. *Exp. Cell Res.*, **84**, 363–366.

Dell'Orco, R. T. and Whittle, W. L. (1978), Unscheduled DNA synthesis in confluent and mitotically arrested populations of aging human diploid fibroblasts. *Mech. Aging Dev.*, **8**, 269–279.

Duffy, P. E. and Kremzner, L. T. (1977), Ornithine decarboxylase activity and polyamines in relation to aging of human fibroblasts. *Exp. Cell Res.*, **108**, 435–439.

Duker, N. J. and Teebor, G. W. (1975), Different ultraviolet DNA endonuclease activity in human cells. *Nature (London)*, **255**, 82–84.

Duncan, M. R., Dell'Orco, R. T. and Guthrie, P. L. (1977), Relationship of heat labile glucose-6-phosphate dehydrogenase and multiple molecular forms of the enzyme in senescent human fibroblasts. *J. Cell. Physiol.*, **93**, 49–56.

Ebeling, A. H. (1913), The permanent life of connective tissue outside of the organism. *J. Exp. Med.*, **17**, 273–285.

Edelstein, S. B., Castiglione, M. and Breakefield, X. O. (1978), Monoamine oxidase activity in normal and Lesch-Nyhan fibroblasts. *J. Neurochem.*, **31**, 1247–1254.

Ephrussi, B., Scaletta, L. J., Stenchever, M. A. and Yoshida, M. C. (1964), Hybridization of somatic cells *in vitro*. In *Cytogenetics of Cells in Culture* (ed. R. J. C. Harris), Academic Press, New York, pp. 13–25.

Epstein, J., Williams, J. R. and Little, J. B. (1973a), Deficient DNA repair in progeria and senescent human cells. *Radiat. Res.,* **55**, 527.

Epstein, J., Williams, J. R. and Little, J. B. (1973b), Rate of DNA repair in human progeroid cells. *Proc. Natl. Acad. Sci. U.S.A.,* **70**, 977–981.

Epstein, J., Williams, J. R. and Little, J. B. (1974), Rate of DNA repair in progeric and normal human fibroblasts., *Biochem. Biophys. Res. Commun.,* **59**, 850–857.

Eriksson, G. and Tavrin, E. (1965), Variations in radiosensitivity during meiosis of pollen mother cells in maize. *Hereditas,* **54**, 156–169.

Ford, C. E., Micklem, H. S. and Gray, S. M. (1959), Evidence of selective proliferation of reticular cell-clones in heavily irradiated mice. *Br. J. Radiol.,* **32**, 280.

Fulder, S. J. (1977), Spontaneous mutations in ageing human cells: Studies using a herpesvirus probe. *Mech. Ageing Dev.,* **6**, 271–282.

Fulder, S. J. and Holliday, R. (1975), A rapid rise in cell variants during the senescence of populations of human fibroblasts. *Cell,* **6**, 67–73.

Fulder, S. J. and Tarrant, G. M. (1975), Possible changes in gene activity during the ageing of human fibroblasts. *J. Exp. Gerontol.,* **10**, 205–211.

Gey, G. O., Svotelis, M., Foard, M. and Bang, F. B. (1974), Long-term growth of chicken fibroblasts on a collagen substrate. *Exp. Cell Res.,* **84**, 63–71.

Goldstein, S. (1971), The role of DNA repair in aging of cultures of fibroblasts from xeroderma pigmentosum and normals. *Proc. Soc. Exp. Biol. Med.,* **137**, 730–734.

Goldstein, S. (1974), Aging *in vitro*. Growth of cultured cells from the Galapagos tortoise. *Exp. Cell Res.,* **83**, 297–302.

Goldstein, S. and Csullog, G. W. (1977), Macromolecular synthesis in human fibroblasts at 37° and 42°C during aging *in vitro*. *Mech. Aging Dev.,* **6**, 185–195.

Goldstein, S. and Haslam, R. J. (1973), Cyclic AMP levels in young and senescent fibroblasts: Effects of epinephrine and prostaglandin E_1. *J. Clin. Invest.,* **52**, 35a.

Goldstein, S., Moerman, E. J., Soeldner, J. S., Gleason, R. E. and Barnett, D. M. (1978), Chronologic and physiologic age effect replicative life-span of fibroblasts from diabetics, prediabetics, and normal donors. *Science,* **199**, 781–782.

Goldstein, S. and Singal, D. P. (1972), Loss of reactivity of HL-A antigens in clonal populations of cultured human fibroblasts during aging *in vitro*. *Exp. Cell Res.,* **75**, 278–282.

Goldstein, S. and Singal, D. P. (1974), Senescence of cultured human fibroblasts: Mitotic versus metabolic time. *Exp. Cell Res.,* **88**, 359–364.

Goldstein, S., Stotland, D. and Cordeiro, R. A. J. (1976), Decreased proteolysis and increased amino acid efflux in aging human fibroblasts. *Mech. Aging Dev.,* **5**, 221–223.

Goldstein, S. and Trieman, G. (1974), Glucose consumption by early and late passage human fibroblasts during growth and stationary phase. *Experientia,* **31**, 177–180.

Good, P. I. (1975), Aging in mammalian cell populations: A review. *Mech. Aging Dev.,* **4**, 339–348.

Good, P. I. and Smith, J. R. (1974), Age distribution of human diploid fibroblasts. *Biophys. J.,* **14**, 811–822.

Green, H., Kehinde, O. and Thomas, J. (1979), Growth of cultured human epidermal cells into multiple epithelia suitable for grafting. *Proc. Natl. Acad. Sci. U.S.A.*, **76**, 5665–5668.

Greenberg, S. B., Grove, G. L. and Cristofalo, V. J. (1977), Cell size in aging monolayer cultures. *In Vitro*, **13**, 297–300.

Grell, R. F. (1978), Time of recombination in the Drosophila melanogaster oocyte: Evidence from a temperature-sensitive recombination-deficient mutant. *Proc. Natl. Acad. Sci. U.S.A.*, **75**, 3351–3354.

Grove, G. L. and Cristofalo, V. J. (1977), Characterization of the cell cycle of cultured human diploid cells: Effects of aging and hydrocortisone. *J. Cell. Physiol.*, **90**, 415–422.

Hakami, N. and Pious, D. A. (1968), Mitochondrial enzyme activity in 'senescent' and virus-transformed human fibroblasts. *Exp. Cell Res.*, **53**, 135–138.

Harley, C. B. and Goldstein, S. (1978), Cultured human fibroblasts: Distribution of cell generations and a critical limit. *J. Cell. Physiol.*, **97**, 509–516.

Harrison, D. E. (1972), Normal function of transplanted mouse erythrocyte precursors for 21 months beyond donor life spans. *Nature (London) New Biol.*, **237**, 220–222.

Harrison, D. E. (1973), Normal production of erythrocytes by mouse marrow continuous for 73 months. *Proc. Natl. Acad. Sci. (U.S.A.)*, **70**, 3184–3188.

Harrison, D. E. (1975), Normal function of transplanted marrow cell lines from aged mice. *J. Gerontol.*, **30**, 279–285.

Hart, R. W. and Setlow, R. B. (1974), Correlation between deoxyribonucleic acid excision-repair and life-span in a number of mammalian species. *Proc. Natl. Acad. Sci. U.S.A.*, **71**, 2169–2173.

Hart, R. W. and Setlow, R. B. (1976), DNA repair in late-passage human cells. *Mech. Aging Dev.*, **5**, 67–77.

Haslam, R. J. and Goldstein, S. (1974), Adenosine 3:5′-cyclic monophosphate in young and senescent human fibroblasts during growth and stationary phase in vitro. *Biochem. J.*, **144**, 253–263.

Hay, R. J. and Strehler, B. L. (1967), The limited growth span of cell strains isolated from the chick embryo. *Exp. Gerontol.*, **2**, 123–135.

Hayakawa, M. (1969), Progressive changes of the growth characteristics of human diploid cells in serial cultivation *in vitro*. *Tohoku J. Exp. Med.*, **98**, 171–179.

Hayflick, L. (1965), The limited *in vitro* lifetime of human diploid cell strains. *Exp. Cell Res.*, **37**, 614–636.

Hayflick, L. (1966), Senescence and cultured cells. In *Perspectives in Experimental Gerontology* (ed. N. Shock), Charles C. Thomas, Springfield, Illinois, pp. 195–211.

Hayflick, L. (1970), Aging under glass. *Exp. Gerontol.*, **5**, 291–303.

Hayflick, L. (1972), Cell senescence and cell differentiation *in vitro*. In *Aging and Development* (eds. H. Bredt and J. W. Rohen), F. K. Schattauer Verlag, Stuttgart.

Hayflick, L. (1975), Cell biology of aging. *Bioscience*, **25**, 629–637.

Hayflick, L. (1976), The cell biology of human aging. *N. Engl. J. Med.*, **295**, 1302–1308.

Hayflick, L. (1977a), The cellular basis for biological aging. In *Handbook of the Biology of Aging* (eds. C. Finch and L. Hayflick), VanNostrand Reinhold Co., New York, pp. 159–179.

Hayflick, L. (1977b), Mislabeling, contamination, and other sins of cultured flesh. *Dev. Biol. Standard.*, **37**, 5–15.

Hayflick, L. and Moorhead, P. S. (1961), The serial cultivation of human diploid cell strains. *Exp. Cell Res.*, **25**, 585–621.

Hellman, S., Botnick, L. E., Hannon, E. C. and Vigneulle, R. M. (1978), Proliferative capacity of murine hematopoietic stem cells. *Proc. Natl. Acad. Sci. U.S.A.*, **75**, 490–494.

Hill, B. T. (1976), A lack of correlation between decline in growth capacity and nuclear RNA synthesizing activity in aging human embryo cells in culture. *Mech. Aging Dev.*, **5**, 267–278.

Hill, B. T. (1977), The establishment of criteria for 'quiescence' in aging human embryo cell cultures and their response to a proliferative stimulus. *Gerontology*, **23**, 245–255.

Hill, B. T., Whelan, R. D. H. and Whatley, S. (1978), Evidence that transcription changes in aging cultures are terminal events occurring after the expression of a reduced replicative potential. *Mech. Aging Dev.*, **8**, 85–95.

Holland, J. J., Koline, D. and Doyle, M. V. (1973), Analysis of virus replication in aging human fibroblasts. *Nature (London)*, **245**, 316–318.

Holliday, R. (1972), Ageing of human fibroblasts in culture: Studies on enzymes and mutation. *Humangenetik*, **16**, 83–86.

Holliday, R., Huschtscha, L. I., Tarrant, G. M. and Kirkwood, T. B. L. (1977), Testing the commitment theory of cellular aging, *Science*, **198**, 366–372.

Holliday, R. and Tarrant, G. M. (1972), Altered enzymes in ageing human fibroblasts. *Nature (London)*, **238**, 26–30.

Horoszewicz, J. S., Leong, S. S., Ito, M., DiBerardino, L. and Carter, W. A. (1978), Aging *in vitro* and large scale interferon production by 15 new strains of human diploid fibroblasts. *Infect. Immun.*, **19**, 720–726.

Houben, A. and Remacle, J. (1978), Lysosomal and mitochondrial heat labile enzymes in ageing human fibroblasts., *Nature (London)*, **275**, 59–60.

Houck, J. C., Sharma, V. K. and Hayflick, L. (1971), Functional failures of cultured human diploid fibroblasts after continued population doublings. *Proc. Soc. Exp. Biol. Med.*, **137**, 331–333.

Howard, B. V., Howard, W. J. and Kefalides, N. A. (1976), Regulation of lipid synthesis from acetate in diploid fibroblast cultures – variation with passage level and stage of cell growth. *J. Cell. Physiol.*, **89**, 325–336.

Johnson, J. E., Jr. (1979), Fine structure of IMR-90 cells in culture as examined by scanning and transmission electron microscopy. *Mech. Aging Dev.* (in the press.)

Kahn, A., Guillouzo, A., Cottreau, D., Marie, J., Bourel, M., Boivin, P. and Dreyfus, J.-C. (1977a), Accuracy of protein synthesis and *in vitro* aging. *Gerontology*, **23**, 174–184.

Kahn, A., Guillouzo, A., Leibovitch, M., Cottreau, D., Bourel, M. and Dreyfus, J. (1977b), Heat lability of glucose-6-phosphate dehydrogenase in some senescent human cultured cells. Evidence for its postsynthetic nature. *Biochem. Biophys. Res. Commun.*, **77**, 760–766.

Kantor, G. J., Mulkie, J. R. and Hull, D. R. (1978), A study of the effect of ultra-violet light on the division potential of human diploid fibroblasts. *Exp. Cell Res.*, **113**, 283–294.

Kapp, L. N. and Klevecz, R. R. (1976), The cell cycle of low passage and high passage human diploid fibroblasts. *Exp. Cell Res.*, **101**, 154–158.

Kay, H. E. M. (1965), How many cell generations? *Lancet*, ii, 418–419.

Kalley, R. O. (1976), Development of the aging cell surface: A freeze-fracture analysis of gap junctions between human embryo fibroblasts aging in culture. *Mech. Aging Dev.*, **5**, 339–345.

Kelley, R. O. and Skipper, B. E. (1977), Development of the aging cell surface: Variation in the distribution of intramembrane particles with progressive age of human diploid fibroblasts. *J. Ultrastruct. Res.*, **59**, 113–118.

Kram, D. and Schneider, E. L. (1978), Reduced frequencies of Mitomycin-C induced sister chromatid exchanges in AKR mice. *Hum. Genet.*, **41**, 45–51.

Kritchevsky, D. and Howard, B. V. (1966), The lipids of human diploid cell strain WI-38. *Ann. Med. Exp. Biol. Fenn. (Helsinki)*, **44**, 343–347.

Krohn, P. L. (1962), Review lectures on senescence. II. Heterochronic transplantation in the study of ageing. *Proc. R. Soc. (London) Ser. B*, **157**, 128–147.

Krohn, P. L. (1966), *Topics in the Biology of Aging.* Interscience Publishers, John Wiley, New York.

Kurtz, M. J. and Stidworthy, G. H. (1969), Enzymatic sulfation of mucopolysaccharides as a function of age in cultured rat gut fibroblasts. *Proc. Int. Congr. Gerontol. 8th,* Washington, D.C., II, 49.

Lee, S.-C., Bemiller, M. P., Bemiller, J. N. and Pappelis, A. J. (1978), Nuclear area changes in senescing human diploid fibroblasts. *Mech. Aging Dev.*, **7**, 417–424.

LeGuilly, Y., Simon, M., Lenoir, P. and Bourel, M. (1973), Long-term culture of human adult liver cells: Morphological changes related to *in vitro* senescence and effect of donor's age on growth potential. *Gerontologia*, **19**, 303–313.

Lesher, S., Fry, R. J. M. and Kohn, H. I. (1961a), Age and the generation time of the mouse duodenal epithelial cell. *Exp. Cell Res.*, **24**, 334–343.

Lesher, S., Fry, R. J. M. and Kohn, H. I. (1961b), Aging and the generation cycle of intestinal epithelial cells in the mouse. *Gerontologia*, **5**, 176–181.

Lesher, S. and Sacher, G. A. (1968), Effects of age on cell proliferation in mouse duodenal crypts. *Exp. Gerontol.*, **3**, 211–217.

Levine, E. M., Burleigh, I. G., Boone, C. W. and Eagle, H. (1967), An altered pattern of RNA synthesis in serially propagated human diploid cells. *Proc. Natl. Acad. Sci. (U.S.A.)*, **57**, 431–438.

Lewis, C. M. and Tarrant, G. M. (1972), Error theory and ageing in human diploid fibroblasts. *Nature (London)*, **239**, 316–318.

Lima, L., Malaise, E. and Macieira-Coelho, A. (1972), Aging *in vitro. Exp. Cell Res.*, **73**, 345–350.

Linn, S., Kairis, M. and Holliday, R. (1976), Decreased fidelity of DNA polymerase activity isolated from aging human fibroblasts. *Proc. Natl. Acad. Sci. U.S.A.*, **73**, 2818–2822.

Lipetz, J. and Cristofalo, V. J. (1972), Ultrastructural changes accompanying the aging of human diploid cells in culture. *J. Ultrastruct. Res.*, **39**, 43–56.

Litwin, J. (1972), The effect of light on the ageing of human diploid fibroblasts. *Exp. Gerontol.*, **7**, 381–386.

Litwin, J., Enzell, C. and Pilotti, A. (1978), The effect of tobacco smoke condensate on the growth and longevity of human diploid fibroblasts. *Acta Pathol. Microbiol. Scand. Sect. A*, **86**, 135–141.

Macek, M., Hurych, J. and Chvapil, M. (1967), The collagen protein formation in tissue cultures of human diploid strains. *Cytologia (Tokyo)*, **32**, 426–443.

Macieira-Coelho, A. (1966), Action of cortisone on human fibroblasts *in vitro. Experientia*, **22**, 390–391.

Macieira-Coelho, A. (1970), The decreased growth potential *in vitro* of human fibroblasts of adult origin. In *Aging in Cell and Tissue Culture* (eds. E. Holečková and V. J. Cristofalo), Plenum Press, New York, pp. 121–132.

Macieira-Coelho, A. (1974), Are non-dividing cells present in ageing cell cultures? *Nature (London)*, **248**, 421–422.

Macieira-Coelho, A. (1977), Kinetics of the proliferation of human fibroblasts during their lifespan *in vitro. Mech. Aging Dev.,* **6**, 341–343.

Macieira-Coelho, A. and Diatloff, C. (1976), Doubling potential of fibroblasts from different species after ionising radiation. *Nature (London),* **261**, 586–588.

Macieira-Coelho, A., Diatloff, C., Billard, M., Fertil, B., Malaise, E. and Fries, D. (1978), Effect of low dose rate irradiation on the division potential of cells *in vitro. J. Cell. Physiol.,* **95**, 235–238.

Macieira-Coelho, A., Diatloff, C. Billard, C., Bourgeois, C. A. and Malaise, E. (1977), Effect of low dose rate ionizing radiation on the division potential of cells *in vitro. Exp. Cell Res.,* **104**, 215–221.

Macieira-Coelho, A., Diatloff, C. and Malaise, E. (1976), Effect of low dose rate irradiation on the division potential of cells *in vitro. Exp. Cell Res.,* **100**, 228–232.

Macieira-Coelho, A. and Pontén, J. (1969), Analogy in growth between late passage human embryonic and early passage human adult fibroblasts. *J. Cell Biol.,* **43**, 374–377.

Macieira-Coelho, A., Pontén, J. and Philipson, L. (1966a), The division cycle and RNA-synthesis in diploid human cells at different passage levels *in vitro. Exp. Cell Res.,* **42**, 673–684.

Macieira-Coelho, A., Pontén, J. and Philipson, L. (1966b), Inhibition of the division cycle in confluent cultures of human fibroblasts *in vitro. Exp. Cell Res.,* **43**, 20–29.

Martin, G. M. (1977a), Cellular aging-clonal senescence. *Am. J. Pathol.,* **89**, 484–530.

Martin, G. M. (1977b), Cellular aging-postreplicative cells. *Am. J. Pathol.,* **89**, 513–530.

Martin, G. M., Sprague, C. A. and Epstein, C. J. (1970), Replicative lifespan of cultivated human cells. Effect of donor's age, tissue, and genotype. *Lab. Invest.,* **23**, 86–92.

Martin, R. (1977), In *Molecular Human Cytogenetics: ICN-UCLA Symposia on Molecular and Cellular Biology,* vol. 7 (eds. R. S. Sparkes, D. E. Comings and C. F. Fox), Academic Press, New York, p. 355.

Mattern, M. R. and Cerutti, P. A. (1975), Age dependent excision repair of damaged thymidine from gamma-irradiated DNA by isolated nuclei from fibroblasts. *Nature (London),* **254**, 450–452.

McHale, J. S., Moulton, M. L. and McHale, J. T. (1971), Limited culture lifespan of human diploid cells as a function of metabolic time instead of division potential. *Exp. Gerontol.,* **6**, 89–93.

Meek, R. L., Bowman, P. D. and Daniel, C. W. (1977), Establishment of mouse embryo cells *in vitro. Exp. Cell Res.,* **107**, 277–284.

Merz, G. S. and Ross, J. D. (1969), Viability of human diploid cells as a function of *in vitro* age. *J. Cell. Physiol.,* **74**, 219–225.

Michl, J. and Svoboda, J. (1967), RNA turnover and the growth potential of human cells in culture. *Exp. Cell Res.,* **47**, 616–619.

Milisauskas, V. and Rose, N. R. (1973), Immunochemical quantitation of enzymes in human diploid cell line WI-38. *Exp. Cell Res.,* **81**, 279–284.

Miller, R. C., Nichols, W. W., Pottash, J. and Aronson, M. M. (1977), *In vitro* aging. *Exp. Cell Res.,* **110**, 63–73.

Mitsui, Y. and Schneider, E. L. (1976a), Increased nuclear sizes in senescent human diploid fibroblast cultures. *Exp. Cell Res.,* **100**, 147–152.

Mitsui, Y. and Schneider, E. L. (1976b), Relationship between cell replication and volume in senescent human diploid fibroblasts. *Mech. Aging Dev.,* **5**, 45–56.

Moore, C. J. and Schwartz, A. G. (1978), Inverse correlation between species lifespan and capacity of cultured fibroblasts to convert benzo(a)pyrene to water-soluble metabolites. *Exp. Cell Res.,* **116**, 359–364.

Murphy, E., Holland, M. J. C. and Cox, R. P. (1978), Adenosine deaminase activity in human diploid skin fibroblasts varies with the age of the donor. *J. Med.,* **9**, 237–244.

Niewiarowski, S. and Goldstein, S. (1973), Interaction of cultured human fibroblasts with fibrin: Modification by drugs and aging *in vitro. J. Lab. Clin. Med.,* **82**, 605–610.

Orgel, L. (1973), Ageing of clones of mammalian cells. *Nature (London),* **243**, 441–445.

Packer, L. and Fuehr, K. (1977), Low oxygen concentration extends the lifespan of cultured human diploid cells. *Nature (London),* **267**, 423–425.

Packer, L. and Smith, J. R. (1974a), Extension of the lifespan of cultured normal human diploid cells by vitamin E. *J. Cell Biol.,* **63**, A255.

Packer, L. and Smith, J. R. (1974b), Extension of the lifespan of cultured normal human cells by vitamin E. *Proc. Natl. Acad. Sci. U.S.A.,* **71**, 4763–4767.

Packer, L. and Smith, J. R. (1977), Extension of the lifespan of cultured normal human diploid cells by vitamin E: A reevaluation. *Proc. Natl. Acad. Sci. U.S.A.,* **74**, 1640–1641.

Paffenholz, V. (1978), Correlation between DNA repair of embryonic fibroblasts and different lifespan of 3 inbred mouse strains. *Mech. Aging Dev.,* **7**, 131–150.

Painter, R. B., Clarkson, J. M. and Young, B. R. (1973), Ultraviolet induced repair replication in aging diploid human cells (WI-38). *Radiat. Res.,* **56**, 560–564.

Parker, R. C. (1961), In *Methods of Tissue Culture,* Harper and Row, New York, p. 258.

Parshad, R. and Sanford, K. K. (1977a), Intermittent exposure to fluorescent light extends lifespan of human diploid fibroblasts in culture. *Nature (London),* **268**, 736–737.

Parshad, R. and Sanford, K. K. (1977b), Proliferative response of human diploid fibroblasts to intermittent light exposure. *J. Cell. Physiol.,* **92**, 481–485.

Pendergrass, W. R., Martin, G. M. and Bornstein, P. (1976), Evidence contrary to the protein error hypothesis for *in vitro* senescence. *J. Cell. Physiol.,* **87**, 3–14.

Pereira, O., Smith, J. R. and Packer, L. (1976), Photosensitization of human diploid cell cultures by intracellular flavins and protection by antioxidants. *Photochem. Photobiol.,* **24**, 237–242.

Petes, T. D., Farber, R. A., Tarrant, G. M. and Holliday, R. (1974), Altered rate of DNA replication in ageing human fibroblast cultures. *Nature (London),* **251**, 434–436.

Pitha, J., Adams, R. and Pitha, P. M. (1974), Viral probe into the events of cellular (*in vitro*) aging. *J. Cell. Physiol.,* **83**, 211–218.

Poiley, J. A., Schuman, R. F. and Pienta, R. J. (1978), Characterization of normal human embryo cells grown to over 100 population doublings. *In Vitro,* **14**, 405–412.

Polgar, P., Taylor, L. and Brown, L. (1978), Plasma membrane associated metabolic parameters and the aging of human diploid fibroblasts. *Mech. Aging Dev.,* **7**, 151–160.

Porter, D. D. (1974), Private communication.

Post, J. and Hoffman, J. (1964), Changes in the replication times and patterns of the liver cell during the life of the rat. *Exp. Cell Res.,* **36**, 111–123.

Press, G. D. and Pitha, J. (1974), Aging changes in uptake of polysaccharides by human diploid cells in culture. *Mech. Aging Dev.*, 3, 323–328.

Razin, S., Pfendt, E. A., Matsumura, T. and Hayflick, L. (1977), Comparison by autoradiography of macromolecular biosynthesis in 'young' and 'old' human diploid fibroblast cultures. *Mech. Aging Dev.*, 6, 379–384.

Regan, J. D. and Setlow, R. B. (1973), DNA repair in human progeroid cells. *Biochem. Biophys. Res. Commun.*, 59, 858–864.

Robbins, E., Levine, E. M. and Eagle, H. (1970), Morphologic changes accompanying senescence of cultured human diploid cells. *J. Exp. Med.*, 131, 1211–1222.

Rosenbloom, A. L., Goldstein, S. and Yip, C. C. (1976), Insulin binding to cultured human fibroblasts increases with normal and precocious aging. *Science*, 193, 412–415.

Rosenthal, M. D. and Geyer, R. P. (1976), Phospholipid acyl group stability in cultured fibroblasts. *Biochim. Biophys. Acta*, 441, 465–476.

Rothfels, K. H., Kupelwieser, E. B. and Parker, R. C. (1963), Effects of X-irradiated feeder layers on mitotic activity and development of aneuploidy in mouse-embryo cells *in vitro. Can. Cancer Conf.*, 5, 191–223.

Rowe, D. W., Starman, B. J., Fujimoto, W. Y. and Williams, R. R. (1977), Differences in growth response to hydrocortisone and ascorbic acid by human diploid fibroblasts. *In Vitro*, 13, 824–830.

Ryan, J. M. and Cristofalo, V. J. (1972), Histone acetylation during aging of human cells in culture. *Biochem. Biophys. Res. Commun.*, 48, 735–742.

Ryan, J. M. and Cristofalo, V. J. (1975), Chromatin template activity during aging in WI-38 cells. *Exp. Cell Res.*, 90, 456–458.

Ryan, J. M., Duda, G. and Cristofalo, V. J. (1974), Error accumulation and aging in human diploid cells. *J. Gerontol.*, 29, 616–621.

Ryan, J. M., Sharf, B. B. and Cristofalo, V. J. (1975), The influence of culture medium volume on cell density and lifespan of human diploid fibroblasts. *Exp. Cell Res.*, 91, 389–392.

Sakagami, H. and Yamada, M. (1977), Failure of vitamin E to extend the lifespan of a human diploid cell line in culture. *Cell Struct. Funct.*, 2, 219–227.

Saksela, E. and Moorhead, P. S. (1963), Aneuploidy in the degenerative phase of serial cultivation of human cell strains. *Proc. Natl. Acad. Sci. U.S.A.*, 50, 390–395.

Saunders, J. W., Jr. (1966), Death in embryonic systems. *Science*, 154, 604–612.

Saunders, J. W. and Fallon, J. F. (1966), Cell death in morphogenesis. In *Major Problems in Developmental Biology* (ed. M. Locke), Academic Press, New York, pp. 289–314.

Schachtschabel, D. O. and Wever, J. (1978), Age-related decline in the synthesis of glycosaminoglycans by cultured human fibroblasts (WI-38). *Mech. Aging Dev.*, 8, 257–264.

Schneider, E. L., Braunschweiger, K. and Mitsui, Y. (1978), The effect of serum batch on the *in vitro* lifespan of cell cultures derived from old and young human donors. *Exp. Cell Res.*, 115, 47–52.

Schneider, E. L. and Chase, G. A. (1976), Relationship between age of donor and *in vitro* lifespan of human diploid fibroblasts. *Interdiscip. Top. Gerontol.*, 10, 62–69.

Schneider, E. L. and Fowlkes, B. J. (1976), Measurement of DNA content and cell volume in senescent human fibroblasts utilizing flow multiparameter single cell analysis. *Exp. Cell Res.*, 98, 298–302.

Schneider, E. L. and Kram, D. (1978), Examination of the effect of ageing on cell replication and sister chromatid exchange. In *DNA Repair Processes* (eds. W. W. Nichols and D. G. Murphy), Symposia Specialists, Inc., Miami, Florida.

Schneider, E. L. and Mitsui, Y. (1976), The relationship between *in vitro* cellular aging and *in vivo* human aging. *Proc. Natl. Acad. Sci. U.S.A.*, **73**, 3584–3588.

Schneider, E. L., Mitsui, Y., Aw, K. S. and Shorr, S. S. (1977), Tissue-specific differences in cultured human diploid fibroblasts. *Exp. Cell Res.*, **108**, 1–6.

Schneider, E. L., Mitsui, Y., Tice, R., Shorr, S. S. and Braunschweiger, K. (1973), Alterations in cellular RNA's during the *in vitro* lifespan of cultured human diploid fibroblasts. II. Synthesis and processing of RNA. *Mech. Aging Dev.*, **4**, 449–458.

Schneider, E. L. and Monticone, R. E. (1978), Aging and sister chromatid exchange. *Exp. Cell Res.*, **115**, 269–276.

Schwartz, A. G. and Moore, C. J. (1977), Inverse correlation between species lifespan and capacity of cultured fibroblasts to bind 7,12-dimethylbenz(a)-anthracene to DNA. *Exp. Cell Res.*, **109**, 448–450.

Shakespeare, V. and Buchanan, J. H. (1976), Increased degradation rates of protein in aging human fibroblasts and in cells treated with an amino acid analog. *Exp. Cell Res.*, **100**, 1–8.

Shakespeare, V. and Buchanan, J. H. (1978), Studies on phosphoglucose isomerase from cultured human fibroblasts: Absence of detectable ageing effects on the enzyme. *J. Cell. Physiol.*, **94**, 105–115.

Shen, T.-F. and Strecker, H. J. (1975), Synthesis of proline and hydroxyproline in human lung (WI-38) fibroblasts. *Biochem. J.*, **150**, 453–461.

Simonovitch, L., Till, J. E. and McCulloch, E. A. (1964), Decline in colony-forming ability of marrow cells subjected to serial transplantation into irradiated mice. *J. Cell. Comp. Physiol.*, **64**, 23–31.

Simons, J. W. I. M. (1967), The use of frequency distributions of cell diameters to characterize cell populations in tissue culture. *Exp. Cell Res.*, **45**, 336–350.

Simons, J. W. I. M. and van den Broek, C. (1970), Comparison of ageing *in vitro* and ageing *in vivo* by means of cell size analysis using a Coulter counter. *Gerontologia*, **16**, 340–351.

Smith, C. A. and Hanawalt, P. C. (1976), Repair replication in cultured normal and transformed human fibroblasts. *Biochim. Biophys. Acta*, **447**, 121–132.

Smith, J. R. and Hayflick, L. (1974), Variation in the life-span of clones derived from human diploid cell strains. *J. Cell Biol.*, **62**, 48–53.

Smith, J. R., Pereira-Smith, O. and Good, P. I. (1977), Colony size distribution as a measure of age in cultured human cells. *Mech. Aging Dev.*, **6**, 283–286.

Srivastava, B. I. S. (1973), Changes in enzymic activity during cultivation of human cells *in vitro*. *Exp. Cell Res.*, **80**, 305–312.

Stein, G. S. and Burtner, D. L. (1975), Gene activation in human diploid cells. *Biochim. Biophys. Acta*, **390**, 56–68.

Stewart, H. L., Snell, K. C., Dunham, L. J. and Schylen, S. M. (1959), Transplantable and transmissible tumors of animals. *Washington, D.C., Armed Forces Institute of Pathology*.

Strehler, B. L. (1977), *Time, Cells and Aging*, 2nd Edn., Academic Press, New York.

Sun, A. S., Aggarwal, B. B. and Packer, L. (1975), Enzyme levels of normal human cells: Aging in culture. *Arch. Biochem. Biophys.*, **170**, 1–11.

Suzuki, N. and Withers, H. R. (1978), Exponential decrease during aging and random lifetime of mouse spermatogonial stem cells. *Science*, **202**, 1214–1215.

Takahashi, S., Seifter, S. and Davidson, A. (1978), Enzymes of the a-glutamyl cycle in 'aging' WI-38 fibroblasts and in HeLa S_3 cells. *Biochim. Biophys. Acta*, **522**, 63–73.

Tan, Y. H., Chow, E. L. and Lundh, N. (1975), Regulation of chromosome 21-directed anti-viral gene(s) as a consequence of age. *Nature (London)*, **257**, 310–312.

Taylor, W. G., Camalier, R. F. and Sanford, K. K. (1978), Density-dependent effects of oxygen on the growth of mammalian fibroblasts in culture. *J. Cell. Physiol.*, **95**, 33–40.

Thompson, K. V. A. and Holliday, R. (1975), Chromosome changes during the *in vitro* ageing of MRC-5 human fibroblasts. *Exp. Cell. Res.*, **96**, 1–6.

Thompson, K. V. A. and Holliday, R. (1978), The longevity of diploid and polyploid human fibroblasts. *Exp. Cell Res.*, **112**, 281–287.

Thrasher, J. D. and Greulich, R. C. (1965), The duodenal progenitor population. I. Age related increase in the duration of the cryptal progenitor cycle. *J. Exp. Zool.*, **159**, 39–46.

Till, J. E., McCulloch, E. A. and Simonovitch, L. (1964), Isolation of variant cell lines during serial transplantation of hematopoietic cells derived from fetal liver. *J. Natl. Cancer Inst.*, **33**, 707–720.

Todaro, G. J. and Green, H. (1963), Quantitative studies of the growth of mouse embryo cells in culture and their development into established lines. *J. Cell Biol.*, **17**, 229–313.

Todaro, G. J. and Green, H. (1964), Serum albumin supplemented medium for long-term cultivation of mammalian fibroblast strains. *Proc. Soc. Exp. Biol. Med.*, **116**, 688–692.

Tomkins, G. A., Stanbridge, E. J. and Hayflick, L. (1974), Viral probes of aging in the human diploid cell strain WI-38. *Proc. Soc. Exp. Biol. Med.*, **146**, 385–390.

Turk, B. and Milo, G. E. (1974), An *in vitro* study of senescent events of human embryonic lung (WI-38) cells. *Arch. Biochem. Biophys.*, **161**, 46–53.

Viceps-Madore, D. and Cristofalo, V. J. (1978), Age associated changes in glutamine synthetase activity in WI-38 cells. *Mech. Aging Dev.*, **8**, 43–50.

Vincent, R. A., Jr. and Huang, P. C. (1976), The proportion of cells labeled with tritiated thymidine as a function of population doubling level in cultures of fetal, adult, mutant, and tumor origin. *Exp. Cell Res.*, **102**, 31–42.

Walford, R. L. (1969), *The Immunologic Theory of Aging*, Williams & Wilkins, Baltimore, p. 169.

Wang, K. M., Rose, N. R., Bartholomew, E. A., Balzer, M., Berde, K. and Foldvary, M. (1970), Changes of enzymatic activities in human diploid cell line WI-38 at various passages. *Exp. Cell Res.*, **61**, 357–364.

Waters, H. and Walford, R. L. (1970), Latent period for outgrowth of human skin explants as a function of age. *J. Gerontol.*, **25**, 381–383.

Whitten, J. M. (1969), Cell death during early morphogenesis: Parallels between insect limb and vertebrate limb development. *Science*, **163**, 1456–1457.

Williamson, A. R. (1972), Extent and control of antibody diversity. *Biochem. J.*, **130**, 325–333.

Williamson, A. R. and Askonas, B. A. (1972), Senescence of an antibody-forming cell clone. *Nature (London)*, **238**, 337–339.

Wolosewick, J. J. and Porter, K. R. (1977), Observations on the morphological heterogeneity of WI-38 cells. *Am. J. Anat.*, **149**, 197–226.

Wright, W. E. and Hayflick, L. (1975a), Use of biochemical lesions for selection of human cells with hybrid cytoplasms. *Proc. Natl. Acad. Sci. U.S.A.*, **72**, 1812–1816.

Wright, W. E. and Hayflick, L. (1975b), Contributions of cytoplasmic factors to *in vitro* cellular senescence. *Fed. Proc. Fed. Am. Soc. Exp. Biol.,* **34**, 76–79.

Yamanaka, N. and Deamer, D. (1974), Superoxide dismutase activity in WI-38 cell cultures: Effects of age, trypsinization and SV-40 transformation. *Physiol. Chem. Phys.,* **6**, 95–106.

Yanishevsky, R. and Carrano, A. V. (1975), Prematurely condensed chromosomes of dividing and nondividing cells in aging human cell cultures. *Exp. Cell Res.,* **90**, 169–174.

Young, L. J. T., Medina, D., deOme, K. B. and Daniel, C. W. (1971), The influence of host and tissue age on the life span and growth rate of serially transplanted mouse mammary gland. *Exp. Gerontol.,* **6**, 49–56.

Yuan, G. C. and Chang, R. S. (1969), Effect of hydrocortisone on age-dependent changes in lipid metabolism of primary human amnion cells *in vitro. Proc. Soc. Exp. Biol. Med.,* **130**, 934–936.

9 Nucleic acids in cell death

RICHARD A. LOCKSHIN, MOIRA ROYSTON,
MICHAEL JOESTEN AND TIMOTHY CARTER

9.1 The basic problem

It would be highly desirable to understand better the role and fate of nucleic acids in cell death. Several contributors have touched on the question, some in considerable depth, and many of the arguments are explored, especially in Chapters 7, 10 and 11. Of the large volume of literature otherwise available, very few articles deal with causal relationships. A mid-1980 MEDLINE search of 'Cell Death and Nucleic Acids' collected a maximum 300 items, most of which involved a pharmacological result of the use of antimetabolites. Fifteen articles dealt with causal relationships in non-pharmacological situations, and four faced the question directly. The indexing of these key words, based on Index Medicus, missed most interesting studies in the field, such as those on glucocorticoid-induced involution of thymocytes (see Chapter 11), the growth and involution of neurons (Chapter 10; Landmesser and Pilar, 1978; Pittman, Oppenheim and Chu-Waing, 1978); embryonic cell death (Chapter 2; Pollak and Fallon, 1976; Beaupain, 1979) and numerous studies concerning the involution of the silk gland in the silk moth *Bombyx mori* (Chapter 3; Matsuura, Morimoto, Nagata and Tashiro, 1968; Okabe, Koyanagi and Koga, 1975; Suzuki, 1977; Fournier, 1979). These studies have been pursued in some depth. There is also an extensive literature on the synthesis of nucleic acids in Phase III fibroblasts, as is summarized in Chapter 8, and a formidable literature on the fate of cultures (rather than individual cells) of prokaryotes which we choose not to tackle at this time. Also of interest is a scattered literature on differentiation-related death, in which RNA content drops during final differentiation (Morrissey and Green, 1978) and a consistent argument that during senescence of organisms the machinery of protein synthesis deteriorates. Depending on the organism, there may be loss of total RNA (Linzen and Wyatt, 1964; Lang, Lau and Jefferson, 1965; Chinzei and Tojo, 1972; Burns and Kaulenas, 1979; Miguel and Johnson, 1979, in insects), some of the polymerases (Johnson and Strehler, 1979, in pea cotyledons) or transfer RNAs (Hosbach and Kubli, 1979a, 1979b, in *Drosophila*). Most of the other literature can be described as 'tail-off' literature, in which the authors study a developmental phenomenon but illustrate in passing the close of the active phase as indication of a rapid decline of the synthesis or total amount of messenger RNA or polysomes or, more rarely, another component of the protein-synthesizing machinery.

287

The most interesting of these studies include the involution of the silk gland of *Bombyx mori*. The several readily identifiable and morphologically simple parts of the glands produce well-defined gene products and, because of both the commercial value of the product and the elegance of the system, the characterization of the protein-synthetic machinery has reached an enviable level. Thus, in spite of the fact that the story is usually abandoned at the point which interests us, we can get an image of the events immediately preceding cell death. In order to review this story, it is worthwhile briefly to summarize current concepts of eukaryotic synthesis of protein (Ochoa and de Haro, 1979).

9.2 Protein synthesis in eukaryotic cells

A priori, the synthesis of protein in eukaryotic cells could be regulated at any stage from initiation of transcription on chromatin by the appropriate RNA polymerase to utilization of completed mRNA in the cytoplasm. The scanty information now available concerning regulation of protein synthesis comes mostly from virus-infected cells and a few developmental systems.

The activities of RNA polymerases I and III will affect protein synthesis over the long term by determining ribosome number and function, and the synthesis of small RNA molecules respectively. Ribosome synthesis will also be affected by changes in activity or specificity of methylating enzymes and nucleases, or by changes in rate of synthesis of structural proteins associated with ribosomes.

More immediate effects on protein synthesis require regulation of mRNA synthesis and function. Synthesis depends upon the activity of RNA polymerase II, which may be altered by chemical modifications affecting polymerase structure, or by regulatory interactions with other cell proteins. Alternatively, accessibility of DNA sequences in chromatin to RNA polymerase binding and initiation may be determined in a positive or negative sense by proteins associated with the chromatin.

Once RNA polymerase has begun to work on a gene, transcription may be terminated prematurely at an 'attenuation site' near the promoter, or may continue to the end of the gene, or even beyond. Response of the polymerase to structural signals on DNA or on the nascent mRNA molecule that determine the frequency of termination at a particular site may be closely regulated, for example, by methylation of the nascent mRNA. Messenger RNA precursor molecules are 'capped' at their 5′ ends soon after synthesis commences. The cap is $5'G^{m7}pppG^m pX^{(m)}p$. . . or a related structure. Failure to modify the molecule in this way almost certainly has a profound effect on translation of the resultant mRNA, and alteration of the structure may play a more subtle role in selection of messages for translation at different stages of cell development.

After termination of transcription of the pre-mRNA (which may be synonymous with HnRNA or a specific subclass of HnRNA molecules) and addition of poly(A) to the 3′ end of the RNA molecule, enzymic removal of non-coding sequences from the interior of the pre-mRNA molecule is a clear candidate for regulatory activity.

This so-called 'splicing' reaction has been shown to play a major role in determining which of several proteins are synthesized from a single DNA sequence, and changes in splicing accompany regulatory changes in the development of certain viruses, including acquisition of the ability to transport nuclear RNA sequences to the cytoplasm. The polyadenylation reaction has not been associated unequivocally with mRNA function. However, poly(A) polymerase is one of the proteins bound to mRNA during transport from nucleus to cytoplasm, and differences between the enzymes from normal and malignant cells have been reported.

Once a messenger RNA molecule has arrived in the cytoplasm its function can be modulated by numerous factors that affect its association with and translation on polysomes. These factors include the chemical structure of the cap and the sequence and secondary structure of the remainder of the mRNA molecule. Messenger RNA structure in turn interacts with cellular factors such as: ionic composition and strength; the concentration and type of initiation, elongation and termination factors; abundance and sequence of charged tRNA molecules; ribosome structure; and nucleases to determine the frequency of use and the ultimate fate (functional lifetime) of the polyribonucleotide. Some viral mutants have mRNAs of differing lifetimes, suggesting regulation by nucleases, but otherwise the evidence for hydrolytic control is not strong.

9.3 Nucleic acids in silk glands

The relationship of RNA of insects to that of vertebrates has been reviewed several times in different contexts (Ilan and Ilan, 1974; Suzuki, 1975; Grossbach, 1977; Suzuki, 1977). For the silk gland, the cycles of synthesis of DNA are known (Perdrix-Gillot, 1978, 1979); fibroin mRNA has been isolated (Suzuki and Brown, 1972), tRNAs have been well studied and characterized (Kawakami, Kakutani and Ishiyuka, 1974), the gene for fibroin has been cloned (Ohshima and Suzuki, 1977), and the structure of the gene analysed (Manning and Gage, 1978). When the posterior (fibroin-synthesizing) part of the silk gland involutes, total RNA decreases rapidly (98 per cent is lost in 5 days — Matsuura *et al.*, 1968). There is no accumulation of degradation products, but at first the degradation of 4S RNA is faster than that of other types, whereas later the amount of 26S RNA falls a bit more rapidly than that of 17S; and 4S RNA, probably including some degradation products, accumulates (Okabe *et al.*, 1978). These authors feel that, although much of the RNA is degraded *in situ*, some may be exported. The production and turnover of RNA is hormone-dependent (Shigematsu, Kurata and Takeshita, 1978). In starvation, which strongly limits the amount of RNA (Royston, unpublished work), the synthesis of tRNA and rRNA stops, whereas the level of aminoacyl-tRNA synthetase and charging levels decrease (Chavancy and Fournier, 1979).

It is otherwise apparent or suggested for several systems that the synthesis of mRNA falls rather early as cells die; such restrictions may be seen if adequate correction for pool size, penetration, and turnover are made, in involuting silk glands and thymocytes. In neurons, the dispersal of polysomes suggests equivalent phenomena. Although early attempts have been made to seek alteration in the composition of histones and nuclear proteins (Pollak and Levitt, 1980), there is no solid evidence for new nuclear products (putative repressors or enzymes) at this time. If such are present, they remain at or below the level of resolution attainable by the several investigators who have looked.

9.4 Limitations of present data

It would be of great interest to know how the decrease in synthesis of RNA and the presumed increase in (autophagic) destruction of rRNA comes about — whether it is brought about by the intervention of specific destructor genes, secondarily, by restriction of import of precursors or energy resources, or by alteration of cytoplasmic conditions, limiting enzyme activity. Several researchers have attempted to resolve these questions, and the earlier literature claimed that the synthesis of RNA, presumably mRNA, was necessary for cell death to occur (Weber, 1969; Lockshin, 1969; Munck, 1971; Chapters 3 and 11). Such a viewpoint is regarded in a more circumspect fashion today (Chapters 3 and 11). The elusive messengers have not been identified (Beckingham-Smith and Tata, 1976), and some of the results of studies *in vitro* have been demonstrated to be artefacts (Chapter 11). Furthermore, it is now recognized that the inhibitors have multiple effects, ranging from secondary stimulation of stress hormones in vertebrates, the stress hormones themselves blocking synthetic pathways, to blockage of far more metabolic functions than those presumed to be primary. The uncertainties range from the now classical discovery that actinomycin D inhibits not only RNA synthesis, but also DNA replication and, at low levels, specifically inhibits rRNA but not mRNA synthesis, to the more recent uncertainties about possible effects of cycloheximide on the structure and abundance of nuclear transcripts and the secondary effects of interruption of complex feedback pathways.

9.5 Future developments

The use of inhibitors of macromolecular synthesis has recently given way to direct measurement of molecules of interest in particular developmental systems. Measurements of abundance and sequence of RNA range from gross changes in populations to the structure of individual transcripts. The latter type of analysis is increasingly accessible as a result of advances in the field of gene isolation and cloning. Direct observation of changes in cellular protein content has been facilitated by twodimensional gel electrophoresis and radioimmune assays. These methodological developments have helped to overcome the uncertainties inherent in interpreting the results of experiments that are dependent on the use of inhibitors to disrupt specific types of macromolecular processes. These techniques have not in general been applied to the question of cell death.

Thus the theoretical argument to which the earlier results gave rise is considerably more shaky than had appeared at the time it was promulgated. The question remains to be resolved. If non-traumatic death is not brought about by specific command from the genome, affecting directly the synthesis of either one or more types of RNA or their respective hydrolases, then it presumably is brought about by the failure of a cell, under a perhaps self-imposed duress, to protect itself. Such a failure could be brought about by lack of precursor, source of energy or substrate, or, owing to the lack of any of several potential signals, it could simply fail to receive the message to re-initiate synthesis. This is the question which must be resolved. In order to achieve this resolution, it will be necessary to document the turnover of the several classes of RNA, preferably *in vitro* where extracellular pool sizes can be controlled, and then ultimately to go to a totally cell-free system in which the presence, activity and turnover of nuclear, messenger, transfer and ribosomal RNA and the ribosomal proteins and chain-initiating factors can be independently controlled and evaluated. A very interesting indication of the type of result to be expected is a recent finding concerning DNA in cell death, by Wyllie (1980). He found that, in glucocorticoid-induced apoptosis, a nuclear endonuclease cuts DNA into nucleosome fragments. Studies along these or similar lines would be a direct complement to the studies currently transforming the field of developmental biology, and would contribute insights into control mechanisms which are far too frequently ignored in that burgeoning field.

References

Beaupain, R. (1979), Cell division and cell death during regression of the chick embryo Müllerian ducts. *Experientia*, **35**, 1380–1382.
Beckingham-Smith, K. and Tata, J. R. (1976), Cell death. Are new proteins synthesized during hormone-induced tadpole tail regression? *Exp. Cell Res.*, **100**, 129–146.
Burns, A. L. and Kaulenas, M. S. (1979), Analysis of the translational capacity of the male accessory gland during aging in *Acheta domesticus*. *Mech. Ageing Dev.*, **11**, 153–169.
Chavancy, G. and Fournier, A. (1979), Effect of starvation on t-RNA synthesis, amino acid pool, t-RNA charging levels and amino acyl-t-RNA synthetase activities in the posterior silk gland of *Bombyx mori* L. *Biochimie*, **61**, 229–243.
Chinzei, Y. and Tojo, S. (1972), Nucleic acid changes in the whole body and several organs of the silkworm *Bombyx mori* during metamorphosis. *J. Insect Physiol.*, **18**, 1683–1689.
Fournier, A. (1979), Quantitative data on the *Bombyx mori* L. silkworm: A review. *Biochimie*, **61**, 283–320.
Grossbach, U. (1977), The salivary gland of *Chironomus* (Diptera): A model system for the study of cell differentiation. In *Biochemical Differentiation in Insect Glands* (ed. W. Beerman), Springer-Verlag, New York, pp. 147–196.
Hosbach, H. A. and Kubli, E. (1979a), Transfer RNA in aging Drosophila: I. Extent of aminoacylation. *Mech. Ageing Dev.*, **10**, 131–140.
Hosbach, H. A. and Kubli, E. (1979b), Transfer RNA in aging *Drosophila*. II. Isoacceptor patterns. *Mech. Ageing Dev.*, **10**, 141–150.
Ilan, J. and Ilan, J. (1974), Protein synthesis in insects. In *The Physiology of Insecta, Vol. 4* (ed. M. Rockstein), Academic Press, New York, pp. 356–422.
Johnson, L. K. and Strehler, B. L. (1979), Developmental restrictions on transcription: Determinants of the developmental program and their role in aging. *Mech. Ageing Dev.*, **9**, 535–552.
Kawakami, M., Kakutani, T. and Ishiyuka, S. (1974), Methionine transfer RNA from insect tissue. *J. Biochem. (Tokyo)*, **76**, 187–190.
Landmesser, L. and Pilar, G. (1978), Interactions between neurons and their targets during *in vivo* synaptogenesis. *Fed. Proc. Fed. Am. Soc. Exp. Biol., Proc.*, **37**, 2016–2022.
Lang, C. A., Lau, H. Y. and Jefferson, D. J. (1965), Protein and nucleic acid changes during growth and aging in the mosquito. *Biochem. J.*, **95**, 372–377.
Linzen, B. and Wyatt, G. R. (1964), The nucleic acid content of tissues of *Cecropia* silkmoth pupae. Relations to body size and development. *Biochim. Biophys. Acta*, **87**, 188–198.

Lockshin, R. A. (1969), Programmed cell death. Activation of lysis by a mechanism involving the synthesis of protein. *J. Insect Physiol.*, **15**, 1505–1516.

Manning, R. and Gage, L. (1978), Physical map of the *Bombyx mori* DNA containing the gene for silk fibroin. *J. Biol. Chem.*, **253**, 2044–2052.

Matsuura, S., Morimoto, T., Nagata, S. and Tashiro, U. (1968), Studies on the posterior silk gland of the silkworm, *Bombyx mori*. II. Cytolytic processes in posterior gland cells during metamorphosis from larva to pupa. *J. Cell Biol.*, **38**, 589–603.

Miguel, J. and Johnson, J. E., Jr. (1979), Senescent changes in the ribosomes of animal cells *in vivo* and *in vitro*. *Mech. Ageing Dev.*, **9**, 247–266.

Morrissey, J. H. and Green, H. (1978), Differentiation-related death of an established keratinocyte line in suspension culture. *J. Cell. Physiol.*, **97**, 469–479.

Munck, A. (1971), Glucocorticoid inhibition of glucose uptake by peripheral tissues: old and new evidence, molecular mechanisms, and physiological significance. *Persp. Biol. Med.*, **14**, 265–289.

Ochoa, S. and de Haro, C. (1979), Regulation of protein synthesis in eukaryotes. *Annu. Rev. Biochem.*, **48**, 549–580.

Ohshima, Y. and Suzuki, Y. (1977), Cloning of the silk fibroin gene and its flanking sequences. *Proc. Natl. Acad. Sci. U.S.A.*, **74**, 5363–5367.

Okabe, K., Koyanagi, R. and Koga, K. (1975), RNA in the degenerating silk gland of *Bombyx mori*. *J. Insect Physiol.*, **21**, 1305–1309.

Perdrix-Gillot, S. (1978), Durée et succession des cycles de synthèse d'ADN dans les noyaux polyploïdes de la glande séricigène de *Bombyx mori* au cours des 2e et 3e stades larvaires. *Biol. Cell.*, **32**, 245–256.

Perdrix-Gillot, S. (1979), DNA synthesis and endomitoses in the giant nuclei of the silk gland of *Bombyx mori*. *Biochimie*, **61**, 171–204.

Pittman, R., Oppenheim, R. W. and Chu-Waing, I. W. (1978), Beta-bungarotoxin-induced neuronal degeneration in the chick embryo spinal cord. *Brain Res.*, **253**, 199–204.

Pollak, R. D. and Fallon, J. F. (1976), Autoradiographic analysis of macromolecular synthesis in prospectively necrotic cells of the chick limb bud. II. Nucleic acids. *Exp. Cell Res.*, **100**, 15–22.

Pollak, R. D. and Levitt, J. B. (1980), Private communication.

Shigematsu, H., Kurata, K. and Takeshita, H. (1978), Nucleic acids accumulation of silk gland of *Bombyx mori* in relation to silk protein. *Comp. Biochem. Physiol.*, **61B**, 237–242.

Suzuki, Y. (1975), Fibroin messenger RNA and its genes. *Adv. Biophys.*, **8**, 33–114.

Suzuki, Y. (1977), Differentiation of the silk gland. A model system for the study of differential gene action. In *Biochemical Differentiation in Insect Glands* (ed. W. Beerman), Springer-Verlag, New York, pp. 1–44.

Suzuki, Y. and Brown, D. (1972), Isolation and identification of the messenger RNA for silk fibroin from *Bombyx mori*. *J. Mol. Biol.*, **63**, 409–429.

Weber, R. (1969), The tadpole tail as a model system for studies on the mechanism of hormone-dependent tissue involution. *Gen. Comp. Endocrinol. Suppl.*, **2**, 408–416.

Wyllie, A. H. (1980), Glucocorticoid-induced thymocyte apoptosis is associated with endogenous endonuclease activation. *Nature*, **284**, 555–556.

10 Mechanism(s) of action of nerve growth factor in intact and lethally injured sympathetic nerve cell in neonatal rodents

RITA LEVI-MONTALCINI AND LUIGI ALOE

10.1 Introduction

The central issue in the study of hormone-target cell interaction is to uncover the mechanism(s) of activation of the receptive cells by the ligand. Simultaneous progress along three different lines of hormone research provided new approaches to this problem. The first was the discovery of the second messenger, identified as the cyclic nucleotides, endowed with the property of transducing different hormonal signals in intracellular biochemical changes responsible for transient and longer-lasting physiological actions. The second was the recognition of the existence of membrane receptors, identified as macromolecules which selectively bind to different hormones and to neurotransmitters. These findings in turn sparked an ever-growing series of investigations directed towards the elucidation of the properties of the hormone-receptor complexes and of the mechanisms by which these complexes, once formed, modify the activities of specific membrane-localized enzymes, transport systems and intracellular events which follow the binding or internalization of these complexes in the target cells. The third was the finding that several polypeptides released *in vivo* by different organs and tissues, and *in vitro* by several normal and neoplastic cell lines, are endowed with a hormone-like activity on other cell types. Among these polypeptides, which became known as specific growth factors because of their main property of activating morphological and metabolic processes in cells at an early stage of their lifecycle, the most intensely investigated, as well as the first to have been discovered, is the Nerve Growth Factor (NGF). A unique and remarkable feature of this protein molecule, and one that is particularly relevant in this book devoted to the study of cell death, is that it elicits its potent growth and differentiative effects not only in its intact target cells, but also in the same cells that have been mortally injured by the most diversified treatments. In some but not all instances, evidence was obtained that death in the experimentally manipulated cells is directly related to experimental depletion of NGF, whether produced by immunological, chemical or surgical interventions. These findings illuminate some of the problems discussed in the first section of this volume, namely the possible causes of the widespread occurrence of 'natural cell death' during ontogenetic processes as well as during metamorphosis in invertebrate and vertebrate species. They raise the question whether natural and experimentally induced cell death could not result from similar, even if not necessarily identical, events occurring in the critical early phases of growth and differentiation, when cells are much more vulnerable than the fully differentiated cells.

In the first part, after a brief historical survey of the phenomenon, we shall consider the characteristics of the growth response elicited by NGF in its target cells and its possible mechanism of action. In the second part we shall report on the 'rescue effect' and even paradoxical overgrowth elicited by NGF in sympathetic nerve cells lethally injured by some pharmacological compounds or by their mechanical disconnection from the end organs.

10.2 Historical survey

The critical event that started these investigations was the finding that two mouse tumours known as mouse sarcomas 180 and 37, grafted into the body wall of 3-day chick embryos, selectively enhanced growth and differentiative processes of sensory (Bueker, 1948) and sympathetic nerve cells (Levi-Montalcini and Hamburger, 1951) innervating the neoplastic tissues. The excessive production and aberrant distribution of sympathic nerve fibres inside the viscera and into the cavity of blood vessels bearing these intra- or extra-embryonic transplants provided the first evidence that these anomalous effects are elicited by a diffusible agent released by the neoplastic cells (Levi-Montalcini, 1952). The next step was the development of an *in vitro* bioassay which consisted of the explantation *in vitro* of embryonic sensory and sympathetic ganglia, dissected from 8-day chick embryos and cultured in proximity to one of the two mouse sarcomas or of other neoplastic tissues devoid of such *in vivo* nerve-growth-promoting activity. The finding that ganglia cultured in close proximity to fragments of mouse sarcomas 180 or 37, but not to other tumours or normal tissues, developed in a 12–24 h incubation period a dense fibrillar halo of nerve fibres (Levi-Montalcini, Meyer and Hamburger, 1954) signalled the turning point of this investigation. The *in vitro* bioassay provided the conditions unattainable in experiments performed in developing embryos, allowing the identification of nerve-growth-promoting factor, which was tentatively identified in 1954 (Cohen, Levi-Montalcini and Hamburger, 1954) in a nucleoprotein fraction, and named after its activity Nerve Growth Factor or more simply NGF. At the same time this *in vitro* technique afforded the opportunity for a rapid screening of other tissues and organic fluids as potential sources of NGF. The discoveries in 1956 and in 1958 that the snake venom (Cohen and Levi-Montalcini, 1956) and the extracts of mouse submaxillary salivary glands (Cohen, 1958; Levi-Montalcini, 1958) are two much richer NGF sources than are mouse sarcomas 180 and 37, and the subsequent identification in the venom and salivary NGFs of two polypeptides remarkably similar to each other by all biochemical and immunological criteria (Cohen, 1959) diverted the attention from the mouse tumours to the snake venom at first, and then to the mouse salivary glands. The latter became the tissue of choice ever since 1958 for all studies directed towards the characterization of this NGF and of the morphological and biochemical changes in its target nerve cells: the sensory embryonic nerve cells during a restricted period of their development, and the sympathetic nerve cells during their entire lifecycle (Levi-Montalcini, 1966). The results of these investigations will be summarized in the following section only to the extent needed to consider the possible mechanism of action of NGF in promoting differentiation and maintenance of its target cells.

10.3 The salivary NGF: morphological and biochemical effects induced in its target cells.

10.3.1 *The 2.5S and the 7S NGFs*

Cohen, who first purified the mouse salivary NGF, evaluated the molecular weight of this polypeptide moiety as 44000 (Cohen, 1960). Subsequent work indicated molecular weights of 30000 (Bocchini and Angeletti, 1969), 28000 and even 14000 (Zanini and Angeletti, 1971). The subunit structure of the NGF first reported in 1968 (Zanini, Angeletti and Levi-Montalcini, 1968) was re-examined by other investigators (Angeletti and Bradshaw, 1971), who found that the salivary NGF, purified according to the procedure devised by Bocchini and Angeletti and reported in the 1969 article mentioned above, represents a dimer of two identical subunits linked together by non-covalent bonds in a tight complex which is difficult to dissociate. This salivary NGF, which became known as the 2.5S, has a molecular weight of 26516; each of the two polypeptide chains contains 118 amino acids and has a molecular weight of 13259. The amino acid sequence and the primary structure of NGF were determined and reported by Angeletti and Bradshaw (1971). Of the other salivary NGF molecular species determined by other investigators, the most extensively studied is a 7S NGF (Varon, Nomura and Shooter, 1967) which consists of the NGF dimer also called the β-subunit, bound in a stable complex with two copies each of two other proteins, designated as the α- and the γ-subunits. The γ-subunit is a specific enzyme which cleaves polypeptide chains only at a point adjacent to the amino acid arginine. The α-subunit has no known biological activity. The NGF β-dimer is endowed with full biological activity, identical with that of the 2.5S NGF and does not require the α- and the γ-subunits to elicit the effects associated with NGF activity. Several hypotheses have been proposed to account for the presence of the α- and γ-subunits in the 7S NGF. Studies by Shooter and co-workers (see review article, Server and Shooter, 1977) led to the suggestion that the NGF chains are derived from larger precursor chains by the proteolytic action of the esteropeptidase. The initial gene product in the synthesis of NGF would be a polypeptide chain, the 7S NGF. Its role would be in biosynthesis, and in conferring stability to the low-molecular-weight NGF, which is more susceptible to degradation than the high-molecular-weight form.

10.3.2 *Morphological effects of NGF*

Daily injections of the purified 2.5S NGF in neonatal mice and rats, at 10 μg/g

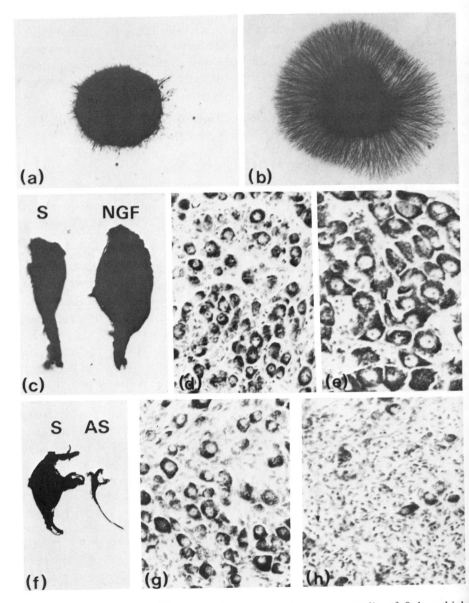

Fig. 10.1 (a) and (b) Photomicrographs of sensory ganglia of 8-day chick embryos after 18 h of culture *in vitro*. (a) Ganglion in a control medium; (b) ganglion in a medium supplemented with 10 ng of NGF/ml. Silver-stained. Magnification 53 ×. (c) Whole mounts of SCGs of 6-day-old rats injected since birth with saline(S) or NGF. Magnification 30 ×. (d) and (e) Transverse sections of SCGs of 6-day-old rats injected since birth with saline(d) or NGF(e). Toluidine Blue-stained. Magnification 220 ×. (f) Whole mounts of SCGs of 14-day-old rats injected since birth with saline (S) or NGF antiserum-

body weight from the day of birth to the end of the second or third postnatal week, result in an increase in volume of 10–12-fold over that of the same ganglia in untreated littermates (Levi-Montalcini, 1966). This overall effect (Fig. 10.1c, 1d and 1e) was traced to four separate processes: increase in nerve cell number, enhanced speed of differentiation, hypertrophic effects which result in the attainment of a size of individual neurons markedly larger than that of fully differentiated untreated nerve cells, and exuberant production of neuropile (stem axons and collaterals) which forces the cells apart and fills all available space among cell bodies in the NGF-treated ganglia. The most debated and still unsettled question is the origin of the supernumerary cells in ganglia of NGF-injected neonatal rodents. This increase was at first attributed to a direct mitotic effect in nerve cell precursors or in these cells at an incipient stage of differentiation (Levi-Montalcini and Booker, 1960a). This hypothesis was, however, refuted when counts of nucleus-labelled neurons in the sympathetic superior cervical ganglia (SCG) of experimental and control rats injected during the first postnatal week with tritiated thymidine did not show a statistically significant increase in number of labelled neurons in the SCGs of NGF-treated rats as compared with those of controls (Hendry, 1977). The prevailing and generally adopted viewpoint, championed by Hendry, is that the NGF hyperplastic effects are instead due to prevention of 'natural cell death' by the exogenous supply of the growth factor which would provide the means of subsistence to nerve cells that failed to establish synaptic connections with their end tissues. Natural cell death in developing nerve centres is in fact considered by the majority of authors (see review articles: Prestige, 1970; Cowan, 1973) as being consequent to a competition race between nerve cells in active growth phase for a limited number of available synaptic sites on the receptor cells. The neurons that fail to establish morphological or, according to the results of more recent studies, functional connections with their end cells (Pittman, Oppenheim and Chu-Waing, 1978; Laing and Prestige, 1978) are doomed.

The hypothesis that prevention of natural cell death by NGF fully accounts for the numerical increase of nerve cells in sympathetic ganglia of neonatal rodents injected with NGF is difficult to accept in view of the following findings: the nerve cell population in the SCGs of 4-day-old untreated rats exceeds by only 17–20% that of the same fully differentiated nerve cells (Table I from Aloe, Mugnaini and Levi-Montalcini; 1975). Other authors evaluate the nerve cell loss which occurs in the SCGs in the first postnatal week to be around 30% (see review article by Mobley, Server, Ishii, Riopelle and Shooter, 1977). The numerical increase in nerve cells in the SCGs of NGF-treated infant rats is, however, 2–3-fold over that of control ganglia (Table 10.1). The discrepancy between the above figures makes it unlikely that the NGF hyperplastic effects result only from prevention of natural cell death. Thus there is a need for a reinvestigation of early events elicited by

(AS). Magnification 10×. (g) and (h) Transverse sections of SCGs of 14-day-old rats injected since birth with saline(g) or NGF antiserum(h). Toluidine Blue-stained. Magnification 250×.

Table 10.1 Average cell number/superior cervical ganglion of rats injected since birth as indicated

Taken from Aloe *et al.* (1975)

Age (days)	NGF+6-OHDA	NGF	6-OHDA	Saline
4	65 020	61 620	11 435	29 400
10	55 413	71 291	4 958	27 230
16	54 100	66 020	3 773	27 027
21	58 060	62 520	2 484	25 314

NGF in sympathetic nerve cell precursors assembled in the still not fully developed ganglia.

The NGF hypertrophic effects investigated *in vivo* at first in the chick embryos and in neonatal and fully developed rodents (Levi-Montalcini, 1966) were recently the object of studies in rat foetuses (Aloe and Levi-Montalcini, 1979a). The *in vitro* investigations pursued in parallel with those *in vivo* analyzed the effects elicited by NGF in three systems: (a) intact sensory and sympathetic ganglia from 8-day chick embryos explanted in semi-solid media; (b) dissociated sensory and sympathetic embryonic nerve cells cultured in liquid media; and (c) cells of a neoplastic pheochromocytoma cell line known as PC12 line grown in the presence or absence of NGF.

The growth response of embryonic sensory nerve cells and of sympathetic nerve cells, from their early inception to full differentiation in adult rodents, is characterized both *in vitro* and *in vivo* by the massive production of nerve fibres, which represents the earliest detectable event seen in the light and the electron microscopes (EM). Increase in microtubules, microfilaments and neurofilaments is already apparent at the EM level 2 h after the addition of NGF to the culture media, and is so extensive as to overshadow other ultrastructural changes such as increase in the rough endoplasmic reticulum and in the Golgi apparatus (Levi-Montalcini, Caramia, Luse and Angeletti, 1968).

The production of the dense fibrillar halo by embryonic sensory and sympathetic ganglia (Fig. 10.1a and 1b), which materializes in the first 10–20 h of culture, is the most spectacular effect evoked by NGF at a concentration of 10 ng/ml of culture (an effect which became known as the growth response elicited by 1 NGF biological unit). Even more revealing of the role played by NGF in the life of these cells than the precocious and massive production of nerve fibres by whole ganglia is the growth response elicited by 1–10 NGF biological units (BU) from embryonic sensory and sympathetic nerve cells dissociated and cultured in liquid media. In the presence of NGF, the dissociated nerve cells survive in excellent condition and produce a dense fibre network which in a short period extends over the entire surface of the culture dish. Daily additions of NGF to the culture medium permit survival and continuous axonal growth for indefinite lengths of time. In control cultures, nerve cells undergo massive degeneration in the first 23 h, while satellite cells and fibroblasts persist even in the absence of serum (Levi-Montalcini and Angeletti, 1963).

Two additional features of the NGF-evoked response, which hardly fit into the conventional concept of 'trophic effects' and add further weight to the notion of the unique role played by this protein molecule in the life of its receptive cells, are: an NGF-transforming effect of neoplastic and immature chromaffin cells in sympathetic nerve cells, and an NGF-neurotropic effect. In view of the interest that attaches to these two newly discovered NGF properties, they will be briefly reported here.

The NGF-transforming effect was first described by Greene and co-workers in a clonal cell line, the PC12 cell line originating from a rat pheochromocytoma (Greene and Tischler, 1976). The PC12 cells exhibit typical morphological and biochemical features of neoplastic chromaffin cells when cultured in NGF-free media. Upon addition of NGF to the culture medium, the cells cease to divide, produce an axon and show all differentiative and biochemical marks of sympathetic nerve cells, but upon withdrawal of NGF from the medium, they revert to the original cell type. This *in vitro* system provided, as we shall see in a following section, an invaluable model by which to explore the earliest events evoked by NGF in the transformed nerve cells, and provided a new lead to the study of its mechanism of action.

A similar NGF-transforming effect *in vitro* was described in dissociated rat adrenal medullary cells cultivated in an NGF-rich medium (Unsicker, Kirsch, Otten and Thoenen, 1978). Experiments *in vivo* in rat foetuses and chick embryos, gave additional impressive evidence of a NGF-transforming effect of normal chromaffin cell precursors in sympathetic nerve cells. NGF injections in 16–17-day rat foetuses, continued after birth until the end of the first or second postnatal week, produce a massive transformation of chromaffin cells of the adrenal medullary cells, carotid bodies and abdominal paraganglia into sympathetic neurons: indistinguishable from genuine nerve cells (Aloe and Levi-Montalcini, 1979a). Daily injections of NGF in chick embryos cultivated in petri dishes from the third to the 17th day of incubation, according to the technique devised by Auerbach (Auerbach, Kubai, Klinghton and Folkman, 1974), showed a similar transformation of chromaffin cell precursors into sympathetic nerve cells. In 11–17-day embryos, the chromaffin cells were entirely replaced by massive sympathetic cell aggregates which filled the abdominal cavity, and surrounded the diminutive adrenal gland which itself consisted only of the cortical tightly packed cell strands. In control embryos, the cortical gland component is loosely interlaced with medullary cordons. The chromaffin cells which underwent transformation into sympathetic cells under the action of NGF remain, outside the gland, assembled in the enormous newly formed ganglia (Levi-Montalcini and Aloe, 1980).

The NGF neurotropic effects came most unexpectedly to our attention in the course of studies exploring whether NGF enhances growth and differentiative processes in intracerebral catecholaminergic systems. Microcapillary NGF injections into the medulla oblongata of neonatal rodents, repeated daily for a 10-day period, did not result in detectable morphological changes in cells of the locus coeruleus or of other catecholaminergic brain centres, but elicited marked hyperplastic and hypertrophic effects in paravertebral sympathetic ganglia. A contingent of nerve fibres growing out from these ganglia gained access inside the spinal cord and brain stem through the dorsal roots of spinal nerves and the sensory motor roots of the lowest cephalic nerves, and settled respectively in the dorsal funiculi of the spinal cord and ventro-lateral cordons of the brain stem. Collaterals emerging from these ectopic fibre bundles entered inside the white and grey matter in close association with blood vessels. Discontinuation of the intracerebral NGF injections resulted in the fading away and final disappearance of this ectopic parasitic system. The hypothesis that the entrance of sympathetic nerve fibres into the central nervous system of intracerebrally NGF-injected rodents is produced by an NGF diffusion gradient from the site of injection in the medulla oblongata, and realized through its transport along sensory and motor roots to the adjacent sympathetic ganglia, is supported by the marked volume increase in these ganglia. There is also a high NGF level in the brain and spinal cord of the injected infant rats, 30 h after the last NGF injection (Menesini Chen, Chen and Levi-Montalcini, 1978). Recent most ingenious experiments *in vitro* (Campenot, 1977; Letourneau, 1978; Gundersen and Barret, 1979) provided additional strong support in favour of an NGF-neurotropic effect on sympathetic and sensory fibres produced by dissociated nerve cells of both types.

10.3.3 *Metabolic effects of NGF*

Experiments performed on sensory and sympathetic ganglia cultured *in vitro* for a 4–8 h period showed that NGF enhances synthetic and metabolic processes. Glucose oxidation is stimulated through a direct oxidative pathway (Angeletti, Liuzzi, Levi-Montalcini and Gandini Attardi, 1964). Lipid biosynthesis is increased, as indicated by the enhanced incorporation of labelled acetate (Liuzzi, Angeletti and Levi-Montalcini, 1965). RNA and protein in sensory ganglia incubated in an NGF-rich medium undergo a net increase during the first 8 h of culture (Angeletti, Levi-Montalcini and Calissano, 1968). The possibility that these and other metabolic effects, result from a combination of better maintenance of ganglia and NGF-stimulated nerve fibre outgrowth as suggested by Larrabee (1972) is to be taken into serious consideration. Recent studies which made use of more sophisticated techniques, such as two-dimensional gel electrophoresis, to detect more than 800 proteins in control and NGF-treated pheochromocytoma PC12 cells showed that none of these proteins were qualitatively repressed after NGF treatment, and no new proteins appeared after NGF treatment resulting in the transformation of the neoplastic chromaffin cell line into a line exhibiting the typical phenotype of sympathetic cells (Garrels and Schubert, 1979). These findings add further to the notion, gathered from earlier experiments with sensory and sympathetic nerve cells *in vitro*, that NGF causes only quantitative modulation of protein synthesis rather than qualitative changes.

10.3.4 *Effects of NGF on catecholamine synthesis*

Early studies showed an increased noradrenaline content in the SCGs of NGF-treated mice as compared with the noradrenaline content of control ganglia (Crain and Wiegand, 1961). The recent development of more sensitive techniques to measure the activity of enzymes involved in the synthesis and metabolic degradation of the noradrenergic neurotransmitter made available a more precise, even if indirect, method of evaluating the effects of NGF on the function of sympathetic nerve cells. Studies by Thoenen, Angeletti, Levi-Montalcini and Kettler (1971) on the levels of these enzymes in the SGCs of infant rats injected with NGF from the day of birth to the tenth postnatal day and of control littermates gave evidence for a marked increase in the specific activity of the two key enzymes, tyrosine hydroxylase and dopamine β-hydroxylase, in the synthesis of noradrenaline in the experimental ganglia, while two other enzymes, dopadecarboxylase and monoamine oxidase, rose only in proportion to the increase in protein content in the NGF-treated ganglia (Thoenen *et al.*, 1971). These and other findings to be reported in a following section support the notion that NGF does not only enhance the cell's metabolic processes resulting in the hypertrophic effects described above, but also markedly stimulates the specific nerve cell function, which is indissolubly bound to the synthesis and release of the noradrenergic neurotransmitter.

10.3.5 *Interaction of NGF and contractile fibrillar proteins*

The earliest morphological event detectable at the E.M. level in the sequence of changes produced by NGF in its target cells consists, as already mentioned, of a massive increase in contractile fibrillar proteins in cell perikaryon and in the axon. These findings focused attention on the possible role of cytoskeletal proteins in mediating the NGF action. Studies in this last decade on nerve cells as well as on several other cell types have implicated these fibrous proteins in most diversified cell functions, such as axonal growth and axoplasmic transport, sensory transduction, cilia movements, hormonal actions on the target cells, and cell-surface changes in different cellular systems (Margulis, 1972; Olmsted, 1976; Tilney, 1971). It will also be remembered that the sudden interest in these contractile proteins, which were practically ignored only two decades ago, developed with the discovery that the basic unit and building block of microtubules, the most extensively studied of these fibrillar structures, also known as tubulin, is characterized by the unique property of binding colchicine. It thus became possible to detect tubulin wherever present in different cell compartments, including the membranes of a variety of eukaryotic cells and synaptosomal membranes of nerve terminals (Feit and Barondes, 1970). The subsequent discovery that microtubules are highly dynamic structures which are capable of rapid assembly and disassembly in response to a variety of drugs and environmental conditions raised the question as to the role of these ubiquitous structures in all processes where they are implicated. According to Shelanski (1973), who pioneered these studies, 'microtubules can be visualized as skeletal structures which because of their ability to readily form and dissolve, provide an ideal metastable cytoskeleton or a key component in a motility transport system in the cell'.

Of the other two contractile fibrillar proteins, microfilaments and neurofilaments, microfilaments are the best known from the structural viewpoint as well as for their functional role, which consists of generating cell locomotion. They are thin (about 5nm) fibrillar proteins, very similar to muscle actin in terms of molecular weight and peptide 'map' (Bray and Thomas, 1976). They subsist in two forms: globular unpolymerized actin (G-actin) and filamentous actin (F-actin). F-actin interacts with myosin protein filaments, functionally similar to but immunologically different from muscle myosin (Kuczmarski and Rosenbaum, 1979). F-actin is restricted in nerve cells to the axon growth cone, where it accounts for the vigorous motility and contractility of the tip of the growing fibres (Yamada and Wessells, 1971; Fine and Bray, 1971).

A direct effect of NGF on tubulin and G-actin *in vitro* was first reported by Calissano and co-workers (Calissano and Cozzari, 1974; Calissano, Monaco, Levi *et al.* 1976; Levi, Cimino, Mercanti *et al.* (1975). These authors found that addition of NGF to a 100 000 *g* supernatant of mouse brain, induces an instantaneous turbidity of the solution. The electrophoretic and amino acid analysis of the precipitated and redissolved pellet showed that NGF interacts and co-precipitates mainly with the precursor proteins of microtubules and microfilaments among all the brain proteins. Subsequent studies by the same group demonstrated that NGF favours polymerization of tubulin to form microtubules, and stabilizes these structures against depolymerizing agents such as vinblastine, both *in vitro* and *in vivo*. They also showed that the NGF–F-actin complex forms paracrystalline structures which activate myosin ATPase activity much more than actin alone (Calissano, Monaco, Castellani, Mercanti and Levi, 1978). These findings *in vitro* raised the question whether an interaction between NGF and the two fibrillar protein precursors, tubulin and G-actin, similar to those which take place *in vitro*, could also occur in intact NGF target cells. Results of experiments to be reported in the second part of this article lend support to this hypothesis.

10.4 Dual access and mechanisms of action of NGF in its target cells

An important and still debated question in neurobiology is whether nerve cells are receptive to the action of exogenous factors only or mainly through their end terminals, or whether these factors also gain entrance through the membrane that envelopes the cell perikaryon. Studies on NGF–target cell interaction provided unequivocal evidence for a dual access of this protein molecule to sympathetic nerve cells through binding to specific receptors distributed on the cell membrane and on the axonal endings. Whether, and to what extent, NGF also binds to the preterminal nerve fibre endings remains to be determined. The results to be summarized here indicate the existence of a specific NGF-receptor system on the target cell perikarya even prior to the production of the axon. In subsequent developmental stages, and throughout the entire lifecycle, NGF binding to membrane receptor sites, co-exists with binding and retrograde NGF transport from axonal endings to the cellular compartments of sympathetic nerve cells.

Evidence for a non-axonal NGF-mediated effect during early development came from experiments *in vivo* and *in vitro* (Levi-Montalcini and Booker, 1960a). Injections of the salivary 2.5S NGF into newborn rodents cause the precocious differentiation of nerve cell precursors in sympathetic nerve cells. Since in chick embryos and in rat foetuses NGF induces, as reported on p. 301, the transformation of immature chromaffin cells to sympathetic nerve cells, this effect can only materialize through the interaction of NGF with membrane receptors, in view of the fact that these cells lack axonal processes. Likewise the formation of a fibrillar halo around sensory and sympathetic ganglia explanted *in vitro* occurs through a direct effect of the growth factor on the cell bodies, since both pre- and post-ganglionic roots were sectioned in the immediate vicinity of the ganglia at the moment of their dissection from the embryos. Even more clear-cut evidence for a NGF binding to the perikaryon of immature nerve cells came from experiments mentioned on p. 301 on the response elicited by NGF in nerve cells dissociated by enzymic or mechanical means. Upon dissociation, the cells inspected with the inverted microscope exhibit a round or ovoidal shape and are entirely devoid of axonal processes. The outgrowth and subsequent elongation of a neurite from the majority of the cultured cells takes place only in NGF-rich media, an effect obviously mediated through direct interaction between the growth factor and the cell perikaryon.

How does NGF gain access to the cells receptive to its action? Work performed in this last decade on polypeptide hormones provided most valuable models and criteria for evaluating the general and specific features of the interaction between these ligands and macromolecules, known as receptor sites, located on the cell membrane. The same criteria followed in the NGF–receptor studies gave definite evidence for the presence in intact 8-day embryonic sensory cells (Herrup and Shooter, 1973) and in plasma-membrane-enriched fractions of rabbit superior cervical ganglia (Banerjee, Snyder, Cuatrecasas and Greene, 1973) of saturable reversible high-affinity binding sites with dissociation constants in the range of 10^{-10} mol. Both groups were able to show that the binding of chemically modified NGF was altered in proportion to its altered biological activity. Failure of other hormones to bind to these receptor sites provided additional evidence for the specificity and biological significance of this interaction. While these findings satisfy most of the basic criteria for establishing the existence of specific NGF receptors, some perplexity is raised by the finding that other cell lines such as neuroblastoma and C-6 glioma cells (unpublished experiments by Calissano and Kimli) and a melanoma cell line (Fabricant, De Larco and Todaro, 1977) also possess NGF-specific binding sites on their cell membranes, and yet no morphological changes are detectable in these cells upon addition of NGF to the culture medium. The question is raised whether the NGF–receptor complexes need some cellular components, lacking in these systems, to mediate their biological action.

The next problem was to uncover the mechanism of action and the sequence of events sparked by NGF binding to its receptors. The question whether this growth factor acts through a second messenger, such as cyclic AMP, cyclic GMP or trans-membrane calcium fluxes directly or indirectly altered by NGF, or whether NGF elicits morphological and biochemical changes in its target cells through a direct action at the transcription or translation levels, remains unanswered. In favour of the second messenger hypothesis are the studies by Roisen, Murphy and Braden (1972), Nikodijevic, Nikodijevic and Wong (1975), Schubert, La Corbiere, Witlock and Stallcup (1978) and Garrels and Schubert (1979). Schubert and co-workers pro-pose the hypothesis that the NGF-increased cyclic AMP level in PC12 cells would lead to a mobilization of calcium ions. This effect would in turn result in structural changes in the plasma membrane, increased cell-substratum adhesion and neurite outgrowth Strong evidence against the biological significance of transient changes in intra-cellular cyclic AMP levels, as well as against the hypothesis of a direct NGF influence on calcium fluxes in NGF target cells, was, however, presented by other groups (Frazier, Ohlendorf, Boyd, Aloe *et al.*, 1973; Hatanaka, Otten and Thoenen, 1978; Landreth, Cohen and Shooter, 1980). These and other investigators showed that NGF undergoes internalization after binding to membrane receptors of embryonic sensory and PC12 cells (Andres, Jeng and Bradshaw, 1977: Yanker and Shooter, 1979; Marchisio, Naldini and Calissano, 1980; Levi, Shechte, Neufeld and Schlessinger, 1980). Translocation of the internalized NGF to the PC12 cell nuclei, isolated with the detergent Triton X-100, was reported by Yanker and Shooter (1979), while Marchisio *et al.* (1980) provided evidence with immuno-fluorescence and radioautographic techniques for a peri- and intra-nuclear location for NGF. Levi *et al.* (1980) showed that NGF shares with several serum proteins and hormones the property of being internalized through the mechanism of receptor-mediated endocytosis, a coupled process by which selected extracellular proteins or peptides are first bound to specific cell-surface receptors and then rapidly internalized by the cell (Goldstein, Anderson and Brown), 1979). Whatever the intracellular distribution of NGF in its receptor cells, these findings favour the view that internalization is instrumental to the mechanism of NGF action. This hypothesis finds support in the demonstration that half-life of NGF upon inter-nalization is 18–24 h, and, even more important, that when it is extracted under suitable conditions, it binds and competes with native NGF for the same receptors, a finding which would prove that NGF does not undergo immediate degrading after internalization, since it is still recognized by highly specific receptors (Calissano and Shelanski, 1980).

While studies on NGF–membrane receptor interaction and on the subsequent internalization of NGF-receptor complexes were almost exclusively done *in vitro*, other studies performed at the same time were instead directed at examining the mechanism whereby NGF is taken up by noradrenergic nerve endings and conveyed retrograded to the perikarya of fully developed sympathetic neurons. Although this analysis *in vivo* could not avail itself of the sophisticated quantitative techniques available to measure NGF-receptor affinity to the membranes of cells cultured *in vitro*, nor could it afford equally good opportunities for visualizing the binding and early internalization processes, it did reveal other facets of the NGF–target cell interaction which are very valuable for the understanding of the role played by NGF in the life of sympathetic nerve cells. Here we shall mention only the most relevant findings of these investigations which were prompted by the demonstration that iodinated NGF injected into the anterior eye chamber of adult mice labels only the small percentage of sympathetic neurons in the SCG that innervate the iris (Hendry, Stöckel, Thoenen and Iversen, 1974). These and subsequent studies (Stöckel, Paravicini and Thoenen, 1974; Stöckel, Schwab and Thoenen, 1975) showed that labelled NGF is taken up by noradrenergic nerve endings by a highly selective saturable mechanism. It is transferred to the cell bodies and enclosed in vesicles and cisternae that exhibit the morphological characteristics of smooth endoplasmic reticulum. In the cytoplasmic compartment they fuse, at least partially, with lysosomes (Schwab, 1977; Schwab and Thoenen, 1977; Thoenen, Barde, Edgar, Hatanaka, Otten and Schwab, 1979). Since the results of E.M. studies on the intracellular localization of NGF coupled with horse-radish peroxidase gave no evidence for a release of intact NGF into the cytoplasm, nor for its transfer into the nuclear compartment, the authors suggest the hypothesis that the retrogradely transported NGF acts as a 'second messenger' either after binding to the membrane structure or after splitting into smaller biologically active fragments (Thoenen *et al.*, 1979). It may, however, be objected that failure to detect labelled NGF in the cytoplasmic or nuclear compartments does not necessarily rule out the possibility that NGF may have gained access to subcellular components in one or both compartments, but remained unidentifiable by the presently available techniques.

10.5 Destruction of immature sympathetic nerve cells by immunochemical, pharmacological and surgical procedures

10.5.1 *Immunosympathectomy*

The selective destruction of sympathetic nerve cells produced in neonatal rodents and other mammals by a specific antiserum to NGF (AS-NGF) first reported in 1960 (Cohen, 1960; Levi-Montalcini and Booker, 1960b) emphasizes the key role played by NGF in the early developmental stages of these cells. Sympathetic 'long adrenergic neurons' located in para- and pre-vertebral sympathetic ganglia undergo massive degeneration, whereas the 'short adrenergic neurons' positioned in close proximity to the innervated tissues, such as the sex organs, the interscapular brown adipose bodies and few other peripherally located sympathetic ganglia (Steiner and Schönbaum, 1972), are not damaged by the specific AS-NGF. This effect, which deprives the animal of about 90 per cent of the entire sympathetic nerve cell population, without interfering with its somatic development and vitality, became known ever since 1961 as 'immunosympathectomy' (Levi-Montalcini and Angeletti, 1966). The precise mechanism by which the AS-NGF produces these dramatic effects is still a matter of discussion (see page 316), even if the lesions induced by

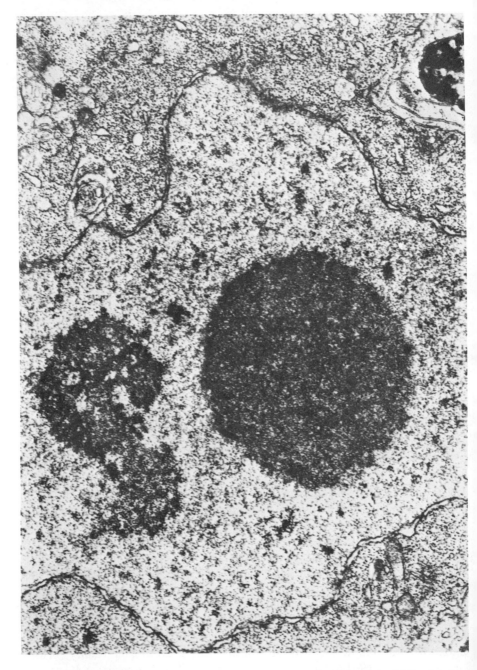

Fig. 10.2 Electron micrograph of a sympathetic nerve cell in SCG of 2-day-old rat injected on the first day of life with NGF antiserum. Magnification 22 000×.

AS-NGF in the immature sympathetic nerve cells have been the object of detailed sequential studies at the structural and ultrastructural levels. Inspection with the light microscope of sympathetic ganglia showed, by 24 h after an AS-NGF injection, the prescence of dead nerve cells and cell debris. Immature, apparently still normal, nerve cells are smaller than in control ganglia and the cytoplasm is almost deprived of ribonucleic acid, as shown by light stainability with basic dyes and the fact that nucleoli are barely visible. In subsequent days, even if the AS-NGF treatment is not repeated (see p. 316) dead cells increase progressively in number, whereas satellite cells are intact and appear at first slightly more numerous than in controls. Subsequently these cells also undergo regressive changes and the ganglia are reduced to diminutive sclerotic nodules barely detectable with the aid of the dissecting microscope (Fig. 10.1f, 1a and 1h). E.M. studies give evidence for the extraordinarily rapid degenerative effects elicited by antibodies to NGF. The most precocious alterations, already apparent 2-3 h after the injection of AS-NGF, consist of marked changes in the fine structure of the nucleoli (Fig. 10.2), condensation of chromatin into large and small masses, and folding of the nuclear envelope. The only cytoplasmic alteration which is apparent in these early stages consists of some disorganization of the endoplasmic reticulum. In immediately subsequent stages, 4 and 12 h after the injection of the antiserum, the nuclear and cytoplasmic material becomes intermixed following rupture of the nuclear membrane. Vacuolization and other signs of cytolysis are seen in the large majority of nerve cells (Levi-Montalcini, Caramia and Angeletti, 1969). In adult rodents the antiserum to NGF induces marked atrophic changes in the ganglionic cell bodies characterized by loss of neurofibrillar material, marked decrease in noradrenaline content and decrease in this catecholamine in peripheral organs (Angeletti, Levi-Montalcini and Caramia, 1971). These effects are, however, of a temporary nature, at variance with those produced in neonatal rodents where the destruction of sympathetic ganglia is irreversible. Recent electrophysiological studies on adult guinea pigs, injected with AS-NGF for a 4-5-day period, showed depression of intracellularly recorded synaptic response within 4-5 days after the end of the antiserum administration. Synapses counted in the E.M. showed only half as many contacts as in control ganglia (Njä and Purves, 1978).

10.5.2 *Chemical sympathectomy*

The possibility of selectively destroying by means of pharmacological agents sympathetic nerve cells in neonatal rodents was first reported from this laboratory in 1970 (Angeletti and Levi-Montalcini, 1970a). It was shown that injections of 6-hydroxydopamine (6-OHDA), a dopamine derivative which in adult animals produces a degeneration of the synaptic vesicles in adrenergic nerve endings and a long-lasting blocking effect of the sympathetic transmission of the nerve impulse (Thoenen and Tranzer, 1968), in neonatal animals produces massive destruction of immature sympathetic nerve cells. This process became known as 'chemical sympathectomy' (Angeletti and Levi-Montalcini, 1970b). The destruction of immature sympathetic nerve cells by 6-OHDA is not due to a direct cytotoxic effect of this compound on the cell perikaryon. E.M. studies showed that upon subcutaneous injections, 6-OHDA accumulates in the axonal endings in the immature as well as in the fully differentiated sympathetic nerve cells and does not diffuse along the axon or gain direct access to the perikaryon through the cell membrane as demonstrated by the fact that both nuclear and cytoplasmic compartments do not show any degenerative signs. Death of immature sympathetic nerve cells following 6-OHDA injection does not result therefore from critical lesions of the cell's vital centre, but from the chemical disconnection of the growing nerve fibre from its end organs. The cytotoxic lesions produced by 6-OHDA in the adrenergic endings prevent the retrograde axonal transport of the trophic factor released by peripheral tissues, thus causing death of the cells depending on NGF for their subsistence during the active growth phase (Levi-Montalcini and Aloe, 1980).

The much greater vulnerability of developing nerve cells compared with fully differentiated neurons in rodents injected with AS-NGF or 6-OHDA in the first postnatal week rather than in adult life was also apparent in treatments with guanethidine, a compound which interferes with the release of noradrenaline, and with the antimitotic alkaloid vinblastine. Guanethidine gains access and accumulates within adrenergic neurons by the same specialized membrane transport system that is responsible for the uptake of noradrenaline (Mitchell and Oates, 1970). It was shown that mitochondria are the first organelles to undergo regressive changes (Fig. 10.3) upon guanethidine administration (Angeletti and Levi-Montalcini, 1971). The conclusion was reached that, upon intracellular accumulation of guanethidine in the sympathetic neuron, the cell is destroyed because of inability to generate ATP and/or sequester calcium below toxic levels (Juul and Sand, 1973). At 8 days practically no intact nerve cells are seen in the ganglia and the residual neuronal population amounts to less than one tenth that of controls (Eränkö and Eränkö, 1971; Burnstock, Evans, Gannon, Heath and James, 1971; Angeletti and Levi-Montalcini, 1972).

A third drug, vinblastine was recently the object of studies *in vitro* and *in vivo* of the ability of the vinca alkaloid to favour microtubule formation in conditions where tubulin dimers are not able to self-assemble, and of the effects of this drug on immature and fully differentiated sympathetic nerve cells in living organisms (Calissano *et al.*, 1976a). Vinblastine prevents microtubule and microfilament polymerization *in vitro* and *in vivo* and in this way it interferes with processes that are essential in the most diversified cell functions such as cell replication, anterograde and retrograde axonal transport (Wilson, Bryan, Ruby and Mazia, 1970). While fully differentiated sympathetic neurons undergo reversible damage, immature nerve cells are destroyed by daily injections of 1.0 nM-vinblastine/g body weight (Calissano *et al.*, 1976a). Studies with the E.M. at first conveyed the impression that the earliest lesions produced by vinblastine in sympathetic nerve cells are localized in the cell nucleus (Menesini Chen, Chen, Calissano and Levi-Montalcini, 1977). A reinvestigation of this process (Aloe, unpublished results) showed that the earliest lesion caused by vinblastine consists instead of marked swelling and

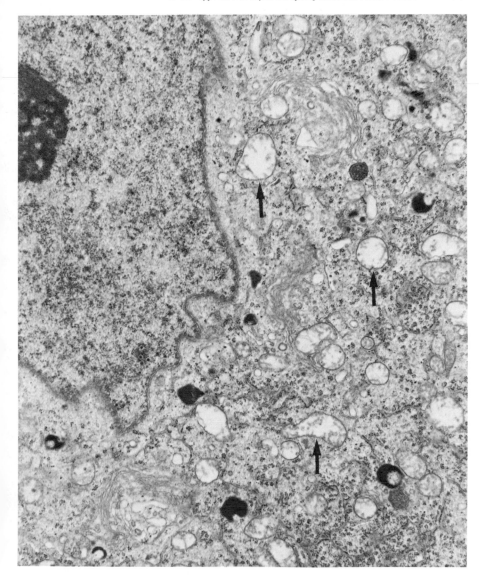

Fig. 10.3 Electron micrograph of a sympathetic nerve cell in SCG of 2-day-old rat injected on the first day of life with guanethidine. Arrows point to dilated and grossly altered mitochondria. Magnification 23 000 ×.

disruption of microtubules (Fig. 10.4). In immediately subsequent stages, alterations in the nuclear compartment, strikingly similar to those that occur in the same cells upon AS-NGF injection, become apparent. In the case of the vinblastine treatment, they follow rather than precede disruption of axonal microtubules.

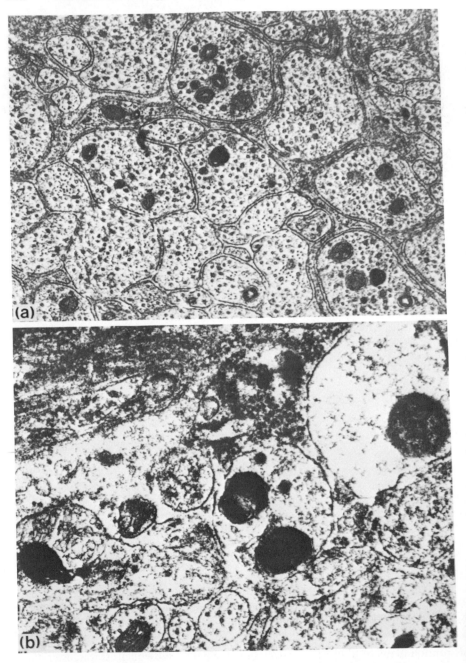

Fig. 10.4 Electron micrographs of cross-sections of postganglionic fibres of SCGs of 5-day-old rats injected at day 4 with saline(a) or with vinblastine(b). Note in (b) massive destruction of microtubules in the swollen axons. Magnification 24 000 ×.

10.6 Surgical axotomy

Surgical axotomy of the two main postganglionic roots in SCGs of neonatal rats results in death of 90 per cent of sympathetic nerve cells (Hendry and Campbell, 1976). The process is irreversible, as shown by the fact that the ganglia persist as diminutive sclerotic nodules in the fully differentiated animal. Light-microscopic and electron-microscopic studies showed the striking similarity in the devastating effects produced by 6-OHDA (chemical axotomy) and surgical axotomy of the superior cervical ganglion. The only obvious difference is that 6-OHDA affects to the same extent all sympathetic para- and pre-vertebral ganglia, whereas the lesions produced by surgical axotomy of postganglionic roots of the SCG are restricted to nerve cells of this ganglion. In both experimental groups the nuclear and cytoplasmic compartments do not show regressive changes at least for the first 40 h subsequent to the intervention. However, in both situations, by 24 h after surgical or chemical axotomy, one sees detachment of apparently intact nerve endings from the nerve cell surface (Aloe, L. and Levi-Montalcini, R., 1979b).

10.7 Protective effects of NGF against 6-OHDA, guanethidine, vinblastine, AS-NGF and surgical axotomy

The protective effects of NGF against the otherwise destructive effects produced by the above treatments differ markedly from each other in the sequence of events which follow the exogenous supply of NGF and in the growth response produced in the 'rescued nerve cells'. Only in the cases of chemical surgical axotomy are the effects similar if not identical. Since the differences are perhaps even more revealing than the similarities in elucidating some aspects of the mechanism of NGF action on the mortally injured nerve cells, we shall at first consider the most significant features of each one of the above combined treatments.

10.7.1 *NGF and 6-OHDA*

The concomitant daily treatment of rats immediately after birth with NGF ($10 \mu g/g$ body weight) and 6-OHDA ($100 \mu g/g$ body weight) results in paradoxical increase in volume of the ganglia. Thus after the combined treatment for 3 weeks the volume of sympathetic ganglia is 30 times that of controls and 2.3 times that of NGF-injected littermates (Levi-Montalcini, Aloe, Mugnaini, Oesch and Thoenen, 1975: Aloe *et al.*, 1975). This effect illustrated in Fig. 10.5(a), is due to hyperplastic and hypertrophic effects only slightly inferior to those evoked by injections of NGF alone. The enormous volume increase in the ganglia is due mainly to an extraordinary production of collateral fibres branching out from the proximal segment of the chemically axotomized nerve fibre. These collaterals find accommodation inside the ganglia where they displace nerve cells and build a thick fibrillar capsule around the same ganglia. The survival in the excellent conditions of sympathetic nerve cells treated with NGF but prevention by 6-OHDA from establishing anatomical and functional connections with their end organs suggests

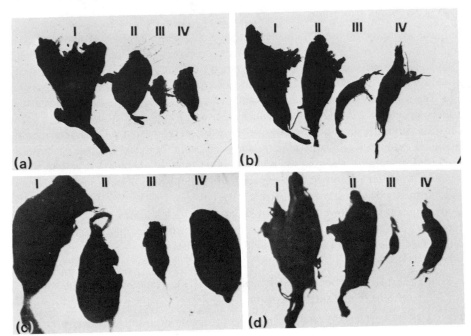

Fig. 10.5 Comparison of whole mounts of SCG of 2-week-old rats injected as follows: (a) NGF + 6-hydroxydopamine (I), NGF alone (II), 6-hydroxydopamine alone (III), saline (IV). Magnification 10×. (From Aloe *et al.*, 1975.) (b) NGF + guanethidine (I), NGF alone (II), guanethidine alone (II), saline (IV). Magnification 12×. (c) NGF + vinblastine (I), NGF alone (II), vinblastine alone (III), saline (IV). Magnification 20×. (From Calissano *et al.*, 1976a). (d) NGF + axotomy (I), NGF alone (II), axotomy (III), saline (IV). Magnification 12×.

that the growth factor finds access to its target cells through the receptors of the cell membrane. The exuberant production of collaterals by the chemically axotomized sympathetic neurons brings to light growth potentialities of these fibres which materialize only upon disconnection of their endings from peripheral tissues.

10.7.2 *NGF and guanethidine*

The combined NGF and guanethidine treatment also results in survival of nerve cells and in a volume increase of sympathetic ganglia only slightly inferior to that elicited by injections of NGF alone (Fig. 10.5b) (Johnson and Aloc, 1974). The protective effect of NGF does not appear to result from inhibition of guanethidine accumulation in the ganglia, since discontinuation of NGF treatment is followed by massive degeneration of nerve cells. In prolonged NGF and guanethidine treatments nerve cells exhibit severe cytoplasmic alterations even in the presence of NGF. The lethal guanethidine effects on immature and mature sympathetic nerve cells could be due to inhibition of calcium uptake into the mitochondria leading to an increase in free Ca^{2+}, which exerts its highly toxic effects in the cells where it accumulates. The ability of NGF to overcome cytotoxicity of guanethidine may be the result of two possible effects: (a) increased extrusion of Ca^{2+}; (b) NGF counteraction of the depolymerizing effects of calcium on contractile proteins. It has in fact been shown that even very low levels of free Ca^{2+} affect several intracellular processes and in particular the assembly and organization of microtubules and microfilaments. NGF completely prevents the effects of Ca^{2+} by preserving preformed microtubules, even in the presence of Ca^{2+} concentrations as high as 1 mmol/litre, a concentration which in 30 min would result in the total disorganization of microtubules (Calissano, unpublished studies).

Since recent studies showed that NGF has no significant influence on rates of calcium uptake and efflux (Landreth *et al.*, 1980), a protective effect of NGF against guanethidine is more likely due to its property of counteracting the depolymerizing effects of calcium on microtubules and microfilaments.

10.7.3 *NGF and vinblastine*

Nerve cells of newborn mice injected simultaneously with NGF and vinblastine exhibit none of the severe axonal and nuclear lesions produced by vinblastine alone. All cell compartments appear intact and microtubules and microfilaments even undergo an increase comparable with that elicited by NGF alone. This growth factor does not exert its protective effects by preventing the access of vinblastine to the ganglia, as shown by the fact that binding of tritiated vinblastine to ganglia in the combined vinblastine and NGF treatment is 4-fold higher in these than in control ganglia in 15-day-old mice (Fig. 10.5c), an effect that correlates with the increase in cell size and number in the former as compared with controls (Menesini Chen *et al.*, 1977). On the basis of the previous studies *in vitro* (Calissano *et al.*, 1976a) it is suggested that NGF prevents death of immature sympathetic nerve cells, by favouring the assembly or organization of microtubules which in the organized state are inaccessible to vinblastine (Wilson *et al.*, 1975). An alternative explanation of the protective effects of NGF against vinblastine, submitted by Johnson and co-workers (Johnson, 1978; Johnson, Macia, Andres and Bradshaw, 1979), is that the vinca alkaloid inhibits retrograde transport of NGF and thereby causes death of sympathetic neurons. Cell death is prevented by exogenous administration of NGF. According to these authors both 6-OHDA and vinblastine would block retrograde axonal transport through a similar mechanism. The markedly different damage produced by these two compounds in different cell compartments of the developing sympathetic nerve cells seem to us difficult to reconcile with such an unitary hypothesis, which also neglects the important role of membrane receptors in making NGF available to its target cells.

A still different situation which suggests an entirely different mechanism of action of the growth factor in counteracting lesions incompatible with the life of immature sympathetic nerve cells will be considered in the following section.

10.7.4 *NGF and AS-NGF*

Immunosympathectomy differs from chemical sympathectomy produced by 6-OHDA, guanethidine and vinblastine, not only in the different localization of the earliest lesions detectable in the immature sympathetic nerve cells, but also in the much more precocious materialization of the degenerative events leading to nerve cell death upon injections of AS-NGF than upon administration of one of the three pharmacological compounds listed above.

Ever since the discovery of the devastating effects produced by AS-NGF, the question was raised whether they were due to a cytotoxic complement-mediated effect or to depletion of circulating NGF by the specific antibodies.

Early studies by Sabatini, De Iraldi and De Robertis (1965) and by our group (Levi-Montalcini *et al.*, 1969) favoured the first alternative, while the prevailing viewpoint of more recent investigators (Ennis, Pearce and Vernon, 1979; Saito and Thoenen, unpublished observations) is that depletion of circulating NGF rather than the cytotoxic effects of AS-NGF accounts for the effects elicited by antibodies to NGF. NGF depletion rather than cytotoxic action is indicated, according to Thoenen (1979), by the finding that injections of NGF up to 24 h after the administration of NGF antibody prevent the destructive effects of AS-NGF. This delayed protective effect would not be compatible with a complement-mediated cytotoxic action of NGF antibodies. In view of the interest that attaches to the elucidation of this problem, we reinvestigated the effect delayed injections of NGF after the AS-NGF treatment. Although the results of these investigations will be considered in detail elsewhere, it is of interest to report on some preliminary findings, since they suggest an entirely different mechanism of action of NGF from those envisioned above. At variance with previous studies, we injected neonatal mice only once on the day of birth with AS-NGF and inspected the SCGs and other sympathetic ganglia on subsequent days, from the end of the first day to the end of the first postnatal week. The decrease in size of the ganglia of injected mice was in the same range as upon more prolonged AS-NGF treatment. Light microscopic studies performed 72 h after the AS-NGF injection showed that the ganglia consisted of highly atrophic densely packed nerve cells and of normal glia and satellite cells. When NGF injections were started 24, 48 or even 72 h after the AS-NGF treatment, the ganglia appeared at the external inspection to be slightly larger than those of untreated littermates, but smaller than ganglia of mice injected since birth with only NGF. At the histological inspection, nerve cells in ganglia of 8-day-old mice submitted to the AS-NGF treatment followed by daily injections of NGF, appeared hypertrophic, but to a lesser extent than neurons of ganglia of littermates injected since birth with NGF.

The findings could be explained on the basis of a 'rescue effect' by NGF on severely damaged but still viable nerve cells. Two alternative hypothesis are however not ruled out: (a) NGF promotes proliferation of nerve cells precursors and the newly formed nerve cells replace the dead nerve cells and this effect accounts for the finding of an even larger than normal neuronal cell population, (b) NGF channels still uncommitted cells toward the neuronal rather than other cell lines. Studies in progress are aimed at providing a definite answer to this long debated but still unsettled question of whether NGF is endowed with a proliferative effect on nerve-cell precursors beside its well-known differentiative effect on immature sympathetic nerve cells.

10.7.5 *NGF and surgical axotomy*

The prevention by NGF of the near-total destruction of immature sympathetic nerve cells of superior cervical ganglia in neonatal rats submitted to surgical transection of the two main postganglionic roots (Hendry and Campbell, 1976) was recently reinvestigated by us (Aloe and Levi-Montalcini, 1976b). These studies brought to light similarities and differences between the effects elicited by NGF in chemically and surgically axotomized ganglia of neonatal rats. Since the differences are perhaps even more revealing than the similarities for the understanding of the mechanism of NGF action in peripherally lesioned nerve cells, we shall focus attention on the former. The volume of sympathetic ganglia of rats submitted to the combined 6-OHDA and NGF treatment for the first 3 postnatal weeks is, as reported on p. 313, about 30-fold that of control ganglia, whereas the maximum volume increase elicited by NGF in surgically axotomized ganglia is eight times that of ganglia of untreated littermates (Fig. 10.5d). Histological and histofluorescence studies explained the cause of this less-prominent growth effect elicited by NGF in surgically rather than chemically axotomized nerve cells. The enormous volume increase produced by NGF in 6-OHDA-injected neonatal rats is to a large extent due to accumulation of collaterals of the chemically axotomized axons, prevented by this compound from undergoing further elongation and innervating the periphery. On the other hand, NGF accelerates the regrowth of the transected fibre in surgically axotomized nerve cells, as indicated by the reinnervation of the iris. The channelling of the newly synthesized fibrillar proteins in the stem axon limits the overproduction of collaterals from the proximal segment of the neurite. In both experimental sets, NGF gains access to the axotomized nerve cells through membrane receptors, and perhaps also through the preterminal non-lesioned axonal segment.

10.8 Some considerations and concluding remarks

... 'the almost universal ability of the experimenter to ignore the possible existence of unknown premises and to argue as if they were dealing with a logically closed system, suggests a lack of respect for the complexity of the nervous system that is rather surprising' (Gaze in *The Formation of Nerve Connections*, p. 116, 1970).

A viewpoint similar to that formulated by Gaze was expressed by another neurologist Luria: 'In any deductive activity, and particularly in science, the law that we can call the 'law of disregard of negative information' holds good: facts that fit into a preconceived hypothesis attract attention, are singled out, and are remembered: facts that are contrary to it are disregarded, treated as 'exceptions and forgotten' (Luria, 1975).

Studies on NGF-target cell interaction illustrate both principles stated above: the general tendency to visualize complex stimulus–response systems as if they were amenable to one rather than multiple processes and at the same time to ignore the evidence that does not fit in the preconceived scheme. In the following we shall comment only on the most controversial issues considered in this article.

The currently prevailing view that the NGF hyperplastic effects can be explained as the result of 'prevention of natural cell death' does not take into consideration the discrepancy between the magnitude of the numerical increase in the neuronal cell population produced by NGF in sympathetic ganglia of neonatal rodents and the relatively moderate nerve cell loss which occurs in the same ganglia during the first postnatal week and has been attributed to the phenomenon known as 'natural cell death'. Pending additional data from experiments in progress, we feel that the information now available suggests that the numerical increase in nerve cells in sympathetic ganglia of neonatal rodents may result from multiple responses to NGF: prevention of 'natural cell death' and enhanced proliferative activity of nerve cell precursors. When NGF is supplied to avian embryos or rodent foetuses, an additional cause for the enormous numerical increase in sympathetic nerve cells may also be traced to a third effect: channeling of common stem cell precursors of chromaffin and sympathetic nerve cells toward the latter rather than the former cell line.

Since NGF is endowed with the property of counteracting natural nerve cell death as well as experimentally induced nerve cell death, as illustrated in the second part of this article, it is of interest to see whether the processes set in motion by NGF in these different experimental conditions are amenable to the same or to different mechanisms. First of all we shall briefly comment on the phenomenon that is generally referred to as 'natural cell death'. Its occurrence during neurogenesis has been the object of extensive studies ever since it came to our attention in the nervous system of chick embryos (Hamburger and Levi-Montalcini, 1949; Levi-Montalcini, 1950). It may consist of the sporadic loss of immature nerve cells in developing sensory ganglia, as described in the first of the two articles quoted above, or it may result from the sudden breakdown of about two-thirds of the neuronal cell population, as in the case of the disappearance of developing nerve cells located in the cervical segment of the spinal cord of chick embryos, as reported in the second of the two above-mentioned articles.

The sporadic loss of immature nerve cells, described in subsequent years in a large number of developing nerve centres in amphibian and avian embryos (Glücksmann, 1951; Hamburger, 1958, 1975; Hughes, 1968; Prestige, 1970; Cowan, 1973; Pittman *et al.*, 1978; Laing and Prestige, 1978), has been attributed by most authors to the failure of all developing nerve cells to establish morphological or functional connections with postsynaptic cells. As such, it would fit well into the concept of 'survival of the fittest' and reflect a probabilistic or epigenetic event. Since nerve cells are produced in excess of available synaptic sites, the losers in this competitive race would be doomed to death. The abrupt and massive death of motor nerve cells in the cervical segment of the spinal cord of chick embryos between the fourth and the fifth day of incubation exhibits entirely different features from those mentioned above, and has been attributed to the disappearance of an abortive visceromotor preganglionic column (Levi-Montalcini, 1950). As such it would fit into the concept of a 'programmed' rather than a probabilistic developmental event. Recent ultrastructural studies give support to the hypothesis that cell death in developing nerve centres may result from entirely different processes. A detailed E.M. study by O'Connor and Wyttenbach (1974) of the degenerative events in nerve cells of the putative visceromotor column of the cervical region of the spinal cord of 4–5-day chick embryos showed that the initial set of degenerative changes in motor nerve cells include a decrease in nuclear size, clumping of the chromatin beneath the nuclear envelope, an increase in electron opacity of the cells, disappearance of Golgi bodies and disaggregation of polysomes. Normally occurring cell death, examined at the ultrastructural level in embryonic chick ciliary ganglion [attributed to the failure of supernumerary cells to establish synapsis with the peripheral target organ (Pilar and Landmesser, 1976)] consists at first of dilation of the rough endoplasmic reticulum, with eventual cytoplasmic disruption, nuclear changes appearing only secondarily. The authors suggest that the failure to form or maintain peripheral synapses could result in the accumulation of transmission-related proteins with consequent cisternal dilation and eventual cell death. These findings would therefore suggest that entirely different processes may account for cell death during neurogenesis. In this connection it is of interest to call attention to the striking similarity between the morphological features exhibited by the massive cell death that occurs in the motor column of the cervical segment of the spinal cord of avian embryos and the similarly dramatic disappearance of entire nerve cell populations in the cerebellum or other cerebral segments in mutant mice, which was the object of extensive and most revealing studies by Sidman and co-workers (Sidman, 1972).

In the second part of this article we reported on the protective effects of NGF and prevention of the massive destruction of immature sympathetic nerve cells resulting from injections of different pharmacological compounds or of specific antibodies to NGF. While the prevailing view expressed by other authors (Hendry *et al.*, 1974; Thoenen *et al.*, 1979; Ennis *et al.*, 1979) is that in all the above instances, with perhaps the exception of the mechanism of action of guanethidine (Johnson *et al.*, 1979), the cause of death is prevention of retrograde NGF transport by these agents, we believe that this mechanism is likely to account for only the protective action by NGF of nerve cells lethally injured by chemical or surgical axotomy. We have already discussed and need not reconsider the putative protective mechanisms of NGF against guanethidine, vinblastine and specific antibodies to this protein molecule endowed with such potent and specific nerve-growth-promoting activity. In this connection it is worth calling attention to the still different mechanism set in motion by NGF in the neoplastic pheochromocytoma cell line PC12. In this instance, NGF does not counteract a lethal effect but channels a malignant neoplastic cell toward a 'normal' rather than deviant cell type. It is planned to examine the possible protective effect of NGF against other equally damaging, if not necessarily lethal, agents such as viral infections in sensory and sympathetic ganglia.

As a final remark it may well be worthwhile calling attention to the most valuable contribution provided by studies of 'natural' or experimentally induced cell death to our knowledge of the mechanisms responsible for the diametrically opposite effect, namely cell growth and differentiation.

Note added in proof

On page 316 we submitted, as one of the alternative hypotheses to NGF prevention of natural death, a NGF proliferative effect on nerve cell precursors in ganglia of neonatal rodents. Thymidine labelled experiments performed after the completion of the manuscript confirm the previous findings by J. Hendry, that NGF does not promote proliferative activity among immature sympathetic nerve cells. We therefore agree with this author concerning the lack of NGF mitotic acticity in sympathetic nerve cell precursors.

Acknowledgements

This investigation was in part supported by the Progetto Finalizzato Biologia della Riproduzione of the Italian National Council of Research (CNR).

References

Aloe, L. and Levi-Montalcini, R. (1979a), Nerve Growth Factor *in vivo* induced transformation of immature chromaffin cells in sympathetic neurons: effects of antiserum to the Nerve Growth Factor. *Proc. Natl. Acad. Sci. U.S.A.,* **76**, 1246–1250.

Aloe, L. and Levi-Montalcini, R. (1979b), Nerve growth factor induced overgrowth of axotomized superior cervical ganglia in neonatal rats: similarities and differences with NGF effects in chemically axotomized sympathetic ganglia. *Arch. Ital. Biol.,* **117**, 287–307.

Aloe, L., Mugnaini, E. and Levi-Montalcini, R. (1975), Light and electromicroscopic studies of the excessive growth of sympathetic ganglia in rats injected daily from birth with 6-OHDA and NGF. *Arch. Ital. Biol.,* **113**, 326–353.

Andres, R. Y., Jeng, I. and Bradshaw, R. A. (1977), NGF receptors: identification of distinct classes in plasma membranes and nuclei of embryonic dorsal root neurons. *Proc. Natl. Acad. Sci. U.S.A.,* **74**, 2785–2789.

Angeletti, P. U. and Levi-Montalcini, R. (1970a), Sympathetic nerve cell destruction in newborn mammals by 6-hydroxydopamine. *Proc. Natl. Acad. Sci. U.S.A.,* **65**, 114–121.

Angeletti, P. U. and Levi-Montalcini, R. (1970b), Specific cytotoxic effect of 6-hydroxydopamine on sympathetic neuroblasts. *Arch. Ital. Biol.,* **108**, 213–221.

Angeletti, P. U. and Levi-Montalcini, R. (1972), Growth inhibition of sympathetic cells by some adrenergic blocking agents. *Proc. Natl. Acad. Sci. U.S.A.,* **69**, 86–88.

Angeletti, P. U., Levi-Montalcini, R. and Calissano, P. (1968), The nerve growth factor: chemical properties and metabolic effect. *Adv. Enzymol. Relat. Areas Mol. Biol.,* **31**, 51–75.

Angeletti, P. U., Levi-Montalcini, R. and Caramia, F. (1971), Analysis of the effects of the antiserum to the nerve growth factor in adult mice. *Brain Res.,* **27**, 343–355.

Angeletti, P. U., Liuzzi, A., Levi-Montalcini, R. and Gandini Attardi, D. (1964), Effects of a nerve growth factor on glucose metabolism by sympathetic and sensory nerve cells. *Biochim. Biophys. Acta,* **90**, 445–450.

Angeletti, R. H. and Bradshaw, R. A. (1971), Nerve growth factor from mouse submaxillary gland: amino acid sequence. *Proc. Natl. Acad. Sci. U.S.A.,* **68**, 2417–2420.

Auerbach, R., Kubai, L., Klinghton, D. and Folkman, J. (1974), A simple procedure for long-term cultivation of chick embryos. *Dev. Biol.,* **41**, 391–394.

Banerjee, S. P., Snyder, S. H., Cuatrecasas, P. and Greene, L. A. (1973), Binding of nerve growth factor receptor in sympathetic ganglia. *Proc. Natl. Acad. Sci. U.S.A.,* **70**, 2519–2523.

Bocchini, V. and Angeletti, P. U. (1969), The nerve growth factor: purification as a 30,000 molecular weight protein. *Proc. Natl. Acad. Sci. U.S.A.,* **64**, 787–794.

Bray, D. and Thomas, C. (1976), Unpolymerized actin in tissue cells. In *Cell Motility, Book B* (eds. R. Goldman, T. Pallard and J. Rosenbaum), Cold Spring Harbor Laboratory, pp. 461–474.

Bueker, E. D. (1948), Implantation of tumors in the hind limb field of the embryonic chick and developmental response of the lumbosacral nervous system. *Anat. Rec.,* **102**, 369–390.

Burnstock, G., Evans, B., Gannon, B. J., Heath, J. W. and James, V. (1971), A new method of destroying adrenergic nerves in adult animals using guanethidine. *Br. J. Pharmacol.,* **43**, 295–301.

Calissano, P. and Cozzari, C. (1974), Interaction of NGF with the mouse-brain neurotubule protein(s). *Proc. Natl. Acad. Sci. U.S.A.,* **71**, 2131–2135.

Calissano, P., Levi, A., Alemà, S., Chen, J. S. and Levi-Montalcini, R. (1976a), Studies on the interaction of the Nerve Growth Factor with tubulin and actin. In *26 Colloquium Mosbach, Molecular Basis of Motility* (eds. L. Heilmeyer, J. C. Ruegg and Th. Wieland), Springer-Verlag, Berlin and Heidelberg, pp. 186–202.

Calissano, P., Monaco, G., Castellani, L., Mercanti, D. and Levi, A. (1978), Nerve growth factor potentiates actinomyosin adenosine triphosphatase. *Proc. Natl. Acad. Sci. U.S.A.,* **75**, 2210–2214.

Calissano, P., Monaco, G., Levi, A., Menesini-Chen, M. G., Chen, J. S. and Levi-Montalcini, R. (1976b), New developments in the study of NGF–tubulin interaction. In *Contractile Systems in Non-Muscle Tissue* (eds. S. W. Perry, A. Margret and R. S. Adelstein), Elsevier/North-Holland, New York, pp. 201–211.

Calissano, P. and Shelanski, M. L. (1980), Interaction of nerve growth factor with pheochromocytoma cells. Evidence for tight binding and sequestration. *Neuroscience*, (in press).

Campenot, R. (1977), Local control of neurite development by nerve growth factor. *Proc. Natl. Acad. Sci. U.S.A.,* **74**, 4516–4519.

Cohen, S. (1958), A nerve growth-promoting protein. In *Chemical Basis of Development* (eds. W. D. McElroy and S. Glass), Johns Hopkins Press, Baltimore, pp. 665–667.

Cohen, S. (1959), Purification and metabolic effects of a nerve growth promoting protein from snake venom. *J. Biol. Chem.,* **234**, 1129–1137.

Cohen, S. (1960), Purification of a nerve growth promoting protein from the mouse salivary gland and its neuro-cytoxic antiserum. *Proc. Natl. Acad. Sci. U.S.A.,* **46**, 302–311.

Cohen, S. and Levi-Montalcini, R. (1956), A nerve growth stimulating factor isolated from snake venom. *Proc. Natl. Acad. Sci. U.S.A.,* **42**, 571–574.

Cohen, S., Levi-Montalcini, R. and Hamburger, V. (1954), A nerve growth stimulating factor isolated from sarcomas 37 and 180. *Proc. Natl. Acad. Sci. U.S.A.,* **40**, 1014–1018.

Cowan, M. V. (1973), Neuronal death as a regulative mechanism in the control of cell number in the nervous system. In *Development and Aging in the Nervous System* (eds. M. Rockstein and L. M. Sussman), Academic Press, New York and London, pp. 19–41.

Crain, S. M. and Wiegand, R. G. (1961), Catecholamine levels of mouse sympathetic ganglia following hypertrophy produced by salivary nerve-growth factor. *Proc. Soc. Exp. Biol. Med.,* **107**, 663–665.

Ennis, M., Pearce, F. L. and Vernon, C. A. (1979), Some studies on the mechanisms of action of antibodies to nerve growth factor. *Neuroscience*, 4, 1391–1398?

Eränkö, O. and Eränkö, L. (1971), Histochemical evidence of chemical sympathectomy by guanethidine in newborn rats. *Histochem. J.*, 3, 451–456.

Fabricant, R. N., De Larco, J. E. and Todaro, G. J. (1977), Nerve growth factor receptors on human melanoma cells in culture. *Proc. Natl. Acad. Sci. U.S.A.*, 74, 565–569.

Feit, H. and Barondes, S. H. (1970), Colchicine-binding activity in particulate fractions of mouse brain. *J. Neurochem.*, 17, 1355.

Fine, E. R. and Bray, D. (1971), Actin in growing nerve cells. *Nature (London) New Biol.*, 234, 115–118.

Frazier, W. A., Ohlendorf, C. E., Boyd, L. F., Aloe, L., Johnson, E. M., Ferrendelli, J. A. and Bradshaw, R. A. (1973) Mechanism of action of nerve growth factor and cyclic AMP on neurite outgrowth in embryonic chick sensory ganglia. Demonstration of independent pathways of stimulation. *Proc. Natl. Acad. Sci. U.S.A.*, 70, 2448–2452.

Garrels, I. J. and Schubert, D. (1979), Modulation of protein synthesis by nerve growth factor. *J. Biol. Chem.*, 254, 7978–7985.

Gaze, R. N. (1970), The formation of nerve connections. A consideration of neural specificity modulation and comparable phenomena. Academic Press, New York.

Glücksmann, A. (1951), Cell death in normal vertebrate ontogeny. *Biol. Rev.*, 26, 59–86.

Goldstein, L. J., Anderson, G. W. R. and Brown, S. M. (1979), Coated pits, coated vesicles, and receptor-mediated endocytosis. *Nature (London)*, 279, 679–685.

Greene, L. A. and Tischler, A. S. (1976), Establishment of a noradrenergic clonal line of rat adrenal pheochromocytoma cells which respond to nerve growth factor. *Proc. Natl. Acad. Sci. U.S.A.*, 73, 2424–2428.

Gundersen, W. R. and Barret, N. J. (1979), Neuronal chemotaxis: Chick dorsal-root axons turn toward high concentrations of nerve growth factor. *Science*, 206, 1079–1080.

Hamburger, V. (1958), Regression versus peripheral control of differentiation in motor hypoplasia. *Am. J. Anat.*, 102, 365–410.

Hamburger, V. (1975), Cell death in the development of the lateral motor column of the chick embryo. *J. Comp. Neur.*, 160, 535–546.

Hamburger, V. and Levi-Montalcini, R. (1949), Proliferation, differentiation and degeneration in the spinal ganglion of the chick under normal and experimental conditions. *J. Exp. Zool.*, 111, 457–501.

Hatanaka, H., Otten, U. and Thoenen, H. (1978), NGF-mediated selective induction of ornithine decarboxylase in rat pheochromocytoma; a cyclic AMP-independent process. *FEBS Lett.*, 92, 313–315.

Hendry, I. A. (1977), Cell division in the developing sympathetic nervous system. *J. Neurocytol.*, 6, 299–309.

Hendry, I. A. and Campbell, J. (1976), Morphometric analysis of rat superior cervical ganglion after axotomy and nerve growth factor treatment. *J. Neurocytol.*, 5, 351–360.

Hendry, I. A., Stöckel, K., Thoenen, H. and Iversen, L. L. (1974), The retrograde axonal transport of nerve growth factor. *Brain Res.*, 68, 103–121.

Herrup, K. and Shooter, M. E. (1973), Properties of the β nerve growth factor receptors of avian dorsal root ganglia. *Proc. Natl. Acad. Sci. U.S.A.*, 70, 3884–3888.

Hughes, A. (1968), Aspects of neural ontogeny. Academic Press, New York.

Johnson, E. M. and Aloe, L. (1974), Suppression of the *in vitro* and *in vivo* cytotoxic effects guanethidine in sympathetic neurons by Nerve Growth Factor. *Brain Res.*, **81**, 519–532.

Johnson, E. M., Macia, A. R., Andres, R. Y. and Bradshaw, A. R. (1979), The effects of drugs which destroy the sympathetic nervous system on the retrograde transport of nerve growth factor. *Brain Res.*, **171**, 461–472.

Johnson, M. J. (1978), Destruction of the sympathetic nervous system in neonatal rats and hamsters by vinblastine: Prevention by concomitant administration of nerve growth factor. *Brain Res.*, **141**, 105–118.

Juul, P. and Sand, O. (1973), Determination of guanethidine in sympathetic ganglia. *Acta Pharmacol. Toxicol.*, **32**, 487–499.

Kuczmarski, R. E. and Rosenbaum, L. (1979), Chick brain actin and myosin. *J. Cell Biol.*, **80**, 341–355.

Laing, N. G. and Prestige, M. C. (1978), Prevention of spontaneous motoneurone death in chick embryos. *J. Physiol. (London)*, **282**, 33–34.

Landreth, G., Cohen, P. and Shooter, M. E. (1980), Ca^{2+} transmembrane fluxes and nerve growth action on clonal cell line of rat pheochromocytoma. *Nature (London)*, **283**, 202–204.

Larrabee, M. G. (1972), Metabolism during development in sympathetic ganglia of chickens: effects of age, Nerve Growth Factor and metabolic inhibitors. In *Nerve Growth Factor and its Antiserum* (eds. E. Zaimis and J. Knight), The Athlone Press of the University of London, pp. 71–88.

Letourneau, P. C. (1978), Chemotactic response of nerve fibre elongation to Nerve Growth Factor. *Dev. Biol.*, **66**, 183–196.

Levi, A., Cimino, M., Mercanti, D., Chen, J. S. and Calissano, P. (1975), Interaction of nerve growth factor with tubulin. Studies on binding and induced polymerization. *Biochim. Biophys. Acta*, **339**, 50–60.

Levi, A., Shechte, Y., Neufeld, J. and Schlessinger, J. (1980), Mobility, clustering and transport of nerve growth factor in the embryonal sensory cells and in a sympathetic neuronal cell line. *Proc. Natl. Acad. Sci. U.S.A.*, **77**, 3469–3473.

Levi-Montalcini, R. (1950), The origin and development of the visceral system in the spinal cord of the chick embryo. *J. Morphol.*, **86**, 253–283.

Levi-Montalcini, R. (1952), Effects of mouse tumor transplantation on the nervous system. *Ann. N. Y. Acad. Sci.*, **55**, 330–343.

Levi-Montalcini, R. (1958), Chemical stimulation of nerve growth. In *Chemical Basis of Development* (eds. W. D. McElroy and B. Glass), Johns Hopkins Press, Baltimore, pp. 646–664.

Levi-Montalcini, R. (1966), The nerve growth factor: its mode of action on sensory and sympathetic nerve cells. *Harvey Lect. Ser.*, **60**, 217–259.

Levi-Montalcini, R. and Aloe, L. (1980), Tropic, trophic and transforming effects of nerve growth factor. In *Histochemistry and Cell Biology of Autonomic Neurons, SIF Cells and Paraneurons* (eds. O. Eränkö, S. Soinila and H. Päivävinta) pp. 3–16.

Levi-Montalcini, R., Aloe, L., Mugnaini, E., Oesch, F. and Thoenen, H. (1975), Nerve growth factor induced volume increase and enhanced tyrosine hydroxylase synthesis in chemically axotomized sympathetic ganglia of newborn rats. *Proc. Natl. Acad. Sci. U.S.A.*, **72**, 595–599.

Levi-Montalcini, R. and Angeletti, P. U. (1963), Essential role of the nerve growth factor in the survival and maintenance of dissociated sensory and sympathetic nerve cells *in vitro*. *Dev. Biol.*, **7**, 655–659.

Levi-Montalcini, R. and Angeletti, P. U. (1966), Immunosympathectomy. *Pharmacol. Rev.*, **18**, 619–628.

Levi-Montalcini, R. and Booker, B. (1960a), Excessive growth of the sympathetic ganglia evoked by a protein isolated from mouse salivary glands. *Proc. Natl. Acad. Sci. U.S.A.*, **46**, 373-384.

Levi-Montalcini, R. and Booker, B. (1960b), Destruction of the sympathetic ganglia evoked by a protein isolated from mouse salivary glands. *Proc. Natl. Acad. Sci. U.S.A.*, **46**, 384-391.

Levi-Montalcini, R., Caramia, F. and Angeletti, P. U. (1969), Alterations in the fine structure of nucleoli in sympathetic neurons following NGF-antiserum treatment. *Brain Res.*, **12**, 54-73.

Levi-Montalcini, R., Caramia, F., Luse, S. A. and Angeletti, P. U. (1968), *In vitro* effects of the nerve growth factor on the fine structure of the sensory nerve cells. *Brain Res.*, **8**, 347-362.

Levi-Montalcini, R. and Hamburger, V. (1951), Selective growth-stimulating effects of mouse sarcoma on the sensory and sympathetic nervous system of the chick embryo. *J. Exp. Zool.*, **118**, 321-362.

Levi-Montalcini, R., Meyer, H. and Hamburger, V. (1954), *In vitro* experiments on the effects of mouse sarcoma 180 and 37 on the spinal and sympathetic ganglia of the chick embryo. *Cancer Res.*, **14**, 49-57.

Liuzzi, A., Angeletti, P. U. and Levi-Montalcini, R. (1965), Metabolic effects of a specific nerve growth factor (NGF) on sensory and sympathetic ganglia. Enhancement of lipid biosynthesis. *J. Neurochem.*, **12**, 705-708.

Luria, R. A. (1975) Neuropsychology: Its sources, principles and prospects. In *Neurosciences: Paths of Discovery* (eds. G. F. Worder, P. J. Swazey and G. Adelman), The MIT Press Cambridge, MA and London, U.K., pp. 335-362.

Marchisio, P. C., Naldini, L. and Calissano, P. (1980), Intracellular distribution of nerve growth factor in the rat pheochromocytoma PC12 cells: evidence for a peri- and intracellular location. *Proc. Natl. Acad. Sci. U.S.A.*, (in press).

Margulis, L. (1972), Colchicine sensitive microtubules. In *International Review of Cytology* (eds. G. H. Bourne and J. F. Danielli), Academic Press, New York, pp. 333-361.

Menesini Chen, M. G., Chen, J. S., Calissano, P. and Levi-Montalcini, R. (1977), Nerve growth factor prevents vinblastine destructive effects on sympathetic ganglia in newborn mice. *Proc. Natl. Acad. Sci. U.S.A.*, **74**, 5559-5563.

Menesini Chen, M. G., Chen, J. S. and Levi-Montalcini, R. (1978), Sympathetic nerve fibres ingrowth in the central nervous systems of neonatal rodents upon intracerebral NGF injection. *Arch. Ital. Biol.*, **116**, 53-84.

Mitchell, R. J. and Oates, A. J. (1970), Guanethidine and related agents. I: Mechanism of selective blockade of adrenergic neurons and its antagonism by drugs. *J. Pharmacol. Exp. Ther.*, **172**, 100-107.

Mobley, C. W., Server, A. C., Ishii, N. D., Riopelle, J. R. and Shooter, M. E. (1977), Nerve growth factor. *N. Engl. J. Med.*, **297**, 1096-1104.

Njä, A. and Purves, D. (1978), The effects of nerve growth factor and its antiserum on synapses in the superior cervical ganglion of the guinea pig. *J. Physiol. (London)*, **277**, 53-75.

Nikodijevic, B., Nikodijevic, O. and Wong Yu, M. Y. (1975), The effect of nerve growth factor on cyclic AMP levels in superior cervical ganglia of the rat. *Proc. Natl. Acad. Sci. U.S.A.*, **72**, 4765-4771.

O'Connor, T. M. and Wyttenbach, R. C. (1974), Cell death in the embryonic chick spinal cord. *J. Cell Biol.*, **60**, 448-459.

Olmsted, J. B. (1976), The role of divalent cations and nucleotides in microtubule assembly *in vitro*. In *Cell Motility, Book C* (eds. R. Goldman, T. Polland and J. Rosenbaum), Cold Spring Harbor Laboratory, pp. 1081-1092.

Pilar, G. and Landmesser, L. (1976), Ultrastructural differences during embryonic cell death in normal and peripherally deprived ciliar ganglia. *J. Cell Biol.*, **68**, 339–359.

Pittman, R., Oppenheim, R. W. and Chu-Wang, I-W. (1978), Beta-bungarotoxin induced neuronal degeneration in the chick embryo spinal cord. *Brain Res.*, **153**, 199–204.

Prestige, M. C. (1970), Differentiation, degeneration and the role of periphery: Quantitative consideration. In *Neurosciences, Second Study Program* (ed. F. Smith), Rockefeller University Press, New York, pp. 73–99.

Roisen, F. J., Murphy, R. A. and Braden, W. G. (1972), Neurite development *in vitro*. I. The effects of adenosine 3′,5′-cyclic monophosphate (cyclic AMP). *J. Neurobiol.*, **3**, 347–368.

Sabatini, M. T., De Iraldi, A. P. and De Robertis, E. (1965), Early effects of the antiserum (AS) against the nerve growth factor (NGF) on the structure of sympathetic neurons. *J. Exp. Neurol.*, **12**, 370–383.

Schubert, D., La Corbiere, M., Witlock, C. and Stallcup, W. (1978), Alterations in the surface properties of cells responsive to nerve growth factor. *Nature (London)*, **273**, 718–723.

Schwab, M. E. (1977), Ultrastructural localization of a NGF-horseradish peroxidase (NGF-HRP) coupling product after retrograde axonal transport in adrenergic neurons. *Brain Res.*, **130**, 190–196.

Schwab, M. E. and Thoenen, H. (1977), Selective trans-synaptic migration of tetanus toxin after retrograde axonal transport in peripheral sympathetic nerves: a comparison with nerve growth factor. *Brain Res.*, **122**, 455–474.

Server, A. C. and Shooter, E. M. (1977), Nerve Growth Factor. In *Advances in Protein Chemistry* (eds. C. B. Anfinsen, J. T. Edsal and F. M. Richards), Academic Press, New York, San Francisco and London, pp. 339–409.

Shelanski, M. L. (1973), Chemistry of the filaments and tubules of brain. *J. Histochem. Cytochem.*, **21**, 529–539.

Sidman, R. L. (1972), Cell interaction in developing nervous system. In *Cell Interaction, Third Lepetit Colloquium* (ed. L. Silvestri), North-Holland, Amsterdam, pp. 1–13.

Steiner, G. and Schönbaum, E. (eds.) (1972), *Immunosympathectomy*, Elsevier, Amsterdam.

Stöckel, K., Paravicini, U. and Thoenen, H. (1974), Specificity of the retrograde axonal transport of nerve growth factor. *Brain Res.*, **76**, 413–421.

Stöckel, K., Schwab, M. and Thoenen, H. (1975), Comparison between the retrograde axonal transport of nerve growth factor and tetanus toxin in motor sensory and adrenergic neurons. *Brain Res.*, **99**, 1–61.

Thoenen, H. (1979), Private communication.

Thoenen, H., Angeletti, P. U., Levi-Montalcini, R. and Kettler, R. (1971), Selective induction by nerve growth factor of tyrosine hydroxylase and dopamine β-hydroxylase in the rat superior cervical ganglia. *Proc. Natl. Acad. Sci. U.S.A.*, **68**, 1598–1602.

Thoenen, H., Barde, A. Y., Edgar, D., Hatanaka, H., Otten, U. and Schwab, M. (1979), Mechanism of action and possible sites of synthesis of nerve growth factor. *Prog. Brain Res.*, **51**, 95–107.

Thoenen, H. and Tranzer, J. P. (1968), Chemical sympathectomy by selective destruction of adrenergic nerve endings. *Arch. Pharmacol. Exp. Pathol.*, **261**, 271–288.

Tilney, L. G. (1971), Origin and continuity of microtubules. In *Origin and Continuity of Cell Organelles* (eds. J. Reinert and H. Ursprung), Springer-Verlag, Berlin, pp. 222–260.

Unsicker, K., Kirsch, B., Otten, U. and Thoenen, H. (1978), Nerve growth factor-induced fiber outgrowth from isolated adrenal chromaffin cells. Impairement by glucocorticoids. *Proc. Natl. Acad. Sci. U.S.A.*, **75**, 3498–3502.

Varon, S., Nomura, J. and Shooter, E. M. (1967), The isolation of the mouse nerve growth factor protein in a high molecular weight form. *Biochemistry*, **6**, 2202–2209.

Wilson, L., Bryan, J., Ruby, A. and Mazia, D. (1970), Precipitation of proteins by vinblastine and calcium ions. *Proc. Natl. Acad. Sci. U.S.A.*, **66**, 807–814.

Yamada, M. K. and Wessells, N. K. (1971), Axonal elongation: Effect of nerve growth factor on microtubule protein. *Exp. Cell Res.*, **66**, 346–352.

Yanker, A. B. and Shooter, M. E. (1979), Nerve growth factor in the nucleus: Interaction with receptors on the nuclear membrane. *Proc. Natl. Acad. Sci. U.S.A.*, **76**, 1269–1273.

Zanini, A. and Angeletti, P. U. (1971), Studies of the nerve growth factor by microcomplement fixations. Effects of physical, chemical and enzymatic treatments. *Biochim, Biophys. Acta*, **229**, 724–729.

Zanini, A., Angeletti, P. U. and Levi-Montalcini, R. (1968), Immunochemical properties of the nerve growth factor. *Proc. Natl. Acad. Sci. U.S.A.*, **61**, 835–842.

11 Glucocorticoid-induced lymphocyte death

ALLAN MUNCK and GERALD R. CRABTREE

11.1 Introduction

Glucocorticoids are the principal steroid hormones produced by the inner zones of the adrenal cortex. The natural glucocorticoids, present in amounts which vary according to species, are cortisol and corticosterone. Corticosterone is generally about half as active as cortisol. Dexamethasone, prednisolone, triamcinolone acetonide and certain other synthetic analogues of glucocorticoids which are widely used therapeutically and experimentally have some advantages over the natural hormones with regard to both activity and specificity. Dexamethasone, for example, is 10 to 100 times more active than cortisol, has about 10 times higher affinity for glucocorticoid receptors, and is a 'purer' glucocorticoid in the sense that it cross-reacts much less with mineralocorticoid receptors. Cortisone and prednisone, once thought to be glucocorticoids, are now known to have no activity in themselves but to require conversion, respectively, to cortisol and prednisolone (cf. Munck and Leung, 1977).

Since much of the research on glucocorticoids and lymphocytes is carried out by adding glucocorticoids to isolated cells, it is worth noting the limitations on use of high steroid concentrations. Many studies, most in the older literature but some quite recent, have been rendered meaningless by the excessively high steroid levels employed. Plasma concentrations of the natural glucocorticoids under physiological conditions are of the order of 100 nmol/l or less, and even under pathological conditions probably do not reach 1 μmol/l. Administered glucocorticoids are unlikely to exceed concentrations of 1 μmol/l except locally, at the site of intake. The consistent experience of several decades of work with isolated systems is that effects that are physiological (in the sense that they have counterparts *in vivo* and are specific for steroids with glucocorticoid activity) can usually be reproduced in such systems with glucocorticoid concentrations well below 1 μmol/l. In fact, these effects give relationships between concentration and response in good agreement with the physicochemically determined relationships between concentration and occupancy of glucocorticoid receptors, and there is every reason to suppose the effects are mediated by receptors. With commonly used glucocorticoids such as those mentioned above, the concentration for half-saturation of receptors (which is equal to K_d, the dissociation constant) lies in the range of 5-50 nmol/l (Munck and Leung, 1977). At 1 μmol/l the glucocorticoids virtually saturate their receptors and produce maximal physiological effects.

At higher concentrations, particularly beyond 10 μmol/l, glucocorticoids do not simply give maximal physiological effects but in isolated systems can produce a plethora of effects that have no known relation to their activities *in vivo*. These effects are non-specific in the sense that they can be obtained as well or better with steroids, such as progesterone, that lack glucocorticoid activity. Non-specific effects are not mediated by glucocorticoid receptors – none, at least, that have been identified to date. They probably result from the mixed polar–non-polar character of steroids that makes them interfacially active, causing them to lodge preferentially at interfaces between polar and non-polar regions (Munck, 1976). At the high concentrations characteristic of non-specific effects, such interfacial regions may become saturated with steroid. Effects in isolated systems that can be produced only with high steroid concentrations are therefore unlikely to have significance for glucocorticoid action.

Glucocorticoids have been known for many years to exercise physiological control over the lymphoid system and immunological reactions (cf. Dougherty and White, 1945; Claman, 1972; Munck and Young, 1975; Fauci, 1978). Their most dramatic effect *in vivo* is on the thymus, which enlarges in the absence of glucocorticoids and may decrease measurably in size within hours of glucocorticoid administration or a stress-induced rise in glucocorticoid blood levels. Schrek (1949), in what was probably the first demonstration of a physiological effect *in vitro* with any steroid hormone, showed that glucocorticoids at physiological concentrations kill isolated rabbit thymus cells. Most lymphocyte pools are influenced by the hormones, but, as will be discussed presently, all are not influenced the same way. Different subpopulations, even within the thymus, respond differently. Superimposed on these differences are significant variations in sensitivity among species.

A curious correlation, of which there were glimmerings already 25 years ago (Krohn, 1954, 1955), holds for several species between their sensitivity to glucocorticoid suppression of lymphoid tissue and the type of glucocorticoid secreted by their adrenal glands. In the species that, for reasons to be given later, are considered resistant, namely man, monkey, guinea pig and ferret (Claman, 1972), the principal glucocorticoid is cortisol (Bush, 1953; Fajer and Vogt, 1963). Among the sensitive species, mouse, rat, rabbit and hamster, the principal glucocorticoid is corticosterone in all but the hamster (Bush, 1953; Schindler and Knigge, 1959; Spackman and Riley, 1977), and the hamster's levels of cortisol seem to be exceptionally low compared with those in resistant species. Thus the available data suggest an association, perhaps owing to chance, of resistance with cortisol and of sensitivity with corticosterone (or with low levels of cortisol).

Cell death, the subject of this chapter, has been thought of in the past as the main physiological consequence of glucocorticoid action on lymphocytes – though what physiological 'purpose' is served by this action has never been clear (Munck and Young, 1975). In recent years many new effects have been detected, on lymphocyte distribution, differentiation, proliferation, expression of immunological activities and other properties (Claman, 1972; Munck and Young, 1975; Fauci, 1978). Physiologically these more subtle modulations of lymphocyte behaviour may be at least as important as the lethal actions. They may also play major roles in the immunosuppressive and anti-inflammatory actions of glucocorticoids, as well as in glucocorticoid therapy of various lymphoproliferative diseases. It has become apparent, furthermore, that glucocorticoids can shorten lifespans and otherwise influence lymphocytes not only through direct actions on the lymphocytes themselves, but through indirect actions that deprive lymphocytes of proliferative and other stimuli emanating from associated cells in the form of hormone-like chemical signals. This type of indirect effect, which may have major significance for glucocorticoid physiology, will be dealt with briefly at the end of this chapter.

Whatever their physiological and therapeutic importance, the direct lethal actions on lymphoid cells have permitted genetic analysis of hormone mechanisms to proceed further with glucocorticoids than with most other hormones. These actions can be applied in simple screening techniques for identifying glucocorticoid-resistant cells. We will be discussing several such studies in detail. It is worth mentioning, however, that resistance is not an all-or-none property, and, at least in normal lymphocytes, resistance to lethal actions of glucocorticoids can co-exist with sensitivity to metabolic and other actions.

One point that will emerge from this chapter is that we know very little about how glucocorticoids kill lymphocytes. We have a great deal of information, outlined in the next section, about initial interactions of glucocorticoids with their receptors and the early metabolic events triggered by the hormones. Despite many efforts and hypotheses, however, we still do not know how these initial events are linked to cell death.

11.2 Glucocorticoid receptors and metabolic effects in lymphocytes

11.2.1 *Rat thymus cells*

Because of their easy availability in large numbers, rat and mouse thymic lymphocytes have been studied more intensively than other lymphocytes. In Fig. 11.1 we summarize the results of many experiments, our own and those of others, on the time course of formation of glucocorticoid-receptor complexes and of appearance of glucocorticoid effects after addition of cortisol to rat thymus cells incubated at 37°C. Dexamethasone gives similar results. References to most original papers are given in Munck and Leung (1977), Crabtree, Gillis, Smith and Munck (1979a) and Young, Nicholson, Guyette, Giddings, Mendelsohn, Nordeen and Lyons (1979), and will not be repeated here.

Glucocorticoids, as other steroid hormones, pass rapidly through the cell membrane, apparently by free diffusion. The first detectable physicochemical reaction in which the hormone molecules participate is that of binding non-covalently to soluble protein receptors to form the so-called 'cytoplasmic' hormone-receptor complexes. These complexes are characterized by the fact that they are found in

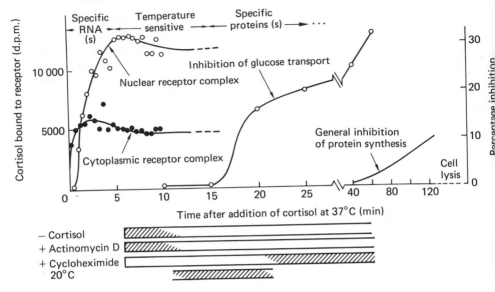

Fig. 11.1 Time course in rat thymus cell suspensions at $37°C$ of cortisol–receptor complex formation, cortisol-induced inhibition of glucose transport, and inhibition of protein synthesis. Kinetics of receptor complex formation were determined with [^{3}H] cortisol at about 0.1 μmol/l. Inhibitory metabolic effects were produced with about 1 μmol/l of cortisol. Glucose transport was measured with 2 min pulses of radioactive hexose, initiated at the times indicated. Shaded segments of the horizontal bars in the lower part of the figure indicate roughly the time intervals during which emergence of the cortisol effect on glucose metabolism can be blocked by treatment with cortexolone (which displaces cortisol from the glucocorticoid receptors), actinomycin D and cycloheximide, and delayed by lowering temperature. Open bars indicate periods during which these treatments have no effect. At the top of the figure is the sequence of steps by which it is hypothesized that the cortisol–receptor complex leads to synthesis of a specific protein that inhibits glucose transport.

cytosol after cells are broken. Binding sites on the receptors have high stereo-chemical specificity and affinity for steroids with glucocorticoid activity such as those mentioned in Section 11.1, with affinities in rough relation to the biological activities of the steroids. They also have affinity for certain closely related steroids such as cortexolone (11-deoxycortisol) and progesterone, which have little glucocorticoid activity and can thus act as anti-glucocorticoids by competition with active steroids.

The cytoplasmic complexes rapidly become 'activated' or 'transformed', acquiring affinity for nuclear structures to which they become bound almost instantaneously. Activation involves some kind of temperature-sensitive conformational rearrangement, perhaps irreversible, accompanied by redistribution of charges. At low temperatures activation is very slow. The nature of the nuclear sites to which the complexes are bound is poorly understood, but at least a substantial fraction of them are associated with the chromatin. Some years ago there were a number of reports that glucocorticoid target cells contain a limited number of high-affinity nuclear sites for glucocorticoid–receptor complexes. These reports seem to have been based on artefacts, and at present there is little evidence for anything but low-affinity nuclear binding with no particular specificity for glucocorticoid versus other steroid hormone–receptor complexes.

After the arrival of the hormone–receptor complexes in the nucleus the first general metabolic change that can be observed in the cell is inhibition of glucose uptake resulting from a decrease in the rate of glucose transport, which, as shown in Fig. 11.1, appears abruptly after a time lag of 15–20 min (Zyskowski and Munck, 1979).

We have recently found (Foley, Jeffries and Munck, 1980) that glucose-dependent acetate incorporation into lipids, particularly nuclear lipids, is also inhibited, with a time course and magnitude similar to that of glucose transport.

A number of other inhibitory metabolic effects then develop more slowly, starting after about 1 h. These include a decrease in amino acid incorporation into proteins (probably a manifestation of general inhibition of protein synthesis), decreases in uridine incorporation into RNA, in ATP levels, and in various transport processes including transport of α-aminoisobutyric acid (AIB), an amino acid analogue. At about this time there is also an increase in so-called 'nuclear fragility', measured by leakage of DNA from nuclei or decreased recovery of nuclei after cells are broken by hypo-osmotic shock. Morphological evidence of cell damage begins to appear after 1–2 h, and measurable loss of viable cells can be detected after 8–12 h.

Removal of cortisol from the glucocorticoid receptors after as little as 5 min by washing or by addition of 10^{-5} mol of cortexolone/l, fails to prevent the appearance subsequently of the inhibitory effect on glucose metabolism. This observation, symbolized by the shaded area in the first horizontal bar in the lower section of Fig. 11.1, indicates that by 5 min the primary hormonal 'message' has already been received and recorded by the cell in such a way that the hormone itself is no longer essential, at least for the rapid effect on glucose metabolism.

The form in which the message is initially recorded is indicated by experiments with actinomycin D and other inhibitors of RNA synthesis. As shown in Fig. 11.1, actinomycin D blocks the early hormone effects if it is present during the first 5 min. These results suggest that the immediate effect of the nuclear hormone–receptor complex is to stimulate transcription of RNA. There has been no direct evidence reported for glucocorticoid stimulation of RNA synthesis in thymus or other lymphoid cells, but Bell and Borthwick (1979a) have detected a substantial increase in RNA polymerase B activity after 10 min.

Experiments with puromycin and cycloheximide, both inhibitors of protein synthesis, have provided circumstantial evidence for synthesis, starting at about 15 min, of a protein mediator of the hormone effect. For example, if cycloheximide is present for 15 min but is then removed, it does not block. If it is present from

15 min on, it blocks. If it is added later it does not affect the already established hormone effect, but does prevent further development. As in the case of RNA, we lack direct evidence for early synthesis of a protein. Finally, between the hypothetical steps involving RNA and protein synthesis there appears to be a separate step that is drastically slowed by lowering the temperature to 20°C (Fig. 11.1).

From these results we have postulated that the nuclear hormone–receptor complex initiates a burst of transcription of specific mRNA that after about 15 min is translated to an effector protein that directly or indirectly inhibits glucose transport. The temperature-sensitive step between RNA and protein synthesis may be associated with translocation of newly synthesized mRNA to the ribosomes.

Similar schemes, qualitatively identical with this one up to the synthesis of the specific effector protein but diverging on the functions of the proteins, have been postulated for all the primary actions not only of glucocorticoids but of all steroid hormones, and may be said to constitute the current dogma of steroid hormone action.

Several of the glucocorticoid effects that begin at about 1 h, including the inhibition of uridine incorporation into RNA, the decrease in ATP levels and the increase in nuclear fragility, are blocked by inhibitors of protein synthesis, and where tested, by inhibitors of RNA synthesis. The glucocorticoid inhibition of protein is blocked by inhibitors of RNA synthesis. These later effects are therefore also thought to depend on prior synthesis of mRNA and effector proteins. How many such proteins there are, and how they are related to each other, is an open question. The temporal separation of steps outlined in Fig. 11.1 is of course something that is seen only immediately after the hormone first impinges on the cells, as each event in turn begins. Later the various processes all take place simultaneously, presumably interacting with each other to give rise to a complex spreading network of effects.

As we will discuss presently, it is unclear in what way, if any, the ultimate lethal effect depends on known earlier events. In this connection it should be pointed out that, although the rapid inhibition of glucose transport in thymus cells can be produced with only 5 min exposure to hormone, such brief exposure will not necessarily suffice to give rise to all later effects. For example, 1 μmol of cortisol/l after 24 h incubation with rat thymus cells will reduce the number of viable cells by about 40 per cent relative to untreated controls (Leung and Munck, 1975), but has no effect on the number of viable cells if it is removed from the cells by washing after 5 min (Leung and Munck, unpublished observations). Cortexolone, present throughout such an incubation at 10 μmol/l, significantly reduces the cytolytic activity of 1 μmol of cortisol/l (Leung and Munck, 1975). To the extent therefore that cell death is a consequence of the earlier hormone-induced changes, it depends on the cumulative effects and perhaps interactions of those changes over an extended period.

11.2.2 *Normal and leukaemic human lymphocytes*

Fig. 11.2 outlines the time course in normal human peripheral lymphocytes of some of the same glucocorticoid-related processes just described for rat thymus cells. A number of effects besides these shown in Fig. 11.2, including inhibition of uridine incorporation into RNA, thymidine incorporation into DNA, AIB transport, mitosis and modification of immunological functions, have also been described (cf. Claman, 1972; Munck and Leung, 1977; Crabtree, Smith and Munck, 1979b).

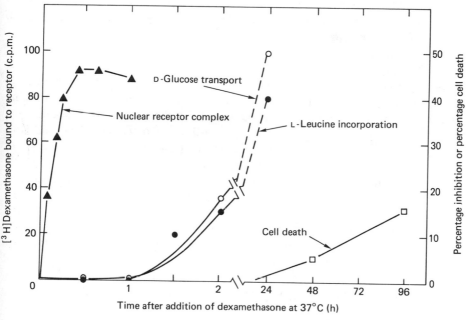

Fig. 11.2 Time course of binding of [^{3}H] dexamethasone to glucocorticoid receptors and subsequent metabolic effects of 100 nmol of dexamethasone/l in human peripheral lymphocytes. Dexamethasone binding and metabolic effects were measured as previously described (Crabtree *et al.*, 1978).

There are both similarities and differences in the responses of rat thymus cells and human lymphocytes. Rates of formation of glucocorticoid–receptor complexes are about the same in both systems. The time course of formation of nuclear complex shown in Fig. 11.2 for human lymphocytes is slower than in Fig. 11.1 for rat thymus cells, but the difference is largely or entirely due to the fact that [^{3}H] cortisol was used for the results in Fig. 11.1 and [^{3}H] dexamethasone for the results in Fig. 11.2. The higher affinity of dexamethasone for glucocorticoid receptors is due to a lower dissociation rate constant, which in turn leads to a slower time course of formation of complexes when compared with cortisol (Munck, 1976). The time course of formation of nuclear dexamethasone–receptor complexes in rat thymus cells is in fact similar to that in Fig. 11.2 (Munck, Crabtree and Smith, 1979).

The rate of inhibition of glucose transport in the human cells is much slower than in rat cells, beginning after more than 1 h, and does not precede inhibition of leucine incorporation. Thymidine incorporation into DNA is inhibited with a similar time course. Furthermore, many experiments have shown that human peripheral and thymic lymphocytes are killed much more slowly than mouse or rat thymus cells (Schrek, 1961; Claman, Moorhead and Benner, 1971). For example, dexamethasone at 1 μmol/l (and 100 nmol/l) leads to 50 per cent decrease in viable rat thymus cells after 24 h (Leung and Munck, 1975) as compared with only a 15 per cent decrease in normal human peripheral lymphocytes after 96 h (Fig. 11.2). Among subpopulations of human lymphocytes there are also differences in sensitivity (Homo, Duval, Thierry and Serrou, 1979).

So far there have been few studies with human lymphocytes of the kind described with rat thymus cells to determine the nature of the initial metabolic steps in hormone action preceding the inhibitory effects, but there is no reason to think that these steps differ in any fundamental way from those in rat thymus cells.

In recent years there have been many investigations of glucocorticoid receptors and glucocorticoid effects in lymphocytes taken from patients with various forms of lymphocytic leukaemia and lymphoma. Much of this work, the general goal of which is to understand the nature of the disorder and to find diagnostic and prognostic tests for use in choosing appropriate therapy, is described in a recent symposium volume (Bell and Borthwick, 1979b). We will not deal with it here in any detail, since the kind of experiment that can be conducted with such cells, coming as they do from individual patients who may be undergoing treatment with a variety of cytotoxic drugs including glucocorticoids, inevitably lacks the degree of control and reproducibility that can be attained with normal cells and cell lines.

On the whole, though, lymphocytes obtained from patients with lymphocytic leukaemia respond to glucocorticoids remarkably like normal human peripheral lymphocytes, showing if anything a generally greater sensitivity to the inhibitory metabolic effects and lethal actions (Schrek, 1964; Crabtree, Smith and Munck, 1978). They have glucocorticoid receptors that can form cytoplasmic and nuclear hormone–receptor complexes. Nuclear binding of the complex is generally followed after an hour or two by inhibition of glucose transport, and reduction in rates of incorporation of precursors of protein, RNA and DNA. By 24–48 h a significant increase in rate of cell death is generally measurable, and by 4–6 days almost all cells may be dead (cf. Bell and Borthwick, 1979b).

11.2.3 *Lymphoid cell lines*

A number of mouse and human lymphoid tumour cell lines have been studied with respect to glucocorticoid receptors and effects. These cells and their mutants are of particular interest in providing models for sensitivity and resistance to the lethal effects of glucocorticoids, as we will discuss later. Some have also yielded a considerable amount of information on metabolic effects.

Results found so far with cell lines are qualitatively similar to those already described for rat thymus cells and human lymphocytes. Time courses resemble most closely those found with human lymphocytes in Fig. 11.2. The mouse lympho-sarcoma P1798 is probably the most extensively studied so far from the point of view of glucocorticoid mechanisms and metabolic effects. We confine ourselves here to results found with the glucocorticoid-sensitive strain of these cells. Results for resistant cells of P1798 and other cell lines are taken up later.

P1798 cells contain glucocorticoid receptors similar to those found in other lymphocytes (Kirkpatrick, Milholland and Rosen, 1971). Rosen, Fina, Milholland and Rosen (1970, 1972) demonstrated that glucocorticoids inhibit uptake of glucose and 2-deoxyglucose by P1798 cells within 1 h, and somewhat more slowly inhibit precursor incorporation into protein, RNA and DNA, and AIB transport. In contrast, Stevens, Stevens, Behrens and Hollander (1973), perhaps working with a different strain of P1798, found that glucocorticoids after 1–2 h inhibited uridine but not 2-deoxyglucose uptake.

The effect on uridine incorporation depends on a series of prior steps similar to those shown for glucose uptake by rat thymus cells in Fig. 11.1, the main difference being a more extended time course (Stevens and Stevens, 1975b). After addition of cortisol at 1 μmol/l, the effect on uridine after 1.5 h persists if cortisol is removed by washing after 30 min. Actinomycin D blocks the effect if it is added initially or after 15 min, but not if it is added after 30 min. Cycloheximide blocks the effect if added initially or after 30 min, but does not block if added initially and then removed after 30 min. Cycloheximide (but not actinomycin D) prevents further development of the effect when added after 1.5 h. These results thus point to a series of steps leading up to the inhibition of uridine incorporation: first a step requiring the hormone, and concurrently a step requiring RNA synthesis; then a step requiring protein synthesis. Such a sequence is entirely consistent with the current view of glucocorticoid action outlined in Section 11.2.1.

Nicholson and Young (1978) have shown that P1798 cells exhibit a gluco-corticoid-induced increase in 'nuclear fragility' similar to that seen in rat thymus cells, beginning after about 2 h exposure to 1 μmol of cortisol/l. The effect is also blocked by cycloheximide, added either initially or after 30 min.

11.3 Lethal effects of glucocorticoids on lymphocytes

11.3.1 *Morphological changes*

Morphological changes that take place in lymphocytes as they die in response to glucocorticoids have been studied by both light and electron microscopy (Dougherty and White, 1945; Schrek, 1949; Cowan and Sorensen, 1964: Burton, Storr and Dunn, 1967; Waddell, Wyllie, Robertson, Mayne, Au and Currie, 1979). The earliest changes that have been observed in rabbit, rat or mouse thymus cells take place in the nucleus. They begin with coarsening, increased osmophilia and con-densation of nuclear chromatin. Then follow gradual loss of structure and dis-solution of the nuclear membrane, and subsequently fragmentation of the cell into membrane-bounded vesicles.

A point already emphasized by Schrek (1949) is that, morphologically, nothing distinguishes the course of glucocorticoid-induced cell death from that of spontaneous cell death or cell death due to a number of other causes including X-irradiation. What glucocorticoids appear to do is accelerate the normal processes of aging and degeneration leading to cell death.

11.3.2 *Species variation in sensitivity to killing by glucocorticoids*

The rapid cytotoxic effect which occurs in the thymus of rats and mice treated with corticosteroids is less striking in the lymphoid tissues of several other species. On the basis of these differences, as well as on variations in immunological sensitivity (cf. Claman, 1972), it has been suggested that there are 'cortisone-resistant' species, such as the guinea pig, monkey, and man (Long, 1957; Schrek, 1961; Claman, Moorehead and Benner, 1971; Claman, 1972; Frenkel and Havenhill, 1963).

Glucocorticoids administered to mice and rabbits can cause a rapid decrease in size of the thymus, associated with the appearance of dying pyknotic cells (Dougherty and White, 1945). The rapidity with which this effect occurs led to the use of the terms 'cytolytic' or 'lympholytic' to characterize this response. In contrast, glucocorticoid treatment of children with large thymic shadows produces only moderate reduction in thymic size, as assessed radiographically after 2–7 days of steroid treatment (Caffey and di Liberti, 1959; Caffey and Silbey, 1960). Furthermore, the thymus, taken at autopsy, of children treated for short periods with high doses of dexamethasone for severe head injury shows little or no cellular necrosis when examined microscopically, even though there is a reduction in the mean thymic weight (Crabtree, unpublished observations). As already noted, studies *in vitro* in which cells are incubated directly with glucocorticoids show that thymus and peripheral lymphocytes from man are killed very slowly compared with rat and mouse cells.

One of the best documented effects of glucocorticoids in man is a reduction in the number of circulating lymphocytes (Dale, Fauci and Wolfe, 1974). Lymphopenia can be observed within 24 h after administration of glucocorticoids (Glasser, Huestis and Jones, 1977). Although this effect was originally attributed to direct cell killing and is frequently referred to as the 'lympholytic' effect of glucocorticoids, recent evidence indicates that it may be due mainly to redistribution of lymphocytes. Two groups of investigators have reported that 4 h after a single dose of glucocorticoid there is redistribution of lymphocytes (primarily T lymphocytes) from the peripheral blood to extravascular sites; by 24 h blood levels return to normal (Fauci and Dale, 1974; Yu, Clements, Paulus, Peter, Levy and Barnett, 1974). Rapid return to normal levels argues strongly against cell killing as the basis for these short-term effects in man.

Glucocorticoids can also affect proliferation of lymphocytes. They produce clear inhibitory effects on mitosis in mitogen-stimulated human lymphocytes (Nowell, 1961) and in bone marrow cells in human volunteers (Mauer, 1965).

Recently, glucocorticoids have been shown to inhibit the production of specific growth factors necessary for T lymphocyte proliferation *in vitro* (Gillis, Crabtree and Smith, 1979a, 1979b).

Thus it seems likely that glucocorticoids can exert large-scale influences on lymphoid tissues *in vivo* in at least three ways: through direct cell killing, through lymphocyte redistribution, and through inhibition of proliferation. Long-term influences are probably due to a combination of all three. Acute changes in man, a resistant species, seem to be due mainly to redistribution; in rats and mice, which are sensitive species, they seem to be due mainly to cell killing.

11.3.3 Variation in sensitivity among subpopulations of lymphocytes

Early studies showed that lymphocytes from different anatomic sites differ in their susceptibility to glucocorticoids. Schrek (1949) noted that rabbit thymus cells were much more sensitive to glucocorticoids *in vitro* than rabbit bone marrow cells. In mice, the thymus medulla persists after cortisone treatment, and lymph nodes and spleen undergo less involution after steroid therapy than the thymus (Dougherty, Berliner, Schneebeli and Berliner, 1964). These results were extended by immunological studies in which antibody production by spleen cells was found to be much more sensitive to glucocorticoid inhibition than antibody production by bone marrow lymphocytes (Levine and Claman, 1970). With the development of specific markers for functional subclasses of lymphocytes it was discovered that the redistribution of lymphocytes that occurred in man after a single dose of glucocorticoids was limited largely to T lymphocytes, (Fauci and Dale, 1974; Yu *et al.*, 1974), and that among T cells there was a greater effect on those with Fc receptors for immunoglobulin M (IgM) than on those with Fc receptors for immunoglobulin G (IgG) (Haynes and Fauci, 1978). Thus there appears to be considerable variation in glucocorticoid sensitivity among the various subclasses of lymphocytes.

In general, T lymphocytes and cell-mediated immune responses have been found to be somewhat more sensitive to glucocorticoids than are B cells and antibody production (Fauci, Dale and Balow, 1976). Although treatment of normal volunteers with glucocorticoids produces a decrease in serum IgG concentration (Butler and Rossen, 1973a, 1973b), there is little or no effect on specific antibody production to injected antigens (Tuchinda, Newcomb and De Vald, 1972). Antibody production by human B cells *in vitro* is actually enhanced by glucocorticoids (Fauci, Pratt and Whalen, 1977), indicating that cell–cell interactions may be necessary to produce the reduction in immunoglobulin levels found *in vivo*.

In addition to these effects on well-defined subpopulations of lymphocytes, glucocorticoids also have differential effects on certain functional classes of lymphocytes. Cells capable of mediating various immunological reactions appear to be remarkably resistant to glucocorticoids (Blomgren and Andersson, 1969). For example, after mice are treated with a massive dose of glucocorticoids, the 5 per cent or so surviving thymic lymphocytes are as effective in mediating a a graft-versus-host reaction as all the thymic lymphocytes from untreated animals (Cohen, Fischback and Claman, 1970).

A note of caution was recently sounded to the usual interpretation of resistant and sensitive subpopulations of thymic lymphocytes. Studies with thymus cells from mice treated with 5 mg of cortisol hemisuccinate for 48 h showed that many of the 3-5 per cent surviving 'cortisone-resistant' cells are still sensitive to killing by glucocorticoids *in vitro* in 20 h cultures (Weissman and Levy, 1975). They are also distinctly sensitive to inhibition of uridine incorporation, as measured in 4 h incubations, though less sensitive than normal thymus cells, and do not constitute a homogeneous cell population by gradient fractionation (Duval, Dardenne, Dausse and Homo, 1977). Aside from the unlikely possibility that glucocorticoids are in some way prevented from entering the thymic medulla, these results suggest that medullary cells are not a completely distinct subpopulation of thymic lymphocytes. They also raise the question of whether cell killing measured *in vivo* and *in vitro* represent the same phenomenon.

11.3.4 *Effects on malignant lymphocytes*

While there is some doubt about the possible significance of cytotoxic effects of glucocorticoids on normal human lymphocytes, it is very likely that the gluco-corticoids produce their therapeutic effects in human lymphoid malignancy by killing cells. Shortly after the institution of steroids in the treatment of acute lymphocytic leukaemia, it was noted that children with large numbers of peripheral lymphoblasts treated with glucocorticoids alone frequently developed increased levels of uric acid concurrent with a rapid fall in the number of peripheral lympho-blasts (Wolff, Brubaker, Murphy, Pierce and Severo, 1967; Ranney and Gellhorn, 1957). These results have been repeatedly confirmed, and indicate that the thera-peutic effects of glucocorticoids are probably due to cell killing rather than inhibition of proliferation.

Lymphoid malignancies appear to vary in their clinical sensitivity to gluco-corticoids. Thus acute lymphocytic leukaemia of children has an overall remission rate of 50-70 per cent with glucocorticoids alone (Wolff *et al.*, 1967; Vietti, Sullivan, Berry, Haddy, Haggard and Blattner, 1965; Leikin, Brubaker, Hartman, Murphy, Wolff and Perrin, 1968), while malignant lymphoma of the histiocytic type has been reported to have only an 18 per cent response rate (Ezdinli, Stutz-man, Aungst and Firat, 1969). Most other lymphoid malignancies appear to be intermediate between these two extremes (Ezdinli, *et al.*, 1969).

11.4 Genetic analysis of glucocorticoid-induced cell death

11.4.1 *Models for studying the lethal actions of glucocorticoids*

The use of models for studying sensitivity and resistance to the lethal actions of glucocorticoids began in 1958 with the observation that some transplantable mouse tumours become resistant to glucocorticoids after treatment of the host animal with the hormones (Lampkin and Potter, 1958). Subsequently it was observed that the mouse lymphoma cell line ML-388 developed stable heritable resistance to glucocorticoids when grown *in vitro* with sublethal concentrations of glucocorticoid (Aronow and Gabourel, 1962). Studies since then on a number of lymphoid cell lines form the basis for much of the current thinking about development of glucocorticoid resistance in human acute lymphocytic leukaemia and lymphoma.

In this section we will describe each of the model systems used for study of glucocorticoid-induced cell death. A discussion of the relevance of these studies to the mechanism of glucocorticoid-induced cell death will be the subject of Section 11.4.2. Table 11.1 lists most of the cell lines that have been used to obtain glucocorticoid-resistant mutants.

Table 11.1 Cell lines used for studying the cytolytic actions of glucocorticoids

Cell line	Origin	Responses to glucocorticoids	Reference
ML-388	Mouse	Cell death	Aronow and Gabourel (1962)
P1798 S	Mouse	Tumour regression	Rosen *et al.* (1970)
SIAT.4	Mouse	Cell death	Baxter *et al.* (1971)
SIAT.8	Mouse	Resistant	Baxter *et al.* (1971)
S49	Mouse	Cell death	Harris *et al.* (1973)
			Sibley and Tomkins (1974a)
AKR-A	Mouse	Resistant	Lippman *et al.* (1974)
CCL-119	Human	Resistant	Lippman *et al.* (1974)
CCL-120	Human	Resistant	Lippman *et al.* (1974)
P288	Mouse	Sensitive	Turnell and Burton (1975)
L1210	Mouse	Resistant	Turnell and Burton (1975)
W7	Mouse	Cell death	Bourgeois and Newby (1977)
CEM	Human	Cell death	Norman and Thompson (1977)
			Harmon *et al.* (1979a)

The mouse lymphosarcoma cell line P1798 is passed serially in mice and exists as a glucocorticoid-sensitive and a glucocorticoid-resistant tumour (Lampkin and Potter, 1958; Stevens, Mashburn and Hollander, 1969; Rosen, Rosen, Milholland and Nichol, 1970). Treatment of intact animals bearing tumours results in inhibition of growth and regression of the tumour only in the sensitive subline.

Resistance in subcutaneous P1798 tumours has been found by Davis, Chan and Thompson (1980) to develop reproducibly as the tumours grow beyond a certain size. They suggest that the cells undergo a non-mutational alteration analogous to differentiation.

The S49 cell line is derived from a glucocorticoid-sensitive lymphoma induced by mineral oil in a Balb/c mouse (Horibata and Harris, 1970; Harris, 1970). These cells are almost all killed by 10 nmol of dexamethasone/l within 48 h (Rosenau, Baxter, Rousseau and Tomkins, 1972). When grown in the presence of 10 nmol of dexamethasone/l, glucocorticoid-resistant cells appear at a frequency of 3.5×10^{-6}/ cell/generation (Sibley and Tomkins, 1974a, 1974b).

A recent addition to the models used for studying glucocorticoid-induced cell death has been the WEHI-7 (W7) cell line first isolated from a radiation-induced thymoma in a Balb/c mouse (Harris, Bankhurst, Mason and Warner, 1973). Wild-type W7 cells are rapidly killed by low concentrations of glucocorticoids (Bourgeois and Newby, 1977). They are more sensitive than S49 cells.

Observations to be described later indicate that the W7 cell line contains two functional copies of the structural gene for the glucocorticoid receptor, while S49 and a partially resistant W7 cell line have only a single functional copy. An additional feature of this cell line is that sublines resistant to thioguanine and 5-bromodeoxyuridine are available, and fusion between various sublines is possible. Thus, with these cells, complementation analysis and studies of dominance can be performed. With S49 cells this has apparently not been possible (Bourgeois, 1979).

Of the different cell lines which have been used for the analysis of the cytotoxic effects of glucocorticoids, the one which perhaps holds the most promise for understanding the therapeutic effects of glucocorticoids in human leukaemia is the CEM cell line. Established from a patient with acute lymphocytic leukaemia, these cells form rosettes with sheep erythrocytes and bear thymic lymphocyte specific antigen (Norman and Thompson, 1977). Addition of glucocorticoids to these cells produces irreversible G_1 arrest and cell death, but the concentrations of glucocorticoid necessary are approximately tenfold higher than for mouse lymphoid cells (Harmon, Norman and Thompson, 1979a; Harmon, Norman, Fowlkes and Thompson, 1979b). Glucocorticoid-induced cell death requires approximately 48 h, and is therefore much slower than cell death in rat or mouse thymus cells or mouse cell lines, but similar to that in fresh human leukaemia cells (Crabtree, Smith and Munck, 1978b).

The CEM cell line has been used to study the interactions between prednisolone and other chemotherapeutic agents. Prednisolone shows slight synergism with methotrexate or vincristine, and some degree of antagonism with 6-mercaptopurine (Norman, Harmon and Thompson, 1978). This cell line may be useful for constructing new chemotherapeutic protocols of acute lymphocytic leukaemia.

11.4.2 *Glucocorticoid receptors in mutants and other cells resistant to glucocorticoid-induced cell death*

Early studies of glucocorticoid-resistant cell lines focused attention on the receptor as the weak link in the chain of processes leading to cell death or inhibition of proliferation. In virtually every case in which wild-type clones have been compared in detail with their resistant progeny, the resistant cells have been found to have either fewer receptors or defective receptors. With mouse fibroblasts it was found that glucocorticoid-resistant cells have fewer glucocorticoid receptors than sensitive cells (Hackney, Gross, Aronow and Pratt, 1970). Similar results were reported with three resistant clones of S49 mouse lymphoma cells (Baxter, Harris, Tomkins and Cohn, 1971) and with the P1978 cell line, a model somewhat more analogous to human lymphoma (Kirkpatrick *et al.*, 1971; Kaiser, Milholland and Rosen, 1974).

These receptor defects in resistant mutants have been studied in detail with the S49 and W7 cell lines. (For review, see Bourgeois, 1979; Sibley and Yamamoto, 1979.) In S49 cells, the transition from glucocorticoid-sensitive to -resistant was found to be stochastic and, as noted above, to occur at a rate of 3.5×10^{-6}/cell/generation (Sibley and Tomkins, 1974a). Among the glucocorticoid-resistant clones, 80 per cent had reduced numbers of cytoplasmic receptors, and 10 per cent had receptors which were unable to translocate to the nucleus. In the remaining 10 per cent no defect was demonstrated in receptor number or translocation (Sibley and Tomkins, 1974b). Receptors defective in nuclear translocation also had altered sedimentation characteristics and bound to DNA–cellulose with lower or higher affinity (Yamamoto, Stampfer and Tomkins, 1974). Thus only 10 per cent of the resistant clones had apparently normal glucocorticoid receptors. However, as pointed out by Bourgeois (1979), the normality of the receptors in these cell lines was not conclusively demonstrated by complementation between receptor-deficient cells and resistant cells possessing 'normal' receptors.

Analysis of resistance in the mouse thymoma cell line W7, the wild form of which is more sensitive to glucocorticoids than S49 and has about twice the number of receptors, revealed that resistance develops by a two-step process. The first step is associated with a 50 per cent reduction in number of receptors and a similar reduction in sensitivity. This step occurs with a frequency between 6.5×10^{-8} and 2×10^{-6}/cell/generation. Totally resistant cells can be produced from these partially resistant clones by growing them in the presence of high concentrations of glucocorticoid. The frequency of conversion to total resistance is 4×10^{-6} to 15×10^{-6}/cell/generation (Bourgeois and Newby, 1977). In 80–90 per cent of the resistant clones, receptors were absent, and in the remaining 10–20 per cent were defective in nuclear translocation. No resistant cells were found with normal receptors. Comparison of these results with similar studies in the S49 cell line led to the conclusion that the W7 cell line was diploid for the structural gene for the glucocorticoid receptor while the S49 cell line, as well as the partially resistant variants of W7, were haploid (Bourgeois, 1979).

Studies with several other cell lines have indicated that abnormal receptors are the most common defect found in glucocorticoid-resistant cell lines. The glucocorticoid-sensitive CEM cell line derived from human acute lymphocytic leukaemia has been used to obtain a large number of resistant clones. Although almost all of these resistant clones have glucocorticoid receptors, the receptors are either reduced in number or have abnormal chromatographic characteristics on DNA–cellulose or DEAE–cellulose (Harmon *et al.*, 1979a; Schmidt, Harmon and Thompson, 1980).

The resistant tumours of P1798 suggested by Davis *et al.* (1980) to arise by non-mutational processes have receptors that are similar in number but appear to have lower affinity for glucocorticoids than those in sensitive cells.

The observation that resistance to glucocorticoids is associated with receptor defects in virtually all of the carefully studied models to date raises the question of why this one step in a complex biological process should apparently be the only one affected. Judging from observed mutation frequencies, the receptor gene does not seem particularly prone to mutation. Regarding the subsequent steps in hormone action, while it is easy to see that mutations affecting the mechanisms of transcription or translation would probably be fatal to cells, it is less clear why mutations in an effector protein of the kind hypothesized to mediate glucocorticoid effects (Section 11.2) could not give resistant but viable cells. Such mutations would not be seen, however, if the protein was essential at non-induced levels for other cellular processes; nor would they be seen except at very low frequencies if more than one effector protein could produce cell death, since in that case several mutations would be required for glucocorticoid resistance.

The fact that normal rat and human lymphocytes all appear to contain glucocorticoid receptors and generally exhibit some degree of sensitivity to glucocorticoids (Duval *et al.*, 1977; Lippman and Barr, 1977; Crabtree *et al.*, 1978; Homo *et al.*, 1979) raises questions about the relevance of receptor defects observed in mutant cell lines to normal cells. It is known that lymphoid tissue in the human and the rat is subject to tonic suppression by normal serum concentrations of glucocorticoid, since adrenalectomy in the rat and Addison's disease in the human are both accompanied by hypertrophy and hyperplasia of lymphoid tissue. This tonic suppression, exerted on the kilogram or more of human lymphoid tissue, would be expected to exert powerful selective pressures for the emergence of resistant clones of lymphoid cells in the course of a human lifespan. The apparent absence of such clones among normal lymphocytes – or for that matter among surviving lymphocytes from mice or humans treated with large doses of glucocorticoids – would seem to indicate that the inhibitory effects of glucocorticoids are beneficial for survival and proliferation of lymphoid cells and that resistant mutants, if they appear *in vivo*, do not have selective advantages over sensitive cells. It is therefore likely that the factors controlling glucocorticoid resistance *in vivo* are more complex than those now documented by studies *in vitro*.

11.5 Mechanisms of glucocorticoid-induced cell death

The molecular pathology of cell death in general, and of glucocorticoid-induced cell death in particular, is poorly understood. As mentioned in Section 11.3.1, the morphological changes that have been observed in lymphocytes dying as a result of exposure to glucocorticoids do not differ markedly from those of cells dying spontaneously, and give no clue as to specific mechanisms that may be activated by glucocorticoids.

Resistant mutants, though potentially holding great promise for identifying individual steps in the sequence of events leading to cell death, have in this respect so far proved disappointing. As we have discussed, virtually all mutants turn out to have defects in their glucocorticoid receptors. Receptors are already recognized to be essential for glucocorticoid activity, and the studies with mutants have disclosed no other essential mechanisms. The absence of other types of mutants does in itself lead to some tentative conclusions drawn in Section 11.4.2. Rephrased somewhat, those conclusions are that the lack of other mutants is consistent with glucocorticoid-induced cell death being mediated on the one hand through several independent processes, each of which is initiated by a different induced protein, or on the other hand, through a single process initiated by a single induced protein, if that process is essential at non-induced levels for normal cell function. One proposed mechanism, according to which glucocorticoids release a cellular 'self-destruct' process (cf. Munck and Leung, 1977), is rendered unlikely by the genetic results, unless the self-destruct process is released rather than blocked by mutations in its components.

A set of observations of considerable importance in connection with mechanisms has been made with CEM cells by Harmon *et al.* (1979a, 1979b). These workers found that glucocorticoids cause irreversible arrest of cells in the G_1 phase of the cell cycle. Furthermore, the glucocorticoid-induced increase in fraction of cells in G_1 arrest, in cell death and in loss of viability measured by decreased capacity for colony formation in 2-week cultures, were all initiated almost simultaneously after 24 h exposure to hormone. They concluded that G_1 arrest may be a consequence of cell death and that cell killing may be confined to G_1.

Another conclusion that may be drawn from these results is that there is no significant latent period between the time a cell is committed to die and the time it dies. This conclusion is consistent with the observation noted earlier that brief exposure of thymus cells to glucocorticoids has no lethal effect on the cells, even though the exposure is sufficient to initiate measurable inhibitory metabolic effects.

Most efforts to analyse the mechanisms of glucocorticoid-induced cell death have relied on biochemical approaches, which we discuss in the next section. Progress in this area has been somewhat faltering, the main accomplishments being the elimination of some of the more easily testable hypothesis. Nonetheless, as we shall try to show in Section 11.6, the results are surprisingly consistent with the genetic and other findings described in this chapter, and together with them lead to a coherent if tentative picture of the lethal actions of glucocorticoids.

11.5.1 *Relations among early glucocorticoid effects and cell death*

Much thinking and experimentation regarding early glucocorticoid effects in lymphoid cells has been focused on cell death as the terminal outcome of these effects. The most widely studied systems have been cells from rat and mouse thymus and lymphosarcoma P1798, and effects considered have usually been among those described in Section 11.2 (see Munck and Leung, 1977, for background).

Perhaps the first concrete hypothesis to be proposed regarding relations among glucocorticoid effects was that the inhibitory effects and cell death were direct consequences of the rapid inhibition of glucose uptake. The plausibility of this hypothesis, which had its roots in the bedrock of glucocorticoid physiology (cf. Munck, 1971), stems from observations that inhibition of glucose uptake is one of the most rapid and widespread effects of glucocorticoids in a number of peripheral tissues that are targets for catabolic effects of glucocorticoids, and from the likelihood that glucose is the principal energy source for lymphocytes *in vivo*.

It is one of the few hypotheses that, at least in its narrowest form, has been disproved. A number of observations with thymus cells at one time pointed to a close dependence of inhibition of uridine uptake and protein synthesis, and lowering of ATP levels, on prior inhibition of glucose uptake. It is now clear, however, that *in vitro* inhibition of glucose uptake is neither necessary nor sufficient to account for these effects, nor for the inhibition of AIB transport or the increase in nuclear fragility (cf. Munck and Leung, 1977; Young *et al.*, 1979). The relation of inhibition of glucose uptake to cell death has not been studied with thymus cells, but with a strain of P1798 cells glucocorticoid-induced cell death *in vitro* is unaffected by the presence or absence of glucose (Stevens and Stevens, 1975a).

The potentially important inhibition of lipid synthesis in thymus cells that we have recently reported (Foley *et al.*, 1980) appears to be more closely coupled to inhibition of glucose uptake than any other effect described so far, but its relation to other inhibitory effects and cell death has not yet been investigated.

A conclusion from these studies is that in thymus cells glucocorticoids probably produce several independent effects. Young *et al.* (1979) have shown that the increase in nuclear fragility and the decreases in glucose transport and mitochondrial ATP production are all initiated via separate pathways. They consider the decreases in uridine transport and protein synthesis to be consequences of the decrease in ATP production, and find that rates of protein synthesis are closely coupled to the cellular energy charge.

Among all these independent effects there is none that has been shown to be either necessary or sufficient to cause cell death. Young *et al.* (1979) argue that the inhibitory actions on ATP production may even protect cells and increase their longevity, the effects on glucose uptake supporting these actions *in vivo*. If, in fact, none of the early effects – all of which are under way in thymus cells long before there is measurable cell death – signal that cells are irreversibly committed to die, that would be in accord with the conclusion from the previous section that in CEM cells there is no significant latent period between the time a cell is committed to die and the time it dies.

Burton and colleagues (Burton *et al.*, 1967; Turnell, Clarke and Burton, 1973; Turnell and Burton, 1974, 1975) have proposed that glucocorticoids kill lymphoid cells by raising the intracellular levels of free fatty acids. They found that glucocorticoid treatment *in vivo* raised free fatty acid levels in mouse thymus and sensitive P1978 cells, and lowered levels in resistant P1798 cells. They also found that *in vitro*, glucocorticoids decreased fatty acid oxidation by thymus cells and sensitive P1798 cells, but if anything stimulated oxidation by resistant cells. Furthermore, they showed that fatty acids, when incubated with thymus and sensitive P1798 cells, caused degenerative nuclear changes and loss of viability comparable with those caused by glucocorticoids; similar changes could be produced in the resistant cells, but required much higher fatty acid levels. Isolated nuclei from both sensitive and resistant P1798 cells were damaged by fatty acids. Burton and colleagues envision a mechanism according to which glucocorticoids stimulate release of free fatty acids from triglycerides, with nuclear damage leading to cell death starting when the intracellular free fatty acids reach a certain level. The resistance of resistant P1798 cells is attributed mainly to their capacity to oxidize free fatty acids.

Although this is an attractive hypothesis, there is no evidence that cell death is preceded by increases in intracellular free fatty acids when glucocorticoids kill thymus and P1798 cells *in vitro*. In what appear to be the only published experiments on this question (Zyskowski, Cushman and Munck, 1979), no changes could be detected in free fatty acid levels of rat thymus cells incubated with 1.3 μmol of cortisol/l for up to 2 h. These experiments were designed to study the relation of free fatty acid levels to known early glucocorticoid effects; longer incubations would be required to determine the relation to cell death. If it turns out that no increase precedes cell death *in vitro* the proposed mechanism must be discarded, at least for lethal actions *in vitro*. The significance of the changes in free fatty acids produced by glucocorticoid treatment *in vivo* also would come into question. Those changes could reflect glucocorticoid action elsewhere in the body – for example in adipose tissue where glucocorticoids are known to stimulate fatty acid release and lipolysis (cf. Munck and Leung, 1977).

Even if the free fatty acid hypothesis is not correct, there would still remain the intriguing observation (Turnell *et al.*, 1973) that the cells respond in somewhat similar ways *in vitro* to cortisol and exogenous fatty acids. The similarity of morphological changes is not surprising, since, as we point out in Section 11.3.1, the same changes accompany spontaneous cell death and are induced by a variety of lethal agents. The fact that the cells that are resistant to killing by glucocorticoids are also resistant to killing by fatty acids, however, means that unless the resistant cells differ from the sensitive cells in a specific mechanism related to fatty acid metabolism, as Burton and colleagues propose, their resistance must be due to some more general ability to withstand injury.

It would be interesting to know, for example, if these cells are more resistant to radiation. Already in his early studies Schrek (1949) noted that thymus cells were sensitive and bone marrow cells were insensitive to the lethal actions of both glucocorticoids and X-rays, and suggested that this parallel reflected common underlying mechanisms.

Young and his colleagues (Nicholson and Young, 1978; Young *et al.*, 1979) have made an explicit suggestion for a mechanism of resistance that is based on their observations on nuclear fragility. Increased nuclear fragility (Giddings and Young, 1974) would seem the most likely among early glucocorticoid effects to be a forerunner of cell death, yet even this effect does not appear to be coupled tightly to lethal glucocorticoid actions. Nicholson and Young (1978) have found that both the sensitive and resistant lines of P1978 cells exhibit increased nuclear fragility on exposure to glucocorticoids *in vitro*, despite the fact that *in vivo* the resistant line is unaffected by doses of glucocorticoid that reduce the weight of tumours of the sensitive line by 70 per cent. The magnitude of the increase in nuclear fragility is about the same in both lines, but in the resistant line it develops more slowly. To account for these observations they suggest that the resistant and sensitive cell lines differ less in their initial response to glucocorticoids than in the ability of their membranes to withstand the later consequences of glucocorticoid actions. This hypothesis might be testable by measurement of resistance to lysis by osmotic shock.

Kaiser and Edelman (1977) found with rat thymus cells *in vitro* that a bivalent cation ionophore had cytolytic effects similar to those of glucocorticoids, similarly correlated to increases in ^{45}Ca uptake, and that removal of Ca^{2+} from the medium blunted the glucocorticoid effect. They suggested that enhanced Ca^{2+} uptake is involved in glucocorticoid-induced cytolysis of lymphocytes. Subsequently, however (Kaiser and Edelman, 1978), they found different results with lymphocytes from rat lymph nodes. They concluded that glucocorticoid-induced cytolysis in these cells does not depend on Ca^{2+} to the extent it does in thymus cells, and that even in thymus cells Ca^{2+} has a facilitating rather than a central role in causing cell death. Nicholson and Young (1978) state that the increase in nuclear fragility due to glucocorticoids occurs in sensitive P1798 cells before any increase in Ca^{2+} uptake, and does not require Ca^{2+}.

11.5.2 *Indirect effects of glucocorticoids on lymphocytes*

Recently several observations with both malignant and normal human cells have suggested that some apparent cytotoxic actions of glucocorticoids may be exerted indirectly through effects on other cells. Glucocorticoids added to lymphocytes in culture taken directly from patients with lymphoid leukaemias and lymphomas generally give only modest cytotoxic effects. For example, we have found that 20–60 per cent of such cells are killed over a 6-day period of culture *in vitro* with 100 nmol of dexamethasone/l (Crabtree *et al.*, 1978, 1979b). These are small effects compared with the more than 99 per cent cell kill which appears to be associated with induction of remission in acute leukaemia and lymphoma. The time course of cell killing in culture also differs from the response *in vivo*: *in vitro* there is little effect on the viability of the cells until 48–72 h; clinically, glucocorticoids used in treating acute lymphocytic leukaemia produce a decrease in the white cell count as well as an increase in the serum uric acid within 24 h (Wolff *et al.*, 1967; Ranney and Gellhorn, 1957). In preliminary studies we have found, furthermore, that cells of several patients who did not respond clinically to glucocorticoids were killed by glucocorticoids *in vitro*. These disparities between the glucocorticoid sensitivity *in vitro* and *in vivo* lead to the conclusion that additional factors may be responsible for the cytotoxic effects observed *in vivo*.

This conclusion is supported by studies from which it has become apparent that proliferation of lymphocytes *in vitro* is controlled by a separate class of hormone-like growth factors. Using one of these factors (T-cell growth factor) it has been possible to maintain normal T lymphocytes as well as antigen-specific cytotoxic T cells in continuous exponential growth for several years (Morgan, Ruscetti and Gallo, 1976; Gillis and Smith; 1977). Since it appears that this factor and others like it may be responsible for the proliferation of lymphocytes *in vivo*, glucocorticoid effects have been sought on the production of T-cell growth factor and the function of cytotoxic cell lines. The results indicate that the growth-inhibitory effects as well as the cytotoxic effect of glucocorticoids may in part be related to inhibition of production of specific growth factors.

Glucocorticoids produce a dose-dependent reduction in the rate of thymidine incorporation in mitogen-stimulated lymphocytes. An example of this response using dexamethasone and human peripheral lymphocytes is shown by the solid circles in Fig. 11.3. Reduction in thymidine incorporation is paralleled by a similar decrease in the production of T-cell growth factor (open circles). That inhibition of production of growth factor, and not direct effects of glucocorticoids, was responsible for the inhibition of thymidine incorporation was shown by the fact that T-cell growth factor added to cultures of human lymphocytes stimulated by concanavalin A completely overcomes the inhibitory effect of dexamethasone (Fig. 11.4). This occurs at concentrations of T-cell growth factor which do not increase the rate of thymidine incorporation in mitogen-stimulated cells in the absence of dexamethasone.

The importance of these studies is that they suggest that glucocorticoids may accelerate cell death as well as inhibit growth by reducing the production of soluble growth factors necessary for growth and survival *in vivo*. It is also apparent that such effects on production of lymphocyte growth factors may be responsible for immunosuppressive actions of glucocorticoids. Thus, by interfering with the production of factors necessary for the clonal expansion of antigen-stimulated lymphocytes, a greater immunosuppressive effect would be produced than could be accounted for by the modest inhibitory effects observed on differentiated cells (Crabtree *et al.*, 1979a; Crabtree, Gillis, Smith and Munck, 1980). While the physiological importance of these inhibitory effects is largely speculative, they suggest new approaches to the study of the immunosuppressive and cytotoxic effects of glucocorticoids.

11.6 Conclusions

In this final section we shall try to weave together into a coherent picture the various strands of observations and ideas on glucocorticoid-induced lymphocytolysis that we have followed through this chapter. The many sources we draw on are referred to in the preceding sections. A theme we have touched on repeatedly is that of the differences between responses to glucocorticoids of lymphocytes *in vitro* and lymphocytes in their natural setting *in vivo*. This theme has come up in

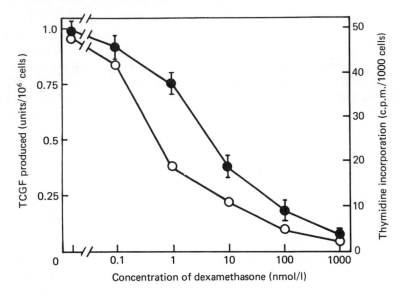

Fig. 11.3 T-cell growth factor production and [³H]thymidine incorporation by human peripheral lymphocytes in culture stimulated for 48 h by phytohaemagglutinin in the presence of various concentrations of dexamethasone. T-cell growth factor (TCGF) production (○) was assayed in the supernatants using a bioassay which measures the ability of the sample to maintain the proliferation of a growth-factor-dependent T-cell line. This assay and the cell line have been described in detail (Gillis and Smith, 1977; Gillis, Ferm, Ou and Smith, 1978). In separate experiments the cell line used to measure T-cell growth factor was found to be relatively resistant to the effects of dexamethasone (Gillis, *et al.*, 1979a, 1979b). [³H]Thymidine incorporation (●) was measured as previously described (Crabtree *et al.*, 1978). (Redrawn from Gillis *et al.*, 1979a.)

connection with the existence of responses *in vivo*, such as redistribution of lymphocytes, that are not associated with cell death (Section 11.3.2); in connection with apparent differences in sensitivity between the same lymphocyte population *in vivo* and *in vitro* (Section 11.3.3); in connection with the apparent absence *in vivo* of clones of resistant lymphocytes such as appear frequently in culture, with the corollary that the inhibitory effects of glucocorticoids must confer selective advantage *in vivo* over cells that are resistant to these effects (Section 11.4.2) – for which a parallel conclusion has been derived from metabolic studies (Section 11.5.1); and finally, in connection with the possibility that some glucocorticoid-induced influences on lymphocytes *in vivo* may be due to direct effects on other cells or tissues (Sections 11.5.1 and 11.5.2). These differences do not imply that the phenomena observed *in vitro* are artefacts (though that may sometimes be the case), but only that in the controlled environment of the culture dish or incubation vessel the interactions and responses of the cells are drastically curtailed. In any case, from

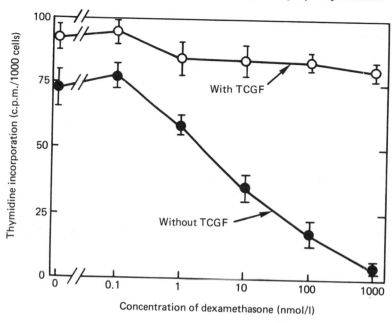

Fig. 11.4 The ability of T-cell growth factor to overcome dexamethasone-induced inhibition of mitogenesis in phytohaemagglutinin-stimulated human peripheral blood lymphocytes. [³H] Thymidine incorporation was measured after 72 h exposure to phytohaemagglutinin and several concentrations of dexamethasone with and without T-cell growth factor (TCGF) at 1.0 unit/ml. (Redrawn from Gillis *et al.*, 1979a.)

the standpoint of understanding the mechanisms of glucocorticoid-induced cell death, we have at present no choice but to rely heavily on observations *in vitro*, recognizing that extrapolation to the whole organism must be done with care.

Regarding the primary mechanisms by which glucocorticoids initiate their actions on lymphocytes, practically all the evidence points to one universal mechanism, the 'central dogma' (Section 11.2) according to which every hormone effect begins with formation of glucocorticoid–receptor complexes, which become 'activated' and are translocated to nuclear sites where they stimulate transcription of certain mRNAs that are then translated to specific effector proteins. The effector proteins directly or indirectly produce the effect.

The essential role of the glucocorticoid receptors, of activation and of nuclear binding is confirmed by the extensive studies with mutants (Section 11.4) that have revealed no glucocorticoid-sensitive mutant that lacks receptors or has receptors that are incapable of forming activated complexes that bind to nuclei. The truly surprising outcome of these studies is that practically all glucocorticoid-insensitive mutants isolated so far have defective receptors.

Secondary glucocorticoid effects that result from these primary mechanisms and precede cell death constitute a fairly extensive repertoire that can be observed with almost any sensitive lymphocyte population after at most a few hours exposure to glucocorticoids. They are generally inhibitory, and include inhibition of glucose uptake, protein synthesis, uridine transport, ATP synthesis and others described in Section 11.2. Several different primary induction processes appear to be involved in producing these effects (Section 11.5.1).

Despite the variety of secondary effects, cell death itself, the ultimate consequence of glucocorticoid action on lymphocytes, is approached through a stereotyped sequence of morphological changes that start in the nucleus (Section 11.3.1), and that are similar to changes that take place when cells die spontaneously or death is induced by certain other agents.

Thus the glucocorticoid actions begin through a single general mechanism, and end in a seemingly general mechanism of cell death. How the beginning is linked to the end is the problem that has stimulated much of the work in this field. An assumption underlying the approach of many researchers is that there is a linear sequence of events, a single path, that once embarked on leads irreversibly to cell death. The search for events along such a path has focused on the secondary inhibitory effects — inhibition of glucose uptake, etc. — and, as we have detailed in Section 11.5.2, has so far been unsuccessful. Not only have particular secondary inhibitory effects like that on glucose uptake been dissociated from cell death, but most of these effects have been dissociated from each other. None of them has been shown to lead irreversibly to cell death, neither the inhibitory metabolic effects nor the increase in 'nuclear fragility', nor even the early morphological changes seen in the nucleus.

Though this mass of negative evidence is hardly conclusive, it does bring to the fore the question of whether any irreversible path at all links the secondary events to cell death. After surveying this subject from the many points of view we have presented in this chapter, we are inclined to answer the question in the negative, and adopt the hypothesis that *cell death is the final response to accumulated defects that can be generated by glucocorticoids in a variety of ways through many sets of paths*.

With this hypothesis, which simply applies to the cell the conventional idea that death of a complex organism can be brought about in many ways, not only does the metabolic evidence become understandable, but a number of otherwise disparate observations which we now summarize can also be reconciled.

As we have already pointed out (Sections 11.4.2 and 11.5), the fact that almost all glucocorticoid-resistant mutants have defective receptors can be accounted for if there are many different paths to cell death. The necessity for an accumulation of defects can explain why brief exposure to hormone, sufficient to initiate early events, will not necessarily lead to cell death (Section 11.2.1). This necessity is consistent furthermore, with the observations with CEM cells that imply that cell death is not predetermined until very shortly before the cells die (Section 11.5). (Since the experiments with CEM cells provide the most direct evidence against an early irreversible commitment to cell death, they should be repeated with other cell lines.) Finally, the necessity for an accumulation of defects is also consistent with the indications that resistance to glucocorticoids may be due not only to an inability to respond to glucocorticoids, as in the case of receptor-defective mutants (Section 11.4.2), but to the ability of some cells to withstand the effects of glucocorticoids without dying (Section 11.5.2).

Acknowledgements

The preparation of this chapter was supported by Research Grants AM03535 and CA17323 from the US Public Health Service.

References

Aronow, L. and Gabourel, J. D. (1962), Development of a hydrocortisone-resistant subline of mouse lymphoma *in vitro*. *Proc. Soc. Exp. Biol. Med.*, **111**, 348–349.

Baxter, J. D., Harris, A. W., Tomkins, G. M. and Cohn, M. (1971), Glucocorticoid receptors in lymphoma cells in culture: relationship to glucocorticoid killing activity. *Science*, **171**, 189–191.

Bell, P. A. and Borthwick, N. M. (1979a), Regulation of transcription in rat thymus cells by glucocorticoids. *J. Steroid Biochem.*, **11**, 381–387.

Bell, P. A. and Borthwick, N. M. (eds.) (1979b), Glucocorticoid Action and Leukemia. Alpha Omega Publishers Ltd., Cardiff, Wales.

Blomgren, H. and Andersson, B. (1969), Evidence for a small pool or immuno-competent cells in the mouse thymus. *Exp. Cell Res.*, **57**, 185–192.

Bourgeois, S. and Newby, R. (1977), Diploid and haploid states of the gluco-corticoid receptor gene of mouse lymphoid cell lines. *Cell*, **11**, 423–430.

Bourgeois, S. (1979), Genetic analysis of glucocorticoid-induced lymphocyto-lysis. In Steroid Receptors and the Mangement of Cancer, vol. II, (eds. E. B. Thompson and M. E. Lippman), CRC Press, Boca Raton, FL, pp. 99–111.

Burton, A. F., Storr, J. M. and Dunn, W. L. (1967), Cytolytic action of cortico-steroids on thymus and lymphoma cells *in vitro*. *Can. J. Biochem.*, **45**, 289–297.

Bush, I. E. (1953), Species differences in adrenocortical secretion. *J. Endocrinol.*, **9**, 95–100.

Butler, W. T. and Rossen, R. D. (1973a), Effects of corticosteroids on immunity in man. *J. Clin. Invest.*, **52**, 2629–2640.

Butler, W. T. and Rossen, R. D. (1973b), Effects of corticosteroids on immunity in man. II. Alterations in serum protein components after methylprednisolone. *Transplant Proc.*, **5**, 1215–1219.

Caffey, J. and di Liberti, C. (1959), Acute atrophy of the thymus induced by adrenocorticosteroids; observed roentgenographically in living human infants. *Am. J. Roentgenol.*, **82**, 530–540.

Caffey, J. and Silbey, R. (1960), Regrowth and overgrowth of the thymus after atrophy induced by the oral administration of adrenocorticosteroids to human infants. *Pediatrics*, **26**, 762–770.

Claman, H. N. (1972), Corticosteroids and lymphoid cells. *N. Engl. J. Med.*, **287**, 388–397.

Claman, H. N., Moorhead, J. W. and Benner, W. H. (1971), Corticosteroids and lymphoid cells *in vitro*. I. Hydrocortisone lysis of human, guinea pig, and mouse thymus cells. *J. Lab. Clin. Med.*, **78**, 499–507.

Cohen, J. J., Fischback, M. and Claman, H. N. (1970), Hydrocortisone resistance of graft vs host activity in mouse thymus, spleen and bone marrow. *J. Immunol.,* **105**, 1146–1150.

Cowan, W. K. and Sorensen, G. D. (1964), Electron microscopic observations of acute thymic involution produced by hydrocortisone. *Lab. Invest.,* **13**, 353–370.

Crabtree, G. R., Gillis, S., Smith, K. A. and Munck, A. (1979a), Glucocorticoids and immune responses. *Arthritis Rheum.,* **22**, 1246–1256.

Crabtree, G. R., Gillis, S., Smith, K. A. and Munck, A. (1980), Mechanisms of glucocorticoid induced immunosuppression: Inhibitory effects on expression of Fc receptors and production of T-cell growth factor. *J. Steroid Biochem.,* **12**, 445–449.

Crabtree, G. R., Smith, K. A. and Munck, A. (1978), Glucocorticoid receptors and sensitivity of isolated human leukemia and lymphoma cells. *Cancer Res.,* **38**, 4268–4272.

Crabtree, G. R., Smith, K. A. and Munck, A. (1979b), Glucocorticoid receptors and *in vitro* sensitivity of cells from patients with leukemia and lymphoma: A reassessment. In *Glucocorticoid Action and Leukemia* (eds. P. A. Bell and N. M. Borthwick), Alpha Omega Publishers Ltd., Cardiff, Wales, pp. 191–204.

Dale, D. C., Fauci, A. S. and Wolfe, S. M. (1974), Alternate day prednisone therapy. Leukocyte kinetics and susceptibility to infections. *N. Engl. J. Med.,* **291**, 1154–1158.

Davis, J. M., Chan, A. K. and Thompson, E. A., Jr. (1980), Nonmutational alteration in glucocorticoid sensitivity of lymphosarcoma P1798. *J. Natl. Cancer Inst.,* **64**, 55–62.

Dougherty, T. F., Berliner, M. L., Schneebeli, G. L. and Berliner, D. L. (1964), Hormonal control of lymphatic structure and function. *Ann. N. Y. Acad. Sci.,* **113**, 825–843.

Dougherty, T. F. and White, A. (1945), Functional alterations in lymphoid tissue induced by adrenal cortical secretion. *Am. J. Anat.,* **77**, 81–116.

Duval, D., Dardenne, M., Dausse, J. P. and Homo, F. (1977), Glucocorticoid receptors in corticosensitive and corticoresistant thymocyte subpopulations. II. Studies with hydrocortisone treated mice. *Biochim. Biophys. Acta,* **496**, 312–320.

Ezdinli, E. Z., Stutzman, L., Aungst, C. W. and Firat, D. (1969), Corticosteroid therapy for lymphomas and chronic lymphocytic leukemia. *Cancer,* **23**, 900–909.

Fajer, A. B. and Vogt, M. (1963), Adrenocortical secretion in the guinea-pig. *J. Physiol. (London),* **169**, 373–385.

Fauci, A. S. (1978), Mechanisms the immunosuppressive and anti-inflammatory effects of glucocorticosteroids. *J. Immunopharmacol.,* **1**, 1–25.

Fauci, A. S. and Dale, D. C. (1974), The effect of *in vivo* hydrocortisone on subpopulations of human lymphocytes. *J. Clin. Invest.,* **53**, 240–246.

Fauci, A. S., Dale, D. C. and Balow, J. E. (1976), Glucocorticoid therapy: mechanisms of action and clinical considerations. *Ann. Int. Med.,* **84**, 304–315.

Fauci, A. S., Pratt, K. R. and Whalen, G. (1977), Activation of human B lymphocytes. IV. Regulatory effects of corticosteroids on the triggering signal in the plaque-forming cell response of human peripheral blood B lymphocytes to polyclonal activation. *J. Immunol.,* **119**, 598–603.

Foley, J. E., Jeffries, M. and Munck, A. (1980), Glucocorticoid effects on incorporation of lipid and protein precursors into rat thymus cell fractions. *J. Steroid Biochem.,* **12**, 231–243.

Frenkel, J.K. and Havenhill, M.A. (1963), The corticoid sensitivity of golden hamsters, rats and mice. Effects of dose, time and route of administration. *Lab. Invest.*, **12**, 1204–1220.

Giddings, S.J. and Young, D.A. (1974), An effect *in vitro* of physiological levels of cortisol and related steroids on the structural integrity of the nucleus in rat thymic lymphocytes as measured by resistance to lysis. *J. Steroid. Biochem.*, **5**, 587–595.

Gillis, S., Crabtree, G.R. and Smith, K.A. (1979a), Glucocorticoid-induced inhibition of T cell growth factor production. I. The effect on mitogen-induced lymphocyte proliferation. *J. Immunol.*, **123**, 1624–1631.

Gillis, S., Crabtree, G.R. and Smith, K.A. (1979b), Glucocorticoid-induced inhibition of T cell growth factor production. II. The effect on the *in vitro* generation of cytotoxic T cells. *J. Immunol.*, **123**, 1631–1635.

Gillis, S., Ferm, M.M., Ou, W. and Smith, K.A. (1978), T cell growth factor: parameters of production and a quantitative microassay for activity. *J. Immunol.*, **120**, 2027–2032.

Gillis, S. and Smith, K.A. (1977), Long term culture of tumor specific cytotoxic T cells. *Nature (London)*, **268**, 154–156.

Glasser, L., Huestis, D.W. and Jones, J.F. (1977), Functional capabilities of steroid-recruited neutrophiles harvested for clinical transfusion. *N. Engl. J. Med.*, **297**, 1033–1036.

Hackney, J.J., Gross, S.R., Aronow, L. and Pratt, W.B. (1970), Specific glucocorticoid binding macromolecules from mouse fibroblasts growing *in vitro*: A possible steroid receptor for growth inhibition. *Mol. Pharmacol.*, **6**, 500–512.

Harmon, J.M., Norman, M.R. and Thompson, E.B. (1979a), Human leukemic cell lines in culture – a model system for the study of glucocorticoid-induced lymphocytolysis. In *Steroid Receptors and the Management of Cancer*, vol. II (eds. E.B. Thompson and M.E. Lippman), CRC Press, Boca Raton, FL, pp. 113–129.

Harmon, J.M., Norman, M.R., Fowlkes, B.J. and Thompson, E.B. (1979b), Dexamethasone induces irreversible G_1 arrest and death of a human lymphoid cell line. *J. Cell. Physiol.*, **98**, 267–278.

Harris, A.W. (1970), Differentiated functions expressed by cultured mouse lymphoma cells. *Exp. Cell Res.*, **60**, 341–353.

Harris, A.W., Bankhurst, A.D., Mason, S. and Warner, N.L. (1973), Differentiated functions expressed by cultured mouse lymphoma cells: II. O Antigen, surface immunoglobin and a receptor for antibody on cells of a thymoma cell line. *J. Immunol.*, **110**, 431–438.

Haynes, B.F. and Fauci, A.S. (1978), The differential effect of *in vivo* hydrocortisone on the kinetics of subpopulations of human peripheral blood thymus-derived lymphocytes. *J. Clin. Invest.*, **61**, 703–707.

Homo, F., Duval, D., Thierry, C. and Serrou, B. (1979), Human lymphocyte subpopulations: effects of glucocorticoids *in vitro*. *J. Steroid Biochem.*, **10**, 609–613.

Horibata, K. and Harris, A.W. (1970), Mouse myelomas and lymphomas in culture. *Exp. Cell Res.*, **60**, 61–77.

Kaiser, N. and Edelman, I.S. (1977), Calcium dependence of glucocorticoid-induced lymphocytolysis. *Proc. Natl. Acad. Sci. U.S.A.*, **74**, 638–642.

Kaiser, N. and Edelman, I.S. (1978), Further studies on the role of calcium in glucocorticoid-induced lymphocytolysis. *Endocrinology*, **103**, 936–942.

Kaiser, N., Milholland, R.J. and Rosen, F. (1974), Glucocorticoid receptors and mechanism of resistance in the cortisol-sensitive and resistant lines of lymphosarcoma P1798. *Cancer Res.*, **34**, 621–626.

Kirkpatrick, A. F., Milholland, R. J. and Rosen, F. (1971), Stereospecific glucocorticoid binding to subcellular fractions of the sensitive and resistant lymphosarcoma P1978. *Nature (London) New Biol.*, **232**, 216–218.

Krohn, P. L. (1954), The effect of ACTH on the reaction to skin homografts in rabbits. *J. Endocrinol.*, **11**, 71–77.

Krohn, P. L. (1955), The effect of ACTH and cortisone on the survival of skin homografts and on the adrenal glands in monkeys (*Macaca Mulata*). *J. Endocrinol.*, **12**, 220–226.

Lampkin, J. M. and Potter, M. (1958), Response to cortisone and development of cortisone resistance in a cortisone-sensitive lymphosarcoma of the mouse. *J. Natl. Cancer Inst.*, **20**, 1091–1108.

Leikin, S. L., Brubaker, C., Hartman, J. R., Murphy, M. L., Wolff, J. A. and Perrin, E. (1968), Varying prednisone dosage in remission induction of previously untreated childhood leukemia. *Cancer*, **21**, 346–351.

Leung, K. and Munck, A. (1975), Long-term incubation of rat thymus cells: cytolytic actions of glucocorticoids *in vitro*. *Endocrinology*, **97**, 744–748.

Levine, M. A. and Claman, H. N. (1970), Bone marrow and spleen: dissociation of immunologic properties by cortisone. *Science*, **167**, 1515–1517.

Lippman, M. E. and Barr, R. (1977), Glucocorticoid receptors in purified subpopulations of human peripheral blood lymphocytes. *J. Immunol.*, **118**, 1977–1981.

Lippman, M. E., Perry, S. and Thompson, E. B. (1974), Cytoplasmic glucocorticoid-binding proteins in glucocorticoid-unresponsive human and mouse leukemia cell lines. *Cancer Res.*, **34**, 1572–1576.

Long, D. A. (1957), The influence of corticosteroids on immunological responses to bacterial infections. *Int. Arch. Allergy*, **10**, 5–12.

Mauer, A. M. (1965), Diurnal variation of proliferative activity in the human bone marrow. *Blood*, **26**, 1–7.

Morgan, D. A., Ruscetti, F. W. and Gallo, R. (1976), Selective *in vitro* growth of T-lymphocytes from normal human bone marrows. *Science*, **193**, 1007–1008.

Munck, A. (1971), Glucocorticoid inhibition of glucose uptake by peripheral tissues: old and new evidence, molecular mechanisms, and physiological significance. *Perspect. Biol. Med.*, **14**, 265–289.

Munck, A. (1976), General aspects of steroid hormone–receptor interactions. In *Receptors and Mechanism of Action of Steroid Hormones, Part I* (ed. J. R. Pasqualini), Marcel Dekker, New York, pp. 1–40.

Munck, A., Crabtree, G. R. and Smith, K. A. (1979), Glucocorticoid receptors and actions in rat thymocytes and immunologically stimulated human peripheral lymphocytes. In *Glucocorticoid Hormone Action* (eds. J. B. Baxter and G. G. Rousseau), Springer-Verlag, Berlin, pp. 341–355.

Munck, A. and Leung, K. (1977), Glucocorticoid Receptors and Mechanisms of action. In *Receptors and Mechanism of Action of Steroid Hormones, Part II* (ed. J. R. Pasqualini), Marcel Dekker, New York, pp. 311–397.

Munck, A. and Young, D. A. (1975), Corticosteroids and lymphoid tissue. *Hand. Physiol. Sect. 7*, **6**, 231–243.

Nicholson, M. L. and Young, D. A. (1978), Effect of glucocorticoid hormones *in vitro* on the structural integrity of nuclei in corticosteroid-sensitive and -resistant lines of lymphosarcoma P1798. *Cancer Res.*, **38**, 3673–3680.

Norman, M. R., Harmon, J. M. and Thompson, E. B. (1978), Use of a human lymphoid cell line to evaluate interactions between prednisolone and other chemotherapeutic agents. *Cancer Res.*, **38**, 4273–4278.

Norman, M. R. and Thompson, E. B. (1977), Characterization of a glucocorticoid-sensitive human lymphoid cell line. *Cancer Res.*, **37**, 3785–3790.

Nowell, P. C. (1961), Inhibition of human leukocyte mitosis by prednisolone *in vitro*. *Cancer Res.*, **21**, 1518–1521.

Ranney, H. M. and Gellhorn, A. (1957), The effect of massive prednisone and prednisolone therapy on acute leukemia and malignant lymphoma. *Am. J. Med.*, **22**, 405–413.

Rosen, J. M., Fina, J. J., Milholland, R. J. and Rosen, F. (1970), Inhibition of glucose uptake in lymphosarcoma P1798 by cortisol and its relationship to the biosynthesis of deoxyribonucleic acid. *J. Biol. Chem.*, **245**, 2074–2080.

Rosen, J. M., Fina, J. J., Milholland, R. J. and Rosen, F. (1972), Inhibitory effect of cortisol *in vitro* on 2-deoxyglucose uptake and RNA and protein metabolism in lymphosarcoma P1798. *Cancer Res.*, **32**, 350–355.

Rosen, J. M., Rosen, F., Milholland, R. J. and Nichol, C. A. (1970), Effects of cortisol on DNA metabolism in the sensitive and resistant lines of mouse lymphoma P1798. *Cancer Res.*, **30**, 1129–1136.

Rosenau, W., Baxter, J. D., Rousseau, G. G. and Tomkins, G. M. (1972), Mechanism of resistance to steroids: glucocorticoid receptor defect in lymphoma cells. *Nature (London) New Biol.*, **237**, 20–24.

Schindler, W. J. and Knigge, K. M. (1959), Adrenal cortical secretion by the golden hamster. *Endocrinology*, **65**, 739–747.

Schmidt, T. J., Harmon, J. M. and Thompson, E. B. (1980), 'Activation-labile' glucocorticoid-receptor complexes of a steroid-resistant variant of CEM-C7 human lymphoid cells. *Nature*, **286**, 507–510.

Schrek, R. (1949), Cytotoxic action of hormones of the adrenal cortex according to the method of unstained cell counts. *Endocrinology*, **45**, 317–334.

Schrek, R. (1961), Cytotoxicity of adrenal cortex hormones on normal and malignant lymphocytes of man and rat. *Proc. Soc. Exp. Biol. Med.*, **108**, 328–332.

Schrek, R. (1964), Prednisolone sensitivity and cytology of viable lymphocytes as test for chronic lymphocytic leukemia. *J. Natl. Cancer Inst.*, **33**, 837–847.

Sibley, C. H. and Tomkins, G. M. (1974a), Isolation of lymphoma cell variants resistant to killing by glucocorticoids. *Cell*, **2**, 213–220.

Sibley, C. H. and Tomkins, G. M. (1974b), Mechanisms of steroid resistance. *Cell*, **2**, 221–227.

Sibley, C. H. and Yamamoto, K. R. (1979), Mouse lymphoma cells: Mechanisms of resistance to glucocorticoids. In *Glucocorticoid Hormone Action* (eds. J. D. Baxter and G. G. Rousseau), Springer-Verlag, Berlin, pp. 357–376.

Spackman, D. H. and Riley, V. (1977), Corticosterone concentrations in the mouse. *Science*, **200**, 87.

Stevens, J., Mashburn, L. T. and Hollander, V. P. (1969), Effect of 9α-fluoro-prednisolone and L-asparaginase on uridine incorporation into ribosomal RNA of P1798 lymphosarcoma. *Biochim. Biophys. Acta*, **186**, 332–339.

Stevens, J. and Stevens, Y. M. (1975a), Cortisol-induced lymphocytolysis of P1798 tumor cells in glucose-free, pyruvate-free medium. *J. Natl. Cancer Inst.*, **54**, 1493–1494.

Stevens, J. and Stevens, Y. W. (1975b), Sequential irreversible, actinomycin D-sensitive, and cycloheximide-sensitive steps prior to cortisol inhibition of uridine incorporation by P1798 tumor lymphocytes. *Cancer Res.*, **35**, 2145–2153.

Stevens, J., Stevens, Y. W., Behrens, U. and Hollander, V. P. (1973), Role of nucleoside transport in glucocorticoid-induced regression of mouse lymphoma P1798. *Biochem. Biophys. Res. Commun.*, **50**, 799–806.

Tuchinda, M., Newcomb, R. W., DeVald, B. L. (1972), Effect of prednisone treatment on the human immune response to keyhole limpet hemocyanin. *Int. Arch. Allergy*, **42**, 533–544.

Turnell, R. W. and Burton, A. F. (1974), Studies on the mechanism of resistance to lymphocytolysis induced by corticosteroids. *Cancer Res.*, **34**, 39–42.

Turnell, R. W. and Burton, A. F. (1975), Glucocorticoid receptors and lymphocytolysis in normal and neoplastic lymphocytes. *Mol. Cell. Biochem.*, **9**, 175–189.

Turnell, R. W., Clarke, L. H. and Burton, A. F. (1973), Studies on the mechanism of corticosteroid-induced lymphocytolysis. *Cancer Res.*, **33**, 203–212.

Vietti, T. J., Sullivan, M. P., Berry, D. H., Haddy, T. B., Haggard, M. E. and Blattner, R. J. (1965), The response of acute childhood leukemia to an initial and a second course of prednisone. *J. Pediat.*, **66**, 18–26.

Waddell, A. W., Wyllie, A. H., Robertson, A. M. G., Mayne, K., Au, J. and Currie, A. R. (1979), Lethal and growth-inhibitory actions of glucocorticoids. In *Glucocorticoid Action and Leukemia* (eds. P. A. Bell and N. M. Borthwick), Alpha Omega Publishers Ltd., Cardiff, Wales, pp. 75–83.

Weissman, I. L. and Levy, R. (1975), *In vitro* cortisone sensitivity of *in vivo* cortisone-resistant thymocytes. *Isr. J. Med. Sci.*, **11**, 884–888.

Wolff, J. A., Brubaker, C. A., Murphy, M. L., Pierce, M. I. and Severo, N. (1967), Prednisone therapy of acute childhood leukemia: prognosis and duration of response in 330 treated patients. *J. Pediat.*, **70**, 626–631.

Yamamoto, K. R., Stampfer, M. R. and Tomkins, G. M. (1974), Receptors from glucocorticoid-sensitive lymphoma cells and two classes of insensitive clones: Physical and DNA-binding properties. *Proc. Natl. Acad. Sci. U.S.A.*, **71**, 3901–3905.

Young, D. A., Nicholson, M. L., Guyette, W. A., Giddings, S. J., Mendelsohn, S. L., Nordeen, S. K. and Lyons, R. T. (1979), Biochemical endpoints of glucocorticoid hormone action. In *Glucocorticoid Action and Leukemia* (eds. P. A. Bell and N. M. Borthwick), Alpha Omega Publishers Ltd., Cardiff, Wales, pp. 53–74.

Yu, D. T. Y., Clements, P. J., Paulus, H. E., Peter, J. B., Levy, J. and Barnett, E. V. (1974), Human lymphocyte subpopulations. Effects of corticosteroids. *J. Clin. Invest.*, **53**, 565–571.

Zyskowski, L. P., Cushman, S. W. and Munck, A. (1979), Absence of an early effect of glucocorticoids on nonesterified fatty acid accumulation in isolated rat thymus cells. *J. Steroid Biochem.*, **11**, 1639–1640.

Zyskowski, L. P. and Munck, A. (1979), Kinetic studies on the mechanism of glucocorticoid inhibition of hexose transport in rat thymocytes. *J. Steroid Biochem.*, **10**, 573–579.

12 The role of the LT system in cell destruction *in vitro*

MONICA W. ROSS, ROBERT S. YAMAMOTO AND
GALE A. GRANGER

12.1 Introduction

When a foreign antigen is introduced into host tissues, the host lymphoid system
can respond in two ways. One population of lymphoid cells, B lymphocytes,
responds by producing humoral antibodies. The second response is cellular, termed
cell-mediated immunity (CMI), and is mediated primarily by non-B lymphoid cell
populations, and it can lead to specific or non-specific tissue destruction. The
principal effectors of these reactions appear to be certain lymphocyte subpopulations and monocytes or macrophages. Situations in which these reactions are
operative are allograft rejection, tumour immunity, autoimmune diseases and
delayed hypersensitivity.

In an effort to understand the molecular mechanism operative in CMI reactions
further, *in vitro* systems have been developed. Extensive research in this area has
revealed that mammalian lymphocytes can mediate four classes of CMI *in vitro*.
One class is T-lymphocyte-mediated killing (CTL) (Goldstein and Smith, 1977;
Henney, 1977; Martz, 1977), which is mediated by thymus-derived (T) lymphocytes preimmunized with antigen *in vivo* or *in vitro*. These T cells possess antigen-
binding receptors specific for the cellular antigen(s) present on the surface of the
sensitizing target cell. Although the antigen-binding receptors are not classical
immunoglobulins, they have been termed IgT, and are only found on the surface
of the T cell. These receptors appear to direct the lytic reaction, causing lysis to be
specific for the sensitizing target cell line. Another class of CMI is natural killing
(NK) (Bonnard and West, 1980), mediated by certain non-B lymphocytes. It is
selective for some target cell lines, but the nature of the recognition mechanism
of natural killing is still unclear. The third class of CMI is antibody-dependent
cellular cytotoxicity (ADCC) (MacLennan, 1972), which may be mediated by
the same subpopulation of lymphocytes which mediates natural killing (Bonnard
and West, 1980). ADCC is initiated and directed by antibody, in particular immunoglobulin G or M (IgG or IgM), bound to the target cell surface. Finally, a system
employed solely for experimental purposes is mitogen-induced cellular cytotoxicity
(MICC) (Holm, 1969). This is non-specific target cell destruction mediated by T or
B lymphocytes. It requires a mitogen to promote target–effector cell adhesions and
to trigger the reaction.

While the nature of the recognition mechanism appears to be different for each class of CMI, there are several features common to the events which follow recognition (Goldstein and Smith, 1977; Henney, 1977; Martz, 1977). Once the lymphocyte has recognized and achieved direct membrane contact with the target cells, a series of lymphocyte-dependent events occur. These are collectively termed programming for lysis. They lead to delivery of the lethal hit, after which the presence of the effector cell is not required for target cell lysis. An additional feature is that the lytic reaction is localized, so that bystander cells are unaffected. Furthermore, the effector cells remain viable, indicating that the lytic process is unidirectional. Still under intense investigation is the nature of the lethal hit. One theory holds that the aggressor lymphocytes cause some type of mechanical damage, either by inserting destructive lymphocyte membrane components into target cells or tearing the target cell membrane. An alternative hypothesis is that molecules released at the point of contact by effector lymphocytes bind to and destroy target cells.

Lymphocytes respond to stimulation by proliferating and by releasing several types of soluble mediators, collectively referred to as lymphokines (LK) (Bloom, 1971; Granger, 1972). The response can be initiated either specifically, with a soluble or cell-bound antigen (Oppenheim, 1968), or non-specifically, with a mitogen (Ling, 1968). The interaction of mitogens, which include various lectins, with lymphocytes apparently bypasses the specific antigen–receptor interactional event. The mitogenic response is polyclonal. Lymphokines (LK) are generally named in accordance with their various effector functions. For example, materials which attract and activate macrophages are termed macrophage chemotactic factor (MCF) and macrophage-activating factor (MAF) respectively. Several additional LKs are blastogenic factor (BF), proliferation-inhibition factor (PIF), colony-inhibition factor (CIF), interferons (IF) and lymphotoxins (LT) (Cohen, Pick and Oppenheim, 1979). Owing to the fact that these LKs are present in supernatants in very small quantities, biochemical characterization and purification has been exceedingly difficult. In addition, the assays developed to detect their presence are cumbersome and difficult to standardize, making the activities difficult to quantify.

One particular lymphokine class, LT, has been proposed as playing a role in the lytic mechanism of CMI. There are several reviews available which contain more in depth information on previous studies of this mediator (Granger, 1972; Granger, Daynes, Runge, Prieur and Jeffes, 1975; Granger, Yamamoto, Ware, and Ross, 1980). This chapter will present more recent information on the molecular features of LTs, the conditions required for their release, and the capacity of LT to mediate various lytic reactions. These studies have led to the formulation of a new concept, namely, that the toxic molecules associate with receptors to become specific and potent mediators of lysis.

12.2 Molecular characteristics of the LT system of cytotoxic effector molecules

Lymphotoxins are cell-lytic molecules which can be released by lymphocytes *in vitro* upon stimulation by a variety of agents. These lytic materials are released by lymphoid cells obtained from both humans and experimental animals (Granger, 1972; Heise and Weiser, 1969; Kolb and Granger, 1968; Peter, Stratton, Stempel, Yu and Cardin, 1973; Ruddle, 1972; Russell, Rosenau, Goldberg and Kunitomi, 1972; Walker and Lucas, 1972). Release can be specifically initiated by interaction of immune cells with antigens, cellular or soluble, or it can be initiated non-specifically when non-immune cells are stimulated with polyclonal activating agents, such as the lectins concanavalin A (Con A) and phytohaemagglutinin (PHA). Present data indicate that LT release is inducible and requires the continued presence of the stimulating agent in association with cell surface receptor(s) to maintain the release of these materials into the supernatant (Granger *et al.*, 1975). Several assays have been developed which quantify the amount of material present in a supernatant on the basis of units of lytic activity (Spofford, Daynes and Granger, 1974). The target cell routinely used to detect lysis *in vitro* is the murine L-929 cell line, because it is highly sensitive to lysis mediated by all classes of LT from all animal species. The basis of these assays is a dilution method which chooses the lowest dilution capable of killing a set number of target murine L-929 cells over an incubation time of 16 h. While the sensitivity of the L cells may fluctuate, this is one of the few lymphokine assays that can be quantified.

Biochemical studies reveal that LT molecules from both humans and experimental animals are heterogeneous and present in supernatants in very small quantities (Ross, Tiangco, Horn, Hiserodt and Granger, 1979). Separation of these materials has been achieved by the use of traditional biochemical methods, such as molecular-sieving chromatography, ion-exchange chromatography, and various electrophoretic techniques. These materials can be divided into five molecular-weight classes, termed complex (>200 000), alpha heavy (α_H, 160 000), alpha (α, 80 000), beta (β, 45 000) and gamma (γ, 25 000) (Ross *et al.*, 1979; Harris, Yamamoto, Crane and Granger, 1980). Additional studies reveal that these molecular-weight classes can be further resolved into subclasses by ion-exchange chromatography and gel electrophoresis, as shown in Fig. 12.1. A number of parameters determine which molecular-weight classes or subclasses will be detected in a supernatant, such as: (a) the type and physiological state of the stimulated cell; (b) the time when the culture fluid is collected; and (c) the stability of the various molecular-weight forms. These parameters will be discussed more thoroughly in a later section. However, the most stable of these materials appear to be those in the α and β classes. The higher-molecular-weight complex, α_H, and the lower-molecular-weight complex, γ, are those detected least often, for they are the most unstable. The basic molecular-weight classes detected to date appear to be common to all animal species examined (i.e. human, mouse, rat, guinea pig, hamster and rabbit). The heterogeneity of these forms was a difficult problem encountered by investigators,

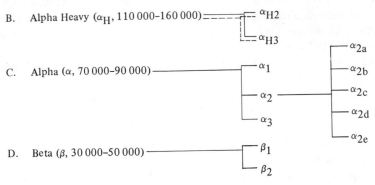

MW Classes (1)   *Ion-Exchange Subclasses* (2)

DEAE-cellulose   Phosphocellulose

Fig. 12.1 Identification of molecular-weight (MW) classes and subclasses of human and murine lymphotoxin. (1) Resolved by molecular-sieving chromatography of supernatants. (2) Resolved by ion-exchange chromatography of a MW class or subclass. ——, Human LT forms. - - - - -, Murine LT forms.

for frequently, different molecular-weight classes were detected which made comparison of materials within and between species difficult. For example, two different investigators studying murine lymphotoxins identified different molecular-weight classes detectable in basically similar supernatants (Kolb and Granger, 1970; Trivers, Baumgart and Leonard, 1976). Present data suggest that these differences appear to be largely due to technical reasons, for cells from many animal species can release identical forms *in vitro* under standardized conditions.

While the story is complex, recent evidence indicates that these are not all different materials, but the various molecular-weight classes are interrelated, and form a subunit system associated by non-covalent bonds (Hiserodt, Yamamoto and Granger, 1978a; Ross *et al.*, 1979; Yamamoto, Hiserodt, Lewis, Carmack and Granger, 1978). The first evidence to support this concept was obtained by immunological studies, for antibodies made against the individual molecular-weight classes or subclasses cross-reacted with the larger and smaller forms. This immunological relationship has been reported for the LT forms obtained from three different animal species, mouse, guinea pig and human (Hiserodt, Tiangco and Granger, 1979c; Ross *et al.*, 1979; Yamamoto *et al.*, 1978). Yet antibodies prepared against forms from one species do not cross-react with those of another species. An example of these immunological relationships can be seen in Table 12.1. Biochemical studies reveal that the complex and α_H forms can dissociate in high-ionic-strength buffers into the smaller α, β and γ subunits. This dissociation has been observed in studies of these larger forms from several different animal species, but most extensively examined with human and murine forms (Ross *et al.*, 1979; Harris, Yamamoto, Crane and Granger, 1980 work; Hiserodt *et al.*, 1978a). While the larger complex and α_H easily dissociate into the smaller α, β and γ forms, the latter cannot be further dissociated. The complex, upon dissociation, gives rise to α_H and the smaller forms. The α_H, upon dissociation, gives rise to only the α, β and γ forms.

Table 12.1 Immunological evidence of relationships between the classes of lymphotoxins of various molecular weights from humans and experimental animals

Source of LT	Antiserum Tested	Molecular-Weight Class of LT Tested				
		Cx	α_H	α	β	γ
A. Human	(a) Rabbit anti-(human α_2)	+	+	+	+	±
	(b) Rabbit anti-(human β_2)	+	+	+	+	±
B. Mouse	Rabbit anti-(mouse α_H)	+	+	+	NT	NT
C. Guinea pig	Rabbit anti-(guinea pig β)	NT	+	+	+	NT
D. Human	(a) Rabbit anti-(guinea pig β)	–	–	–	–	–
	(b) Rabbit anti-(mouse α_H)	–	–	–	–	–

+ Indicates the capacity of the antisera to neutralize the cell lytic activity *in vitro* of the individual LT class. – Indicates no effect. NT indicates not tested.

Recent evidence indicates an exciting new concept, namely that the complex and α_H classes are associated with other non-lytic materials (Harris, Yamamoto, Christensen and Granger, 1980; Hiserodt, *et al.*, 1978a, 1979c; Ware, Klostergaard, Toth, and Granger, 1981). The first of these is an antigen-binding receptor(s), which is not a classical immunoglobulin. Evidence to support this concept first came from immunological studies which revealed that rabbit antisera reactive with the antigen-binding portion of the human IgG molecule (anti-Fab) would neutralize a percentage of the cell-lytic material in the Cx and α_H classes, but had no effect on the smaller α, β and γ classes. Moreover, these LT forms were not neutralized by antibodies directed at the non-antigen-binding heavy chains of various human immunoglobulins. Furthermore, they were not affected in a non-specific fashion by the presence of antigen–antibody complexes. A summary of these data can be seen in Table 12.2. More convincing evidence which soon followed these observations was that these two forms do indeed associate with or contain specific antigen-binding receptors when released by immune cells. It was demonstrated that lectin-activated human or murine cells immune to either soluble or cellular antigens would release LT forms into the supernatant. These LT forms were capable of selectively binding to the specific antigen. During the process of these studies, it became clear that to obtain maximum levels of receptor-associated lytic proteins (RALP) from immune cells, special culturing techniques

Table 12.2 Immunological reactivity of anti-immunoglobulin serum with various human LT classes

Antiserum Tested	LT Molecular Weight Class				
	Cx	α_H	α	β	γ
(a) Rabbit anti-F(ab')$_2$ (IgG)	+	+	−	−	−
(b) Goat anti-F(ab')$_2$ (IgG)	+	+	−	−	−
(c) Rabbit anti-(H chain) sera (μ, γ, α, D, E)	−	−	−	−	−
(d) Rabbit anti-(bovine serum albumin)	−	−	−	−	−
(e) Ag-Ab complexes, bovine serum albumin–anti-(bovine serum albumin)	−	−	−	−	−

+ Indicates the capacity of the antisera to neutralize the cell lytic activity of the individual LT class *in vitro*. − Indicates no effect.

were required. Cells were stimulated with antigen *in vitro* to boost intracellular LT pools, then release was stimulated with lectin. We found that if specific antigens were employed as the stimulus for the final release phase, the amount of RALP detected in the supernatant was low. Additional studies revealed that the low amount of RALP appears to be due to the binding of the RALP forms with the residual stimulating antigens in the supernatant. When this binding occurs, the receptor-bearing forms are not lytically active and therefore undetectable. Further biochemical studies on both murine and human forms revealed that the antigen-binding forms are restricted to the Cx and α_H molecular-weight classes. Moreover, studies in murine systems reveal that these forms are released by 'T cells'. Additional immunological studies reveal that the Cx and α_H human LT molecular-weight classes express other alloantigens. These alloantisera may be recognizing immunological determinants which are expressed on gene products which are associated with regulation of the immune response (Ware, Klostergaard, Toth and Granger, 1981). Finally, extensive studies with monospecific sera directed at various human complement components reveal that antibodies reactive with Clq have the capacity to neutralize these same two human forms. In summary, these findings indicate a very exciting concept, namely, that cell-lytic LT materials form a system of toxins that can associate with receptors, alloantigens, and perhaps the Clq component(s) of complement.

Biochemical and immunological studies on forms released *in vitro* by lectin-activated purified human T cell populations reveal the association of these forms with receptors and the interrelationships of the various molecular-weight LT classes (Harris, Yamamoto, Crane and Granger, 1980). The first form released by these cells *in vitro* is precursor alpha heavy (Pα_H). This form of α_H is stabilized in a supernatant by association with a 12 000-molecular-weight non-toxic component, termed precursor factor (Pf). Upon molecular sieving or mild treatment with agents that disrupt hydrophobic bonds, Pf dissociates from the $P\alpha_H$ form, leaving the α_H. The $P\alpha_H$ form does not express α LT antigenic determinants, but the α_H does, thus dissociation of Pf exposes the α determinant. The α_H form can spontaneously dissociate to some degree into the smaller α, β and γ forms in a supernatant or this process can be enhanced during molecular sieving in high-ionic-strength buffers. Immunological and functional studies reveal the $P\alpha_H$ and α_H, but not the smaller α, β and γ forms, express receptor activity. An exciting concept followed from this work, for when the Pf dissociates from precursor α_H, the α_H may polymerize with itself to form Cx molecules, or it can dissociate into the smaller LT subunits. The former reaction may indicate that Cx forms are multimeric forms of the α_H, and those α_H forms which do not polymerize dissociate into non-active subunits.

12.3 Cellular processes involved in LT release by unstimulated (primary) and stimulated (secondary) human lymphocytes

Several factors influence which LT forms will be detected in supernatants from human lymphocytes. One major consideration is the type of lymphocyte population being stimulated. When unseparated populations of human adenoid or peripheral blood lymphocytes are stimulated with a lectin, the supernatants contain mainly α, β and small amounts of Cx and γ forms (Granger, Yamamoto, Fair and Hiserodt, 1978; Walker and Lucas, 1972) of LT. However, when lymphocytes are separated into subpopulations by standard techniques and then stimulated, important differences can be noted. Populations enriched for T lymphocytes release all of the LT classes upon lectin stimulation, but the majority of the activity resides in the $P\alpha_H$ class (Harris, Yamamoto, Crane and Granger, 1980). On the other hand, B-lymphocyte-enriched populations secrete only α when stimulated with a lectin (Yano and Lucas, 1978). In addition, continuous human B cell lines spontaneously secrete low levels of α (Fair, Jeffes and Granger, 1979; Granger, Moore, White, Matzinger, Sundsmo, Shupe, Kolb, Kramer and Glade, 1970). Therefore, it appears that B cells do not associate α LT with receptor molecules.

In addition to the class of lymphoid cells, the state of differentiation and activation of these cells is also important in determining which LT forms are present in the supernatant. Whereas the work described above utilized supernatants from lectin-activated fresh unstimulated lymphocytes *in vitro*, LT production systems have been developed which utilize preactivated lymphocytes (Granger *et al.*, 1975; Granger *et al.*, 1980). These systems involve 'priming' human adenoid or peripheral blood lymphocytes for 3–5 days prior to the 'secondary' lectin stimulation. This serves two purposes: (1) to expand the population of specifically immune cells; and (2) to build high levels of intracellular LT pools which can be rapidly released upon receiving a secondary stimulation (Hiserodt, Prieur and Granger, 1976; Hiserodt, Yamamoto and Granger, 1978b; Hiserodt, Ford, Harris and Granger, 1979a; Yamamoto, Hiserodt and Granger, 1979). There are several methods employed for priming lymphocytes: (1) culturing for 3–5 days with a lectin (non-specific stimulus); (2) culturing for 5–7 days with a specific soluble antigen; or (3) culturing for 5–7 days in a one-way mixed lymphocyte culture (MLC, stimulation by cellular antigens). After priming, the cells are washed and restimulated with a lectin. Supernatants obtained from unseparated populations of lymphocytes receiving a secondary stimulation contain all forms of LT. However, most of the activity in these supernatants is in the form of $P\alpha_H$ (Ware, Yamamoto and Granger, unpublished work). Populations enriched for T cells which have been primed and restimulated give similar results (Harris, Benvenuti and Granger, unpublished work). There are two advantages to using primed cells for generating LT. One advantage is that these cells are powerful effectors of several classes of CMI, including CTL and NK. Thus they may be more likely to secrete those forms of LT which are important in the various cytolytic effector functions of lymphocytes. Furthermore, through the use of this technique, immune lymphocytes can be generated against a soluble or cellular antigen(s) and then stimulated in the absence of the sensitizing antigen. This facilitates the production of RALPs and prevents them from binding to the sensitizing antigen and thus becoming inactive in the L cell assay.

The kinetics of release of LT are affected by parameters similar to those described above for determining the classes of LT to be released. That is, the kinetics of release are determined by the particular type of lymphocyte population and the state of activation of the cells. When unseparated populations of primary lymphocytes are stimulated with a lectin, intracellular levels of LT in these cells plateau at 3 days. However, supernatant LT appears at 2–3 days, and maximum levels are reached within 4 or 5 days, suggesting that the majority of activity is released from pools of material within activated cells (Lewis, Yamamoto, Carmack, Lundak and Granger, 1976). Primary cell populations enriched for T cells produce higher levels of LT than unseparated populations, and moreover, the levels rise rapidly and peak at 2 days after lectin stimulation (Harris, Yamamoto, Crane and Granger, 1980). On the other hand, B-cell-enriched populations produce low levels of α LT activity, and these levels are attained only after 7 days (Yano and Lucas, 1978). When unseparated or T-cell-enriched populations are primed and then given a secondary stimulus, the major portion of LT activity is rapidly released within 5 h, yet slow release continues for 2 days (Granger *et al.*, 1980; Harris, Yamamoto, Christensen and Granger, 1980). These results further illustrate the concept that LT can be rapidly released by activated cells which contain high levels of intracellular LT activity.

An additional important factor in determining the kinetics and levels of LT release is the type of stimulus applied to the responsive lymphocyte. This situation is most clearly illustrated in the case of LT released by murine lymphoid cells *in vitro*. Low levels of LT activity are released by lectin-activated unstimulated immune cells; however, high levels can be released within 5 h if the same cells are co-cultured with lectin-coated allogeneic cells or allogeneic cells alone (Hiserodt, Tiangco and Granger, 1979b). Thus the appropriate signal triggered rapid release of these materials from cells previously considered incapable of releasing significant levels of LT *in vitro*. This method has been used with similar results in other animal species as well (Ross *et al.*, 1979). However, human lymphocytes do not respond as well to this stimulus, and the human LT activity produced under these conditions is highly unstable.

Important information on the cellular processes involved in the production and release of LT by human lymphocytes has been gained by use of various inhibitory agents. These include inhibitors of protein synthesis, secretory processes and protease activity. An examination of the effect of emetine and cycloheximide, inhibitors of protein synthesis, on LT production yields two different results depending upon the state of activation of the cells (Hiserodt *et at.*, 1979a). LT production and release by primary cells requires protein synthesis, whereas secondary cells secrete LT in the absence of protein synthesis. Furthermore, colchicine and cytochalasin B, inhibitors of secretory processes, inhibit release from primary cells but not from secondary cells (Hiserodt *et al.*, 1979a). Although the actual mechanism of release is unclear, it appears that release is concomitant with the expression of α on lymphocyte membranes.

The effect of protease inhibitors on LT release by activated human lymphoid cells has also been examined (Harris, Benvenuti and Granger, unpublished work; Ross, Ware and Granger, unpublished work). Serine esterase inhibitors, including *p*-tosyl-L-arginine methyl ester HC1 (TAME) and 7-amino-1-chloro-3-L-tosylamido-heptan-2-one *N*-α-*p*-tosyl-L-lysine chloromethyl ketone HC1; TLCK), inhibit LT release from primed cells. Furthermore, soybean trypsin inhibitor, a naturally occurring protease inhibitor which apparently does not enter cells, also blocks LT secretion. These results suggest that a cell surface protease may be required at some stage in the release or activation of LT molecules by primed lymphocytes.

Additional studies revealed that LT release by lectin-stimulated preactivated cells has a Ca^{2+}-requiring step(s) (Ross and Granger, 1980). The results of studies employing EGTA, a Ca^{2+}-chelating agent, indicate that there is an absolute requirement for Ca^{2+}, but not for Mg^{2+}, in the early phases of release of LT from preactivated lymphocytes. However, the Ca^{2+}-requiring step is at an early stage in the release process, for once initiated, secondary release does not require Ca^{2+}. This may reflect the presence of a classical Ca^{2+}-dependent secretory process; however, as mentioned earlier, other agents which inhibit classical secretion, colchicine and cytochalasine B, do not affect LT release from primed cells. This finding is of particular importance, since a similar cation requirement has been demonstrated in the programming for lysis stage of T-cell-mediated cytolysis.

In summary, these data suggest the following model. First, primary lymphocytes respond to antigenic or mitogenic stimulation by producing and storing large pools of intracellular LT activity. Upon secondary stimulation, these pools of LT are rapidly released into the medium. The secondary stimulation and release process appears to be independent of microtubule or microfilament movement, but it does require Ca^{2+}. Finally, a protease(s) may be required at the surface of the lymphocyte to facilitate release of intracellular and/or membrane-associated LT.

12.4 Types of lytic reactions induced by lytic molecules of various molecular weights *in vitro*

Studies examining the mechanism of LT-induced cytolysis *in vitro* have employed whole supernatants or fractions of LT from lyphocytes stimulated with lectin or antigens. However, most studies have examined the effects of α and β LT forms. The target cell routinely used to detect lysis *in vitro* is the murine L-929 cell line, because it is highly sensitive to lysis mediated by all classes of LT from all animal species. There appears to be a wide variation in sensitivity *in vitro* of cells from various tissues and different animal species to these molecules. Within the L cell population there are resistant substrains to LT (Kramer and Granger, 1975a; Namba and Waksman, 1976). The L-cell-induced lysis by LT is usually slow (occurs in 12-24 h) and concentration-dependent. Moreover, while one concentration can induce lysis, intermediate levels cause growth inhibition, and low levels only transient growth inhibition (Granger, 1972). Lysis can also be enhanced by using inhibitors of energy production and macromolecular synthesis on the target L cell (Granger *et al.*, 1975). It has also been shown that a certain portion of human LT molecules in whole supernatants can rapidly bind to the L cell (Hessinger, Daynes and Granger, 1973; Kramer and Granger, 1975a). In addition, a small amount of material in the human α LT class has the ability to bind to the L cell surface (Tsoukas, Rosenau and Baxter, 1976). This binding *in vitro* is rapid and temperature-independent. The LT remains on the L cell surface, for within the first few hours lysis can be reversed or inhibited by limited proteolysis or neutralized by the addition of an anti-LT serum (Kramer and Granger, 1975b). It is still not clear whether the LT remains on the L cell surface or later enters the cytoplasm to cause actual lysis. Studies by other investigators suggest that cell lysis does not result from inhibition of cellular macromolecular synthesis or energy metabolism (Granger *et al.*, 1975). Studies by Friend and Rosenau (1977) with human α LT suggest the molecule acts on the L cell surface. Kobayshi, Sawada and Osawa (1978) have demonstrated that guinea pig β LT binds to a small oligosaccharide on the L cell surface. Okamoto and Mayer (1978) found that this form affects membrane ion transport, and they suggest that cell lysis was due to osmotic imbalances.

We have demonstrated in our laboratory that human α LT appears to cause lysis of the L cell without directly entering the target cell cytoplasm (Granger, Hiserodt and Ware, 1979). We have shown this by covalently linking with cyanogen bromide the α LT molecule to Sepharose 4B beads and then incubating these beads with L cells for 12 h. The data from Table 12.3 suggest that the α LT that is bound to beads induces cytolysis *in vitro* by direct action on the L cell membrane. However, there is the possibility that proteolytic action by the L cell may cause removal of the α LT from the beads, therefore enabling it to either bind to the cell surface or enter the cytoplasm to cause lysis of the target cell. We have also shown that unbound human α LT is actually inactive and requires an L-cell-associated protease to cause cell lysis (Ross and Granger, 1980). The lytic activity of human α LT class can be blocked by reversible and irreversible protease inhibitors that affect trypsin and chymotrypsin-like proteases. These protease inhibitors do not affect the α LT molecule itself but a target-cell-associated enzymic step(s). These studies suggest the α LT subunits are lytically inactive. This may explain how the effector cell can synthesize, handle, and express the molecule on the cell surface without causing damage to itself.

Table 12.3 Cytolysis of the L target by α LT covalently linked to sepharose beads *in vitro*

Treatment	Bead: Target Cell Ratio	Percentage Cell Viability
		(*L Cells*)
α LT beads*	40:1	8
α LT beads + 100 μl of anti-α LT sera	40:1	78
Control BSA bead	40:1	80
Supernatant from L-cells† after 24 h at 37°C with α LT beads (40:1)		96

*Ultrogel fractions of α LT activity on bovine serum albumin (BSA, 2mg/ml of beads) were covalently linked to Sepharose 4B beads by the CNBr methods. α LT or control BSA beads (0.1 ml) were then added to monolayers of L cells (10^5 cells) to give the indicated bead-to-target-cell ratios. After 24 h incubation at 37°C, the beads were washed away, and the remaining adherent cells counted. The percentage cell viability is relative to non-treated controls.
†The supernatant from L cells incubated in the presence of α LT beads after 24 h at 37°C (8% L-cell viability) was tested (undiluted) on a second L cell monolayer for an additional 24 h at 37°C.

Experiments with fractions obtained from molecular-sieving columns revealed that the different human molecular-weight LT classes possess different cell-lytic activities when assayed on various target cells *in vitro* (Yamamoto, Hiserodt and Granger, 1979). The high-molecular-weight LT class, Cx, expressed more lytic activity than the lower-molecular-weight LT classes, α, β, γ, when assayed on allogeneic target cells. HeLa (carcinoma of the cervix) and Chang (carcinoma of the liver), two lymphoblastoid B cell lines, RPMI-1788 and WI-L2, and a myeloid cell line, K562. Extensive studies revealed that the α and β LT classes were weakly lytic, whereas, the Cx LT class was from 5 to 100 times more effective in causing rapid 'non-specific' lysis (6–10 h) of allogeneic target cells. Furthermore, we found that supernatants obtained from lectin-stimulated lymphocytes 'immune' to RPMI-1788 or WI-L2 contain target-cell-specific RALP forms, and these forms caused enhanced lysis of the specific sensitizing target cell. We found that the RALP was from 5 to 10 times more lytically effective on the specific lymphoblastoid target cell than was Cx obtained from lectin-stimulated non-immune cells, making the RALP 50–1000 times more lytically effective than the α or β LT class. Immunological studies with antibodies made against the lower-molecular-weight human LT classes and subclasses demonstrated that these antisera could neutralize the cell-lytic capacities of the immune and non-immune Cx LT forms (Yamamoto *et al.*, 1979). It has also been demonstrated that a rabbit anti-human $F(ab')_2$ serum could neutralize the various cell-lytic activities of the immune and non-immune Cx forms, indicating that this form is in association with an Ig-like molecule.

Studies examining the lytic activities of these complexes on the K562 myeloid cell line indicate that the K562 cell is equally sensitive to all LT complexes from lectin-stimulated immune or non-immune lymphycytes (Yamamoto *et al.*, 1979). This is an interesting finding, because it has been demonstrated by other investigators studying natural killing that the K562 cell is the most sensitive target cell to human natural killers (Bonnard and West, 1980). Owing to the fact that we are working with supernatants obtained from lectin-stimulated immune and non-immune lymphocytes, it is possible that the natural killer population is being expanded and giving rise to LT forms that have natural killer functions and are independent from the specific receptor cell-lytic forms.

We have performed experiments to demonstrate that the RALP is not unique to the human system. Functional studies of murine LT molecules obtained from non-immune and alloimmune T cells indicate that the high-molecular-weight LT classes have an increased cell-lytic potential *in vitro*, whereas, the lower-molecular-weight LT classes (α and β) are only weakly lytic (Hiserodt, Tiangco and Granger, 1979d). The high-molecular-weight LT molecules from alloimmune T cells can cause rapid and 'specific' target cell lysis. We found that Cx LT from non-immune T cells can cause non-specific lysis (10–30%) in a 16–24 h assay, whereas the Cx LT molecules from alloimmune T cells cause a rapid (6–10 h) 'specific' lysis (5–39%) of the specific sensitizing allogeneic target cell. It was established by several different criteria that the alloimmune T cell was responsible for the release of the RALP.

12.5 Conclusions

CMI cell-lytic reactions are complex, yet they all seem to share certain common steps (Goldstein and Smith, 1977; Henney, 1977; Martz, 1977). The first is recognition of the target cell by the aggressor cell, which is facilitated by contact. It is during this contact phase that initiation of aggressor cell processes occur which lead to delivery of the lethal hit to the target cell. Whatever these processes are, they occur primarily in the microenvironment at membrane contact points between killer and target cell. Once programming for lysis has occurred, and the lymphocyte has delivered the lethal hit, the presence of the aggressor cell is no longer necessary for target cell lysis. This has been termed the lymphocyte-independent phase. The lethal hit is unidirectional, and the actual mechanism is not yet known. One possible explanation is that molecules of the LT system may be involved as effectors of lysis.

Biochemical and functional characteristics of LT molecules in humans and in experimental animals reveal a new concept, namely, that these materials appear to form a 'system' of cell toxins. The smaller weakly lytic forms can assemble together with themselves, receptor(s) and other components to form highly lytic high-molecular-weight forms. Moreover, the cell-lytic capacity of the large forms can be directed by the receptor(s). These studies indicate that both the recognition function and cell-lytic components are associated together in a single molecular unit, termed RALP. However, RALP is unstable when free in suspension, and the components rapidly dissociate into the smaller non- or weakly lytic subunits. While the nature of the receptors in these forms is not yet clear, it appears that these are not traditional immunoglobulins and originate from populations of cell-lytic effector cells. The most exciting recent finding is the concept that certain of these receptor forms coming from non-immune lymphoid cells have selective binding capacities for NK-sensitive target cells. These findings suggest that different types of effectors, NK and T cells, may employ receptors with different specificities in association with the same LT components to mediate lysis. If this proves to be the case, it suggests a lytic mechanism which may be common to both NK and T killing. These findings also suggest that if LT molecules are involved in lytic reactions, the major killing mediated by these materials may, under normal circumstances, be all receptor-directed.

While the mechanism of lymphocyte-mediated cell lysis is not yet understood, it is possible to propose a theoretical model of these lytic reactions employing the presently known features of molecules of the LT system. Immunologically responsive or 'immune' cells presumably have been preactivated and possess intracellular pools of LT, which can be mobilized and released in the absence of protein synthesis. The contact and recognition phases serve to initiate delivery of receptor-associated LT forms from the killer to the target cell. This delivery occurs in an as yet unknown fashion; however, proteases and Ca^{2+} are somehow involved in the early phases of release. There are several possibilities of how these molecules get from the aggressor to the target cell: (a) the smaller forms are released as individuals and assemble on the target cell surface; or (b) they are released together as a molecular aggregate. More likely is the possibility that the precursor α_H forms are released as subunits at cellular contact points, which, via receptor, bind to the target cell surface and there assemble into complex forms. Once assembled into complexes on the target cell surface, the lethal hit has been delivered, the materials become lytically active, and the lymphocyte could leave the area with the activated cell-lytic molecules remaining on the target cell surface. Unidirectional lysis then could occur in the absence of the aggressor cell. Those cell-lytic materials released which do not make physical contact with the target cell dissociate in the supernatant into weakly lytic components, and are effectively disarmed. Thus the effective radius of these forms would be restricted to the immediate area around the effector cell(s). The above model, while hypothetical, correlates the findings in the LT system with the observed features of cell-mediated lysis. However, the challenge in future studies will be to demonstrate that this model is indeed valid.

References

Bloom, B. (1971), *In vitro* approaches to the mechanism of cell-mediated immune reactions. *Adv. Immunol.,* **13**, 102–106.

Bonnard, G. D. and West, W. H. (1980), Cell-mediated cytotoxicity in humans. A critical review of experimental models and clinically oriented studies. In *Immunodiagnosis of Cancer* (eds. R. B. Herberman and K. R. McIntire), Marcel Dekker, New York (in press).

Cohen, S., Pick, E. and Oppenheim, J. J. (1979), The lymphokine concept. *Biology of Lymphokines.* Academic Press, New York.

Fair, D. S., Jeffes, E. W. B. and Granger, G. A. (1979), Release of LT molecules with restricted physical heterogeneity by a continuous human lymphoid cell line *in vitro. Mol. Immunol.,* **16**, 186–192.

Friend, D. S. and Rosenau, W. (1977), Target-cell membrane alterations induced by lymphotoxin. *Am. J. Pathol.,* **86**, 149–162.

Goldstein, P. and Smith, E. T. (1977), Mechanism of T-cell-mediated cytolysis: The lethal hit stage. *Contemp. Top. Immunobiol.,* **7**, 273–300.

Granger, G. A. (1972), Review. Lymphokines – The Mediators of Cellular Immunity. *Ser. Haematol.,* **4**, 8–40.

Granger, G. A., Daynes, R. A., Runge, P. E., Prieur, A.-M. and Jeffes, E. W. B. (1975), Review – Lymphocyte effector molecules and cell-mediated immune reactions. *Contemp. Top. Mol. Immunol.,* **4**, 205–241.

Granger, G. A., Hiserodt, J. C. and Ware, C. W. (1979), Cytotoxic and growth inhibitory lymphokines. In *Biology of the Lymphokines* (eds. M. Landy and S. Cohen), Academic Press, New York, pp. 141–163.

Granger, G. A., Moore, G. E., White, J. G., Matzinger, P., Sundsmo, J. S., Shupe, S., Kolb, W. P., Kramer, J. J. and Glade, P. R. (1970), Production of lymphotoxin and migration inhibitory factor by established human lymphocytic cell lines. *J. Immunol.,* **103**, 1476–1485.

Granger, G. A., Yamamoto, R. S., Fair, D. S. and Hiserodt, J. C. (1978), The human LT system. I. Physical–chemical heterogeneity of LT molecules released by mitogen activated human lymphocytes *in vitro. Cell. Immunol.,* **38**, 388–402.

Granger, G. A., Yamamoto, R. S., Ware, C. F. and Ross, M. W. (1980), *In vitro* detection of human lymphotoxin. In *Manual of Clinical Immunology* (eds. N. R. Rose and H. Friedman), ASM Publications, Washington, D.C., pp. 267–274.

Harris, P., Yamamoto, R. S., Christensen, C. and Granger, G. A. (1980), The Human LT system XI: cell-lytic effectiveness and association of LT forms with antigen-binding receptor(s) released by T-enriched lymphocytes, *J. Immunol.,* (in press).

Harris, P., Yamamoto, R. S., Crane, J. and Granger, G. A. (1980), The Human LT system X: The first cell-lytic molecule released by T-enriched lymphocytes is a 150 000 MW form in association with small non-lytic components, *J. Immunol.*, (in press).

Heise, E. R. and Weiser, R. S. (1969), Factors in delayed sensitivity: Lymphocyte and macrophage cytotoxins in the tuberculin reaction. *J. Immunol.*, **103**, 570–576.

Henney, C. S. (1977), T-cell mediated cytolysis: An overview of some current issues. *Contemp. Top. Immunobiol.*, **7**, 245–271.

Hessinger, D. A., Daynes, R. A. and Granger, G. A. (1973), The binding of human lymphotoxin to target cell membranes and its relation to cell mediated cytodestruction. *Proc. Natl. Acad. Sci. U.S.A.*, **11**, 3082–3086.

Hiserodt, J. C., Prieur, A.-M. and Granger, G. A. (1976), *In vitro* lymphocyte cytotoxicity. I. Evidence of multiple cytotoxic molecules secreted by mitogen activated human lymphoid cells *in vitro*. *Cell. Immunol.*, **24**, 277–288.

Hiserodt, J. C., Ford, S. C., Harris, P. C. and Granger, G. A. (1979a), The human LT system. VII. Release of soluble forms and delivery to target L-929 cells by lectin-preactivated human lymphocytes in the absence of protein synthesis and secretory processes. *Cell. Immunol.*, **47**, 32–45.

Hiserodt, J. C., Yamamoto, R. S. and Granger, G. A. (1978a), The human LT system. III. Characterization of a large molecular weight LT class (Complex) composed of the various smaller LT classes and subclasses in association with Ig-like molecules. *Cell. Immunol.*, **38**, 417–433.

Hiserodt, J. C., Yamamoto, R. S. and Granger, G. A. (1978b), The human LT system. IV. Studies on the large MW LT complex class: Association of these molecules with specific antigen binding receptor(s) *in vitro*. *Cell. Immunol.*, **41**, 380–396.

Hiserodt, J. C., Tiangco, G. J. and Granger, G. A. (1979b), The LT system in experimental animals. I. Rapid release of high levels of lymphotoxin (LT) activity from murine lymphocytes during the interaction with lectin-treated allogeneic or xenogeneic target cells *in vitro*. *J. Immunol.*, **123**, 311–316.

Hiserodt, J. C., Tiangco, G. J. and Granger, G. A. (1979c), The LT system in experimental animals. II. Physical and immunological characteristics of molecules with LT activity rapidly released by murine lymphoid cells activated on lectin coated allogeneic monolayers *in vitro*. *J. Immunol.*, **123**, 317–324.

Hiserodt, J. C., Tiangco, C. J. and Granger, G. A. (1979d), The LT system in experimental animals. IV. Rapid specific lysis of allogeneic target cells mediated by highly unstable high molecular weight lymphotoxin–receptor complexes released by alloimmune murine T lymphocytes *in vitro*. *J. Immunol.*, **123**, 332–341.

Holm, G. (1969), Cytotoxic effects of lymphoid cells *in vitro*. *Adv. Immunol.*, **11**, 117.

Kobayashi, Y., Sawada, J.-I. and Osawa, T. (1978), Isolation and characterization of an inhibitory glycopeptide against guinea pig lymphotoxin from the surface of L cells. *Immunochemistry*, **15**, 61–68.

Kolb, W. P. and Granger, G. A. (1968), Lymphocyte *in vitro* cytotoxicity: Characterization of human lymphotoxin. *Proc. Natl. Acad. Sci. U.S.A.*, **61**, 1250–1255.

Kolb, W. P. and Granger, G. A. (1970), Lymphocyte *in vitro* cytoxicity: Characterization of mouse lymphotoxin. *Cell. Immunol.*, **1**, 122–132.

Kramer, J. J. and Granger, G. A. (1975a), The role of lymphotoxin in target cell destruction by mitogen-activated human lymphocytes. *Cell. Immunol.*, **15**, 57–68.

Kramer, J. J. and Granger, G. A. (1975b), Role of lymphotoxin in target cell destruction induced by mitogen-activated human lymphocytes *in vitro*. II. The correlation of temperature and trypsin-sensitive phases of lymphotoxin-induced and lymphocyte-mediated cytotoxicity. *J. Immunol.*, **116**, 562–567.

Lewis, J. E., Yamamoto, R. S., Carmack, C., Lundak, R. L. and Granger, G. A. (1976), Lymphocyte effector molecules: An *in vitro* production method for obtaining liter volumes of supernatants from mitogen-stimulated human lymphocytes. *J. Immunol. Methods*, **11**, 371–383.

Ling, N. R. (1968), *Lymphocyte Stimulation*, Vol. I, John Wiley and Sons, New York.

MacLennan, I. C. M. (1972), Antibody in the induction and inhibition of cytotoxicity. *Transplant. Rev.*, **13**, 67–75.

Martz, E. (1977), Mechanism of specific tumor-cell lysis by alloimmune T lymphocytes: Resolution and characterization of discrete steps in the cellular interaction. *Contemp. Top. Immunobiol.*, **7**, 245–273.

Namba, Y. and Waksman, B. H. (1976), Regulatory substances produced by lymphocytes. III. Evidence that lymphotoxin and proliferation inhibitory factor are identical and different from the inhibitor of DNA synthesis. *J. Immunol.*, **116**, 1140–1144.

Okamoto, M. and Mayer, M. M. (1978), Studies on the mechanism of action of guinea pig lymphotoxin II. Increase of calcium uptake rate in LT-damaged target cells. *J. Immunol.*, **120**, 279–285.

Oppenheim, J. J. (1968), Relationship of *in vitro* lymphocyte transformation to delayed hypersensitivity in guinea pigs and man. *Fed. Proc. Fed. Am. Soc. Biol.*, **27**, 21.

Peter, J. B., Stratton, J. A., Stempel, K. E., Yu, D. and Cardin, C. (1973), Characteristics of a cytotoxin ('Lymphotoxin') produced by stimulation of human lymphoid tissue. *J. Immunol.*, **111**, 770–782.

Ross, M. W. and Granger, G. A. (1980), The human LT system VIII: A target cell dependent enzymatic activation step required for the expression of the cytotoxic activity of human lymphotoxin. *J. Immunol.*, **25**, 719–724.

Ross, M., Tiangco, G. J., Horn, P., Hiserodt, J. C. and Granger, G. A. (1979), The LT system in experimental animals. III. Physical–chemical characteristics and relationships of lymphotoxin (LT) molecules released *in vitro* by activated lymphoid cells for several animal species. *J. Immunol.*, **123**, 325–331.

Ruddle, N. H. (1972), Approaches to the quantitative analysis of delayed hypersensitivity. *Curr. Top. Microbiol. Immunol.*, **57**, 75–93.

Russell, S. W., Rosenau, W., Goldberg, M. L. and Kunitomi, M. L. (1972), Purification of human lymphotoxin. *J. Immunol.*, **109**, 784–790.

Spofford, B., Daynes, R. A. and Granger, G. A. (1974), Cell mediated immunity *In Vitro*: A highly sensitive assay for human lymphotoxin. *J. Immunol.*, **112**, 2111–2116.

Trivers, G., Baumgart, D. and Leonard, E. J. (1976), Mouse lymphotoxin. *J. Immunol.*, **117**, 130–135.

Tsoukas, C. D., Rosenau, W. and Baxter, J. D. (1976), Cellular receptors for lymphotoxin: Correlation of binding an cytotoxicity in sensitive and resistant target cells. *J. Immunol.*, **116**, 184–187.

Walker, S. M. and Lucas, Z. J. (1972), Cytotoxic activity of lymphocytes. II. Studies on mechanism of lymphotoxin-mediated cytotoxicity. *J. Immunol.*, **109**, 1233–1244.

Ware, C. F., Klostergaard, J., Toth, M. K., and Granger, G. A. (1981), The Human LT System IX. Serological identification of F(ab')$_2$, major histocompatibility complex, and Clq-like components associated with the cytotoxic activity of the large MW LT classes. *Cell. Immunol.*, (in press).

Yamamoto, R. S., Hiserodt, J. C. and Granger, G. A. (1979), The human LT system V: A comparison of the relative lytic effectiveness of various MW human LT classes on ^{51}Cr labeled allogeneic target cells *in vitro*. Enhanced lysis by LT complexes associated with Ig-like receptor(s). *Cell. Immunol.*, **45**, 261–275.

Yamamoto, R. S., Hiserodt, J. C., Lewis, J. E., Carmack, C. E. and Granger, G. A. (1978), The human LT system II: Immunological relationships of LT molecules released by mitogen activated human lymphocytes *in vitro*. *Cell. Immunol.*, **38**, 403–416.

Yano, K. and Lucas, Z. J. (1978), Cytotoxic activity of lymphocytes VII: Cellular origin of α-lymphotoxin. *J. Immunol.*, **120**, 385–394.

13 Techniques for demonstrating cell death

I. D. BOWEN

'We have only a hardened bit of the past the bygone life under our observation'.
(D. H. Lawrence, 1918)

13.1 Introduction

Cell death is as difficult to define as clinical death. The research scientist, like the medical practitioner, often finds himself probing the remains. Cell death itself is a dynamic process, and the end point or the point of commitment may be both fleeting and controversial. Much of the scientific evidence is still largely histological or cytological so that as D. H. Lawrence (1918) surmized, 'We have only a hardened bit of the past the bygone life under our observation'. In such a situation we have to select profiles of previously dying cells from amongst a host of artificially killed or fixed cells.

A broad analysis of the published evidence leads us to the conclusion that there are two basic kinds of cell death. There may be a planned programmed or physiological cell death often involving an ordered series of events, usually genetically predetermined, or there may be an induced provoked necrosis. The latter would include acute injury and could also be the result of disease.

13.1.1 *Programmed cell death*

As a matter of principle, it is important to regard programmed cell death as a mechanism for survival rather than destruction. It is a paradox that cell death is intimately involved in the birth and development of life as we know it. Cell death occurs during the formation of eggs and sperm and at many stages during embryological and adult development. Like autophagy and cell atrophy, cell death may also play a conservative role under conditions of stress. In these circumstances the products of cell death may be recycled for the benefit of surviving cells (Bowen, 1980) so that cell destruction has an economic and nutritive role.

Cell death exerts a homoeostatic function in relation to tissue kinetics and may be an important factor in the control of size and shape in normal tissues, organs, appendages and whole organisms. Here, morphometric data may be useful to illustrate the results of cell deletion and tissue involution. A theoretical basis for

379

a cytokinetic model of cell growth and death in heterogeneous cell populations has been presented by Shackney (1973) (see also Chapters 4 and 6). Cell death is also an important element in the control of abnormal growth and many therapeutic agents exploit this factor to induce regression in cancers and tumours.

In programmed cell death, as in differentiation, there appears to be a point of commitment, and in general this is followed by irreversible degenerative change. From the commitment point onwards, the cell fails to maintain itself in a highly organized state and disintegrates rapidly.

From the technical point of view, early indicators of cell death could be sought in particular DNA sequences in the case of genetically mediated cell death. This DNA could give rise to a messenger RNA which in turn would produce enzymes such as hydrolases or autolysins, or, more subtly, affect metabolic pathways (dehydrogenases, ATPases, adenylate cyclase) or otherwise alter cell function (by altering membrane permeability, inhibiting specific enzymic pathways, disrupting cell cytoskeleton or intracellular transport).

Evidence of hydrolase involvement in cell death is extensive: acid phosphatase (Lockshin and Williams, 1965c; Brandes, Bertini and Smith, 1965; Bowen and Ryder, 1974; Lockshin and Beaulaton, 1974a, b and c), β-glucuronidase (Slater, Greenbaum and Wang, 1963; Weber, 1969; Ogawa, 1975; Zeligs, Janoff and Dumot, 1975), cathepsin (now considered to be a group of intracellular proteases – Weber, 1963, 1969; Seshimo, Ryuzaki and Yoshizato, 1977), and other proteases or peptidases (Sylven and Niemi, 1972; Gossrau, 1977, 1978). N-Acetyl-β-glucosaminidase (Derby and McEldoon, 1976), esterases (Oxford and Fish, 1979), collagenase (Weber, 1969, 1977; Beckingham-Smith and Tata, 1976) and other hydrolases have been implicated in the degradative or lytic phases of cell death. Autolysin involvement in cell death has been demonstrated by Tomasz (1974). Changes in dehydrogenase activity in association with cell death have been studied for some time (Hammar and Mottet, 1971; Fallon, Brucker and Harries, 1974) and more recently by Bidlack and Lockshin (1976) and Gartner, Hiatt and Provenza (1978a, 1978b). Transient increases in anaerobic glycolysis have also been reported by Lockshin, Bidlack and Beaulaton (1975). Schlichtig, Lockshin and Beaulaton (1977) have demonstrated increases in several ATP-digesting enzymes in degenerating insect muscles.

With regard to enzymic changes preceding cell destruction, there seems to be either an increased production or activity of lytic enzymes such as hydrolases or appropriate changes in the enzymes involved in energy production and transfer, causing an eventual drop in ATP production. Whilst enzymic involvement often occurs at an irreversible stage of cell death, enzymic changes still precede morphological destruction and are useful markers. Ultimately enzymes, of course, dealing with the mechanics of cellular shut-down and self-disposal make cell death possible.

During the destructive phase many potential indicators of cell death become available as the initial products of cell breakdown accumulate. There is an increase in free amino nitrogen as the lysis of protein proceeds, and there may be changes in ionic balance (Trump, Laiho, Mergner and Arstila, 1974; Farber and El Mofty,

1975; El Mofty, Scrutton, Serroni, Nicolini and Farber, 1975), or release of low-molecular-weight DNA (Williams, Little and Shipley, 1974) and low-molecular-weight RNA. Any or all of these parameters are amenable to biochemical analysis. Some of these changes, however, can be seen as the result of sublethal damage, for example transient changes in ionic concentrations and protein leakage.

Specific proteins may be synthesized in relation to cell death [suggested by several authors and sought by Beckingham-Smith and Tata (1976) who could not, however, identify any] and these technically could be employed as labels of cell death. Some proteins may even act as triggers of programmed cell death. The range of triggers may well be as varied as the number of cell types. In this context, the role of hormonal triggers in cell death is well documented (Lockshin and Williams, 1965a, b and c; Weber, 1963, 1969).

Where homogeneous synchronous cultures of cells are available, physiological parameters such as reproductive viability and respiration can be used to predict death. Loss of respiratory function on the face of it would seem to be a good indicator of cell death, but may not always be the case. There may be a switch from oxidative to anaerobic respiration, and at the structural level mitochondria often seem relatively robust and survive well even under adverse conditions. Indeed, changes in cytochemically demonstrated mitochondrial dehydrogenases have been quite poor indicators of impending cell deletion (Fallon *et al.*, 1974).

Although morphological changes are obviously secondary to biochemical shifts, much of the published data on cell death depend on histological, cytological and fine-structural studies. Histological interpretations are often based on changes in membrane permeability, which result in differential staining patterns. Thus apparently crude histological assessments may be based on quite subtle membrane and pH changes, and the continuing utility of such an empirical approach should not be underestimated. Similarly, fine-structural changes observed at electron-microscope level reflect underlying biochemical changes. Here again, the status of the plasma membrane and cytomembranes can give reliable and early clues as to the probability of cell death. Farber and El Mofty (1975) argue strongly that liver cell death is not primarily caused by disturbances of RNA and protein synthesis, mitochondrial function or release of lysosomal enzymes; they implicate the plasma membrane and its role in the maintenance of calcium balance.

At fine-structural level, distinct and useful patterns of cellular behaviour such as apoptosis (Kerr, Wyllie and Currie, 1972) may be revealed. Nuclear pyknosis, endoplasmic vesiculation, mitochondrial, lysosomal and Golgi changes have all been recorded in relation to cell death as well as the more general cellular budding or fragmentation.

Senescence has a bearing on the probability of cell death (see Chapter 8). Cells from different tissues may age at different rates, but death is thought by some to be brought about either by the accumulation of specific faulty (or waste) products or by the accumulation of random genetic errors. In aging cells, specific accumulations such as lipofuscin granules — related to the peroxidation of lipid — can thus form the basis of a technique for pin-pointing old or dying cells. Changes in the degrees of mineralization of various granules can also be useful in this respect. Genetic errors could potentially be traced either at source (inaccurate DNA sequences) or at later stages when they would give rise to faulty messenger RNA molecules, faulty proteins, enzymes or other products (or lack of products). See also Chapter 9.

13.1.2 *Acute cell death*

Acute cell death presents a somewhat different case in that the initial agent of injury is always external and impinges directly on the well being of the cell (see Chapter 7). The cellular response to lethal and sublethal injury has been reviewed by Trump and Mergner (1974). The resultant changes depend to a large extent on the harmful agent, but in a lethal situation the final outcome is cell disruption and disintegration.

Induced cell death, in a mechanistic way, often mimics programmed cell death (Schweichel and Merker, 1973), particularly in terms of fine-structural changes. Because of these similarities most of the parameters outlined earlier can be used to measure loss of viability (Zeligs *et al.*, 1975). Thus injured cells often exhibit membrane changes, respiratory changes, ionic changes, ribosomal changes, synthetic changes, loss of ATP production and cell autolysis. Even sublethal injury can, however, result in leakage of proteins and enzymes, as well as osmotic and ionic shifts.

Evidence of the importance of plasma membrane injury and related changes in calcium ion concentrations to cell death comes from the work of Chien, Abrams, Pfau and Farber (1977), El Mofty *et al.* (1975) and Farber and El Mofty (1975) on ischaemia. Jones and Trump (1975) also describe the cellular and subcellular effects of ischaemia.

Biochemical aspects of cell death following injury have been recently reviewed from a forensic point of view by Gevers (1975).

13.2 Microscopical

13.2.1 *Histological*

The histological basis of the phenomenon of cell lysis and death was recognized long ago by Looss (1889), who worked on sarcolysis in amphibian metamorphosis. An extensive review of cell death in normal vertebrate ontogeny was presented by Glücksmann (1951), in which he lists the sites of cell death and also classifies cell degenerations according to their developmental functions, i.e. morphogenetic, histiogenetic and phylogenetic. Leblond and Walker (1956), who attempted to look at the subject dynamically, deal with cell loss and renewal in mammalian systems. A histologically based review of the biology of cell death in tumour has been presented by Cooper (1973), and more recently Gruenwald (1975) has dealt with pathogenetic aspects of abnormal embryological development and reports on teratogenic and induced cell death.

(a) *Vital dyes and histological stains.* An important and critical review of cell death in embryonic systems was given by Saunders (1966), in which he discussed the mechanical utility of morphogenetic death, possible mechanisms and control of developmental degeneration and also the disposal of products of cell death.

Saunders, Gasseling and Saunders (1962) introduced very useful vital and supravital staining procedures for demonstrating regions of cell death. The technique consists of exposing the embryo or living tissue for 15 to 30 min at room temperature (or at 37°C) to a 1:10 000 solution of Nile Blue Sulphate (Grublers) in Ringer's saline. The tissues are then rinsed with Ringer's solution which removes the Nile Blue Sulphate from normal cells. Degenerating cells, however, retain the stain, indeed the dye appears to be concentrated in dying cells.

Embryos may also be stained *in ovo* by using Nile Blue Sulphate at concentrations of 1:40 000 in Ringer's saline. For parallel histological studies, Saunders *et al.* (1962) advised fixation in Bouin's solution and staining in iron/haematoxylin or Feulgen reagent. Microscopic examination of supravitally stained tissue revealed 'numerous degenerating figures' ranging from small cells possessing one or several blue 'degeneration granules', thought to be materials lost from the nucleus, to very large bodies containing many granules apparently formed by the fusion of several disintegrating cells. The same bodies are initially haematoxylin- and Feulgen-positive, although Feulgen staining decreases eventually as chromatin is lost. It was concluded that many of the degenerations observed were the result of nuclear pyknosis.

Nuclear pyknosis is described in detail by Sandritter and Riede (1975). It appears to start with aggregation of chromatin at the nuclear membrane (hyperchromatosis). This is a very sensitive indicator of anoxic cell death and occurs only 15 min after loss of oxygen supply in the liver. After 2 h, maximum chromatin aggregation can be observed (granular hyperchromatic nucleus), followed by dilations of the nuclear membrane and chromatin condensation typical of pyknosis. In due course chromatin progressively becomes less visible and is dispersed or lost (chromatolysis, karyolysis). It is thought that changes in nuclear membrane permeability and dissociation of nucleoprotein lead to nuclear staining with Nile Blue Sulphate. Other vital dyes such as Acridine Orange and Neutral Red behave in a similar fashion. A review of the effects of vital dyes on living organisms and cells has been undertaken by Barbosa and Peters (1971). They reported evidence that Neutral Red at a concentration of 1-2 per cent inhibits acid phosphatase activity, and that Methylene Blue, Nile Blue A and others reduce the availability of ATP. Evidence that Methylene Blue inhibits acetylcholinesterase was also cited.

Nuclei of phagocytosed dead cells are intensively chromophilic (Chang, 1939; Ballard, 1965) and can readily be identified in iron/haematoxylin-stained sections. This hyperchromasia of pyknotic cell nuclei may be explained by the condensation of chromatin and also by the exposure of charged groups. For example, pyknotic nuclei also bind more eosin, owing to the release of basic histone groups (Sandritter and Riede, 1975).

Ballard and Holt (1968) report that in sections of foetal rat foot stained in

Baker's acid haematin (without pyridine extraction) and counterstained with Neutral Red, the pyknotic nuclei of dead cells stain dark red and are enveloped with black-stained material not observed in viable cells, nor in pyridine-extracted controls. Similar dark granules were also demonstrable in sections stained with Heidenhain's haematoxylin. According to Ballard (1965) these cells of the inter-digital tissue of the mouse embryo limb consist of transformed macrophages containing several phagocytosed dead cells. Ballard and Holt (1968) also report that in sections stained with Methyl Green and pyronin the viable mesenchymal cells have pale-green nuclei, dark-green chromatin and pink cytoplasm, whereas dead mesencymal cells have uniformly blue–green pyknotic nuclei and red con-densed cytoplasm.

Pexieder (1972a) noted formations characterized as degenerating or dying cells in haematoxylin- and eosin-stained preparations of developing chick embryo heart. These cells were also Feulgen-positive initially, although with advancing degener-ation and DNA loss there was a slow decrease in reaction intensity and eventually a sudden loss of staining. Arees and Åström (1977) studying cell death in the optic tectum of the developing rat found that dead or dying cells were identifiable by their shrunken, dark and often fragmented nuclei, whilst the cytoplasm of such cells was 'liquified'. They employed haematoxylin and eosin or Cresyl Violet as histological stains and concluded that cell death in the tectum of the rat was histogenic, occurring normally as the neurons differentiated. Chu-Wang and Oppen-heim (1977) described cell death of mononeurons in the chick embryo spinal cord. Here, nuclear changes were observed histologically by employing haematoxylin and Orange G. as stains. Electron-microscopical evidence was also presented.

Vital and supravital staining for cell death has in general been quite productive following its introduction by Saunders *et al.* (1962). Hinchliffe and Ede (1967) describe the application of Nile Blue Sulphate for the localization of cell death in development of the polydactylous talpid mutant of the fowl and again in the demonstration of cell death and the development of limb form and skeletal pattern in normal and wingless chick embryos (Hinchliffe and Ede, 1973). Areas of necrosis were found corresponding to areas taking up vital stain; these were also recognized by the presence of macrophages ('giant cells') containing chromophilic granules. Pexieder (1972a) described cell death during the morphogenesis of chick heart and also found that macrophages were identifiable in the bulbar cushions stained supravitally with Nile Blue Sulphate and Neutral Red (Pexieder 1972b).

Fallon and Cameron (1977) investigating cell death in turtle and skink embryos, employed 1:10 000 (w/v) of Neutral Red or Brilliant Cresyl Blue in tissue culture medium. These topically applied dyes stained dying cells. Interestingly, Fallon (1972), using Nile Blue Sulphate as a marker, demonstrated that Janus Green B could be used to prevent cell death in chick foot interdigital regions. Cameron and Fallon (1974) also noted the normal absence of interdigital cell death in developing *Xenopus laevis*. Using Nile Blue Sulphate as an indicator, Scott, Ritter and Milson (1977) studied delayed appearance of ectodermal cell death in rat embryo hind-limbs as a mechanism of induction of polydactyly.

Spreij (1971) in an extensive study of cell death during the development of imaginal discs in *Calliphora erythrocephala* indicated that the distribution of degenerating cells was readily revealed by their ability to bind vital dyes such as Nile Blue Sulphate, Neutral Red, Bismark Brown and Acridine Orange. For insect work he recommended a 1:100 000 solution of Nile Blue Sulphate in Ringer, to be applied for 15 to 20 min. Degenerating cells then appeared as dark blue dots against a pale-bluish background, although the intensity of the reaction faded with time. Vijverberg (1974) presented an exhaustive study of the cell-proliferation patterns in the imaginal discs of the same species. Using a range of histological techniques, he demonstrated cell death as a major quantitative and qualitative factor in the morphogenesis of the imaginal discs. Starre- van der Molen and Otten (1974) working on the embryology of *Calliphora* demonstrated cell death in the central nervous system during the later stages of development. Degenerating cells were small and rounded and contained pyknotic nuclei. They were found to accumulate vital dyes such as Nile Blue Sulphate and Acridine Orange selectively. The latter fluorescing dye was found to be more satisfactory, since the Nile Blue Sulphate was reported to diffuse out of the tissue.

(b) *Trypan Blue exclusion* Trypan Blue-exclusion techniques can be useful in estimating cyto-toxicity, injury and cell death. Brus and Glass (1973) presented an estimation of cytotoxic injury to gastric parietal cells by Trypan Blue-exclusion test followed by haematoxylin/eosin counterstaining of fixed smears. This technique was adapted by Hinsull, Bellamy and Franklin (1977) for determinations *in vivo* of non-viable cells in the thymus. They demonstrated a gradual increase in the percentage of dead cells with age whilst the rate of cell division declined. Interestingly it was also found that there existed a density-dependent distribution pattern of mitotic and dead cells. Drake, Ungoro and Mardiney (1972) used formalin-fixed preparations as standards for use in an automated Trypan Blue cytotoxic assay. They confirmed that any injury to the cell membrane caused by physical, chemical or enzymic damage results in the loss of Trypan Blue exclusion, leading to the selective staining of non-viable cells.

(c) *Radioautography and cell counts* Some studies combine radioautographic incorporation techniques for detecting mitosis with histological staining for cell death (see also page 413). Other more general estimates of cell death depend on an assessment of total cell loss under particular circumstances as compared with a normal or control situation. Lewis (1975), estimating cell death in the germinal layers of the postnatal rat brain, employed radioautographic incorporation of [³H] thymidine to calculate mitotic indices, and used routine haematoxylin and eosin staining to estimate pyknotic indices. This was backed up by Feulgen staining for nuclear DNA, where pyknotic nuclei were shown to have a lower than diploid DNA content. The results indicated that 3.2 per cent or less of newly acquired neurons and glia were lost by cell death. In undernourished rats, more pyknotic nuclei were found in the germinal layers of the brain at 12 days when loss of new cells was 3.7 to 7.8 per cent. Smadja-Joffe, Klein, Kerdiles, Feinendegen and Jasmin (1976) using radioautographic techniques demonstrated massive cell death in Friend leukaemic spleen.

Hamburger's (1975) estimates of cell death in the development of the lateral motor column of the chick embryo were dependent on nuclear and nucleolar counts (corrected) as compared with the number of distinctly pyknotic cells. The results indicated that at least 5 to 6 per cent of the total population was in a process of degeneration at any particular time. Arguments were presented to support the hypothesis that naturally occurring cell death in the lateral motor column was due to a failure of axons to survive in competition for peripheral contact sites (see Chapter 10).

Sohal (1976) estimated a low level of cell death in the trochlear nucleus of developing duck brain and compared this level with the massive cell death (85 per cent) observed in the nucleus after removal of the optic primordium. In this instance sections were stained with 2 per cent silver nitrate or with 1 per cent thionin. Sohal and Holt (1977) have described a similar study of cell death during normal development of the abducens nucleus of white Peking duck embryos.

Butterworth (1971, 1972) and Butterworth and Lanczy La Tendresse (1973) have used apparent cell disappearance as a criterion of cell death. The rate of cell death in the larval fat body of *Drosophila* was determined by counting the number of cells remaining in animals of various ages and plotting the number of cells remaining against time.

(d) *Morphology* Various morphological studies attempt to classify types of cell death. Baroldi (1975) demonstrated the different morphological types of myocardial cell death in man [a tissue that has been investigated histochemically by Musy and Haag (1977)]. Other morphological studies draw from electron-microscopical as well as histological data. Becker and Lane (1966) refer to the early preparation of the hepatocyte for mitosis following 70 per cent hepatectomy, where they observed an impressive increase in autophagocytosis. Schweichel and Merker (1973) describe various types of induced and physiological cell death in prenatal mouse tissues. They distinguished three basic types:

(i) Condensation and fragmentation of cells undergoing phagocytosis (with lysosomal disintegration of the fragments in neighbouring cells).

(ii) Primary formation of lysosomes in dying cells (with activation and subsequent destruction and phagocytosis of the fragments by neighbouring cells).

(iii) Disintegration of cells into minute fragments.

Many studies, including investigations into the concept of apoptosis (Kerr *et al.*, 1972), draw on a mixture of optical and electron-microscopical techniques. Sandritter and Riede (1975) conclude that, from a morphological point of view, the first changes on the way to cell death are degenerative swelling, attributable to the failure of the sodium pump and ATP activity which in turn causes inflow of sodium and calcium and outflow of potassium into the extracellular space. Basically this means a change in cell membrane permeability which at light microscope level is reflected in the cell's reactions to dyes.

13.2.2 *Fine-structural*

(a) *Apoptosis* Fine-structural studies on both induced and programmed or physiological cell death abound, and attempts to rationalize the literature are to be welcomed. One such attempt has been made by Kerr and colleagues, who have put forward a general concept of apoptosis (Kerr *et al.*, 1972). The concept attempts to clarify the previously known process of fragmentation and degeneration and demonstrates how these fragments are phagocytotically disposed of by neighbouring cells. The most powerful aspect of the concept, however, relates to the idea of a controlled cell deletion which is complementary and opposite to mitosis and which is thus aptly named apoptosis (see Chapter 1). This broad and useful functional concept draws largely from morphological evidence and offers little in terms of our detailed biochemical understanding of the mechanisms of cell death.

The thesis of apoptosis stems from the earlier work of Kerr (1965, 1971), on 'shrinkage necrosis' in diseased liver, and Kerr and Searle (1972), who looked at cell loss in basal cell carcinoma of human skin. Apoptosis has now been described in a wide range of tissues and situations and has gained acceptance in the field of pathology. Searle, Collins, Harmon and Kerr (1973) describe a distinctive type of cell necrosis occurring in squamous carcinomas of the uterine cervix. In the initial stages cells become condensed and fragmented into compact membrane-bound bodies containing intact and relatively well preserved organelles. The fragments were seen to be phagocytosed by neighbouring epithelial cells or by histiocytes and finally degraded within heterophagic lysosomes. The results indicated a considerable cell loss in cervical carcinomas, which obviously influenced the rate of tumour growth. A similar cell deletion by apoptosis during castration-induced involution of the rat prostrate was described (Kerr and Searle, 1973). Whilst the incidence of autophagy in epithelial atrophy was confirmed in their paper, the problems of differentiating between autophagic vacuoles and apoptotic bodies were stressed. It was this type of problem which prompted Ericsson (1969) to introduce the term cytosegresome which would cover both types of bodies.

Wyllie, Kerr and Currie (1973a) and Wyllie, Kerr, Macaskill and Currie (1973b) presented evidence of changing incidence of adrenocortical apoptosis in the rat during the newborn period and further demonstrated that adrenocorticotropin (ACTH) withdrawal in the rat is followed by deletion of adrenocortical cells.

The morphological data presented in studies on apoptosis are reminiscent of the earlier work of Saunders *et al.* (1962) and Bellairs (1961), who dealt with cell death in vertebrate ontogeny; the processes of pyknosis, cellular fragmentation and phagocytosis by neighbouring cells are common elements in all the investigations.

Apoptosis has been described in a developmental situation by Kerr, Harmon and Searle (1974), who studied cell deletion in tadpole tail during spontaneous metamorphosis. Special attention was given to apoptosis of striated muscle fibres. The signs of cell death observed in the epidermis included aggregation of chromatin beneath the nuclear envelope, breakup of nucleus and cytoplasm, condensation and budding. These activities produced well-preserved membrane-bound fragments, some containing bits of nucleus. The fragments were subsequently phagocytosed and degraded by neighbouring viable cells and macrophages. The authors claim that deletion of striated muscle in tadpole tail is accomplished by a process akin to 'classical apoptosis'. Dilation and fusion of elements of the endoplasmic reticulum or sarcoplasm leads first to internal fragmentation accompanying nuclear pyknosis. This fragmentation is followed by degradation of the muscle remnants within neighbouring cells or within macrophages. A similar process was described by Crossley (1968) in the intersegmental muscles of *Calliphora*. Degenerative changes involving fracture of myofibrils and sarcoplasmic reticulum preceded phagocytosis by haemocytes.

With regard to fine-structural studies on muscle cell death, Ketelsen (1977) reviewed the changes involved in dystrophic skeletal muscle and showed that different forms of the disease displayed different cell reaction patterns. Webb (1972, 1974, 1977) sees muscular dystrophy as a derangement in the normal process of developmentally based cell death. Webb (1974), studying cell death in human embryos, combined electron microscopy with a range of histochemical tests. He reported that muscle cell death was found in all foetuses between 10 and 16 weeks and suggests that disintegration or disruption of the cell membrane is a crucial factor. He also concluded that acid phosphatase activity and phagocytosis of cell debris was not the main means of disposal. Lesch (1977) clearly indicates that release of lysosomal proteases is not of primary importance in the pathogenesis of autolytic ischaemic muscle cell death.

With regard to apoptosis, the mechanism of cell fragmentation appears to be unclear. No hydrolase activity can apparently be seen in fragmenting muscle, and Searle *et al.* (1973) reported no diffuse acid phosphatase activity in extracellular fragments. Lysosomal digestion is brought about by invading macrophages. Hurle and Hinchliffe (1978), however, report autophagic and diffuse acid phosphatase activity in fragmenting chick mesenchyme (see Section 3.2). Here, acid phosphatase activity appeared to be prevalent in cytoplasmic fragments containing no nuclear elements. They concluded that autophagy accompanied by participation of acid phosphatase occurred at an early stage of degeneration of chick limb mesoderm.

Olson and Everett (1975) describe epidermal apoptosis or cell deletion by phagocytosis in skin following sunburn, during Bowen's disease and in basal cell carcinoma. Here, dyskeratotic cells become phagocytosed intact or in fragments by keratinocytes. The authors conclude that the mass of a tissue is related to the balance between mitosis and cell deletion or apoptosis. Cell loss is also an additional factor in skin and epidermal tissues. Hopwood and Levison (1976) demonstrated atrophy and apoptosis in the cyclical human endometrium. They divided apoptotic bodies into 'early' and 'late' types. 'Early apoptotic bodies', can be recognized as membrane-bound fragments containing condensed cytoplasm with recognizable organelles and sometimes nuclear fragments. These bodies appeared to be broken down by other cells forming dense 'late apoptotic bodies' or residual bodies. They also stress that if a body contains no nuclear fragment it could either be an apoptotic body or an autophagic vacuole.

(b) *Cell deletion* Cell deletion does not always take the form of apoptosis. Bowen and Ryder (1974) reported a selective cell autolysis and deletion in the normal adult planarian which did not result in the formation of apoptotic bodies. In the parenchyma of adult planaria, cells exhibiting varying degrees of lysis, ranging from peripheral degeneration to complete cellular breakdown, can be seen, next to normal well-fixed elements. Morphological signs of lysis appear first in the peripheral prolongations of cells where a characteristic release and concentration of ribosomes occurs. Aggregation of free ribosomes often occurs at boundaries between degraded and more normal looking cytoplasm (Fig. 13.1). Cytoplasmic vesiculations become obvious and there appears to be a concomitant increase in electron opacity. In general, dying cells appear to take up osmium and counterstain more readily, leading to an increase in contrast and the formation of so-called 'dark cells'. Degraded areas spread throughout the whole cell and finally the nucleus is destroyed and deleted (Fig. 13.2). Intermediate stages of cellular breakdown can be seen, the final appearance of a lysed area taking the form of a space lined with highly electron-dense lipid-like material and associated broken membranes (Fig. 13.3). Many cells, in particular gland cells, maintain a well-developed and distended endoplasmic reticulum during the earlier phases of this physiological cell deletion. A characteristic pattern of extracisternal or exoplasmic acid phosphatase activity can also be seen at this stage (Section 3.2). Expansion of endoplasmic reticulum prior to cell death has also been observed by Cantino and Daneo (1972), Pilar and Landmesser (1976), Chu-Wang and Oppenheim (1977) and Reichel (1976). Increases in the ribosomal population as a response to sublethal cellular injury has been previously observed in the neurons of *Periplaneta americana* (Young *et al.*, 1970).

Starvation in planarian worms increases the amount of cell death (Bowen, Ryder and Dark, 1976), as does regeneration (work in progress). Indeed over a 5-week period of starvation there is on average a 32 per cent decrease in the size of the organism. The initial fine-structural response to starvation appears to be increased autophagy within the parenchymal cells. Examples of muscle lysis can also be seen at approximately 7 days of starvation. Here, there is a loss of cellular matrix and a reduction in glycogen content, leaving empty patches in the sarcoplasm. Some muscle blocks become completely lysed, muscle fibres condense and consolidate into compact masses (Fig. 13.4). Profiles of cell lysis and death can be observed with increasing frequency both in the gut and the parenchyma as starvation proceeds, leaving large vacuities (Fig. 13.3). After 14 days of starvation evidence of gross histolysis can be seen.

Cell deletion in the normal adult planarian means that the great majority of cells in the organism are being replaced. Undifferentiated cells (neoblasts) in the adult animal are undergoing a fairly constant rate of mitosis. Cell production is

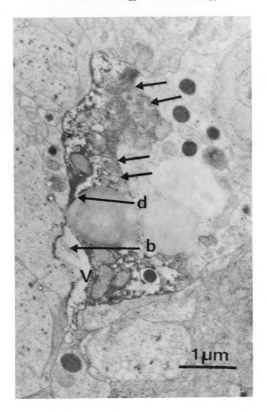

Fig. 13.1 Part of a planarian parenchymal cell. A dense concentration of ribosomes (arrows) can be seen bordering a lysed area. In the lysed area vacuities (V), dense material (d) and broken membranes (b) can be seen.

counterbalanced by cell deletion, since surface loss is minimal and the adult is not growing. Cell deletion probably takes place by a programmed cell lysis.

Selective cell deletion is accelerated during starvation and the animal shrinks. Here, cell death plays a conservative role, the products of death probably being made available to the remaining survivor cells some of which retain a regenerating capability. Paradoxically, therefore, cell death may be an important factor in enabling the organism to survive adverse conditions. It also reduces the finite mass of cells to be maintained and reduces the organism's total requirements whilst increasing or maximizing the available nutrient. The factors which control cell death in planaria are not known. It is possible that each cell type has a genetically programmed lifespan which is short compared with that of the animal as a whole. Clearly the lysis of entire cells followed by their recycling may, like autophagy, be a factor in the ability of planaria to survive starvation.

(c) *Neuronal cell death* In the case of physiological or normal cell death,

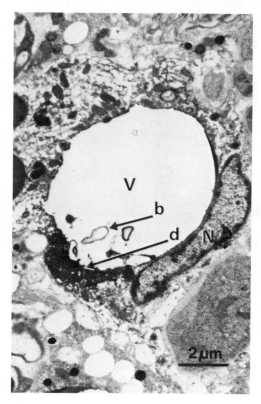

Fig. 13.2 The nucleus (N) of a lysing cell from planarian parenchyma. A large vacuity (V), dense material (d) and broken membranes (b) can be seen.

Levi-Montalcini has proposed that usually, during the development of the central nervous system (CNS), an excess of neurons is produced, these neurons send out more axons than the periphery can support and only those cells that establish contact with the peripheral target survive (see Chapter 10). Pilar and Landmesser (1976) and Landmesser and Pilar (1976) showed that neurons in the chick ciliary ganglion peripherally deprived by excision of the optic vesicle underwent a different type of cell death from that of 'normal' neurons. The main difference was that the normal cells developed an extensive rough endoplasmic reticulum whose dilation indicated the onset of degeneration, whilst peripherally deprived cells did not develop such a reticulum and here cell death was marked by nuclear changes followed by release of ribosomes from polyribosomal systems. Pilar and Land-messer (1976) concluded that the difference between natural and induced cell death in this case stemmed from the greater degree of differentiation that occurred in the neurons of the normal ganglia as a result of their having made some contact with the target zone. Other possibilities could be that the distension of the endo-plasmic reticulum was due to synthesis of specific proteins or enzymes associated

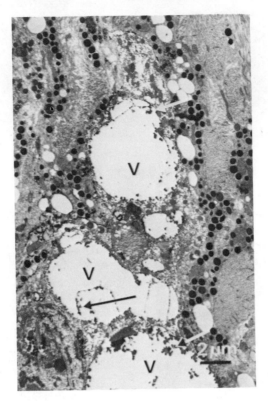

Fig. 13.3 The planarian parenchyma after 35 days starvation. Note the large vacuities (V) containing fragmenting cytoplasm and broken membranes. Many dense lipid-like droplets can also be seen (arrows).

with programmed cell death, or that proteins accumulated as a result of failure to export. Chu-Wang and Oppenheim (1977) also indicate the occurrence of two basic types of cell death in developing chick spinal cord. In one type polyribosomes become dissociated and appear free in the cytoplasm. Ribosomes are also released from the endoplasmic reticulum and mitochondria degenerate. These changes are accompanied by nuclear pyknosis. Another type of cell degeneration showed dilation of the endoplasmic reticulum, nuclear envelope and Golgi. Here, ribosomes remain in rosette formation with little or no mitochondrial degeneration. Dilated membranes eventually break down into numerous vesicles still bearing ribosomes which may be released at a later stage. They disagree, however, with Pilar and Landmesser's interpretation of induced versus 'normal' cell death and indicate that both types of degeneration occur under induced circumstances. Chu-Wang and Oppenheim (1977) suggest that the different patterns of cell death may take place at different stages of cell maturity or may even be associated with functionally different motoneurons. Both types of cell death lead to disintegration by means of

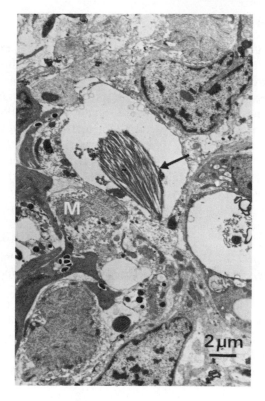

Fig. 13.4 Muscle lysis (arrow) in the planarian parenchyma after 14 days starvation. Note condensation of muscle fibres. Other less-lysed areas of muscle can also be seen (M).

autolysis, since phagocytosis occurs only in the final stages. Autolysis in spinal sensory neurons of chick is thought to involve lysosomal acid phosphatase (Pannese, Luciano, Iurato and Reale, 1976).

O'Connor and Wyttenbach (1974) recorded detailed fine-structural changes associated with cell death of visceromotor neurons of embryonic chick spinal cord. These changes included nuclear pyknosis and disaggregation of polyribosomes. On the other hand Cantino and Daneo (1972), studying developing chick optic tectum, claimed that nuclear changes were not a prominent feature of neuroblast cell death. They described several fine-structural changes in three cell types. These changes ranged from enlargement of the endoplasmic cisternae and membrane disruption to more prominent changes involving gross cytoplasmic disruption. There appeared to be no involvement of macrophages.

(d) *Cell death, development and metamorphosis* There have been many electron-microscopic investigations of cell death in chick development. Comparative studies were presented by Bellairs (1961) and Jurand (1965). Dawd and Hinchliffe (1971) and Hurle and Hinchliffe (1978) specialized on the necrotic zones of the chick limb, and Ojeda and Hurle (1975) and Pexieder (1972a, 1972b) presented data on cell death during the development of the tubular heart of the embryo. These studies commonly referred to nuclear pyknosis, increased cell density, vacuolar or autophagic changes and phagocytosis by macrophages. The same elements appeared again in studies on mammalian morphogenesis (Bellairs, 1961; Jurand, 1965; Schlüter, 1973; Price, Donahoe, Ito and Hendren, 1977). Firth and Hicks (1970) gave an interesting account of differentiation and cell death in transitional epithelium of the urinary bladder of developing rats, and later (Firth and Hicks, 1972) presented evidence of a synchronous death of the superficial cells of the transitional urothelia. Much of the fine-structural work on insect and amphibian cell death relates to metamorphosis (see Chapter 3) and is often presented in an experimental context as part of a cytochemical, biochemical or interdisciplinary study (Weber, 1969; Fox, 1972a, 1972b, 1973; Kerr *et al.*, 1974; Lockshin and Beaulaton, 1974a, 1974b). Michinomae (1977) described biochemical and fine-structural studies on cell death of developing bar eye discs in *Drosophila melanogaster*.

Areas studied from a largely fine-structural point of view include aspects of differentiation of the opercular integument of *Rana* (Reichel, 1976), cell death in the ovarian chambers of *Drosophila* (Giorgi and Deri, 1976), the anatomy of cell death in the epidermis of *Rhodnius* (Goldsmith, 1966) and morphogenetic cell death in the wing hypodermis of *Lucilia cuprina* (Seligman, Filschie, Day and Crossley, 1975).

(e) *Induced cell death* Studies on induced cell death include an investigation into the cellular and subcellular effects of ischaemia by Jones and Trump (1975) and also an interesting paper on the prevention by chlorpromazine of ischaemic liver cell death (Chien *et al.*, 1977). Jones and Trump report dilation of endoplasmic reticulum, swelling and degeneration of mitochondria and cell autolysis in ischaemic pancreatic acinar cells. Chien *et al.* (1977) suggest that chlorpromazine may exert a protective effect by inhibition of the increased calcium entry into injured cells, thus preventing lethal membrane injury. Mittnacht and Farber (1978) confirm these results with liver cell mitochondria. The importance of membrane injury in cell death is further demonstrated by El Mofty *et al.* (1975), who showed early reversible plasma membrane injury in galactosamine-induced liver cell death. Galactosamine-treated liver cells show a 40 per cent reduction in 5'-nucleotidase activity. Administration of uridine prevents loss of enzyme activity and early alterations in the plasma membrane, including damaging changes in calcium ion concentration. Galactosamine-induced liver cell death has been reviewed by Farber and El Mofty (1975) and by Decker (1975).

The course and nature of acinar cell death following pancreatic ligation in guinea pig is described in an interesting paper by Zeligs *et al.* (1975). The fine-structural changes observed included swelling and vesiculation of rough endoplasmic reticulum, a change in osmotic and ionic balance, increase of autophagic activity, and increased activity of acid hydrolases.

A large volume of work is available on the cellular and subcellular effects of damaging radiation of all kinds, and some of this will be discussed in relation to DNA changes (Section 3.5). With regard to histological or morphological damage, autonomous and non-autonomous cell death in the metamorphosis of the epidermis of *Drosophila* is described by Poodry (1975), and u.v.-induced epidermal cell death or apoptosis in mouse skin is documented by Searle and Patten (1977). An apoptotic type of cell death was also observed in relation to lymphocytically induced cell death (Don, Ablett, Bishop, Bundesen, Donald, Searle and Kerr, 1977) and in response to the administration of cancer chemotherapeutic agents (Searle, Lawson, Abbott, Harmon and Kerr, 1975). In the latter paper a comparison of the morphological features of apoptosis and coagulative necrosis is presented (see Chapter 1). A comparison between 'normal' or physiological cell death and chemically or drug-induced cell death is also presented by Wendler (1973) and Schweichel and Merker (1973) (see page 386).

(f) *Scanning electron microscopy* Developmental biologists have traditionally found the scanning electron microscope (SEM) of great value in morphogenetic studies, where changes of shape and form often need to be resolved. Modern sophisticated instrumentation and preparative procedures such as critical point drying have also meant that cells and subcellular components can now confidently be identified on topographical grounds. Histochemical and radioautographic confirmation by means of X-ray microanalysis at SEM level is also now becoming available (Hodges and Muir, 1975; Bowen and Lewis, 1980b). Hodges and Muir (1975) described measurements of mass of developed silver grains in radioautograms. They applied the scanning method to the study of protease-induced cell division in [^{3}H] thymidine-labelled mouse bladder cells. Quantitative data were obtained showing increased ^{3}H incorporation in protease-induced cultures. In our own laboratory we have combined this type of approach with a histochemical technique (Bowen *et al.*, 1976) for demonstrating acid phosphatase in cryostat sections. By scanning emulsion-coated cryosections of [^{3}H] thymidine-labelled specimens prepared to show acid phosphatase we can obtain a simultaneous indication of cell division and cell lysis. Silver grains can be detected topographically and by X-ray microanalysis. Similarly the lysing acid phosphatase-rich cells give a characteristic signal for bromine, which is a constituent of the azo dye reaction product. Technically, the method compares with that presented optically by Bowen and Lewis (1979) and Lewis, Bowen and Bellamy (1979).

Hurle and Hinchliffe (1978) were able to recognize dying mesenchymal cells in developing chick limb by means of topographical features. Dying cells were found to bud and fragment. Similar conclusions may be found in papers by Case and Hendrix (1976), Pexieder (1975) and Garcia-Porrero and Ojeda (1979). A low-resolution SEM study of the effect of 5-bromodeoxyuridine on mouse limb development was recently presented by Skalko, Perrins and Niles (1977).

(g) *Conclusions* In conclusion, commonly occurring features of fine-structural change can be identified in dying cells, although these may not be general to all types of cells. Changes cited in the literature are shown in Table 13.1.

Table 13.1 Commonly occurring features of fine-structural change in dying cells

	Source of change		Result of change
(i)	Membrane	Plasma	Leakiness
		Microsomal	Loss of ribosomal binding
		Mitochondrial	Influx of Ca^{2+}
		Nuclear	Dilation
		Lysosomal	Leakage of enzymes
(ii)	Nuclear pyknosis	Margination of chromatin	
		Hyperchromatosis	Functional DNA changes
		Chromatolysis	DNA loss
(iii)	Cytoplasmic condensation and fragmentation	Vesiculation	Loss of synthetic capacity
		Autophagy	Degradation of macro-molecular structure
		Autolysis	
(iv)	Phagocytosis	By neighbouring cell	Disposal of debris
		By macrophage	

No strict temporal significance should be read into the sequence shown, although it does appear likely that the earliest and most sensitive changes would be in membrane, particularly permeability of plasma membrane. A very common feature following such change appears to be nuclear pyknosis and various stages of chromatolysis, resulting eventually in cell fragmentation. The mechanisms of cell fragmentation still remain unclear, although presumably antecedent membrane and possibly lysosomal or autophagic changes could be contributory factors.

13.3 Cytochemical and biochemical

13.3.1 *Markers of phagocytosis*

Phagocytosis is a secondary response to cell death: in complex metazoa dead or dying cells are often engulfed by wandering amoebocytes, histiocytes, macrophages or, in the case of apoptosis, by neighbouring viable cells. This fact has traditionally been made use of by developmental biologists to mark the sites of 'necrosis', although in effect they would be witnessing the scavenging aftermath of cell death. In recent times the phenomenon of phagocytosis has become closely allied with the lysosome concept (Straus, 1967a and b), although the vacuolar and digestive basis of the process was clearly understood and formulated early on (Metchnikoff, 1892, 1901). The term phagocytosis is generally applied to the vacuolar uptake of solid particles by cells, although Straus (1959) employs the term 'phagosome' to describe particles which may include material taken in by pinocytosis or phago-cytosis. Such a broad definition would encompass autophagy, which could be considered as a specialized kind of phagocytosis. Arguments have also been put forward for considering apoptosis as a special kind of autophagy, since it often involves phagocytosis of identical cell types (Olson and Everett, 1975).

Phagocytosis appears to occur in response to both induced and physiological cell death. Schweichel and Merker (1973) indicate that after application of cyclo-phosphamide, 6-mercaptopurine and actinomycin to early embryos, surviving organisms showed an increase in cell fragmentation and phagocytosis by neigh-bouring cells.

Early embryological studies (Saunders *et al.*, 1962) refer to remains of phago-cytosed cell nuclei which were chromogenic and stained with Nile Blue Sulphate and Feulgen reagent. This phagocytosis was seen to occur in localized zones of necrosis. Salzgeber and Weber (1966) indicate that increased digestive activity probably occurs in the macrophages of regressing chick mesonephros. Ballard and Holt (1968) confirm that cell death and digestion occurs within phagocytes in the developing rat foot where they describe large macrophages containing numerous dead cells. Some of these phagocytes were found to be mitotically active. Dawd and Hinchliffe (1971) used the Gomori acid phosphatase test to highlight sites of macrophage activity associated with necrotic zones in the developing chick limb.

Reference to phagocytosis of dead or dying cells or fragments of cells abound in the embryological literature. The role of phagocytosis in insect cell death is not entirely clear. Giorgi and Deri (1976) outline the phagocytic function of nurse cells in the ovarian chambers of *Drosophila*, and Crossley (1968) indicates that fragments of muscle cells are phagocytosed by haemocytes. Lockshin and Beau-laton (1974a) and Seligman *et al.* (1975), however, did not find active involvement of haemocytes at necrotic sites. Macrophage activity, in relation to removal of dead cells, appears to be an important and consistent factor in the development of amphibia (Michaels, Albright and Patt, 1971; Reichel, 1976) and in mammalian ontogeny (Price *et al.*, 1977), although involvement of macrophages in brain- and CNS-associated cell death is not obvious (Schlüter, 1973). Cantino and Daneo (1971) comment on the lack of phagocytic activity in the developing chick optic tectum. This is contrary to the observations of O'Connor and Wyttenbach (1974) on the embryonic chick spinal cord, where phagocytosis by glial cells was recorded. Avian ontogeny, however, generally displays either macrophage activity or phago-cytosis by neighbouring cells (Pexieder, 1972a and b; Hinchliffe and Ede, 1973; Hinchliffe, 1974; Hurle and Hinchliffe, 1978). Phagocytosis of 'apoptotic bodies' or cell fragments, usually by neighbouring viable cells forms an essential feature of the process of apoptosis (Wyllie *et al.*, 1973a, 1973b; Kerr and Searle, 1973; Searle *et al.*, 1973, 1975; Kerr *et al.*, 1974; Hopwood and Levison, 1976; Olson and Everett, 1975; Elmes, 1975, 1977).

Phagocytosis is an important element in acute cell death, where cells engulf and kill other cells. Rowley and Ratcliffe (1976a, 1976b) have described the fate of bacterial cells within the haemocytes of *Calliphora erythrocephela*. Infectious cells may alternatively bring about cell death within the tissues which they infect. Dorn (1977) describes the cellular breakdown of *Pseudomonas*-infected *Oncopeltus* fat body.

Phagocytic uptake of cells and cell fragments usually leads to their ultimate digestion within secondary lysosomes (Straus, 1967b). It is generally thought that phagosomes are supplied with digestive hydrolases via the primary lysosomes with which they fuse to form secondary lysosomes or digestive vacuoles. A considerable degree of admixture of autophagic and heterophagic material can occur, however, and secondary lysosomes may fuse with each other (Bowen, Coakley and James, 1979).

The measurement of viability of engulfed or phagocytosed cells can be used as a yardstick for digestion (Bowen *et al.*, 1979). Primary lysosomes containing acid phosphatase activity have been identified in *Acanthamoeba* within minutes of feeding with yeast cells, and these vacuoles accumulate and fuse with phagocytic vacuoles. Levels of acid phosphatase in the digestive vacuoles appeared highest at 20 to 40 min. Yeast digestion was observed and yeast viability began to decline at this time. At least half of the population of yeast cells was still viable after 90 min. Much of the acid phosphatase and other hydrolase activity observed in metamorphosing and resorbing tissues is thought to be involved in similar digestive events (Brandes *et al.*, 1965; Salzgeber and Weber, 1966; Weber, 1969; Ballard and Holt, 1968; Derby and McEldoon, 1976; Michinomae, 1977; Hurle and Hinchliffe, 1978; Price *et al.*, 1977).

What factors control phagocytes is not clear. Certain cells have been shown to be selective in the susbtrates they will phagocytose. Chapman-Andresen and Holter (1955) demonstrated that the amoeba (*Chaos chaos*) which does not normally pinocytose glucose could be induced to do so in the presence of bovine serum albumin. Ryder and Bowen, (1977a, 1977b), studying aspects of phagocytes in the slug foot, showed that epithelial cells could be induced to phagocytose materials (including pesticides) not normally taken up when the slug was presented with ribonuclease, peroxidase or bovine serum albumin. Several authors have speculated whether compounds released during cell death in some way activate or attract macrophages or induce phagocytosis in neighbouring cells. Certainly potentially stimulating proteins and enzymes would be released in necrotic areas. On the other hand Slivinsky and Alepa (1973) refers to phagocytosis-inhibiting factors arising in the serum of patients with rheumatoid arthritis.

Autophagy has been shown to occur in normal cells (Ashford and Porter, 1962; Novikoff and Schin, 1964; de Duve and Wattiaux, 1966) and seems to be enhanced in cells under stress (Novikoff and Essner, 1962; Napolitano, 1962; Couch and Mills, 1968; Bowen, 1968). Several authors have observed that an increase in autophagy occurs as a prelude to cell death (Lockshin, 1969). Autophagy may indeed be an important initial response to adverse conditions but may well be in some instances coincidental with ultimate cell death. Autophagy in general is thought of as a conservative phenomenon (Ericsson, 1969) geared to economy measures and the maintenance of cell life rather than the promotion of cell death.

13.2.2 *Markers of hydrolysis*

Many authors have employed and still employ acid hydrolases as markers of cell death. The main impetus for this approach sprang from de Duve's early concept of the lysosome as a 'suicide bag' (de Duve, 1963). According to this hypothesis, particle-bound acid hydrolase can be released into the surrounding cytoplasm under appropriate conditions and bring about its ultimate destruction. Such a response might occur under programmed or physiological conditions or following cellular injury. Interpretation of biochemical data in this area has been made difficult by the fact that solubilization or release of hydrolase from lysosomal particles may occur by accident during the preparative technique of homogenization. Electron-microscopical results on the other hand are often inconclusive in dynamic terms, hence the development of intravital and fluorimetric lines of research (Allison and Young, 1969).

The investigations of Weber (1963, 1969) into 'lysosomal enzymes' during anuran metamorphosis typically illustrates the initial problems encountered in this field. The total activity of various hydrolases such as cathepsin, deoxyribonuclease, β-glucoronidase and acid phosphatase was found to increase considerably during tail involution. Although some of this activity appears to be non-sedimentable and possibly non-lysosomal, the evidence presented indicated that much of the increased activity observed was encountered in invading macrophages. The specific source or sources of non-sedimentable hydrolase in metamorphosing tadpole tail remained unclear at this stage. Reviewing the situation, Weber (1969) listed increases in activity, recorded by other investigators, of acid and alkaline proteases, cathepsins, collagenase, di- and tri-peptidase, acid and alkaline phosphatases, ribonuclease, deoxyribonuclease and β-glucoronidase. Concomitant decreases were encountered in enzymes related to energy metabolism, for example, Mg^{2+}-dependent ATPase, lactate dehydrogenase, succinate dehydrogenase and alanine aminotransferase (glutamate–pyruvate transaminase).

Referring to changes in the tail of *Amblystoma maculatum* at different stages of metamorphosis, Derby and McEldoon (1976) indicate increases in the specific and total activity of acid phosphatase and *N*-acetyl-β-glucosaminidase. The activities of both the enzymes increased in the fins, which degenerate, and muscular portions of the tail, which do not. The specific activity for acid phosphatase in the fins doubles during resorption and there is approximately a 60 per cent increase in the muscular portion. Similar results were obtained with *N*-acetyl-β-glucosaminidase in the muscular part of the tail. Lesch (1977) has, however, indicated that release of cathepsin is not a causative factor in ischaemic cell death of rabbit myocardium. Colon, Dorsey and Lockshin (1978) are also investigating muscle proteolysis during programmed cell death.

Marked increases in acid phosphatase and β-glucuronidase activity were recorded by Zeligs *et al.* (1975) in response to acinar cell death in guinea-pig pancreas. Most of this activity appeared to be associated with autophagic lysosomes.

A critical appraisal of the 'suicide bag' theory in relation to induced cell death has been presented by Trump and Mergner (1974). The consensus view is that lysosomes in general are relatively stable organelles which fragment and release their contents after cell death rather than before.

Some evidence in favour of the 'suicide bag' theory has been put forward by Soberman, Hoffstein and Weissmann (1973), who cinematographically demonstrated lysosomal disruption in response to monosodium urate uptake. Allison (1969) has put forward the possibility that crystals such as sodium urate and silica perforate lysosomal membranes, thus effecting release of hydrolase into the cytoplasm; he extended this idea into an hypothesis of carcinogenesis where the released hydrolases are thought to induce chromosomal damage.

Reviewing lysosomal changes during mammary involution, Slater *et al.* (1963) drew attention to evidence of increasing cathepsin, β-glucuronidase, acid ribonuclease and β-galactosidase activity linked to the enhancement of autophagy in this tissue. Histochemical studies by Wiley and Esterley (1975) demonstrated a decrease in respiratory enzymes and glucose 6-phosphatase during involution of the corpus luteum, but involvement of lysosomal enzymes was unclear, although the staining of these enzymes was prominent in macrophages. Cajander and Bjersing (1975) have presented some evidence in support of acid phosphatase leakage from lysosomes in rabbit germinal epithelia prior to ovulation. They suggest that this leakage may be promoted by sex steroids and prostaglandins.

Unnecessary confusion arises in the literature from the habit of referring to acid hydrolases as 'lysosomal enzymes'. Hydrolases have now been shown to occur in association with the Golgi and/or GERL (Golgi Endoplasmic Reticulum Lysosomes), (Novikoff *et al.*, 1964; Novikoff, 1964, 1976), endoplasmic reticulum (Borgers and Thoné, 1976) extracisternal spaces and plasma membrane (Bowen and Davies, 1971; Bowen and Ryder, 1974; Beadle, Dawson and Amos, 1976). Fishman *et al.* (1967) have clearly demonstrated the dual localization of β-glucuronidase in lysosomes and endoplasmic reticulum, and ribosomal localizations have been claimed for glycosidase (Hash and King, 1958; Kihara, Hu and Halvorson, 1961).

Referring to inflammatory cells, Van Lancker (1964) proposed that acid hydrolases were directly transferred from the endoplasmic reticulum into foci of cytoplasmic degeneration. In an erudite review of the role of hydrolases in cell death, Van Lancker (1975) concludes that when cells die, areas of the cytoplasm become segregated for autolysis and that at some stage these areas are supplied with hydrolases, either by primary lysosomes or direct from the endoplasmic reticulum. The latter presumably refers to the direct supply of autophagic vacuoles via the cytomembrane system. In our own laboratory we have observed accumulation of free acid phosphatase around ribosomes in the extracisternal space of the endoplasmic reticulum. This early extracisternal build-up appears as a prelude to cell autolysis and death in planarian tissues (Bowen and Ryder, 1974), and in our opinion is not derived from the lysosomes but generated and released at the ribosomes. It could be that the 'non-sedimentable' hydrolase activity observed in earlier biochemical studies on tadpole tail and other tissues represents an extracisternal free hydrolase produced *de novo* as a prelude to cell death, rather than a technical artefact.

Many authors have commented on the significance of free hydrolase in relation to cell death and tissue destruction (Schin and Clever, 1965; Brandes *et al.*, 1965; Kalina and Bubis, 1968; Lockshin and Williams, 1965b; Michaels *et al.*, 1971). Norkin and Ouellette (1976) demonstrated virus-induced release of *N*-acetyl-β-glucosaminidase in monkey kidney cells and indicated that lysosomal hydrolytic enzymes are released into the cytoplasm during certain viral infections. Oxford and Fish (1979) has implicated free esterase activity with cell death in starved *Cepaea nemoralis* (Mollusca). Bowen and Davies (1971) observed free acid phosphatase in the digestive gland of *Arion hortensis* (Mollusca) which was apparently linked with cellular lysis and replacement.

Ogawa (1975) has shown that the release of hydrolases such as β-glucuronidase and cathepsin plays a significant role in the pathogenesis of refractory shock. Splanchnic ischaemia during circulatory shock was found to result in 'extralysossomal' and extracellular release of active hydrolases.

Acid hydrolase release into the cytoplasm also seems to be a consistent feature of cell injury and death in plants (Pitt and Coombes, 1969; Mansfield and Sexton, 1974; Oliveira, 1975; Hislop, Keon and Fielding, 1977, 1979). Most of these studies imply a lysosomal or vacuolar source of the free hydrolase. Current investigations into the actual sources of free hydrolase would be very welcome, especially in view of the increasing evidence of non-lysosomal localization.

The mere presence or release of hydrolase in itself is not evidence for its direct involvement in the degradative process. In some instances, enzymes occur in cells as inactive forms or zymogens. Where active hydrolytic enzyme is demonstrable cytochemically or biochemically we must assume that the enzyme exerts a catalytic role in digestion and degradation of the polymer substrates that constitute cell structure, unless of course latent hydrolase activity is in some way accidentally activated by experimental technique.

(c) *Acid phosphatase* Acid phosphatase activity is often employed to help characterize lytic activities within cells, especially during embryological or larval development. Areas studied include insect metamorphosis (Lockshin and Williams, 1964; 1965a, 1965b, 1965c; Schin and Clever, 1965; Crossley, 1968; Spreij, 1971; Lockshin and Beaulaton, 1974a, 1974b; Larsen, 1976; Michinomae, 1977), amphibian metamorphosis (Weber, 1969; Robinson, 1970; Michaels *et al.*, 1971; Decker, 1976; Derby and McEldoon, 1976; Pannese *et al.*, 1976), chick morphogenesis (Salzgeber and Weber, 1966; Dawd and Hinchliffe, 1971; Hinchliffe and Ede, 1973; Allenspach, 1976; Shamsuddin, Bhattacharya and Medda, 1976; Hurle and Hinchliffe, 1978) and mammalian morphogenesis (Jurand, 1975; Ballard and Holt, 1968; Price *et al.*, 1977; Campbell, Dinsdale and Fell, 1977). Cell death and injury in plants has also received attention from this point of view (Pitt and Coombs, 1969; Mansfield and Sexton, 1974; Oliveira, 1975; Hislop *et al.*, 1979).

The role ascribed to acid phosphatase activity in these studies is varied, but falls into two basic types, lysosomal (including heterophagic and autophagic) and non-lysosomal free hydrolase leading to cell autolysis. Whereas some components are digested within the lysosomes others appear to be destroyed by the action of soluble acid phosphatase acting free in the cytoplasm. Lockshin and Beaulaton (1974b) have shown that whilst mitochondria and other organelles are removed by autophagy, erosion of intersegmental muscle fibres in metamorphosing silk moths occurs external to lysosomal membranes, and the mechanism of muscle degradation in this instance remains unclear. Schin and Clever (1965) demonstrated that in normal salivary gland cells of Chironomid larvae, acid phosphatase is restricted largely to lysosomal elements, whereas in degenerating tissue the enzyme is freely distributed in the cytoplasm. Weber (1963) demonstrated a 20 per cent increase in soluble acid phosphatase during the growth phase of *Xenopus* development, and Salzgeber and Weber (1966) showed a 44 per cent increase in soluble non-lysosomal acid phosphatase during regression of chick embryo mesonesphros.

Such increases in soluble acid phosphatase would necessitate synthesis of the enzyme *de novo* as a prelude to cell degeneration. Many of the reports cited imply or assume that soluble or free acid phosphatase must be released from the lysosomes. The concept of synthesis of soluble-phase nascent hydrolase (Bowen and Ryder, 1974, 1976; Bowen, 1980) is not entertained. In this respect the work of Robinson (1970) on *Xenopus laevis* tail metamorphosis is of interest. She consistently achieved higher acid phosphatase activity with *p*-nitrophenyl phosphate as substrate than with sodium β-glycerophosphate. She demonstrated that specific activity of whole-tail homogenates remained constant during early metamorphosis and rose sharply as the rate of tail degeneration accelerated. Total phosphatase activity of whole-tail homogenates was found to rise linearly during development and early metamorphosis and to maintain a high level during the first half of tail regression before declining.

Bowen and Ryder (1974, 1975, 1976) and Ryder and Bowen (1975a) have recommended *p*-nitrophenyl phosphatase as a good marker for cell death. The *p*-nitrophenyl phosphatase method can be used to advantage in demonstrating intracellular sites of lysis. The substrate, *p*-nitrophenyl phosphate, has a broad specificity (which can be controlled) and can pick up a wide range of isoenzymes. Substrates such as sodium β-glycerophosphate and α-naphthyl phosphate are not as effective in this respect (Ryder and Bowen, 1975b). Other investigators have found *p*-nitrophenyl phosphate to be an advantageous substrate (Borgers and Thoné, 1976; Beadle *et al.*, 1976; Harrison, Borgers and Thoné, 1979). Luppa, Radon, Weiss and Martin (1970) introduced the substrate for the localization of non-specific acid and alkaline phosphatase in rat tissue. Butterworth (1971), who used the method for the histochemical demonstration of acid phosphatase in *Ascaris*, reported that the rate of hydrolysis of *p*-nitrophenyl phosphate was twice that of β-glycerophosphate. Luppa, Weiss and Martin (1972) and Hoffman and Di-Pietro (1972) have employed the method at the level of electron microscopy. The latter who demonstrated a plasma-membrane and lysosomal localization of *p*-nitrophenyl phosphatase in human placenta, concluded that *p*-nitrophenyl phosphate was a more favourable substrate than β-glycerophosphate. More recently the method was refined and applied to mouse kidney cells by Miyayama, Soloman, Sasak, Lin and Fishman (1975).

In our own laboratory we have found *p*-nitrophenyl phosphate to be a useful substrate for the cytochemical demonstration of plasma-membrane acid phosphatase, heterophagy, autophagy, cell autolysis and deletion and also for general biochemical assays. The method can be employed for both plant (Bowen and Bryant, 1978) and animal tissues (Bowen, 1980). Since it readily demonstrates a wide range of acid phosphatases, the technique is also a good broad marker of cell death (Bowen and Lewis, 1979; Lewis *et al.*, 1979). It has been employed to advantage in estimating cell death in normal and involuting thymus (Bowen and Lewis, 1980a). Recently Jones and Bowen (1979a) have used the method for localizing sites of cellular lysis in embryonic and newly hatched grey field slugs.

The method has one general disadvantage; *p*-nitrophenyl phosphate under certain conditions may act as a substrate for a wide range of enzymes, including ATPase, alkaline phosphatase and possibly glucose 6-phosphatase and nucleoside diphosphatases. This difficulty, however, can be overcome by employing appropriate incubating conditions and controls (Miyayama *et al.*, 1975; Borgers and Thoné, 1975, 1976; Jones and Bowen, 1979b). Appropriately controlled experiments show that some plasma-membrane staining and more importantly extracisternal (exoplasmic) activity revealed by the test does in fact represent acid phosphatase.

Using the *p*-nitrophenyl phosphatase method, Bowen and Ryder (1974) described selective cell lysis and deletion in the normal adult planarian. This localization is illustrated in Figs. 13.5, 13.6 and 13.8, and close examination shows it to be extracisternal (or exoplasmic). Identical extracisternal localization has also been observed in slug pore cells Fig. 13.7. Evidence of progressive morphological disintegration accompanied by increasing free acid hydrolase activity may be taken as an indication of cell death. The appearance of extracisternal *p*-nitrophenyl phosphatase activity may be a relatively early signal of cell death; it always precedes any morphological signs of lysis and gradually extends throughout the exoplasmic space (Fig. 13.5b) and eventually the whole cell.

Use of the *p*-nitrophenyl phosphate method for the demonstration of acid phosphatase during starvation and cell autolysis in *Polycelis tenuis* (Planaria) was described by Bowen and Ryder (1975). Starvation-induced acid phosphatase activity was cytochemically demonstrated in relation to autophagy and crinophagy in gland cells, autophagy, autolysis and cell death in parenchymal and gastrodermal cells, and basement-membrane lysis. In a broader study Bowen *et al.* (1976) present a biochemical assay of β-glycerophosphatase activity over a 5-week starvation period. Four peaks of activity could be resolved, which correlated well with parallel cytochemical data. The first peak was associated with phagocytosis and feeding, the second at day 7 with enhanced autophagy and sarcolysis, the third with lysis of intestinal cells, and the fourth with cell deletion in the reproductive system.

With regard to sarcolysis in starving planaria, free acid phosphatase makes a distinctive appearance between 7 and 14 days of starvation in the sarcoplasm of selected muscles. Enzyme activity can initially be demonstrated in the pools of glycogen which occur in the sarcoplasm. The glycogen appears to be progressively cleared, leaving evidence of lysis such as subcellular fragments and broken membranes.

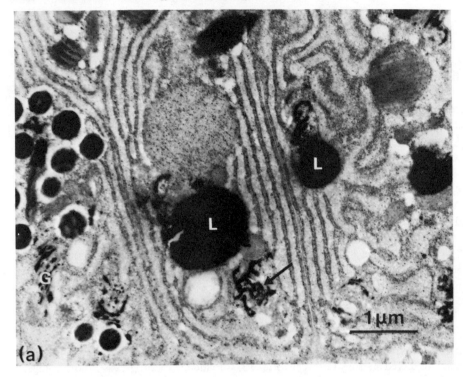

Fig. 13.5 (a) Acid phosphatase (*p*-nitrophenyl phosphatase) activity in acidophil and cyanophil gland cells from a normal planarian. Note Golgi (G), GERL (arrow) and lysosomal (L) localizations of electron-dense reaction product. (b) Acid phosphatase (*p*-nitrophenyl phosphatase) activity in a dying cyanophil gland cell from a 14 day-starved planarian. Note evidence of extracisternal, nascent or free acid phosphatase activity (arrows), as well as the more usual Golgi (G) and lysosomal (L) sources.

By 14 days, acid phosphatase activity can be detected in the fibrillar structure of the muscle, and the complete breakdown of certain muscles occurs by the third week of starvation. Some muscles, especially in the pharynx, however, completely escape this wave of destruction. The selection of muscles for deletion must, therefore, be fully programmed or controlled.

At about the 14th day of starvation a considerable increase in the amount of free acid phosphatase activity in the gut cells was observed. This increase accompanied an accelerated cell deletion in the gastrodermis. Examination of the cytochemical picture reveals that the free activity is extracisternal (Fig. 13.8). The cells are eventually destroyed and completely removed (Fig. 13.9).

In the planarian parenchyma autophagic acid phosphatase appears in most cell types. This phenomenon is enhanced during the early phases of starvation in the 'fixed parenchymal cells' and gland cells. Acid phosphatase activity in these cells extends to specific types of secretions. Profiles indicating the involvement of

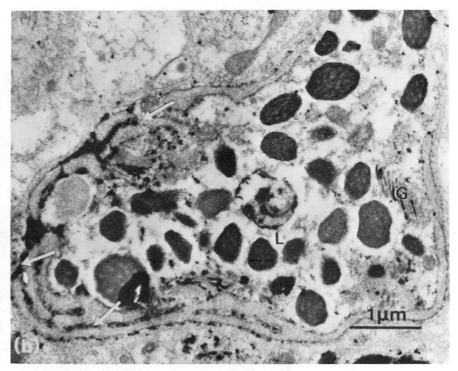

Fig. 13.5 (b)

pigment granules, acidophil, cyanophil and mucous secretions in a controlled lysis have been observed. A process similar to that described as crinophagy (de Duve, 1969) appears to occur, where lysosomes fuse directly with secretory granules. A similar enhancement of autophagic activity has been observed in the gut epithelial cells of starved locust (Bowen, 1968). The role of lysosomes in autophagy and cell death in insect metamorphosis has been reviewed by Lockshin (1969). Coward, Bennett and Hazelhurst (1974) have observed the occurrence of cell death in regenerating planaria and also stressed the importance of autophagy and lysosomal enzymes in the dedifferentiation of specialized cells at the blastema.

(b) *Nascent or extracisternal acid phosphatase* The consistent pattern of exoplasmic localization in lysing cells (Bowen and Ryder, 1974, 1975, 1976; Jones and Bowen, 1979a, 1979b) deserves closer attention. According to the GERL hypothesis (Novikoff, 1974, 1976), enzyme might be expected within the cisternal space but not outside it. Bowen and Ryder (1974) have suggested that extracisternal acid phosphatase may represent *nascent* enzyme newly formed at the ribosomes which occur outside the endoplasmic cisternae. This would mean that such enzymes produced specifically for cell lysis need not pass into the Golgi or lysosomes. de Duve (1963), Brandes *et al.* (1965), Schin and Clever (1965) all implicate

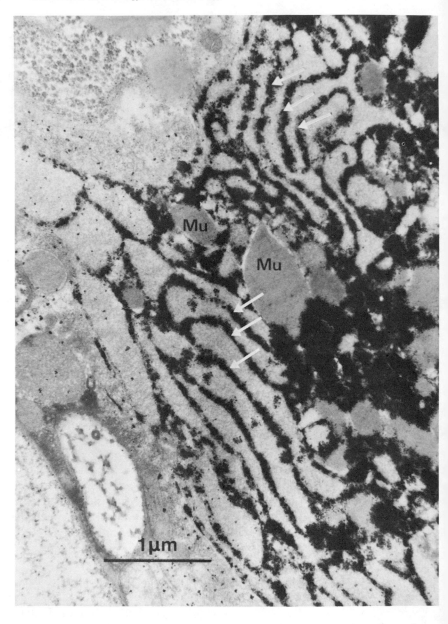

Fig. 13.6 Part of a lysing or dying mucous cell from a normal planarian. Reaction product denoting intense acid phosphatase (*p*-nitrophenyl phosphatase) activity can be seen outside the endoplasmic cisternae (arrows). The endoplasmic cisternae themselves are negative as are the mucous droplets (Mu).

lysosomes in the release of cytoplasmic acid phosphatase. Ballard and Holt (1968) and Weber (1964), on the other hand, present some evidence of cell lysis and death occurring without the involvement of specifically lysosomal hydrolase.

The occurrence of non-lysosomal acid phosphatase is now well established cytochemically (Miyayama *et al.*, 1975; Borgers and Thoné, 1976; Beadle *et al.*, 1976; Harrison *et al.*, 1979).

With regard to the possible functions of extracisternal or nascent hydrolase in cell death, it might be useful to review current ideas on the fate of newly synthesized proteins. Lodish and Rothman (1979) have summarized the events surrounding the translation of proteins. Translation begins with the coupling of messenger RNA with small ribosome subunits, to which large subunits are added. During this period the ribosomes are free in the extracisternal or exoplasmic space and unattached to any cytomembranes. Apparently, up to 40 amino acids need to be assembled before the growing polypeptide starts to emerge from the ribosome. After the addition of a further 30 amino acids the initial sequence of amino acids becomes fully exposed. A part of this exposed chain is now thought of as a signal sequence, since a particular pattern of amino acids is necessary to bring about recognition and attachment between the growing polypeptide and adjacent endoplasmic reticulum. In cells producing a lot of protein the cisternae typically become distended. Enzyme so produced would follow the cytomembrane into the Golgi and hence into the lysosomes, as predicted by the original GERL theory (Novikoff, Essner and Quintana, 1964). This is regarded as the usual passage for *export* and is the probable route for membrane proteins and secreted proteins. Soluble proteins which do not display similar signal sequences, do not become membrane-associated and can remain free in the exoplasmic or extracisternal space. A comparison of the possible pathways taken by acid hydrolases during programmed cell death is shown in Fig. 13.10.

It would be instructive to demonstrate whether the nascent acid phosphatase produced for cell death differs from the conventional acid phosphatase in terms of its amino acid signal sequence. If so, one might expect a lower molecular weight for the cell-death-signal-free enzyme or at least a different set of precursor messengers. In theory thus it might be possible to isolate specific mRNA molecules produced especially to synthesize nascent or free acid hydrolase, as opposed to the membrane-binding variety. It would be useful to search for such 'messengers of death'. The problem is difficult in that it would also involve the identification and isolation of a range of isoenzymes in normal and dying cells.

Ultimately, the best procedure would probably be to make antibodies to the different enzymes or isoenzymes, and test for cross-reactivity. Alternatively, if the enzymes can be isolated reasonably cleanly, isoelectric focusing can detect a difference of one amino acid. Evidence for different species of mRNA encoding for membrane-bound and free secretory protein (IgM μ chains) has recently been presented by Singer, Singer and Williamson (1980).

Robinson (1970) indicated the importance of investigating the electrophoretic separation of acid phosphatase from individual tadpole tails and reports the

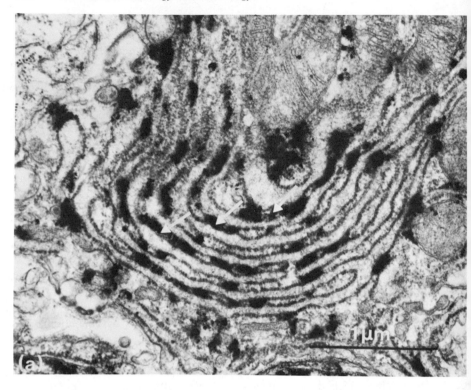

Fig. 13.7 (a) Extracisternal build-up of acid phosphatase (*p*-nitrophenyl phosphatase) activity (arrows) in a pore cell from a slug embryo (*Deroceras reticulatum*). Note the nascent enzyme on the ribosomal or outer surface of the endoplasmic reticulum. (b) A profile of a similar cell to that shown in (a). This pore cell appears to be at a more advanced stage of lysis, and free acid phosphatase (*p*-nitrophenyl phosphatase) activity is extending throughout parts of the extracisternal space. Some morphological evidence of lysis in the form of broken membranes (arrows) can also be seen. (By courtesy of G. W. Jones)

separation of two distinct forms, one of which (a *p*-nitrophenyl phosphatase) increases dramatically in amount during tail regression.

In cells undergoing programmed lysis by means of extracisternal acid phosphatase the rough endoplasmic reticulum initially becomes well developed and many distended cisternae typical of protein-producing cells can be observed (Bowen and Ryder, 1974; Jones and Bowen, 1979b). Pilar and Landmesser (1976) and Chu-Wang and Oppenheim (1977) have observed embryonic chick neurons developing a well-organized rough endoplasmic reticulum before cell death, and several workers have suggested that specific proteins may be synthesized as a prelude to cell death (Tata, 1966; Beckingham-Smith and Tata, 1976; Lockshin and Beaulaton, 1974c).

Cytochemical evidence for extracisternal or free hydrolase can be seen in Figs.

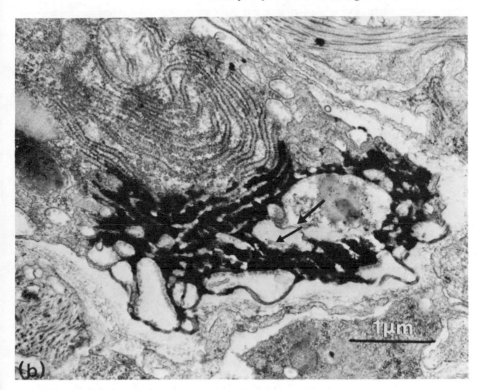

Fig. 13.7 continued

13.5 to 13.8, representing cells from normal and starved planaria, in Fig. 13.11, from old thymus, and in Figs. 13.7a and 13.7b, from developing slug embryos.

A range of distribution patterns can be seen, extending from a slight and limited activity associated with ribosomes to an extensive extracisternal distribution. In the latter stages before cell disruption, acid phosphatase activity may occur throughout the cytosol (Figs. 13.9 and 13.11) as well as in the lysosomes and Golgi.

Conventional wisdom regarding the involvement of acid hydrolase in cell death follows the GERL hypothesis in that hydrolase destined to mediate cell death is thought to pass from endoplasmic reticulum to Golgi and hence into lysosomes. Free hydrolase is traditionally envisaged as being released from broken or fragile lysosomes. Our cytochemical results lend support to the idea of a more direct and economic distribution of destructive acid hydrolase (see Fig. 13.10). Where programmed cell death occurs, it would make greater economic sense to produce new active species of hydrolase capable of sustained action in the cytosol rather than transport enzyme into the lysosomal system for eventual release into the cytosol. No accurate data are available on the operational pH of enzymes within the various subcellular compartments. Nevertheless, the operational pH of newly released hydrolase need not be optimal to be effective.

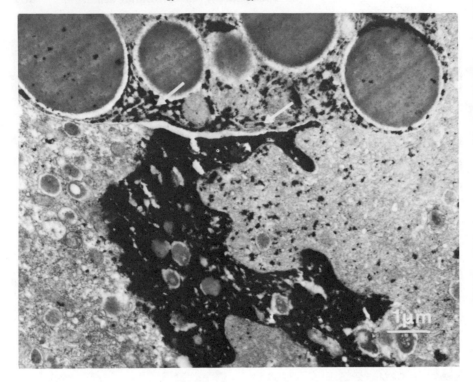

Fig. 13.8 Free acid phosphatase (*p*-nitrophenyl phosphatase) in the planarian gastrodermal cells of a 14 day-starved animal. Note evidence of the extra-cisternal build-up of enzyme activity (arrows) in the Minotian gland cells, and free acid phosphatase activity in the adjacent lysing phagocytic cell.

Michaels *et al.* (1971), working on cell death in the external gills of *Rana pipiens*, have also observed acid phosphatase reaction product present throughout degenerating cells and have suggested that the cells produce hydrolytic enzymes for their own degeneration.

Bernelli-Zazzera (1975) points out that ribosomes from ischaemically damaged liver cells bind less readily with endoplasmic membrane. This change would effectively prevent protein transfer into the endoplasmic reticulum. It would not of necessity, however, inhibit protein synthesis. Protein or enzyme synthesized under such conditions would presumably be deposited or released into the extracisternal space.

To date, specifically extracisternal acid phosphatase has been demonstrated in secretory cells or cells which synthesize much protein. It may be that such secretory or active cells undergo a regular turnover. It is possible, although unlikely, that nascent acid phosphatase specifically acts in clearing specialized or senescent endoplasmic reticulum and is in some way involved in cellular differentiation, which may not always lead directly to cell death (Jones and Bowen, 1979b). It is

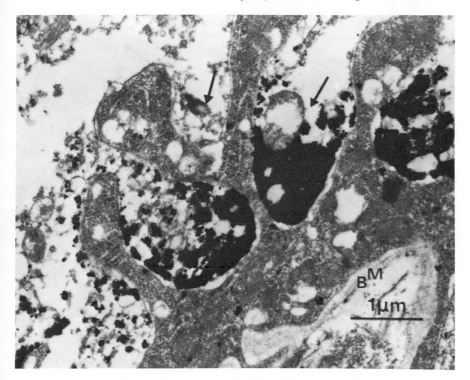

Fig. 13.9 Remains of part of the planarian gastrodermis after 36 days starvation. Gastrodermal cells have been deleted. Hardly any subcellular structures remain in the lysed areas (arrows) which are infused with residual acid phosphatase (*p*-nitrophenyl phosphatase) activity. One thin phagocytic cell remains above the basement membrane (BM).

interesting to note that the mitochondria survive for some time (Fig. 13.8), even when surrounded by free acid phosphatase activity. Cellular integrity is also retained initially (Figs. 13.6 and 13.7), arguing for a controlled sequence of events rather than an immediate shutdown.

Metazoan cells, exhibiting lysosomal or free acid phosphatase, but little or no morphological disruption, may be engulfed by macrophages or surrounding cells. Alternatively the gradual spread or build-up of extracisternal acid phosphatase and other hydrolases may cause cellular fragmentation before ultimate phagocytosis by neighbours, as described by Kerr *et al.* (1972).

The popularity of acid phosphatase as a marker is related to the fact that it is one of the few acid hydrolases which is easily demonstrable at electron-microscope level. This means that other hydrolases, such as the cathepsins and proteases, probably more central and important in terms of cellular degradation, have been neglected cytochemically. However, it is probably safe to ascribe a lytic role to acid phosphatase (EC3.1.3.2). The enzyme is classed biochemically with the

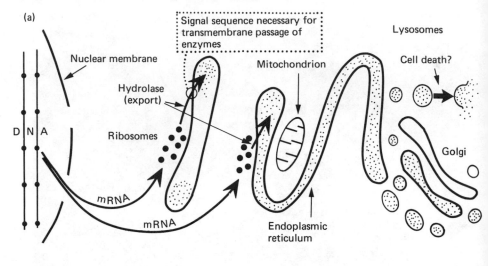

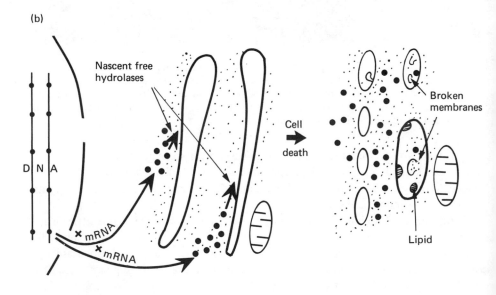

Fig. 13.10 Nascent hydrolase and programmed cell death. (a) Normal pathway for hydrolases. (b) Hypothetical pathway for nascent hydrolases.

phosphoric monoester hydrolases and catalyses the formation of an alcohol and the release of orthophosphate from such esters.

(c) *Simultaneous estimation of mitosis and cell death* Bowen and Lewis (1979) and Lewis *et al.* (1979) have developed a method for the simultaneous estimation of mitosis and cell death. The technique involves the detection of [6-^{3}H] thymidine and acid phosphatase activity in cryostat sections and establishes the principle of combined radioautography and histochemistry. Currently the approach has been modified for the application of X-ray microanalysis and scanning electron microscopy (Bowen and Lewis, 1980b).

It has been shown that in the thymus from 8-week-old mice, acid phosphatase-positive cells represent 1.23±0.06 per cent of the total population, and 8.4±0.27 per cent of the cells incorporate tritiated thymidine. Free acid phosphatase activity was used to estimate cell autolysis and death (Fig. 13.12). This simple technique was based on forming a separate barrier of 5 per cent glycerine albumin or Euparal in xylene between the histochemically reacted tissue section and the photographically sensitive emulsion. Combining the two tests in this manner had no effect on the pattern of acid phosphatase activity. The addition of a separating layer, however, does depress the number of cells demonstrating thymidine incorporation. Interestingly, the percentage of acid phosphatase-positive cells tends to be highest at the lowest total cell densities and the percentage of mitotically active cells tends to decrease at the higher cell densities. Hinsull *et al.* (1977) indicate that 1.17±0.04 per cent of the thymus cells from 8-week-old rats are Trypan Blue-positive. This measure of cell death is very close to our estimate for acid phosphatase-positive cells in mouse thymus. If cells showing free acid phosphatase activity are counted, a good estimate of the levels of cell death can be obtained. This has been confirmed recently by Bowen and Lewis (1980a), who described the fine-structural distribution of acid phosphatase in mouse thymus. Here, using *p*-nitrophenyl phosphate as substrate we demonstrated free acid phosphatase in the cytoplasm of lysing thymic lymphocytes (Fig. 13.11). Vacuolar or lysosomal sites of acid phosphatase activity have been demonstrated in macrophages which appear to dispose of lymphocytes. In 8-week-old mice acid phosphatase-positive cells represent 1.23±0.06 per cent of the total population, whereas in 42-week-old mice, showing involution of the thymus, acid phosphatase-positive cells represent 2.40±0.17 per cent of the total population (see Figs. 13.13a and 13b).

Occasionally, profiles demonstrating free acid phosphatase throughout the cytoplasm of thymic lymphocytes can be seen (Fig. 13.11). The cells at this stage may appear structurally intact. Commonly, in thymus from old mice, lymphocytes showing a greater degree of disintegration can be seen. This cellular debris and associated acid phosphatase activity may subsequently be engulfed by macrophages. Macrophages can also be seen taking up intact lymphocytes containing no free acid phosphatase activity. Such lymphocytes typically possess marginated chromatin and some contain vacuolar acid phosphatase.

Overall the results show that the level of cell death in the thymus from a

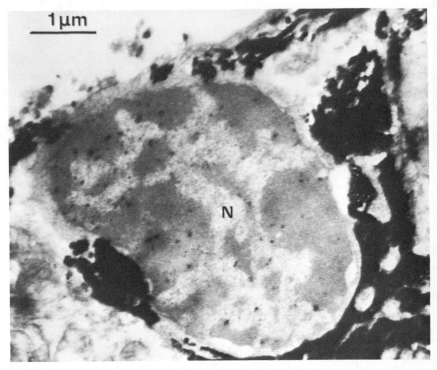

Fig. 13.11 Evidence of free acid phosphatase (*p*-nitrophenyl phosphatase) activity surrounding the nucleus (N) of a small thymic lymphocyte from the thymus of a 42-week-old mouse.

42-week-old mouse is virtually double that in the thymus from an 8-week-old mouse. In areas of low cell density the difference is even greater (Fig. 13.13). Care should be exercised in the interpretation of acid phosphatase-positive cells observed at the level of the optical microscope, since the electron cytochemical results show that macrophages, epithelial reticular cells and lymphocytes contain vacuolar sources. At the level of the optical microscope, examination under oil immersion lenses is preferable for the identification of free and particulate (or vacuolar) acid phosphatase, the former being the primary site of cell lysis. In thymus, however, heterophagy by the macrophages appears to be the main fate of dead or dying cells so that many of the vacuolar sources of acid phosphatase activity in this context are ultimately involved in cell lysis. Indeed, heterophagy in degenerating tissues has been exploited by embryologists as a general marker for cell death.

One interesting and valuable experiment which is now technically possible would be to label thymocytes by short pulse-labelling with tritiated thymidine, transplanting into a new host if necessary, and following them sequentially, looking for incidence of both labelling and free acid phosphatase activity. Such an experiment would give a measure of the mean lifespan of thymocytes and an estimate of mean length of time identifiable as dying.

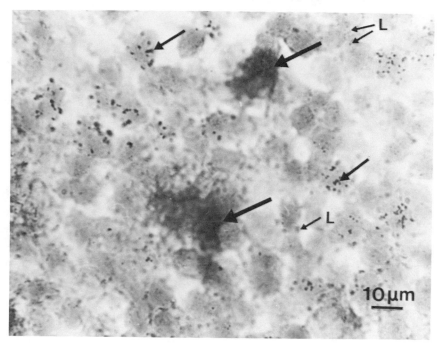

Fig. 13.12 Part of a cryostat section (a histochemical radioautograph) of the thymus of a 42-week-old mouse. Dead or dying cells can be located as sources of free acid phosphatase activity (large arrows). Some lysosomal or particulate sources can also be seen in normal cells (L). Silver grains can be seen immediately above nuclei that have incorporated [6^3H]thymidine (small arrows). The combined technique shown here can give a simultaneous estimate of cell death and division. (By courtesy of G. H. J. Lewis)

13.3.3 *Respiratory enzymes*

Weber (1969) reviews the evidence that tissue involution coincides with a general decrease in activity of enzymes related to energy metabolism. The enzymes investigated included glyceraldehyde 3-phosphate dehydrogenase, lactate dehydrogenase, malate dehydrogenase, succinate dehydrogenase, glutamate dehydrogenase, alanine aminotransferase (glutamate–pyruvate transaminase), aspartate aminotransferase (glutamate-oxalate transaminase), and hydroxyacyl-CoA dehydrogenase. Comparative studies on these enzymes in *Xenopus* larval tails show a 50–70 per cent decrease in specific activity, although the proportions of various enzymes remained unchanged. Wiley and Esterley (1975) report a marked drop in lactate dehydrogenase, succinate dehydrogenase, cytochrome oxidase and glucose 6-phosphatase late in the involution of corpeus luteum.

On the basis that, 'it is logical to assume that mitochondrial and other cellular enzymes are lost from the dying cells or are changed in such a way as to be enzymatically inactive', Hammar and Mottet (1971) proposed that cytochemical

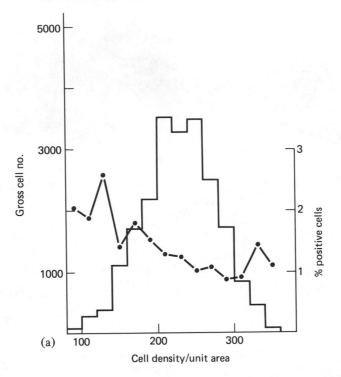

Fig. 13.3 (a) The percentage of free acid phosphatase-positive cells (●) in the thymus of an 8-week-old mouse expressed in relation to density/unit area. Gross cell numbers are also shown (——). (Compare with b). (b) The percentage of free acid phosphatase-positive cells (●) in the thymus of a 42-week-old mouse expressed in relation to density/unit area. Gross cell numbers are also shown (——). (Compare with a). (By courtesy of G. H. J. Lewis)

studies of loss of dehydrogenase activities would be useful in mapping areas of cell death. Specifically they studied necrosis in the interdigital areas of developing hindlimb buds of normal chick embryos. Using succinic acid as a substrate and the tetrazolium salt, Nitro Blue Tetrazolium, they localized succinate dehydrogenase activity in normal cells. They reported that interdigital necrotic areas exhibited minimal levels of succinate dehydrogenase 2 days prior to morphological evidence of degeneration. Such loss of activity was thought to lead to a subsequent fall in cellular ATP, leading to cell death.

The work of Hammar and Mottet (1974) was strongly criticized by Fallon *et al.* (1974). They concluded that succinate dehydrogenase activity cannot be detected using intact chick feet incubated in the Hammar and Mottet medium. They further confirmed that succinate oxidation was demonstrable in frozen sections of chick feet or in interdigit and digit homogenates. They showed that succinate dehydrogenase and associated enzymes such as succinate-cytochrome *c* reductase and NADH-cytochrome *c* reductase were functional and present in digital and necrotic

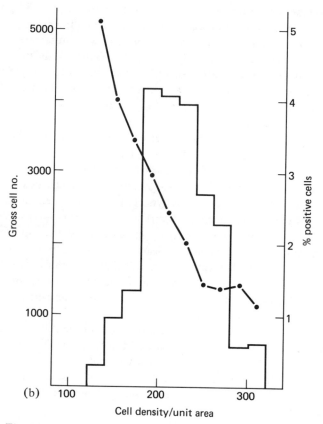

(b)

Fig. 13.3 continued

areas. They indicated that succinate oxidation does not diminish to extinction before cells die and suggested that the results obtained by Hammar and Mottet (1971) were due to histochemical artefact, the non-specific deposition of black formazan crystals.

In one sense, the recent work of Gartner *et al.* (1978a, 1978b) confirms the findings of Fallon *et al.* (1974) in that they also demonstrate the persistence of succinate dehydrogenase activity in the palatal shelf epithelium during the development of normal and X-irradiated mice and also in the Mongolian gerbil. Palatal fusion is a process understood to involve cell death (Pratt and Martin, 1975). Gartner and colleagues interpret the persistence of succinate dehydrogenase activity as evidence against a 'programmed cell death' and in favour of 'epithelial stretching'. Dehydrogenase activity as shown by Fallon *et al.* (1974), however, can persist in necrotic areas and its presence does not automatically exclude cell death.

In an interesting paper by Wrogmann and Pena (1976) it is suggested that the mechanism of muscle cell necrosis is explained by an increase in the influx of calcium into cells which triggers mitochondrial calcium overloading and energy depletion. The mitochondrial problems include defects in oxidative phosphorylation with NAD-linked substrates. For example, the mitochondria from patients with Duchenne dystrophy exhibit depressed respiration rates with glutamate.

With regard to useful markers of necrosis, Woodling, Warner, Hall and Puffer (1973) report briefly on the demonstration of recent and old infarcts in the gross placenta, using a succinate dehydrogenase assay. Van Waes and Lieber (1977) demonstrated that in alcoholics, serum glutamate dehydrogenase reflects the presence and degree of liver cell necrosis.

Lockshin *et al.*, (1975) reported, in abstract, evidence for a transient increase in anaerobic glycolysis in the intersegmental muscles of developing *Manduca sexta*. Bidlack and Lockshin (1976) presented more extensive data on the related evolution of lactate dehydrogenase isoenzymes during programmed insect cell death. They reported that the total activities of isocitrate dehydrogenase, malate dehydrogenase, catalase and lactate dehydrogenase decline at various rates during intersegmental muscle involution. Lactate dehydrogenase activity increases severalfold during the final phases of involution. The enzyme also appears in degenerating silk glands at the time of most rapid involution. The authors conclude that limitations of oxidative metabolism plays an important role in the involution of the intersegmental muscles.

Ischaemia causes a decrease in the oxidative and glycolytic activities of cells (Bernelli-Zazzera and Gaja, 1964; Gaja, Bernelli-Zazzera and Sorgato, 1965; Ragnotti, Gaja and Bernelli-Zazzera, 1968). There is a drop in the ATP and energy charge of ischaemic liver cells and a great decrease in the $[NAD^+]/[NADH]$ ratio, although these effects are reversible depending on the extent of ischaemia. Norkin and Ouellette (1976) noted the release of lactate dehydrogenase and glutamate dehydrogenase from cells killed by Simian virus 40, which they interpreted as being indicative of plasma-membrane damage.

13.3.4 *Other enzymes*

Schlichtig *et al.* (1977) report on the measurement of several 'ATP-digesting' enzymes in degenerating insect muscles. The enzymes were biochemically extracted from the muscles of pharate adults of *Antheraea polyphemus* and *Manduca sexta*. They report the occurrence of myosin ATPase, mitochondrial ATPase, two membrane-bound ATPases and an acid ATPase. They also identified lysosomal acid phosphatase capable of digesting several nucleoside phosphates and indicated that its activity increased threefold at the onset of muscle degeneration.

In an interesting paper on the role of autolysins in cell death, Tomasz (1974) indicates that the cytocidal activity of bacterolytic antibiotics depends largely on the co-operation of a special class of bacterial enzymes, the autolysins, and not simply on their inhibitory effect on bacterial cell wall synthesis.

Giovanni Marzullo (1972) attributes to cell death *in vitro* reduction of the enzyme phosphoglucose isomerase, following administration of high concentrations of 5-bromodeoxyuridine. Burger and Potgieter (1978) comment on the changes in activation of intracellular aspartate aminotransferase by pyridoxyl 5-phosphate after cell death. They concluded that changes in the degree of saturation of aspartate aminotransferase by the coenzyme pyridoxyl 5-phosphate occur intracellularly after cell death or injury and before release into the circulation. Ying, Sarma and Farber (1978) monitored rat liver cell necrosis by determining sorbitol dehydrogenase and alanine aminotransferase levels in serum and by histological examination.

Pratt and Martin (1975) and Hassell and Pratt (1977) have shown that the initiation of epithelial cell death and adhesiveness in the palatal shelf of developing rats may be mediated through increase in cyclic AMP. These findings suggest that there may be underlying changes in the enzyme adenylate cyclase. Coffino, Bourne and Tomkins (1975) propose a mechanism of lymphoma cell death induced by cyclic AMP. They demonstrated that a mouse lymphoma tissue culture line was killed by isoproterenol, cholera toxin and prostaglandin E, all inducers of cyclic AMP. The growth inhibition and cytolysis observed were also produced by administration of the analogue dibutyryl cyclic AMP. Genetic and other evidence presented indicated that both growth inhibition and cell death were mediated by cyclic AMP. Mutant subclones deficient in cyclic AMP-dependent protein kinase were resistant to killing. The authors discussed the possible physiological role of cyclic AMP-induced cell death in thymic lymphocyte differentiation.

13.3.5 *DNA*

(a) *Genetic cell death* The concept of programmed cell death implies the existence of a plan (probably in the shape of DNA) for the normal imposition of cell death during development. Since during development not all cells die, the activation of such a plan must depend on environmental (which may include cytoplasmic) triggers. Since cell death is usually selective, the triggers must also operate selectively either in response to cell type differences or perhaps to differences dependent on cell phase, maturity, density or even spatial position.

Whatever the relationship between the triggers and the plan, programmed cell death must be catered for initially by the existence of particular DNA base sequences, unless the system is controlled entirely at the translational level. Specific lethal DNA sequences are well known to biologists, indeed the majority of genetic mutations fall into this category in the evolutionary sense.

While it is outside the scope of this chapter to catalogue the occurrences of lethal genes, it is nevertheless instructive to draw upon a few examples. Fristrom (1972) describes the chemical modification of cell death in the bar eye of *Drosophila*. The author describes the temporal and spatial distribution of cell death during the development of bar eye imaginal discs and indicates that presumptive eye cells in this mutant usually die during early differentiation, i.e. the stage of organization into groups, corresponding to ommatidia. The author demonstrates changes in phenotype when acetamide and cytosine are applied. Acetamide increases facet number by inhibiting bar-specific cell death, and cytosine delays cell degeneration to a later stage. The paper confirms that imaginal disc cell death is responsible for the bar phenotype.

A substantial review of cell death and autonomous gene action in lethals affecting *Drosophila* imaginal discs was presented by Murphy (1974); lethal gene activity in the form of imaginal cell death was histologically and fluorimetrically demonstrated. The techniques used are the Acridine Orange method of Spreij (1971) and also a Toluidine Blue stain for RNA, which revealed degenerating cells and cell fragments as dark material containing chromatin and ribosomal remains. By these means a total of 13 different *Drosophila melanogaster* mutants were shown to have extensive cell death in the imaginal discs. Three of these mutants showing imaginal cell death had previously been classified as 'normal' on the basis of dissection. The author suggests that discs from these three mutants may be able to respond to the hormone ecdysone, but may be deficient in a later response leading to death of the tissue. Theiler, Varnum, Nadeau, Stevens and Cagianut (1976) describe a new allele of ocular retardation resulting in reduced cell death in the eye cup of an inbred strain of mouse during early development.

Studies on 'lethal genes' clearly show that certain mutations can bring about a catastrophic cell death. Such mutations need not instantaneously affect the phenotype, but do so when a particular DNA sequence is called on and fails. A physiologically or naturally programmed cell death could presumably operate in a similar fashion, with genes lethal for specific cell types within the organism being selected for in view of their morphogenetic utility.

(b) *DNA damage* Lieberman (1972) in a lucid review has dealt with DNA metabolism in relation to induced cell death and chemotherapy. He indicated that agents which physically damaged DNA kill cells more effectively than those which inhibit DNA synthesis by indirect means. It also appears possible to prevent cell death induced by treatment with a variety of inhibitors of DNA synthesis by preventing protein metabolism.

X-irradiation and alkylating agents act on DNA directly, producing physical damage which often only partially inhibits DNA synthesis. Indirect inhibitors which block metabolic pathways often result in complete inhibition of DNA synthesis; these include compounds such as hydroxyurea, arabinofuranosylcytosine (Ara-C) and amethopterin.

There appear also to be tissue differences in terms of response to harmful agents. Directly and indirectly acting agents are lethal for lymphoid tissues, whilst only agents which damage DNA physically are lethal for intestinal crypt cells. Non-dividing cells are resistant to all these agents.

Inhibitors of protein synthesis have been found to give protection from cell death. Thus cycloheximide and tenuazonic acid protect intestinal crypts from damage induced by X-irradiation, nitrogen mustard and Ara-C. One interpretation is that damage to DNA or interference with DNA synthesis results in the production of 'lethal proteins', and that the action of cycloheximide and tenuazonic acid prevents this. It does, however, suggest that cell death is not a passive process, but one that requires active participation of cellular metabolism, as indeed is implied by the term programmed cell death.

Cells appear to be especially sensitive to anti-DNA agents during the S phase. This is an important factor, since, in tissues, cells generally divide asynchronously. Application of DNA-synthesis inhibitors would block some cells at G1 or at G1–S boundaries, and cells which would otherwise be sensitive would be protected at this phase. Cells entering S phase would be killed. Most work to date has been done on DNA synthesis or inhibition of synthesis. In addition to replication, DNA synthesis is essential for repair. The relationship between failure to repair DNA and cell death needs further investigation.

From the point of view of establishing techniques for the demonstration of cell death, Williams, Little and Shipley (1974) present a useful possibility. They associate mammalian cell death with specific endonucleolytic degradation of DNA. They show that human and rodent cells treated with several agents resulting in cell death produce distinct patterns of DNA degradation, as assayed in an alkaline sucrose gradient. They conclude that similar patterns of DNA degradation are common to various mammalian cells after DNA damage, cell membrane disturbance or general toxicity. DNA fragments around 10^6 and 10^7 daltons are obtained, apparently as a result of endonucleolytic incision.

Petes, Farber, Tarrant and Holliday (1974) comment on a reduced rate of DNA replication and chain elongation in aging fibroblasts. Ito and Kobayashi (1977) draw attention to chromosomal DNA damage and membrane injury in cell death. They conclude that in Toluidine Blue-sensitized yeast cells, u.v.-irradiation probably induces cell death primarily through damage to the cell membranes or surface.

Strauss (1977), in a review of the molecular biology of the response of cells to radiation and radiomimetic chemicals, indicates that unrepaired DNA lesions may lead to chromosome aberrations and cell death. Thus, since only viable cells can produce tumours, post-replication repair is important to the primary events of carcinogenesis. DNA lesions which result in strand breaks, however, lead to cell death. Painter (1977), drawing from an investigation of *xeroderma pigmentosum* cells which are unusually sensitive to u.v.-irradiation, also stresses the relationship between unrepaired DNA damage and cell death.

Frankfurt (1976) describes acute cell death induced by inhibition of DNA synthesis by hydroxyurea in stomach and intestinal tissues. He concludes that gastric cells are more resistant to inhibition of DNA synthesis. Scott *et al.* (1977) induced polydactyly in rat embryo hindlimbs by the administration of Ara-C and 5-fluorodeoxyuridine. In this instance the inhibitors of DNA synthesis probably killed the mesodermal cells that normally induce interdigital cell death in overlying ectodermal cells. Using Nile Blue Sulphate and Neutral Red as cell death markers, Webster and Gross (1970) review the effects of a range of chemicals on nuclear DNA in an attempt to validate DNA 'crosslinking' as the basis of cell death.

In conclusion, cell death could be the result of blockage of DNA synthesis or in the case of programmed cell death, the activation of particular lethal genes.

(c) *Tumour cell death* Tumour cell death is dealt with in Chapters 4 and 6. A brief discussion of the techniques used for the demonstration of tumour cell death is included here, since most studies on the tissue kinetics of tumours involve thymidine incorporation. This reflects the importance ascribed to DNA and cell division in the control of tumour growth.

The biology of cell death in tumours has been briefly reviewed by Cooper (1973); the occurrence of cell death in the form of apoptosis has been observed by Searle *et al*. (1973) and the importance of induced cell death is implied in many studies on the effects of X-irradiation and chemotherapeutic agents (Lieberman, 1972). An obvious therapeutic aim is to increase the rate of cell death selectively within tumours and cancers. If the rate of cell death exceeds the mitotic rate, then regression will occur.

Claesson (1972) described the studies on cell death in the peripheral lymphoid organs of mice. The author aimed to determine the percentage of degenerating cells arising from both short- and long-lived members of the lymphocyte population. The techniques employed included repeated injections with [^{3}H] thymidine, nigrosin dye exclusion, cell smears and radioautographic procedures. The percentage of dying or non-viable cells was calculated from nigrosin-positive counts. The results showed that decaying spleen and lymph node cells are derived from both short- and long-lived lymphocytes. There are some ratio differences between spleen and lymph nodes, and the findings suggest that the time of cell degeneration depends on the lifespan of specific progenitor cells. Okumura (1977) investigated the kinetics of tumour cell death by hyperthermic treatment and X-irradiation, using a Trypan Blue-exclusion technique to estimate cell viability. Using a radioautographic technique involving the incorporation of tritiated thymidine into dividing cells, Lala (1971) demonstrated an age-specific elimination of non-cycling cells and a low level of 'mitotic death' in Ehrlich ascites tumour. The study was subsequently extended (Lala, 1972) employing measurements of tumour weight, proportions of necrotic tissues, cell number and thymidine incorporation.

A technical development was proposed by Hofer and Hofer (1971), who labelled tumour cells with [^{125}I] iododeoxyuridine (^{125}IdUrd) as well as tritiated thymidine. They emphasized the problems of obtaining accurate assessments of tumour cell migration and tumour cell death. They reported that after incorporation into DNA, ^{125}IdUrd iodine remained bound within the tumour cells until they died. When labelled cells died, it was claimed that deiodination and excretion of the DNA-breakdown products limited the reutilization of radioactive label by other cells. The authors attempted to assess the rate of ascites-tumour cell death *in vivo* by monitoring the rate of ^{125}I excretion.

Kumar, Hoshino, Wheeler, Baker and Wilson (1974) indicated that, although loss of ^{125}I, from IdUrd incorporated into DNA may be a good indicator of cell death after certain treatments, they were doubtful of its utility for the study of dead tumour cell removal in solid tumours. Russell, Pietro, Laurence and Le Gendre (1974) presented evidence of polyamine depletion and spermidine elevation in the sera of tumour-bearing rats. They employed these biochemical markers as indicators of regression (and related cell death) in mammary tumours. Heby and Andersson (1978) indicated that extracellular polyamines such as putrescine and spermidine accumulate during tumour growth and following radio- and chemo-therapy, as a result of tumour cell death.

13.3.6 *RNA and synthetic changes*

(a) *RNA* Farber (1972) has broadly reviewed the role of RNA, or rather 'translation', in induced cell death. Other investigators dealing with programmed cell death have pointed out that RNA may be released from autophagic vacuoles (Matsuura, Morimoto, Nagata and Tashiro, 1968) or rapidly degraded (Okabe, Koyanagi and Koga, 1975). Chu (1971) demonstrated the incorporation of Ara-C into 2S-7S RNA before cell death. He showed that murine leukemic cells exposed to Ara-C during exponential growth undergo exponential cell death and that increases in cell death correlated with a linear increase in the incorporation of the analogue into RNA. Incorporation of ^{3}H-labelled Ara-C was greatest into the 2S-16S RNA isolated by a sucrose gradient, and incorporation of [^{3}H] uridine into 2S-7S RNA was markedly decreased by pretreatment with Ara-C. The author suggested that Ara-C incorporation into 2S-7S RNA resulted in acute cell death.

Young, Ashhurst and Cohen (1970) observed increases in ribosomal counts as a response to cell injury. Bowen and Ryder (1974) also observed concentrations of ribosomes in areas of cell lysis. Several authors (O'Connor and Wyttenbach, 1974; Pilar and Landmesser, 1976; Chu-Wang and Oppenheim, 1977) have commented on the release of ribosomes from endoplasmic reticulum at particular stages leading to cell death. This question of ribosomal detachment in relation to cell injury and death was reviewed by Bernelli-Zazzera (1975). He showed that ischaemia results in a decrease in the number of ribosomes and in a reduction in the number of polyribosomes. He also showed that stripped rough endoplasmic reticulum from ischaemic cells bound fewer added ribosomes than did reticulum from normal cells. He concluded that there is a 'loosened relationship' between endoplasmic membranes and ribosomes in ischaemic livers.

It is also possible that programmed cell death could be controlled at the cytoplasmic level. Cell death could theoretically follow as a consequence of erroneous transcription of DNA into mRNA, damage to mRNA or erroneous translation, especially if the RNA message itself coded for a vital protein. In actual fact no tissues other than lymphoid cells show any evidence of cell death as a consequence of pure inhibition of protein synthesis, as evidenced by the administration of cycloheximide or tenuazoic acid (Farber, 1972).

The specific synthesis of 'death proteins' or at least regulatory proteins has also been investigated (Beckingham-Smith and Tata, 1976). Several authors indicate that there may be premonitory increases in protein synthesis as a prelude to cell death. Bowen *et al.* (1976) showed an early sharp rise in RNA synthesis in starving planaria which subsequently exhibited an increased level of cell death. Den Hollander and Bowen (unpublished results) have obtained a similar response in regenerating planaria, where there is an enhanced cell death in the blastema.

(b) *Protein* Many studies on changes in protein synthesis in relation to cell death, deal largely with enzymic changes (see Sections 13.3.2, 13.3.3 and 13.3.4). Others deal with the release of the products of autolysis. Kasahara, Kametaka and Kandatsu (1976) describe a Folin-phenol-based technique for demonstrating material released during autolysis. A significant release of Folin-phenol-positive material was observed during both aerobic and anaerobic incubations of liver slices from protein-depleted rats as well as during anaerobic incubation of tissue slices from normal rats. Much of the positive reaction produced appeared to be due to the accumulation and release of xanthine rather than to the products of proteolysis.

Popp, Shinozuka and Farber (1974) tested the hypothesis that protein synthesis is important during the induction of necrosis. They tested the effect of cycloheximide, a known inhibitor of protein synthesis, on the induction of cell death by selected hepatotoxic agents and concluded that cycloheximide gave protection against cytotoxic effects – a finding consistent with the hypothesis that cell necrosis is possibly an active process requiring the induction of specific protein(s).

Pratt and Greene (1975, 1976) describe the inhibition of palatal epithelial cell death by altered protein synthesis. They indicated that cell death was prevented in the presence of 6-diazo-5-oxo-L-norleucine (DON), 2-deoxyglucose (DOG) or cycloheximide. Administration of actinomycin D or Ara-C, inhibitors of RNA and DNA synthesis respectively, did not inhibit cell death, and the authors suggest that inhibition of cell death by DON was most likely due to blockage of glucosamine synthesis. They concluded that palatal epithelial cell death is not a passive event, but an active process necessitating the programmed synthesis of specific proteins.

The possibility that tadpole tail regression was initiated by synthesis of new proteins was carefully investigated by Beckingham-Smith and Tata (1976). Changes in protein synthesis in hormone-induced cultured *Xenopus* tadpole tails were studied using gel electrophoresis. A new technique developed to analyse small differences in distribution of two radioactive isotopes, $[^3H]$ methionine and $[^{35}S]$-methionine, demonstrated that no significant changes occurred in the synthesis of tail proteins. The authors pointed out, however, that this still left open the possibility of changes in the synthesis of regulatory proteins. Having also demonstrated tri-iodothyronine activation of tadpole tail collagenase, the authors discuss the role of hormone-induced activation of proteolytic 'cascades' in tail regression (see Chapter 3).

13.3.7 *Methods related to immune killing*

The topic of immune killing is dealt with in Chapter 12, and discussion in this section will be limited to technical procedures which can be made use of for the demonstration of cell death. Many cell-mediated cytotoxicity tests are based on the release of radioactive ^{51}Cr from previously labelled target cells (Brunner, Mavel, Cerottini and Chapius, 1968; Perlmann, Perlmann, Muller-Eberhard and Manni, 1969; Henney, 1973) or on the release of $^{125}IdUrd$.

Sanderson (1976) presented an important review of the release of different cell components during T-cell-mediated cytotoxicity of mastocytoma target cells. He demonstrated that the radioactive isotope marker (^{51}Cr) is held in the cell as a small molecule rather than bound to large macromolecules as previously thought. He showed that the release of proteins labelled with radioactive leucine or methionine and RNA labelled with uridine show the same kinetics as that for the release of phosphocholine, phosphorylated 3-O-methylglucose, sucrose and chromium. There was no lag phase and release was linear with time, indicating that the kinetics were not related to molecular size. Rubidium and nicotinamide showed some spontaneous and effector-cell-mediated release. Rapid loss of rubidium occurred within 15 min following T cell contact rather than lysis. Some 10 per cent of nicotinamide release also appeared to be due simply to T cell contact. DNA labels such as thymidine and IdUrd were released following a lag phase of about 30 min, implying that nuclear membrane damage occurs later than plasma membrane damage. The author concluded that target cell death is accompanied by sudden release of the cytoplasmic contents, following contact by T cells.

Todd, Stultung and Berke (1973) described a mechanism of blocking lymphocyte-mediated cytolysis of allogenic tumour cells by means of hyperimmune serum. Assay of lymphocyte-mediated cytolysis was by measurement of released radioactivity from cells previously labelled with ^{51}Cr. Newlands (1974) presented two radiolabel assays of immunity to a xenograft in mice based on the IdUrd method of Cohen, Burdick and Ketcham (1971). In the first, target cells were labelled before effector cells were added and loss of radioactivity taken as a measure of direct cytotoxicity. Secondly IdUrd is added as a pulse to the mixed culture of target and effector cells; inhibition of uptake of label is measured here as a cytostatic response. Perlmann *et al.* (1969) reported the cytotoxic effects of leucocytes triggered by complement bound to target cells. Fakhri and Hobbs (1972) describe the phenomenon of target cell death without added complement after co-operation of 7S antibodies with non-immune lymphocytes.

Goldstein and Singal (1974) testing the HL-A antigen reactivity of cultured human fibroblasts presented evidence that continuously replicating cells lose viability before mitotically inhibited but actively metabolizing cohorts. They suggested that factors which increased cellular turnover accelerated senescence.

Norkin and Ouellette (1975) have described variation in pattern of lysosomal enzyme release, cellular enzyme release and cell death during SV40 infection in normal and SV40-transformed liver cells. They reported release of lysosomal *N*-acetyl-β-glucosaminidase into the cytoplasm of rhesus kidney and T-22 (SV40-transformed green-monkey kidney) cells to a degree equal to that obtained during infection of rapidly killed normal green-monkey kidney cells. However, damage to plasma membrane, as demonstrated by the release of lactate dehydrogenase and aspartate aminotransferase into the surrounding medium, occurred to a greater extent in normal cells than in the rhesus kidney or T-22-transformed cells.

Cell death by apoptosis following attachment of allergized lymphocytes is described by Don *et al.* (1977).

13.3.8 Accumulated products

Accumulated products as markers or indicators of impending cell death would include the enzyme and protein increases outlined in earlier sections (13.2.2, 13.3.6), but might also include more immediate ionic changes such as calcium overload, in addition to the more long-term lipofuscin build-up within lysosomes and in specific cases of cell injury the accumulation of particular products such as triglyceride observed in fatty degeneration of the liver.

(a) *Calcium* One definitive change associated with cell death appears to be accumulation of calcium. Farber and El-Mofty (1975) illustrate the importance of calcium overload as a response to plasma-membrane damage, indeed they suggest that cell decline and death is mediated by increases in the concentration of intracellular calcium. Trump (1974) refers to hydroxyapatite deposits within mitochondrial matrices as an indicator of plasma-membrane injury, and this concept is further extended by Chien *et al.* (1977), where control of mitochondrial calcium was shown to be a vital component in the control of ischaemic liver cell death. Recent work by Bank, Buchner and Hunt (1980a), Bank, Emerson, Buchner and Hunt (1980b), Schanne, Kane, Young and Farber (1979) and others also emphasize the importance of calcium entry into dying cells (see Chapter 7).

(b) *Lipid* Interference with protein synthesis or phospholipid formation in liver prevents the final preparation of triglyceride for secretion. Characteristically, this in due course leads to the accumulation of more and more triglyceride within the hepatocytes – producing the so-called 'fatty liver' described by pathologists. Accumulation of various products can be identified in diseased and injured liver cells (Keppler, 1975).

(c) *Lipid peroxidation* The accumulation of lipid-rich material in telolysosomes or residual bodies has been documented. These vacuoles have been termed lipofuscin granules or alternatively 'age pigment'. Such granules typically accumulate with age and are particularly common in cells such as neurons which do not commonly exocytose. Lipofuscin contains both lipid and protein and also products such as haem groups released as a result of autophagic mitochondrial degradation. It has been suggested that haem groups may catalyse the peroxidation of unsaturated lipids in degenerating mitochondrial membranes. The peroxidized lipid molecules may thus cross-link with each other and with denatured protein, forming indigestible lipofuschin. Stoward (1977), discussing muscular dystrophy and aging, has suggested that early lesions in such muscle might well be the result of increased lipid peroxidation of mitochondrial and sarcoplasmic reticulum membranes. He suggests that such peroxidation could result from abnormally high levels of mitochondrial xanthine oxidase or depressed levels of superoxide dismutase or endogenous peroxidase. In leucocytes, xanthine oxidase catalyses the conversion of oxygen into highly injurious superoxide. Superoxide dismutase then converts harmful superoxide into less toxic hydrogen peroxide. Superoxide radicals O_2^- may be formed directly from a flavoprotein hydroxyl interaction and may be transformed under certain conditions into either singlet oxygen or hydroxyl radicals. These strong oxidizing reagents may initiate harmful membrane peroxidation. The biochemical details of the role of lipid peroxidation in cell injury are reviewed by Slater (1975).

Sheldrake (1974) discusses the involvement of lipid peroxidation in the aging, growth and death of cells. He emphasizes that lipid peroxidation can occur in all functional cell membranes and does not always result in the formation of visible lipofuscin granules. He further suggests that the accumulation of such products for example in the Golgi, endoplasmic reticulum, nuclear or lysosomal membranes could be harmful and lead to further lipid peroxidation, resulting in accelerated senescence and even cell death. Interestingly he indicates that lipid peroxidation can be inhibited by lipid-soluble antioxidants, such as vitamin E, and accelerated by vitamin E deficiency, ionizing radiation, chloroform, ethanol poisoning and hyperbaric treatments.

13.4 Conclusions

Laboratories use numerous techniques to evaluate cell death in widely differing situations. There is no one method which is satisfactory for all experiments, since technical or practical limitations will always intervene. For the convenience of the reader, the basic approaches are summarized in Table 13.2 along with pertinent comments and references chosen for their accessibility, generality and clarity of descriptions.

Table 13.2 Means of evaluating cell death

Technique	Advantages	Disadvantages	Comments	Examples
Histological	Routine	Time-consuming	Correct interpretation requires corroborative evidence	Ballard, (1965), Hamburger (1975), Arees and Åström (1977)
Vital dyes	Simple, inexpensive and rapid	At gross level, does not distinguish between phagocytes and dead cells	Dependent on membrane permeability changes	Saunders et al. (1962), Starre-van der Molen and Otter (1974), Fallon and Cameron (1977)
Trypan Blue-exclusion	"	"	"	Drake et al. (1972), Brus and Glass (1973), Hinsull et al. (1977)
Fine-structural	Detailed, visual, subcellular data	Prone to artefacts induced by pretreatment	Useful, but allows little dynamic interpretation	Kerr and Searle (1972), Schweichel and Merker (1973), Pilar and Landmesser (1976)
Markers of phagocytosis	Can be used at the level of the optical and electron microscope	A secondary response to cell death. Limited to tissues containing phagocytically active cells	Demonstrates 'disposal' rather than cell death	Salzgeber and Weber (1966), Ballard and Holt (1968), Hurle and Hinchliffe (1978)
Markers of hydrolysis (a) Biochemical	A good measure of degradative activity	Strictly accurate only in synchronous and homogeneous cell	May be a secondary response to cell death	Weber (1969) Robinson (1970), Derby and McEldoon (1976)

Markers of hydrolysis

(b) Cytochemcial (esp. acid phosphatase)	Can identify lysing cells and subcellular patterns of lysis	Prone to histo-chemical artefacts	Needs a tight experimental control but can be combined with radioautography for kinetic studies	Bowen and Ryder (1974), Lewis et al. (1979), Bowen and Lewis (1980a and b)
Respiratory enzymes	Good biochemical marker of reduced viability and/or metabolic activity	Often not very useful cytochemically	Not always indicative of cell death	Weber (1969), Fallon et al. (1974), Bidlack and Lockshin (1976)
Other enzymes	Wide range available, e.g. ATPases, autolysins, phosphoglucose isomerase, aspartate amino-transferase, adenylate cyclase	Preliminary and need further confirmation	Results vary in different cell types	Tomasz (1974), Coffino et al. (1975), Schlichtig et al. (1977) (see Section 13.3.4)
DNA (release)	Damage to cells detectable as released DNA fragments	Probably a secondary response, following pyknosis	Applicable where there is endonucleolytic degradation	Williams et al. (1974), Ito and Kobayashi (1977)
^{125}IdUrd clearance	Claimed to achieve a dynamic quantitative assessment in some tumours	Of doubtful value in some tissues where label is reutilized	Depends on clearance and excretion of ^{125}I from dead cells	Hofer and Hofer (1971), Kumar et al. (1974)

Table 13.2 continued

Technique	Advantages	Disadvantages	Comments	Examples
RNA (damage)	Direct inhibitory effect on protein synthesis. RNA release from damaged cells	Largely applicable to induced cell death	Experimentally induced by cycloheximide, tenuazoic acid etc.	Chu (1971), Farber (1972), Matsuura et al. (1968)
Protein	Enhanced protein synthesis seen as a prelude to cell death. Release of protein (including enzymes) from damaged cells	No clear evidence for 'death proteins'	Needs further investigation	Pratt and Greene (1975, 1976), Popp et al. (1974), Norkin and Ouellette (1975), Kasahara et al. (1976)
Release of loaded ^{51}Cr	Rapid cytotoxic assay	Indicates cell disruption only	Gives little information on mechanisms of cell death	Brunner et al. (1968), Henney (1973), Sanderson (1976)
Calcium overload	Calcium influx across plasma membrane and increase in mitochondrial calcium indicative of impending cell death	Data on irreversibility unclear	A good early indicator	Farber and El-Mofty (1975), Chien et al. (1977), Bank et al. (1980a, b) (see Chapter 7.)
Lipid	Visible accumulation of lipid product leading to necrosis, e.g. triglyceride	Pathological in diseased liver	Of limited applicability	Keppler (1975)
Lipofuscin	Easily observed as pigmented	An indicator of age rather than cell death	Causal relationship with cell death not established	Sheldrake (1974), Slater (1975)

References

Allenspach, A. L. (1976), Acid phosphatase activity in embryonic chick oesophagus lacking 'programmed cell death' (crooked neck dwarf mutant). *Cytobiologie*, **12**, 356–362.

Allison, A. C. (1969), Lysosomes and cancer. In *Lysosomes in Biology and Pathology*, Vol. 2 (eds. J. T. Dingle and H. B. Fell), North-Holland Publishing Co., Amsterdam and London, pp. 178–204.

Allison, A. C. and Young, M. R. (1969), Vital staining and fluorescence microscopy of lysosomes. In *Lysosomes in Biology and Pathology*, Vol. 2 (eds. J. T. Dingle and H. B. Fell), North-Holland Publishing Co., Amsterdam and London, pp. 600–628.

Arees, E. A. and Åström, K. E. (1977), Cell death in the optic tectum of the developing rat. *Anat. Embryol.*, **151**, 29–34.

Ashford, T. P. and Porter, K. R. (1962), Cytoplasmic components in hepatic cell lysosomes *J. Cell. Biol.*, **12**, 198.

Ballard, K. J. (1965), Studies on the participation of hydrolytic enzymes in physiological autolysis in degenerating regions of rat foetus. Ph.D. Thesis, University of London.

Ballard, K. J. and Holt, S. J. (1968), Cytobiological and cytochemical studies on cell death and digestion in the foetal rat foot: the role of macrophages and hydrolytic enzymes *J. Cell Sci.*, **3**, 245–261.

Bank, H. L., Buchner, L. and Hunt H. (1980a), A statistical design for estimating functional survival. *Cryobiology* (in press).

Bank, H. L., Emerson, D., Buchner, L. and Hunt, H. L. (1980b), Cryogenic preservation of rat polymorphonuclear leukocytes. *Blood cells* (in press).

Barbosa, P. and Peters, T. M. (1971), The effects of vital dyes on living organisms with special reference to Methylene Blue and Neutral Red. *Histochem. J.*, **3**, 71–93.

Baroldi, G. (1975), Different morphological types of myocardial cell death in man. *Recent Adv. Stud. Card. Struct. Metab.*, **6**, 383–397.

Beadle, D. J., Dawson, A. L. and Amos, S. (1976), The demonstration of acid phosphatase in cultured 3T3 mouse cells. *Histochemistry*, **48**, 161–166.

Becker, F. F. and Lane, B. P. (1966), Regeneration of mammalian liver. IV. Evidence on the role of cytoplasmic alterations in preparation for mitosis. *Am. J. Pathol.*, **49**, 227–237.

Beckingham-Smith, K. and Tata, J. R. (1976), Cell death: are new proteins synthesized during hormone induced tadpole tail regression? *Exp. Cell Res.*, **100**, 129–146.

Bellairs, R. (1961), Cell death in chick embryos as studied by electron microscopy. *J. Anat.*, **95**, 54–60.

Bernelli-Zazzera, A. (1975), Ribosomes in dying liver cells. In *Pathogenesis and Mechanisms of Liver Cell Necrosis* (ed. D. Keppler), MTP Press, Lancaster, pp. 103–111.

Bernelli-Zazzera, A. and Gaja, G. (1964), Some aspects of glycogen metabolism following reversible or irreversible liver ischemia. *Exp. Mol. Pathol.*, **3**, 351–368.

Bidlack, J. M. and Lockshin, R. A. (1976), Evolution of lactate dehydrogenase isoenzymes during programmed cell death. *Comp. Biochem. Physiol.*, **55B**, 161–166.

Borgers, M. and Thoné, F. (1975), The inhibition of alkaline phosphatase by L-*p*-bromotetramisole. *Histochemistry*, **44**, 277–280.

Borgers, M. and Thoné, F. (1976), Further characterization of phosphatase activities, using non-specific substrates. *Histochem. J.*, **8**, 301–317.

Bowen, I. D. (1968), Electron-cytochemical studies on autophagy in the gut epithelial cells of the locust *Schistocerca gregaria. Histochem. J.*, **1**, 141–151.

Bowen, I. D. (1980), Phagocytosis in *Polycelis tenuis*. In *Nutrition in Lower Metazoa* (eds. D. C. Smith and Y. Tiffon), Pergamon Press, Oxford, pp. 1–14.

Bowen, I. D. and Bryant, J. A. (1978), The fine structural localization of *p*-nitrophenyl phosphatase activity in the storage cells of pea (*Pisum sativum* L.) cotyledons. *Protoplasma*, **97**, 241–250.

Bowen, I. D., Coakley, W. T. and James, C. J. (1979), The digestion of *Saccaromyces cerevisiae* by *Acanthamoeba castellanii. Protoplasma*, **98**, 63–71.

Bowen, I. D. and Davies, P. (1971), The fine structural distribution of acid phosphatase in the digestive gland of *Arion hortensis* (Fer.). *Protoplasma*, **73**, 73–81.

Bowen, I. D. and Lewis, G. H. J. (1979), A method for the simultaneous estimation of mitosis and cell death. *Suppl. Proc. R. Microsc. Soc.*, **14**, A4.

Bowen, I. D. and Lewis, G. H. J. (1980a), Acid phosphatase and cell death in mouse thymus. *Histochemistry*, **65**, 173–179.

Bowen, I. D. and Lewis, G. H. J. (1980b), Histochemical applications of X-ray microanalysis: the simultaneous assessment of mitosis and cell death using an X-ray microanalytical method in the scanning electron microscope. *SEM (1980)*, IV SEM Inc., AMF O'Hare, Illinois, pp. 179–187.

Bowen, I. D. and Ryder, T. A. (1974), Cell autolysis and deletion in the planarian *Polycelis tenuis* Iijima. *Cell Tissue Res.*, **154**, 265–274.

Bowen, I. D. and Ryder, T. A. (1975), Acid phosphatase as an indicator of cell death. *Proc. R. Microsc. Soc.*, **10**, 271.

Bowen, I. D. and Ryder, T. A. (1976), Use of the *p*-nitrophenyl phosphate method for the demonstration of acid phosphatase during starvation and cell autolysis in the planarian *Polycelis tenuis* Iijima. *Histochem. J.*, **8**, 319–329.

Bowen, I. D., Ryder, T. A. and Dark, C. (1976), The effects of starvation on the planarian worm *Polycelis tenuis. Cell Tissue Res.*, **169**, 193–209.

Brandes, D., Bertini, F. and Smith, E. W. (1965), Role of lysosomes in cellular lytic processes. II. Cell death during holocrine secretion in sebaceous glands. *Exp. Mol. Pathol.*, **4**, 245–265.

Brunner, K. T., Mavel, J., Cerottini, J. C. and Chapius, B. (1968), Quantitative assay of the lytic action of immune lymphoid cells in ^{51}Cr-labelled allogeneic target cells *in vitro. Immunology*, **14**, 181.

Brus, I. and Glass, G. B. J. (1973), Estimation of cytotoxic injury to gastric parietal cells by Trypan blue exclusion test, followed by haematoxylin eosin counterstaining of fixed smears. *Stain Technol.*, **48**, 127–132.

Burger, F. J. and Potgieter, G. M. (1978), The changes in activation of intracellular aspartate aminotransferase by pyridoxal 5-phosphate after cell death. *Clin. Chim. Acta*, **84**, 199–202.

Butterworth, F. M. (1971), Programmed cell death in the fat body of *Drosophila*. *Am. Zool.*, **11**, 690.

Butterworth, F. M. (1972), Adipose tissue of *Drosophila melanogaster*. V. Genetic and experimental studies on an extrinsic influence on the rate of cell death in the larval fat body. *Dev. Biol.*, **28**, 311–325.

Butterworth, F. M. and Lanczy La Tendresse, B. (1973), Quantitative studies of cytochemical and cytological changes during cell death in the larval fat body of *Drosophila melanogaster*. *J. Insect Physiol.*, **19**, 1487–1499.

Cajander, S. and Bjersing, L. (1975), Fine structural demonstration of acid phosphatase in rabbit germinal epithelium prior to induced ovulation. *Cell Tissue Res.*, **164**, 279–289.

Cameron, J. A. and Fallon, J. F. (1974), The absence of interdigital cell death during normal and regenerative forelimb development in *Xenopus laevis*. *Anat. Rec.*, **178**, 320.

Campbell, R. M., Dinsdale, D. and Fell, F. G. (1977), Localization of acid phosphatase activity in the liver of pregnant rats. *Histochem. J.*, **9**, 43–46.

Cantino, D. and Daneo, L. S. (1972), Cell death in the developing chick optic tectum. *Brain Res.*, **38**, 13–25.

Case, D. E. and Hendrix, M. J. (1976), Scanning electron microscopy of atrial septation. *Anat. Rec.*, **184**, 485.

Chang, T. K. (1939), The development of polydactylism in a special strain of *Mus musculus*. *Peking Nat. Hist. Bull.*, **14**, 119–132.

Chapman-Andresen, C. and Holter, H. (1955), Studies on the ingestion of ^{14}C glucose by pinocytosis in the amoeba *Chaos chaos*. *Exp. Cell Res. Suppl.*, **3**, 52–63.

Chien, K. R., Abrams, J., Pfau, R. G. and Farber, J. L. (1977), Prevention by chlorpromazine of ischemic liver cell death. *Am. J. Pathol.*, **88**, 539–558.

Chu, M-Y. (1971), Incorporation of arabinosyl cytosine into 2–7S ribonucleic acid and cell death. *Biochem. Pharmacol.*, **20**, 2057–2063.

Chu-Wang, I-W. and Oppenheim, R. W. (1977), Cell death of motoneurones in the chick embryo spinal cord. 1. A light and electron microscopic study of naturally occurring and induced cell loss during development. *J. Comp. Neurol.*, **177**, 33–58.

Claesson, M. H. (1972), Cell proliferation and cell death in peripheral lymphoid organs of the mouse. *Acta Pathol. Microbiol. Scand. Sect. B*, **80**, 475–477.

Coffino, P., Bourne, H. R. and Tomkins, G. M. (1975), Mechanisms of lymphoma cell death induced by cyclic AMP. *Am. J. Pathol.*, **81**, 199–204.

Cohen, A. M., Burdick, J. F. and Ketcham, A. S. (1971), Cell mediated cytotoxicity: an assay using ^{125}I-iododeoxyuridine-labelled target cells. *J. Immunol.*, **107**, 895.

Colon, A. D., Dorsey, A. and Lockshin, R. A. (1978), Muscle proteolysis during programmed cell death. *Fed. Proc. Fed. Am. Soc. Exp. Biol.*, **37**, 899.

Cooper, E. H. (1973), The biology of cell death in tumours. *Cell Tissue Kinet.*, **6**, 87–95.

Couch, E. F. and Mills, R. R. (1968), The midget epithelium of the American cockroach: acid phosphomonoesterase activity during the formation of autophagic vacuoles. *J. Insect Physiol.*, **14**, 55.

Coward, S. J., Bennett, C. E. and Hazelhurst, B. L. (1974), Lysosomes and lysosomal enzyme activity in the regenerating planarian; evidence in support of dedifferentiation. *J. Exp. Zool.*, **189**, 133–145.

Crossley, A. C. (1968), The fine structure and mechanism of breakdown of larval intersegmental muscles in the blowfly *Calliphora erythrocephala*. *J. Insect Physiol.*, **14**, 1389–1407.

Dawd, D. S. and Hinchliffe, J. R. (1971), Cell death in the opaque patch in the central mesenchyme of the developing chick limb: a cytological, cytochemical and electron microscopic analysis. *J. Embryol. Exp. Morphol.*, **26**, 401–424.

Decker, K. (1975), Quantitative aspects of biochemical mechanisms leading to cell death. In *Pathogenesis and Mechanisms of Liver Cell Necrosis* (ed. D. Keppler), pp. 45–56, MTP Press Ltd., Lancaster.

Decker, R. S. (1976), Influence of thyroid hormones on neuronal death and differentiation in larval *Rans pipiens*. *Dev. Biol.*, **49**, 101–118.

de Duve, C. (1963), The lysosome concept. In *Lysosomes* (eds. A. V. S. de Reuck and M. P. Cameron), J. and A. Churchill Ltd., London, pp. 1–35.

de Duve, C. (1969), Intracellular localization. In *The Phosphohydrolases: Their Biology, Biochemistry and Clinical Enzymology* (ed. W. H. Fishman), *Ann. N. Y. Acad. Sci.*, **166**, 602–603.

de Duve, C. and Wattiaux, R. (1966), Functions of lysosomes. *Ann. Rev. Physiol.*, **28**, 435–492.

Derby, A. and McEldoon, W. (1976), Changes in the tail of *Amblystoma maculatum* at different stages in metamorphosis: observations on tissue remodelling and its relationship to hydrolytic enzymes. *J. Exp. Zool.*, **196**, 205–214.

Don, M. M., Ablett, G., Bishop, C. J., Bundesen, P. G., Donald, K. J., Searle, J. and Kerr, J. F. R. (1977), Death of cells by apoptosis following attachment of specifically allergized lymphocytes *in vitro*. *Aust. J. Exp. Biol. Med. Sci.*, **55**, 407–418.

Dorn, A. (1977), Studies on the fat body of *Oncopeltus fasciatus* invaded by *Pseudomonas aeruginosa*. *J. Invert. Pathol.*, **29**, 347–353.

Drake, W. P., Ungoro, P. C. and Mardiney, M. R. (1972), Formalin fixed cell preparations as standards for use in the automated trypan blue cytotoxic assay. *Transplantation*, **14**, 127–130.

Elmes, M. E. (1975), Controlled cell deletion in the small intestine of zinc deficient rats. In *Trace Substances* (ed. D. D. Hemphill), Environmental Health – IX Symposium, University of Missuri, Colombia, pp. 419–421.

Elmes, M. E. (1977), Apoptosis in the small intestine of zinc deficient and fasted rats. *J. Pathol.*, **123**, 219–224.

El-Mofty, S. K., Scrutton, M. C., Serroni, A., Nicolini, C. and Farber, J. L. (1975), Early reversible plasma membrane injury in galactosamine induced liver cell death. *Am. J. Pathol.*, **79**, 579–596.

Ericsson, J. L. E. (1969), Mechanism of cellular autophagy. In *Lysosomes in Biology and Pathology*, Vol. 2 (eds. J. T. Dingle and H. B. Fell), North-Holland Publishing Co., Amsterdam and London, pp. 345–394.

Fakhri, O. and Hobbs, J. R. (1972), Target cell death without added complement after cooperation of 7S antibodies with non-immune lymphocytes. *Nature New Biol.*, **235**, 177–178.

Fallon, J. F. (1972), The morphology and fate of the apical ectodermal ridge in the normal and Janus Green B treated chick foot. *Am. Zool.*, **12**, 701–702.

Fallon, J. F., Brucker, R. F. and Harries, C. M. (1974), A re-examination of succinic dehydrogenase activity and its association with cell death in the interdigit of the chick foot. *J. Cell Sci.*, **15**, 17–29.

Fallon, J. F. and Cameron, J. (1977), Interdigital cell death during limb development of the turtle and lizard with an interpretation of evolutionary significance. *J. Embryol. Exp. Morphol.*, **40**, 285–289.

Farber, E. (1972), The pathology of translation. In *The Pathology of Transcription and Translation* (ed. E. Farber), Marcel Dekker Inc., New York, pp. 123–158.

Farber, J. L. and El-Mofty, S. K. (1975), The biochemical pathology of liver cell necrosis. *Am. J. Pathol.*, **81**, 237–250.

Firth, J. A. and Hicks, R. M. (1970), Differentiation and cell death in transitional epithelium urinary bladder of foetal and suckling rats. *J. Anat.*, **107**, 192–194.

Firth, J. A. and Hicks, R. M. (1972), Membrane specialization and synchronized cell death in developing rat transitional epithelium. *J. Anat.*, **113**, 95–107.

Fishman, W. H., Goldman, S. S. and De Lellis, R. (1967), Dual localization of β-glucuronidase in endoplasmic reticulum and in lysosomes. *Nature (London)*, **213**, 457–460.

Fox, H. (1972a), Tissue degeneration: An electron microscopic study of the tail skin of *Rana temporaria* during metamorphosis. *Arch. Biol.*, **83**, 373–394.

Fox, H. (1972b), Sub-dermal and notochordal collagen degeneration in the tail of *Rana temporaria*: An electron microscopic study. *Arch. Biol.*, **83**, 395–405.

Fox, H. (1973), Degeneration of the tail notochord of *Rana temporaria* at metamorphic climax. Examination by electron microscopy. *Z. Zellforsch. Mikrosk. Anat.*, **138**, 371–386.

Frankfurt, O. S. (1976), Acute cell death induced by inhibition of DNA synthesis in various parts of the gastrointestinal tract. *Bull. Exp. Biol. Med.*, **80**, 889–891.

Fristrom, M. (1972), Chemical modification of cell death in the bar eye of *Drosophila. Mol. Gen. Genet.*, **115**, 10–18.

Gaja, G., Bernelli-Zazzera, A. and Sorgato, G. (1965), Glycolysis in dying liver cells. *Exp. Mol. Pathol.*, **4**, 275–281.

Garcia-Porrero, J. A. and Ojeda, J. L. (1979), Cell death and phagocytosis in the neuroepithelium of the developing retina. A TEM and SEM study. *Experientia*, **35**, 375–377.

Gartner, L. P., Hiatt, J. L. and Provenza, D. V. (1978a), Succinic dehydrogenase activity during palate formation in the Mongolian gerbil. *J. Anat.*, **125**, 133–136.

Gartner, L. P., Hiatt, J. L. and Provenza, D. V. (1978b), Palatal shelf epithelium: a morphologic and histochemical study in X-irradiated and normal mice. *Histochem. J.*, **10**, 45–52.

Gevers, W. (1975), Biochemical aspects of cell death. *Forensic Sci.*, **6**, 25–29.

Giorgi, F. and Deri, P. (1976), Cell death in ovarian chambers of *Drosophila melanogaster. J. Embryol. Exp. Morphol.*, **35**, 521–533.

Glücksmann, A. (1951), Cell deaths in normal vertebrate ontogeny. *Biol. Rev. Cambridge Philos. Soc.*, **26**, 59–86.

Goldsmith, H. (1966), The anatomy of cell death. *J. Cell Biol.*, **31**, 41A.

Goldstein, S. and Singal, D. P. (1974), Senescence of cultured human fibroblasts mitotic vs. metabolic time. *Exp. Cell Res.*, **88**, 359–364.

Gossrau, R. (1977), Ageing and death of intestinal cells during pre- and post-natal development. *Proc. R. Microsc. Soc.*, **12**, 216.

Gossrau, R. (1978), Zur Verteilung der Stereocilimenzyme im Nebenhodengang von Ratten. *Histochemistry*, **57**, 145–159.

Gruenwald, P. (1975), Embryological basis of abnormal development with special reference to cell death. In *The Mammalian foetus* (ed. E. S. E. Hafez), Springfield, Illinois, pp. 251–281.

Hamburger, V. (1975), Cell death in the development of the lateral motor column of the chick embryo. *J. Comp. Neurol.*, **160**, 535–546.

Hammar, S. P. and Mottet, N. K. (1971), Tetrazolium salt and electron-microscopic studies of cellular degeneration and necrosis in the interdigital areas of the developing chick limb. *J. Cell Sci.,* **8,** 229–231.

Harrison, J. D., Borgers, M. and Thoné, F. (1979), Some observations on the phosphatase cytochemistry of the submandibular gland of cat. *Histochem. J.,* **11,** 311–320.

Hash, J. H. and King, K. W. (1958), Some properties of an aryl-β-glucosidase from culture filtrates of *Myrothecium verrucaria. J. Biol. Chem.,* **232,** 395.

Hassell, J. R. and Pratt, R. M. (1977), Elevated levels of cyclic AMP alter the effect of epidermal growth factor *in vitro* on programmed cell death in the secondary palatal epithelium. *Exp. Cell Res.,* **106,** 55–62.

Heby, O. and Andersson, G. (1978), Tumor cell death: The probable cause of increased polyamine levels in physiological fluids. *Acta. Pathol. Microbiol. Scand. Sect. A,* **86,** 17–20.

Henney, C. S. (1973), Studies on the mechanism of lymphocyte-mediated cytolysis. II. The use of various target cell markers to study cytolytic events. *J. Immunol.,* **110,** 73–84.

Hinchliffe, J. R. (1974), Experimental modification of cell death and chondrogenesis in insulin-induced micronuclia of the developing chick limb. An autoradiographic analysis of $^{35}SO_4$ uptake into chondroitin sulfate. *Teratology,* **9,** 263–274.

Hinchliffe, J. R. and Ede, D. A. (1967), Limb development in the polydactylous *talpid*[3] mutant of the fowl. *J. Embryol. Exp. Morphol.,* **17,** 385–404.

Hinchliffe, J. R. and Ede, D. A. (1973), Cell death and the development of limb form and skeletal pattern in normal and wingless (*ws*) chick embryos. *J. Embryol. Exp. Morphol.,* **30,** 753–772.

Hinsull, S. M., Bellamy, D. and Franklin, A. (1977), A quantitative histological assessment of cellular death in relation to mitosis in rat thymus during growth and age involution. *Age Ageing,* **6,** 77–84.

Hislop, E. C., Keon, J. P. R. and Fielding, A. H. (1977), Effects of fungal pectin lyase on the localization of acid phosphatase in cultured apple cells. *Proc. R. Microsc. Soc.,* **12,** 211–212.

Hislop, E. C., Keon, J. P. R. and Fielding, A. H. (1979), Effects of pectin lyase from *Monilinia fructigena* on viability, ultrastructure and localization of acid phosphatase of cultured apple cells. *Physiol. Plant Pathol.,* **14,** 371–381.

Hodges, M. G. and Muir, M. D. (1975), Quantitative evaluation of autoradiographs by X-ray spectroscopy. *J. Microsc.,* **104,** 173–178.

Hofer, K. G. and Hofer, M. (1971), Kinetics of proliferation, migration and death of L 1210 Ascites cells. *Cancer res.,* **31,** 402–408.

Hoffman, L. H. and Di-Pietro, D. L. (1972), Subcellular localization of human placental acid phosphatases. *Am. J. Obstet. Gynecol.,* **114,** 1087–1096.

Hopwood, D. and Levison, D. A. (1976), Atrophy and apoptosis in the cyclical human endometrium. *J. Pathol.,* **119,** 159–166.

Hurle, J. and Hinchliffe, J. R. (1978), Cell death in the posterior necrotic zone (PNZ) of the chick wing-bud: a stereoscan and ultrastructural survey of autolysis and cell fragmentation. *J. Embryol. Exp. Morphol.,* **43,** 123–136.

Ito, T. and Kobayashi, K. (1977), Chromosomal damage and membrane damage in cell death. *J. Radiat. Res.,* **18,** 17.

Jones, G. W. and Bowen, I. D. (1979a), The fine structural localization of acid phosphatase in molluscan pore cells. *Suppl. Proc. R. Microsc. Soc.,* **14,** A6–A7.

Jones, G. W. and Bowen, I. D. (1979b), The fine structural localization of acid phosphatase in pore cells of embryonic and newly hatched *Deroceras reticulatum* (Pulmonata : Stylommatophora). *Cell Tissue Res., 204,* 253–265.

Jones, R. T. and Trump, B. F. (1975), Cellular and subcellular effects of ischemia on the pancreatic acinar cell *in vitro* studies of rat tissue. *Virchows Arch. B Cell Pathol., 19,* 325–336.

Jurand, A. (1965), Ultrastructural aspects of early development of the fore-limb buds in the chicken and the mouse. *Proc. R. Soc. London Ser. B, 162,* 387–405.

Kalina, M. and Bubis, J. J. (1968), Histochemical studies on the distribution of acid phosphatases in neurones of sensory ganglia; light and electron microscopy. *Histochemie, 14,* 103–112.

Kasahara, T., Kametaka, M. and Kanadatsu, M. (1976), Folin-phenol reagent positive material released during autolysis of tissue slices from rats. *Agric. Biol. Chem., 40,* 337–346.

Keppler, D. (1975), Pathogenesis and mechanisms of liver cell necrosis. MPT Press Ltd., Lancaster.

Kerr, J. F. R. (1965), A histochemical study of hypertrophy and ischaemic injury of rat liver with special reference to changes in lysosomes. *J. Pathol. Bacteriol., 90,* 419–455.

Kerr, J. F. R. (1971), Shrinkage necrosis: a distinct mode of cellular death. *J. Pathol., 105,* 13–20.

Kerr, J. F. R., Harmon, B. and Searle, J. (1974), An electron-microscope study of cell deletion in the anuran tadpole tail during spontaneous metamorphosis with special reference to apoptosis of striated muscle fibres. *J. Cell Sci., 14,* 571–585.

Kerr, J. F. R. and Searle, J. (1972), The digestion of cellular fragments within phagolysosomes in carcinoma cells. *J. Pathol., 108,* 55–58.

Kerr, J. F. R. and Searle, J. (1973), Deletion of cells by apoptosis during castration-induced involution of the rat prostate, *Virchows Arch. Abt. B. Zellpathol., 13,* 87–102.

Kerr, J. F. R., Wyllie, A. H. and Currie, A. R. (1972), Apoptosis: a basic biological phenomenon with wide-ranging implications in tissue kinetics. *Br. J. Cancer, 26,* 239–257.

Ketelsen, U-P. (1977), Ultrastructure of dystrophic skeletal muscle. *Isr. J. Med. Sci., 13,* 107–120.

Kihara, H. K., Hu, A. S. B. and Halvorson, H. O. (1961), The identification of a ribosomal-bound-β-glucosidase. *Proc. Natl. Acad. Sci. U.S.A., 47,* 489.

Kumar, A. R. V., Hoshino, T., Wheeler, K. T., Baker, M. and Wilson, C. B. (1974), Comparative rates of dead tumor cell removal from brain, muscle, subcutaneous tissue and peritoneal cavity. *J. Natl. Cancer Inst., 52,* 1751–1755.

Lala, P. K. (1971), Evaluation of the mode of cell death in Ehrlich ascites tumor. *Cancer, 29,* 261–266.

Lala, P. K. (1972), Age specific changes in proliferation of Ehrlich ascites tumor cells grown as solid tumors. *Cancer Res., 32,* 628–636.

Landmesser, L. and Pilar, G. (1976), Fate of ganglionic synapses and ganglion cell axons during normal and induced cell death. *J. Cell Biol., 68,* 357–374.

Larsen, W. J. (1976), Cell remodelling in the fat body of an insect. *Tissue Cell, 8,* 73–92.

Leblond, C. P. and Walker, B. E. (1956), Renewal of cell populations. *Physiol. Rev., 36,* 255–275.

Lieberman, M. W. (1972), DNA metabolism, cell death and cancer chemotherapy. In *The Pathology of Transcription and Translation* (ed. E. Farber), Marcel Dekker, Inc., New York, pp. 37–53.

Lesch, M. (1977), Kinetics of solubilization of cathepsin D in autolyzing rabbit myocardium. *Circ. Suppl.*, **56**, 111–209.

Lewis, G. H. J., Bowen, I. D. and Bellamy, D. (1979), Combined autoradiography and histochemistry: the simultaneous detection of [6-³H] thymidine and acid phosphatase activity in cryostat sections. *J. Microsc.*, **117**, 225–259.

Lewis, P. D. (1975), Cell death in the germinal layers of the brain. *Neuropathol. Appl. Neurobiol.*, **1**, 21–29.

Lockshin, R. A. (1969), Lysosomes in insects. In *Lysosomes in Biology and Pathology* (eds. J. T. Dingle and H. B. Fell), North-Holland Publishing Co., Amsterdam and London, pp. 363–391.

Lockshin, R. A. and Beaulaton, J. (1974a), Programmed cell death. Cytochemical evidence for lysosomes during the normal breakdown of the intersegmental muscles. *J. Ultrastruct. Res.*, **46**, 43–62.

Lockshin, R. A. and Beaulaton, J. (1974b), Programmed cell death. Cytochemical appearance of lysosomes when death of the intersegmental muscles is prevented. *J. Ultrastruct. Res.*, **46**, 63–78.

Lockshin, R. A. and Beaulaton, J. (1974c), Programmed cell death. *Life Sci.*, **15**, 1549–1565.

Lockshin, R. A., Bidlack, J. M. and Beaulaton, J. (1975), Programmed cell death. Evidence for premonitory increase in anaerobic glycolysis. *J. Cell Biol.*, **67**, 247A.

Lockshin, R. A. and Williams, C. M. (1964), Programmed cell death. II. Endocrine potentiation of the breakdown of the intersegmental muscles of silkmoths. *J. Insect Physiol.*, **10**, 643–649.

Lockshin, R. A. and Williams, C. M. (1965a), Programmed cell death. III. Neural control of the breakdown of the intersegmental muscles of silkmoths. *J. Insect Physiol.*, **11**, 605–610.

Lockshin, R. A. and Williams, C. M. (1965b), Programmed cell death. IV. The influence of drugs on the breakdown of the intersegmental muscles of silkmoths. *J. Insect Physiol.*, **11**, 803–809.

Lockshin, R. A. and Williams, C. M. (1965c), Programmed cell death. V. Cytolytic enzymes in relation to the breakdown of the intersegmental muscles of silkmoths. *J. Insect Physiol.*, **11**, 831–844.

Lodish, H. F. and Rothman, J. E. (1979), The assembly of cell membranes. *Sci. Am.*, **240**, 38–53.

Looss, A. (1889), Über Degeneration serchein ungen im Tierreich, besonders über die Reduktion des Forschlarvenschwanzes und die im Verlaug disselben auftertenden histolytischen. *Proz. Preisschr. Jablonowsk Ges.*, **10**.

Luppa, H., Radon, M., Weiss, J. and Martin, D. (1970), Histochemische Darstellung unspezifischer alkalischer und unspezifischer saurer Phosphatase mit *p*-nitrophenyl phosphat ab Substrat. *Acta Histochem.*, **37**, 197–199.

Luppa, H., Weiss, J. and Martin, D. (1972), Ein Bietrag zur elektronenmikroskopisch-histochemischem Dorstellung der unspezifischen saurer Phosphatase mit *p*-nitrophenyl phosphat. *Acta Histochem.*, **43**, 184–188.

Mansfield, J. W. and Sexton, R. (1974), Changes in the localization of β-glycerophosphatase activity during the infection of *Phaseolus vulgaris* by *Colletotrichum lindemuthianum*. *Ann. Bot.*, **38**, 711–718.

Marzullo, G. (1972), Regulation of cartilage enzymes in cultured chondrocytes and the effect of 5-bromodeoxyuridine. *Dev. Biol.*, **27**, 20–26.

Matsuura, S., Morimoto, T., Nagata, S. and Tashiro, Y. (1968), Studies on the posterior silk gland of the silk worm *Bombyx mori*. II. Cytolytic processes in posterior silk gland cells during metamorphosis from larva to pupa. *J. Cell Biol.*, **38**, 589–603.

Metchnikoff, E. (1892), *Pathologie Comparee de l'Inflammation*, Masson, Paris.

Metchnikoff, E. (1901), *L'Immunite dans les Maladies Infectienses*, Paris. [English translation by F. G. Binnie (1905) *Immunity in infective disease*, Cambridge University Press, Cambridge].

Michaels, J. E., Albright, J. T. and Patt, D. I. (1971), Fine structural observations on cell death in the epidermis of the external gills of the larval frog *Rana pipiens. Am. J. Anat.*, **132**, 301–318.

Michinomae, M. (1977), Fine structural and biochemical studies on cell death of the developing Bar eye discs in *Drosophila melanogaster. Jpn. J. Genet.*, **51**, 315–326.

Mittnacht, S. and Farber, J. L. (1978), Reversibility of ischemic mitochondrial injury. *Fed. Proc. Fed. Am. Soc. Exp. Biol.*, **37**, 403.

Miyayama, H., Solomon, R., Sasak, M., Lin, C. and Fishman, W. H. (1975), Demonstration of lysosomal and extralysosomal sites for acid phosphatase in mouse kidney tubule cells with *p*-nitrophenyl phosphate lead-salt technique. *J. Histochem. Cytochem.*, **23**, 439–451.

Murphy, C. (1974), Cell death and autonomous gene action in lethals affecting imaginal discs in *Drosophila melanogaster. Dev. Biol.*, **39**, 23–36.

Musy, J. P. and Haag, J. R. (1977), Diagnostic de necroses myocardiques au stade initial par des colorations simples, Essai d'interpritation histochimique. *Amn. Anat. Pathol.*, **22**, 97–116.

Napolitano, L. (1962), Cytoplasmic bodies containing mitochondria, ribosomes, and rough surfaced endoplasmic reticulum in the small intestine of new born rats. *J. Cell Biol.*, **18**, 478–481.

Newlands, E. S. (1974), Two methods of detecting different aspects of the immune response to a xenograft in mice. *Immunology*, **27**, 543–552.

Norkin, L. C. and Ouellette, J. (1976), Cell killing by Simian virus 40: Variation in the pattern of lysosomal enzyme release, cellular enzyme release and cell death during productive infection of normal and Simian virus 40-transformed Simian cell lines. *J. Virol.*, **18**, 48–57.

Novikoff, A. B. (1964), GERL, its form and function in the neurons of rat spinal ganglia. *Biol. Bull.*, **127**, 358.

Novikoff, A. B. (1976), The endoplasmic reticulum: A cytochemist's view (A review). *Proc. Natl. Acad. Sci. U.S.A.*, **73**, 2781–2787.

Novikoff, A. B. and Essner, E. (1962), Cytolysosomes and mitochondrial degradation. *J. Cell Biol.*, **15**, 140.

Novikoff, A. B., Essner, E. and Quintana, N. (1964), The Golgi apparatus and lysosomes. *Fed. Proc. Fed. Am. Soc. Exp. Biol.*, **23**, 1010–1022.

Novikoff, A. B. and Schin, W. Y. (1964), The endoplasmic reticulum in the Golgi zone and its relations to microbodies, Golgi apparatus and autophagic vacuoles in rat liver cells. *J. Microsc.*, **3**, 187.

O'Connor, T. M. and Wyttenbach, C. R. (1974), Cell death in the embryonic chick spinal cord. *J. Cell Biol.*, **60**, 448–459.

Ogawa, R. (1975), The role of lysosomal hydrolases in the pathogenesis of refractory shock. *Kitakanto Med. J.*, **25**, 71–84.

Ojeda, J. L. and Hurle, J. M. (1975), Cell death during the formation of tubular heart of the chick embryo. *J. Embryol. Exp. Morphol.*, **33**, 523–534.

Okabe, K., Koyangi, R. and Koga, K. (1975), RNA in degenerating silk gland of *Bombyx mori. J. Insect Physiol.*, **21**, 1305–1309.

Okumura, Y. (1977), Kinetics of tumor cell death by hyperthermic treatment and X-ray irradiation. *Gaun*, **68**, 837–840.

Oliveira, L. (1975), Self-degeneration of mitochondria in the root cap cells of

Triticale. Its contribution to the development of the vacuolar apparatus and significance for senescence. *Caryologia*, **28**, 511–523.

Olson, R. L. and Everett, M. A. (1975), Epidermal apoptosis: cell deletion by phagocytosis. *J. Cut. Pathol.*, **2**, 53–57.

Oxford, G. S. and Fish, L. J. (1979), Ulstrastructural localization of esterase and acid phosphatase in digestive gland cells of fed and starved *Cepaea nemoralis* (L.) (Mollusca: Helicidae). *Protoplasma*, **101**, 181–196.

Painter, R. B. (1977), The relationship between unrepaired DNA damage and cell death. *Mut. Res.*, **46**, 145–146.

Pannese, E., Luciano, L., Iurato, S. and Reale, E. (1976), Lysosomes in normal and degenerating neuroblasts of the chick embryo spinal ganglia. A cytochemical and quantitative study by electron microscopy. *Acta Neuropathol.*, **36**, 209–220.

Perlmann, P., Perlmann, H., Muller-Eberhard, H. J. and Manni, J. A. (1969), Cytotoxic effects of leukocytes triggered by complement bound to target cells. *Science*, **163**, 937–939.

Petes, T. D., Farber, R. A., Tarrant, G. M. and Holliday, R. (1974), Altered rate of DNA replication in ageing human fibroblast cultures. *Nature (London)*, **251**, 434–436.

Pexieder, T. (1972a), The tissue dynamics of heart morphogenesis. I. The phenomena of cell death. A. Identification and morphology. *Z. Anat. Entwickl. Ges.*, **137**, 270–284.

Pexieder, T. (1972b), The tissue dynamics of heart morphogenesis. I. The phenomena of cell death. B. Topography. *Z. Anat. Entwickl. Ges.*, **138**, 241–253.

Pexieder, T. (1975), Scanning electron microscopic investigations on physiological cell death in chick embryo heart. *Experientia*, **31**, 745.

Pilar, G. and Landmesser, L. (1976), Ultrastructural differences during embryonic cell death in normal and peripherally deprived ciliary ganglia. *J. Cell Biol.*, **68**, 339–356.

Pitt, D. and Coombes, C. (1969), Release of hydrolytic enzymes from cytoplasmic particles of solanum tuber tissues during infection by tuber-rotting fungi. *J. Gen. Microbiol.*, **56**, 321–329.

Poodry, C. A. (1975), Autonomous and non-autonomous cell death in the metamorphosis of the epidermis of *Drosophila*. *Wilhelm Roux's Arch. Dev. Biol.*, **178**, 333–336.

Popp, J. A., Shinozuka, H. and Farber, E. (1974), Possible role of protein synthesis in the pathogenesis of cell necrosis. *Am. J. Pathol.*, **74**, 58A.

Pratt, R. M. and Greene, R. M. (1975), Inhibition of programmed cell death in the developing rat secondary palate by diazo-oxonorleucine. *J. Cell Biol.*, **67**, 344A.

Pratt, R. M. and Greene, R. M. (1976), Inhibition of palatal epithelial cell death by altered protein synthesis. *Dev. Biol.*, **54**, 135–145.

Pratt, R. M. and Martin, G. R. (1975), Epithelial cell death and cyclic AMP increase during palatal development. *Proc. Natl. Acad. Sci. U.S.A.*, **72**, 874–877.

Price, J. M., Donahoe, P. K., Ito, Y. and Hendren, W. H. (1977), Programmed cell death in the Müllerian duct induced by Müllerian inhibiting substance. *Am. J. Anat.*, **149**, 353–376.

Ragnotti, G., Gaja, G. and Bernelli-Zazzera, A. (1968), Decline and restoration of glycolysis in homogenates. *Exp. Mol. Pathol.*, **9**, 148–159.

Reichel, P. (1976), Differentiation of the opercular integument in *Rana pipiens*. *Differentiation*, **5**, 75–83.

Robinson, H. (1970), Acid phosphatase in the tail of *Xenopus laevis* during development and metamorphosis. *J. Exp. Zool.*, **173**, 215–224.

Rowley, A. F. and Ratcliffe, N. A. (1976a), An ultrastructural study of the *in vitro* phagocytosis of *Escherichia coli* by the haemocytes of *Calliphora erythrocephala. J. Ultrastruct. Res.,* **55**, 193–202.

Rowley, A. F. and Ratcliffe, N. A. (1976b), The intracellular fate of bacteria and latex particles in insect blood cells. *Eur. Congr. Electron Microsc. 6th,* 301–303.

Russell, D. H., Pietro, M. G., Laurence, J. M. and Le Gendre, S. M. (1974), Polyamine depletion of the MTW9 mammary tumor and subsequent elevation of spermidine in the sera of tumor-bearing rats as a biochemical marker of tumor regression. *Cancer Res.,* **34**, 2378–2381.

Ryder, T. A. and Bowen, I. D. (1975a), A method for the fine structural localization of acid phosphatase activity using *p*-nitrophenyl phosphate as substrate. *J. Histochem. Cytochem.,* **23**, 235–237.

Ryder, T. A. and Bowen, I. D. (1975b), The fine structural localization of acid phosphatase activity in *Polycelis tenuis* (Iijima). *Protoplasma,* **83**, 79–90.

Ryder, T. A. and Bowen, I. D. (1977a), Endocytosis and aspects of autophagy in the foot epithelium of the slug *Agriolimax reticulatus* (Müller). *Cell Tissue Res.,* **181**, 129–142.

Ryder, T. A. and Bowen, I. D. (1977b), Studies on transmembrane and paracellular phenomena in the foot of the slug *Agriolimax reticulatus* (Mü). *Cell Tissue Res.,* **183**, 143–152.

Salzgeber, B. and Weber, R. (1966), La regression du mesonephros chez l'embryon de poulet. *J. Embryol. Exp. Morphol.,* **15**, 397–419.

Sanderson, C. J. (1976), The mechanism of T cell mediated cytotoxicity. I. The release of different cell components. *Proc. R. Soc. London Ser. B,* **192**, 221–239.

Sandritter, W. and Riede, U. N. (1975), Morphology of liver cell necrosis. In *Pathogenesis and Mechanisms of Liver Cell Necrosis,* 1st edn. (ed. D. Keppler), MTP Press, Lancaster, pp. 1–14.

Saunders, J. W. (1966), Death in embryonic systems. *Science,* **154**, 604–612.

Saunders, J. W., Gasseling, M. T. and Saunders, L. C. (1962), Cellular death in morphogenesis of the avian wing. *Dev. Biol.,* **5**, 149–178.

Schanne, F. A. X., Kane, A. B., Young, E. E. and Farber, J. L. (1979), Calcium dependence of toxic cell death: A final common pathway. *Science,* **206**, 700–702.

Schin, K. S. and Clever, U. (1965), Lysosomal and free acid phosphatase in salivary glands of *Chironomus tetans. Science,* **150**, 1053–1055.

Schlichtig, R., Lockshin, R. A. and Beaulaton, J. (1977), Programmed cell death measurement of several ATP digesting enzymes in degenerating insect muscles. *Insect Biochem.,* **7**, 327–336.

Schlüter, G. (1973), Ultrastructural observations on cell necrosis during formation of the neural tube in mouse embryos. *Z. Anat. Entwickl. Ges.,* **141**, 251–264.

Schweichel, J. U. and Merker, H.-J. (1973), The morphology of various types of cell death in prenatal tissues. *Teratology,* **7**, 253–266.

Scott, W. J., Ritter, E. J. and Milson, J. G. (1977), Delayed appearance of ectodermal cell death as a mechanism of polydactyly induction. *J. Embryol. Exp. Morphol.,* **42**, 93–104.

Searle, J. and Patten, C. S. (1977), Distribution and morphology of UV induced epidermal cell death. *Pathology,* **9**, 73.

Searle, J., Collins, D. J., Harmon, B. and Kerr, J. F. R. (1973), The spontaneous occurrence of apoptosis in squamous carcinomas of the uterine cervix. *Pathology,* **5**, 163–169.

Searle, J., Lawson, T.A., Abbott, P.J., Harmon, B. and Kerr, J.F.R. (1975), An electron microscope study of cell death induced by cancer chemotherapeutic agents in populations of prolifering normal and neoplastic cells. *J. Pathol.*, **116**, 129–138.

Seligman, I.M., Filschie, B.K., Day, F.A. and Crossley, A.C. (1975), Hormonal control of morphogenetic cell death of the wing hypodermis in *Lucilia cuprina. Tissue Cell*, **7**, 281–296.

Seshimo, H., Ryuzaki, M. and Yoshizato, K. (1977), Specific inhibition of triiodothyronine-induced tadpole tail-fin regression by cathepsin D-inhibitor pepstatin. *Dev. Biol.*, **59**, 96–100.

Shackney, S.E. (1973), A cytokinetic model for heterogeneous mammalian cell populations. Part 1. Cell growth and cell death. *J. Theor. Biol.*, **38**, 305–333.

Shamsuddin, M., Bhattacharya, A. and Medda, J.N. (1976), Histochemical and biochemical study on acid phosphatase activity in differentiating liver of chick. *Acta Histochem.*, **56**, 215–221.

Sheldrake, A.R. (1974), The ageing, growth and death of cells. *Nature (London)*, **250**, 381–385.

Singer, P.A., Singer, H.H. and Williamson, A.R. (1980), Different species of messenger RNA encode receptor and secretory IgM μ chains differing at their carboxy termini. *Nature (London)*, **285**, 294–300.

Skalko, R.G., Perrins, N.M. and Niles, A.M. (1977), The effect of 5-bromodeoxyuridine on mouse limb development: a scanning electron microscope study. *Acta Embryol. Exp.*, **3**, 323–334.

Slater, T.F. (1975), The role of lipid peroxidation in liver injury. In *Pathogenesis and Mechanisms of Liver Cell Necrosis* (ed. D. Keppler), MTP Press, Lancaster, pp. 209–223.

Slater, T.F., Greenbaum, A.L. and Wang, D.Y. (1963), Lysosomal changes during liver injury and mammary involution. In *Lysosomes* (eds. A.V.S. de Renck and M.P. Cameron), Ciba Foundation Symposium, J. and A. Churchill Ltd., London, pp. 311–334.

Slivinski, A. and Alepa, F.P. (1973), A phagocytosis inhibiting factor(s) in serum from patients with rheumatoid arthritis (RA). *Arthritis Rheum.*, **16**, 132.

Smadja-Joffe, F., Klein, B., Kerdiles, C., Feinendegen, L. and Jasmin, C. (1976), Study of cell death in Friend Leukemia. *Cell Tissue Kinet.*, **9**, 131–145.

Soberman, R.J., Hoffstein, S. and Weissmann, G. (1973), Direct evidence for suicide sac hypothesis of lysosomal enzyme release by mono sodium urate. *Arthritis Rheum.*, **16**, 132–133.

Sohal, G.S. (1976), An experimental study of cell death in the developing trochlear nucleus. *Exp. Neurol.*, **51**, 684–698.

Sohal, G.S. and Holt, R.K. (1977), Cell death during normal development of the abducens nucleus. *Exp. Neurol.*, **54**, 533–545.

Spreij, T.E. (1971), Cell death during development of the imaginal discs of *Calliphora erythrocephala. Neth. J. Zool.*, **21**, 221–264.

Starre-van der Molen, L.G. and Otten, L. (1974), Embryogenesis of *Calliphora erythrocephala* Meigen. Cell death in the central nervous system during late embryogenesis. *Cell Tissue Res.*, **151**, 219–228.

Stoward, P.J. (1977), Muscular dystrophy and the ageing process. *Proc. R. Microsc. Soc.*, **12**, 212.

Straus, W. (1959), Rapid cytochemical identification of phagosomes in various tissues of the rat and their differentiation from mitochrondria by the peroxidase method. *J. Biophys. Biochem. Cytol.*, **5**, 193–204.

Straus, W. (1967a), Changes in intracellular location of small phagosomes (micropinocytic vesicles) in kidney cells and liver cells in relation to time after injection and dose of horse-radish peroxidase. *J. Histochem. Cytochem.*, **15**, 281–293.

Straus, W. (1967b), Lysosomes, phagosomes and related particles. In *Enzyme Cytology* (ed. D. B. Roodyn), Academic Press, London and New York, pp. 239–319.

Strauss, B. S. (1977), Molecular biology of the response of cells to radiation and to radiomimetic chemcials. *Cancer*, **40**, 471–480.

Sylven, B. and Niemi, M. (1972), Histochemical evidence of cell death in transplanted tumours. *Virchows. Arch. Abt. B. Zellpathol.*, **10**, 127–133.

Tata, J. R. (1966), Requirement for RNA and protein synthesis for induced regression of the tadpole tail in organ culture. *Dev. Biol.*, **12**, 77–94.

Theiler, K., Varnum, D. S., Nadeau, J. M., Stevens, L. C. and Cagianut, B. (1976), A new allele of ocular retardation in early development and morphogenetic cell death. *Anat. Embryol.*, **150**, 85–97.

Todd, R. F., Stultung, R. D. and Berke, G. (1973), Mechanism of blocking by hyper-immune serum of lymphocyte mediated cytolysis of allogeneic tumor cells. *Cancer Res.*, **33**, 3203–3208.

Tomasz, A. (1974), The role of cell autolysins in cell death. *Ann. N. Y. Acad. Sci.*, **235**, 439–447.

Trump, B. F., Laiho, K. A., Mergner, W. J. and Arstila, A. U. (1974), Studies on the pathophysiology of acute lethal cell injury. *Beitr. Pathol. Bd.*, **152**, 243–271.

Trump, B. F. and Mergner, W. J. (1974), Cell injury. In *The Inflammatory Process* (eds. B. W. Zweifach, L. Grant and R. T. McCluskey), Academic Press, New York, pp. 115–257.

Van Lancker, J. L. (1964), Concluding remarks, pathology symposium on lysosomes. *Fed. Proc. Fed. Am. Soc. Exp. Biol.*, **23**, 1009–1052.

Van Lancker, J. L. (1975), Hydrolases and cellular death. In *Pathogenesis and Mechanisms of Liver Cell Death* (ed. D. Keppler), MTP Press, Lancaster, pp. 25–35.

Van Waes, L. and Lieber, C. S. (1977), Glutamate dehydrogenase: a biochemical test of alcoholic hepatitis. *Clin. Res.*, **25**, 517A.

Vijverberg, A. J. (1974), A cytological study of the proliferation patterns in imaginal discs of *Calliphora erythrocephala* during larval and pupal development. *Neth. J. Zool.*, **24**, 171–217.

Webb, J. N. (1972), The development of human skeletal muscle with particular reference to muscle cell death. *J. Pathol.*, **106**, 221–228.

Webb, J. N. (1974), Muscular dystrophy and muscle cell death in normal fetal development. *Nature (London)*, **252**, 233–234.

Webb, J. N. (1977), Cell death in developing skeletal muscle histochemistry and ultrastructure. *J. Pathol.*, **123**, 175–180.

Weber, R. (1963), Behaviour and properties of acid hydrolases in regressing tails of tadpoles during spontaneous and induced metamorphosis *in vitro*. In *Lysosomes* (eds. A. V. S. de Renck and M. P. Cameron), Ciba Foundation Symposium, J. and A. Churchill Ltd., London, pp. 282–310.

Weber, R. (1964), Ultrastructural changes in regressing tail muscles of *Xenopus* larvae at metamorphosis. *J. Cell Biol.*, **22**, 481–487.

Weber, R. (1969), Tissue involution and lysosomal enzymes during anuran metamorphosis. In *Lysosomes in Biology and Pathology, Vol. 2* (eds. J. T. Dingle and H. B. Fell), North-Holland Publishing Co., Amsterdam, pp. 437–461.

Weber, R. (1977), Biochemical characteristics of tail atrophy during anuran metamorphosis. *Colloq. Int. C.N.R.S.,* **266**, 137–146.

Webster, D. A. and Gross, J. (1970), Studies on possible mechanisms of programmed cell death in the chick embryo. *Dev. Biol.,* **22**, 157–184.

Wendler, R. (1973), Cell death in normogenesis and teratogenesis. *Teratology,* **8**, 241.

Wiley, C. A. and Esterley, J. R. (1975), Histochemical studies of the corpus luteum. *Am. J. Pathol.,* **78**, 60A.

Williams, J. R., Little, J. B. and Shipley, W. U. (1974), Association of mammalian cell death with a specific endonucleolytic degradation of DNA. *Nature (London),* **252**, 754–756.

Woodling, B. A., Warner, N. E., Hall, T. D. and Puffer, H. W. (1973), The demonstration of recent and old infarcts in the gross placenta, using a succinic acid dehydrogenase assay. *Anat. Rec.,* **175**, 472–473.

Wrogmann, K. and Pena, S. D. J. (1976), Mitochondrial calcium overload: a general mechanism for cell necrosis in muscle diseases. *Lancet,* **i**, 672–673.

Wyllie, A. H., Kerr, J. F. R. and Currie, A. R. (1973a), Cell death in the normal neonatal rat adrenal cortex. *J. Pathol.,* **111**, 255–261.

Wyllie, A. H., Kerr, J. F. R., Macaskill, I. A. M. and Currie, A. R. (1973b), Adrenocortical cell deletion, the role of ACTH. *J. Pathol.,* **111**, 85–94.

Ying, T. S., Sarma, D. S. R. and Farber, E. (1978), Inhibition of diethyl nitrosamine induced acute liver cell necrosis by diethyldithiocarbamate, a possible site of action after the activation step. *Fed. Proc. Fed. Am. Soc. Exp. Biol.,* **37**, 402.

Young, D., Ashhurst, D. E. and Cohen, M. J. (1970), The injury response of the neurones of *Periplaneta americana. Tissue Cell.,* **2**, 387–398.

Zeligs, J. D., Janoff, A. and Dumot, A. E. (1975), The course and nature of acinar cell death following pancreatic ligation in the guinea-pig. *Am. J. Pathol.,* **80**, 203–226.

Author index

Subject index